Nelson Yuan-sheng Kiang

IMPEDANCE MEASUREMENTS IN BIOLOGICAL CELLS

IMPEDANCE MEASUREMENTS IN BIOLOGICAL CELLS

OTTO F. SCHANNE

ELENA RUIZ P.-CERETTI

Department of Biophysics
University of Sherbrooke
Sherbrooke, Quebec

A WILEY–INTERSCIENCE PUBLICATION

JOHN WILEY & SONS, New York • Chichester • Brisbane • Toronto

Library of Congress Cataloging in Publication Data:

Schanne, Otto F
Impedance measurements in biological cells.

"A Wiley-Interscience publication."
Bibliography: p.
1. Impedance, Bioelectric—Measurement.
2. Nerves. 3. Muscle. 4. Epithelium. 5. Cell suspensions. I. Ruiz P.-Ceretti, Elena, 1933- joint author. II. Title.

QP341.S26 591.1′9127 77-17434
ISBN 0-471-76530-9

Printed in the United States of America

10 9 8 7 6 5 4 3 2 1

"Abandon all hope, all ye who enter here!"

Student banner across the entrance of the Department of Physiology, University of Heidelberg, when results of examination were announced in Spring 1963.

Borrowed from Dante's *Divine Comedy*

Preface

Since impedance-measurement techniques attained technically acceptable standards by the middle of the last century, these methods have been applied to biological structures. The history of this branch of electrobiology represents one of great effort, greater hopes, deep frustrations, and some unique contributions to our present biological knowledge.

The great challenge and also an important disadvantage of biological impedance measurements has been the multidisciplinary nature of electrobiology. Since World War II, two camps of "impedance measurers" have developed in parallel with practically no communication between them. One of these groups consisted of researchers with physics or engineering background who employed sophisticated methods and consequently preferred an ac analysis of the biological structures by using mainly extracellular electrodes; however, their biological interest was limited. In the other camp were researchers with biological background using electrical analysis as a tool to solve physiological problems; these investigators preferred the more direct square-pulse analysis. They also became the principal users of the glass microelectrode. By the beginning of the nineteen sixties, some engineers had learned enough biology and some biologists had mastered enough engineering knowledge that the two camps began to communicate, and now it seems that they will finally merge. This development has produced some significant contributions to biological knowledge such as the study of the role of the electrical properties of the transverse tubular system in excitation-contraction coupling in muscle and the application of impedance measurements to the elucidation of ionic transfer mechanisms. Consequently, it is probable that we are at present experiencing a new boom for biological impedance measurements.

When, several years ago, we ventured into writing a summary of biological impedance measurements, we felt that an attempt should be made to show that results obtained by the two groups of impedance measurers complement each other, and that the most relevant biologic information is obtained when the results of electrical measurements are combined with those acquired in other fields of biological research. The discussion of these results is limited here to representative examples of measurements performed at the cellular level and on epithelia since we feel that there lies the greatest future potential of these methods. Some interesting topics like measurements on plant cells, the electric organs, and artificial membranes are not treated, mostly for lack of space within the confines of one book.

Chapter 1 discusses the technical and theoretical tools used in the field of impedance measurements at the cellular level; in the following chapters the results obtainable with those tools are reported. At the end of each chapter, there is a section entitled "Summary and Outlook," which serves as a quick orientation for the reader about the main topics covered in the chapter and indicates the importance of these topics for the future development of the field.

We have tried to insist on quantitative results and to furnish the basic data so that the reader can judge how these results were obtained. We did not try to cover completely the available literature but to select representative examples of studies dealing with impedance problems. Therefore, it is obvious that inclusion or omission of a paper does not mean a judgment of its scientific value.

This book could serve as an introduction to advanced electrobiological measurements for graduate students in biophysics, physiology, and bioengineering who have a reasonable background in electronics and biology. It can also serve as a basis for a graduate course on selected topics of impedance measurements. For the active researcher in the field it should be equally useful as a source of comparative reference data.

During preparation of the manuscript we profited from comments of many colleagues who sacrificed their time in reading parts of it, especially Jean-Pierre Caillé, University of Sherbrooke; Hans G. Haas, University of Bonn; Ignacio Reisin, Centro de Investigaciones Médicas Alberto Einstein, Buenos Aires; William Sleator, University of Illinois; and Merrill Tarr, University of Kansas. We also acknowledge the help of many others who typed the manuscript, prepared the figures, or compiled the references, especially Denis Chartier, Louise Denault, Maryse Godbout, Gertrud Schanne, Roswitha Schanne, and Irene Sjodin. Finally, we thank Beatrice Shube, our editor at Wiley-Interscience, and our referee for their help, understanding, and constructive criticism. During our work on this manuscript, our laboratory research was supported by the Medical Research Council of Canada, the Quebec Heart Foundation, and a training grant from the Ministère de l'Education du Québec.

Otto F. Schanne
Elena Ruiz P.-Ceretti

Sherbrooke, Quebec, Canada
January 1978

Contents

List of Symbols

Complex quantities are identified by boldface capitals and their magnitude is shown by capital italics. The complex dielectric constant is represented as ϵ and its magnitude by ε.

A	Surface, area
A	Amplitude of periodic function
$\mathcal{A}$	Electrochemical activity
a	Cell radius
$\boldsymbol{a}$	Ionic activity
α	Power factor
α	Rate constant in Hodgkin–Huxley theory
b	Relative permeability
b	Phase constant
β	Partition coefficient
β	Rate constant in Hodgkin–Huxley theory
C	Concentration
C	Capacitance
C^*	Specific capacitance
$C^*_{100\%}$	Specific capacitance of a suspension with $\rho' = 100\%$
C_i	Concentration in intracellular medium
C_m	Membrane capacitance
C_o	Concentration in extracellular medium
C_{pol}	Polarization capacitance
$\hat{C}_{\text{pol}}$	Polarization capacitance at $\omega = 1$
C'_{pol}	Bulk polarization capacitance
C_s	Capacitance of surface membrane
C_w	Capacitance per unit area of tubular membrane
$\overline{C}_w$	Capacitance of tubular wall per unit volume of muscle fiber

c_o	Total ion concentration in extracellular medium
c_m	Membrane capacitance per unit length
c_w	Ion concentration in cell wall
c_w^f	Total fixed charge concentration in cell wall
D	Diffusion constant
d	Activation constant of slow inward current
d	Membrane thickness
d	Distance
d	Cell diameter
E_i	Equilibrium potential for ion species *i*
ϵ	Dielectric constant
ε'	Real part of complex dielectric constant
ε''	Imaginary (dissipative) part of dielectric constant
ε_0	Limiting dielectric constant at very low frequencies
ε_∞	Limiting dielectric constant at very high frequencies
ε_r	Relative dielectric constant
ε_v	Dielectric constant of free space
F	Faraday constant
F	Function
f	Inactivation variable of slow inward current
f	Frequency
f	Activity coefficient
f	Form factor
Φ	Function
Φ	Phase angle of polarization element
φ	Phase angle
G	Galvanometer
G	Slope conductance
G_L	Specific luminal conductivity in transverse tubules
$\overline{G}_L$	Effective radial conductivity of tubular lumen
G_m	Membrane conductance
G_s	Conductance of the surface membrane
G_w	Conductance per unit area of tubular membrane
$\overline{G}_w$	Conductance per unit volume of tubular wall
g	Chord conductance
$\bar{g}_i$	Limiting conductance for ion species *i*
γ	Form factor
γ	Propagation constant
h	Inactivation variable of fast inward current
I	Current
I_{ep}	End-plate current
I_{sc}	Short circuit current
i	Instantaneous current

i_m	Instantaneous current density
i_m	Membrane current per unit length
i_r	Radial current
J	Current density
J_m	Density of membrane current
J_c	Density of capacitive current
k	Surface to volume ratio
L	Inductance
Λ	Generalized space constant
Λ	Equivalent ionic conductance
λ	Equivalent ionic conductivity
λ	Length constant in linear cable
λ_T	Length constant in tubule
M	Net flux
M_{in}	Influx
M_{out}	Outflux
m	Activation constant for fast inward current
m	Tan Φ
m	In linear cable model $r_i r_e/(r_i + r_e)$
μ	Local electrochemical potential
N	Constant of proportionality
n	Activation variable for potassium current
n	Number of ions
n	Number of particles per cm^3
ω	Sign of fixed charges in membrane
ω	Angular frequency
$\bar{\omega}$	Characteristic frequency
$\underline{P}$	Ionic permeability
$\bar{P}$	Limiting ionic permeability
Q	Charge
R	Gas constant
R_0	Limiting resistance at low frequency
R_0^*	Limiting resistivity of a suspension at low frequency
R_∞	Limiting resistance at high frequency
R_∞^*	Limiting resistivity of a suspension at high frequency
R_{app}	Apparent resistance
R_d	Resistance of intercalated discs per cm^2 of fiber surface
R_d	Dendritic resistance
R_e	Resistivity of extracellular medium
R_{El}	Electrode resistance
R_i	Resistivity of cytoplasm
R_{inp}	Input resistance
R_{int}	Interaction resistance

R_m	Membrane resistance
R_p	Ionic shunt resistance
R_{pol}	Polarization resistance
R'_{pol}	Bulk polarization resistance
R_s	Soma resistance
R_s	Resistance of surface membrane
r	Donnan ratio
r	Coupling ratio of electrogenic pump
r	Radial distance
r_d	Resistance of intercalated discs per cm fiber length
r_i	Resistance of cytoplasm per unit fiber length
r_m	Membrane resistance per unit fiber length
r_o	Resistance of extracellular medium per unit fiber length
ρ	Resistivity of a solution
ρ'	Volume concentration of particles in a suspension
S	Power density
σ	Conductivity, general
σ	Conductivity of a suspension
σ_0	Limiting conductivity at low frequency
σ_1	Conductivity of suspending medium
σ_2	Conductivity of suspended phase
σ_∞	Limiting conductivity at high frequency
σ_e	Conductivity of suspending medium
σ_{inp}	Input conductivity
σ_w	Conductivity of cell wall
T	Tension
T	Temperature
T^+, T^-	Macroscopic transport numbers for cation and anion
t	Time
τ	Time constant
τ_m	Membrane time constant
τ_{inp}	Input time constant
τ_f	Time constant of foot of action potential
τ_r	Time constant of a relaxation system
U	Voltage
U	Voltage drop
u	Ionic mobility
V	Potential
V	Voltage, general
V_D	Donnan potential
V_m	Membrane potential
v	Voltmeter

v	Conduction velocity
v	Volume
X	Reactance
X	Fixed charge density in membrane
X_m^*	Specific membrane reactance
x	Interelectrode distance
ξ	Planck parameter
ξ	Volume to surface ratio of the tubules
$\mathbf{Y}$	Admittance
$\mathbf{Z}$	Impedance
$\mathbf{Z}_{\text{app}}$	Apparent impedance
$\mathbf{Z}_{\text{inp}}$	Input impedance
$\mathbf{Z}_{\text{int}}$	Interaction impedance
$\mathbf{Z}_m^*$	Specific membrane impedance
z	Number and sign of ionic charge

Technical and Theoretical Aspects of Biological Impedance Measurements

In 1913, Galler reported electrical resistance measurements across the body of a frog. He found that the resistance value differed between living and dead bodies, but one can rightly ask, What kind of relevant conclusion can be drawn from such an experiment? This example typifies the problems encountered in biological impedance measurements. It suggests that a prerequisite for a meaningful measurement is a clear understanding of what kind of electrical quantity is to be analyzed: the transverse resistance across the frog or the alteration of skin resistance at death. This latter quantity is of potential biological significance, but it is not directly accessible to measurement in the chosen system; the former quantity can be measured directly, but it is not very relevant from a biological point of view.

It is the aim of this introductory chapter to provide the elements necessary to a meaningful approach to biological impedance measurements and a critical evaluation of published results.

1.1 OBJECTIVE OF BIOLOGICAL IMPEDANCE MEASUREMENTS

The aim of a biological impedance measurement is the determination of the so-called electrical cell constants, the best known of which are the membrane resistance (R_m) expressed in Ω cm^2, the membrane capacitance (C_m, expressed in μF/cm^2), and the resistivity of the cytoplasm (R_i, expressed in Ω cm). The dimensions of these cell constants show that they are related to the geometry of biological structures and therefore cannot be obtained directly with a simple impedance measurement, for such a measurement results in a quantity with the dimension of Ω.

There are two advantages to determining the electrical cell constants: They permit comparisons of results obtained from structures of different size and geometry (for instance, a comparison of the membrane resistance of an erythrocyte with that of the squid giant axon); and they allow comparison of the electrically determined quantities with data obtained using other methods (for instance, comparison of membrane resistance for a given ionic channel with the same quantity calculated from tracer measurements).

Unfortunately, a direct measurement of the electrical cell constants is generally not possible, and it is useful to analyze the factors that influence the determination of such cell constant. Figure 1.1a shows the simplest case of a bioimpedance measurement. Terminals 1 and 2 are connected to two electrodes of equal surfaces, and between the electrodes is a block of tissue with tissue impedance Z_t. Contact between the metallic surface of the electrodes and the tissue is made by a film of saline solution. An impedance Z_{12} is measured between the electrodes. This impedance is generally higher than Z_t because of the additional tissue–electrode impedances Z_{1t} and Z_{2t}. Since Z_t is the only impedance of biological interest, methods have been devised to experimentally separate the tissue impedance from the tissue–electrode impedances (see Secs. 1.3.1, 1.3.2, and Appendix 1).

Unfortunately, Z_t itself varies with the geometry of the biological preparation and the position of the electrodes in contact with it. For specified conditions, Z_t is related to the electrical cell constants by mathematical models (see Sec. 1.3.1), which are crucial for the interpretation of biological impedance measurements. These models treat the biological structure (cell or tissue) as a two-terminal black box; the impedance measured between the terminals is called input impedance (Z_{inp}), a term that will be used in this book to characterize Z_t. The resistive component of Z_{inp} is the input resistance R_{inp}. In the case of whole tissues, this resistance has also been termed gross tissue resistance (Cole, 1933). At the cellular level, several terms such as electrotonic resistance (Katz, 1948), polarization resistance (Weidmann, 1952), and effective resistance (Tasaki, 1959a) are used for it. Moreover, we call the impedance measured between terminals 1 and 2 (Fig. 1.1a) "apparent impedance," Z_{app}. To arrive at a useful expression for the electrode–tissue impedances Z_{1t} and Z_{2t}, it is expedient to distinguish two components: Z^0_{El}, the impedance of both electrodes under specified reference conditions, and Z_{int} (interaction impedance), an impedance that arises during the process of measuring because of an interaction between the electrodes and the medium in

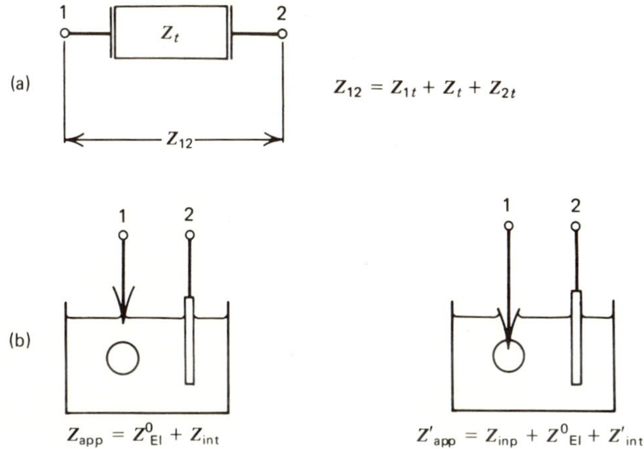

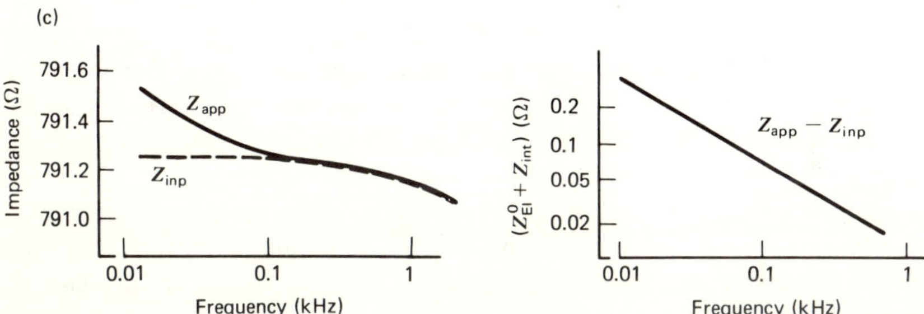

Figure 1.1. (a) Simplest case of a bioimpedance measurement: Two electrodes are connected to a block of tissue via thin films of saline. The impedance is measured between the terminals 1 and 2. (b) Determination of the input impedance of a biological cell with a microelectrode. Left, measurement of Z^0_{El}; right, the impedance between terminals 1 and 2 represents the input impedance when Z'_{int} can be measured independently or $Z_{inp} \gg Z_{int}$. (c) Left, relation between input impedance (dashed curve) and apparent impedance (solid curve) of a blood sample; right, electrode impedance $Z^0_{El} + Z_{int}$ as a function of frequency; note double logarithmic scale. Figure 1.1c modified after Schwan (1963). For details see text.

contact with them (electrode surface and saline in Fig. 1.1a). Now the relation between the apparent impedance and the input impedance can be formalized as follows:

$$Z_{app} = Z_{inp} + Z_{El}^0 + Z_{int} \tag{1.1}$$

In this equation, the terms Z_{El}^0 and Z_{int} comprise contributions from both electrodes.

The following example shows the application of Eq. 1.1 to an experimental situation: When the input resistance of a biological cell is determined by using a single microelectrode in a Wheatstone bridge (method 5 in Table 1.3), the bridge is balanced with the microelectrode resting in the extracellular medium. Under this condition, the quantity measured is $Z_{app} = Z_{El}^0 + Z_{int}$, where Z_{El}^0 is a value of the microelectrode impedance under specified reference conditions and Z_{int} represents the interaction impedance caused by the interaction of some parameters of the microelectrode and the resistivity of the medium surrounding its tip (Fig. 1.1b). Then the cell is impaled, the bridge is balanced again, and one obtains $Z'_{app} = Z_{inp} + Z_{El}^0 + Z'_{int}$. The difference between the two readings ($Z'_{app} - Z_{app}$) is often taken as Z_{inp}, but this interpretation is only valid when $\Delta Z_{int} = Z'_{int} - Z_{int}$ is either zero or small compared to Z_{inp}. However, the resistance of a microelectrode increases when its tip is transferred from the low-resistive extracellular medium into the more resistive cytoplasm (Schanne, Kawata, Schäfer, and Lavallée, 1966; see also Sec. 1.2), and this additional resistance component represents an interaction impedance Z_{int}. Consequently, the difference of the two bridge readings is not equal to the input impedance unless it can be shown that $Z_{inp} \gg Z_{int}$ or Z'_{int} can be measured directly.

Another application of Eq. 1.1 is shown in Fig. 1.1c. This measurement, reported by Schwan (1963), was performed in a measuring cell with platinum electrodes on a sample of blood. The apparent impedance measured at different frequencies is shown as the solid curve in the left figure, whereas the broken line corresponds to the input impedance. The difference of the two curves, which represents the polarization impedance of the electrodes, is shown at the right. According to Eq. 1.1, the polarization impedance can be identified with the term ($Z_{El}^0 + Z_{int}$). These examples are instructive because they show the general applicability of Eq. 1.1.

Input impedance and apparent impedance have the same dimension. In a given experimental situation, it is sometimes difficult to decide whether an apparent impedance or an input impedance has been measured. This question can be decided only with a thorough knowledge of the method applied or by control measurements with an independent method. Some confusion in the literature can be traced to a confounding of Z_{app} with Z_{inp}.

The terminology proposed in this book is general enough to analyze microelectrode measurements and those obtained with cell suspensions, and Eq. 1.1 permits description of important practical interactions between a measuring system and a biological structure that exist in microelectrode measurements.

1.2 ELECTRODE PROPERTIES AND CIRCUIT CONFIGURATIONS

It has been shown in the development of Eq. 1.1 that knowledge of electrode properties can be crucial for accurate measurement of input impedance. The discussion will be limited here to the two most important types of electrodes used in biological impedance measurements: the platinized electrode and the glass microelectrode. Since the configuration of the measuring circuit can also influence the analysis of the results, the more important measuring circuits will be discussed at the end of this section.

1.2.1 The Platinized Electrode

Usually, the platinized electrode is made of a sheet of platinum or a platinum wire that is electrolytically covered with platinum black to increase its effective area, which results in a lower electrode–electrolyte impedance. For a review, see Ferris (1974). This electrode is used mainly for ac measurements in suspensions with frequencies above 10 Hz. As frequency is increased, the electrode's impedance becomes very low compared to that of reversible electrodes, whose domain are measurements in the dc or low-frequency range. The impedance of a platinized electrode can be represented by a resistance and a capacitance in series, which approximates well its electrical properties at a given frequency and current density.

In the literature, the resistance is called polarization resistance R_{pol}, and the capacity is the polarization capacity C_{pol}. The term polarization is used here in a rather broad sense, implying that the nature of this impedance depends upon the not fully understood characteristics of the electrode–electrolyte interface in an ac field. The components of the polarization impedance, R_{pol} and C_{pol}, are linear only up to a current density of approximately 1 mA/cm^2, at a frequency of 1 kHz for a platinum black electrode (Schwan, 1963). Another characteristic of the polarization impedance is that R_{pol} and C_{pol} both depend upon the frequency of measurement. The polarization resistance and capacitance are power functions of the frequency. The exponent for the change of R_{pol} is about -0.5, whereas the exponent of the frequency for a change of C_{pol} is variable between -0.3 and -0.5. Moreover, these exponents depend on the electrode surface, electrolyte species, and current density.

In summary, the platinized electrode can be used for impedance measurements in macroscopic samples using frequencies above 10 Hz. For measurements with lower frequencies or direct current, reversible electrodes are preferable (Cole and Curtis, 1950); therefore, quantitative data given here are valid only for electrodes with a surface larger than 1 mm^2.

The factors influencing the polarization impedance are numerous and difficult to control. Apart from the frequency, the geometry of the electrode, the surface roughness (coat of platinum black), the current density, and the composition of the electrolyte, it has been found, for instance, that the concentration of erythrocytes in a sample of blood also changes the polarization impedance (Schwan, 1963). It is difficult, therefore, to give absolute values for polarization resistance and polarization capacity. For platinized electrodes, values of 5 Ω/mm^2 for R_{pol} and 2 $\mu F/mm^2$ for C_{pol} have been found in actual measurements with a frequency of 10 kHz. Under favorable conditions the value of the polarization impedance can be neglected when high frequencies are used and the resistance of the measured sample is large compared to R_{pol}. If a higher accuracy is desired, or if the measurement must be performed at a low frequency, the polarization impedance should be taken into account in obtaining the true impedance of the sample. Several methods have been proposed to separate the polarization impedance from the sample impedance. These methods use conductivity cells with a variable cell constant (variable electrode distance; Schwan, 1951), impedance measurements over a range of frequencies, substitution of an electrolyte of known conductivity for the sample, and the four-terminal method (see Appendix 1 and Schwan, 1963).

1.2.2 Glass Microelectrodes

Most modern electrophysiological studies on isolated biological cells are performed with glass microelectrodes consisting of a glass capillary tube with a small open tip (micropipet). Micropipets are filled with a highly concentrated electrolyte solution. The small tip is mainly responsible for the electrical properties of the microelectrode. In general, micropipets are drawn from glass capillaries made of glass with a high softening point (Hebert, 1969), to a diameter smaller than 0.5 μm at the tip. Although described by many, there is not complete agreement as to who was the first to use the glass microelectrode, and a history of it can be found in Geddes (1972).

A uniform nomenclature has been developed for the parts of a microelectrode as illustrated in Fig. 1.2. The largest part of uniform diameter is

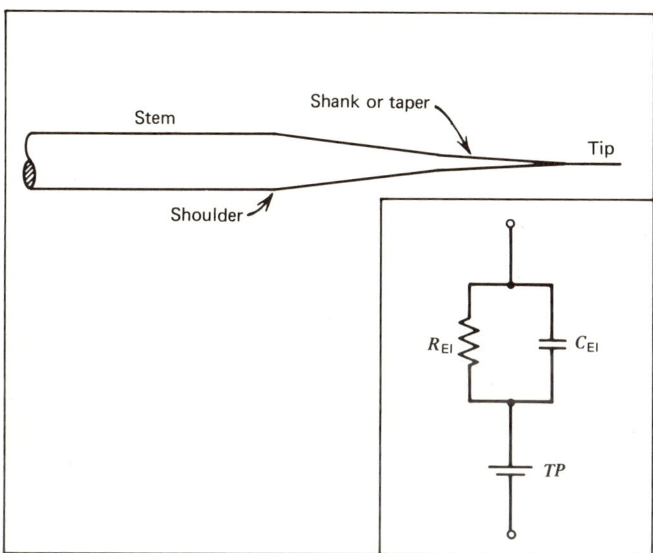

Figure 1.2. Nomenclature of parts of a microelectrode. Inset: Equivalent circuit of a microelectrode: R_{EI}, microelectrode resistance; C_{EI}, microelectrode capacity; and *TP*, tip potential.

the stem, and the part with a continuously decreasing diameter is the shank or taper. The transition between stem and shank is the shoulder; the terminal part is the microelectrode tip.

1.2.2.1 Electrical Properties of Glass Microelectrodes. A glass microelectrode is essentially an active nonlinear electrical element that can be roughly approximated by the circuit shown in the inset of Fig. 1.2: a resistance element in parallel with a capacitance in series with a source of voltage, the tip potential (*TP*). This circuit shows only the electrical parameters that are of practical importance: the tip resistance and the capacitance, which is situated across the wall of the part of the electrode immersed in the solution. This is a frequency-independent capacitance, in contrast to a polarization capacitance, which is of no practical importance in glass microelectrodes. In addition, the voltage source shown in the inset of Fig. 1.2 represents only the tip potential and neglects the small contribution of the diffusion potential (see below). Moreover, this equivalent circuit is an oversimplification in several aspects; it neglects (1) the nonlinear properties of the microelectrode impedance, (2) the fact that the microelectrode represents a network made up of distributed parameters, and (3) the practically negligible electrode–electrolyte impedance of the microelectrode. The electrical parameters of a microelectrode are not constant, but for specified conditions approximate values can be given. For instance, a micropipet made of Pyrex code 7740 glass and filled with 3 M KCl, having a tip diameter of 0.5 μm, and immersed 2 mm in a solution of 0.1 M KCl typically has the following electrical parameters: $R_{\text{El}} = 10$ MΩ; $C_{\text{El}} = 2\mu\,\mu$F, and a tip potential of 10 mV, inside negative.

(*a*) *The tip potential.* A fluid junction separating a 0.1 M KCl solution from 3 M KCl has a junction potential of about 0.6 mV. Therefore, the value of -10 mV actually found in microelectrodes cannot be attributed to a simple junction potential. However, if the tip is enlarged, the tip potential disappears and only the liquid junction potential is measured. Apart from the tip diameter, the parameters that in general determine the tip potential are the surface charge of the glass (Agin, 1969; Lavallée and Szabo, 1969) and the composition (Lev, 1969), concentration (Adrian, 1956; Kern, 1966), and pH of the solutions inside the electrode and surrounding its tip (Lavallée, 1964; Kostyuk, Sorokina, and Kholodova, 1969). The latter point is illustrated by Fig. 1.3, which shows the tip potential as a function of different concentrations of NaCl and KCl solutions.

Borosilicate glasses generally used for the production of microelectrodes have relatively low surface charges, and filling the microelectrode with a highly concentrated electrolyte further reduces the influence of surface charges. Küchler (1964) reported that the tip potential of a standard microelectrode made of Pyrex glass did not change between pH 6 and 8.

(*b*) *The microelectrode resistance.* The resistance of a glass microelectrode depends upon the same parameters as the tip potential. Figure 1.3 shows the microelectrode resistance as a function of different concentrations of KCl and NaCl solutions, the resistivity function of a microelectrode. This experimental result shows that the microelectrode resistance varies with the resistivity of the solution in which the microelectrode is immersed (ρ_e). This property has been used to estimate cytoplasmic resistivity (Schanne and Ceretti, 1965; Schanne, 1969; Law and Atwood, 1972).

Various theories have been proposed to explain the resistivity function of a microelectrode, and none has been supported unequivocally by experimental evidence. There are three types of theories: (1) The resistivity function does not depend on ρ_e (Fatt, 1961); (2) the microelectrode resistance is a linear function of ρ_e (Zablow, 1958; Snell, 1969; Engel et al., 1972), and (3) the microelectrode resistance depends in a complicated way on ρ_e, but for a certain range of resistivities it is a linear function of the logarithm of ρ_e (Lanthier and Schanne, 1966; Schanne, 1969). The first hypothesis has been abandoned in the light of experimental results. Theory 2 is based on a phenomenon known by electrobiologists as convergence

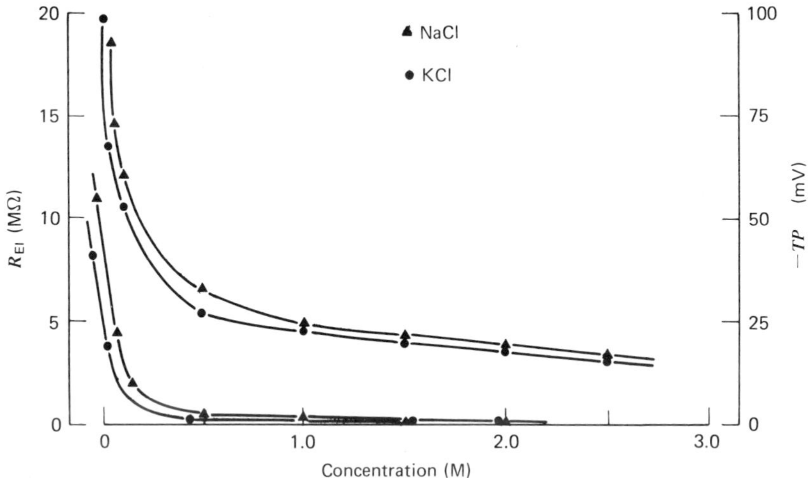

Figure 1.3. Dependence of microelectrode resistance (R_{EI}, upper curves) and tip potential (TP, lower curves) upon the concentration of the solution in which the microelectrode is immersed.

resistance or end effect (Cole and Guttman, 1942). However, this hypothesis is in apparent contradiction with the results of Fig. 1.3, which favor hypothesis 3. A solution to this apparent problem has been proposed by Schanne, Ceretti, and Rivard (1973): The microelectrode resistance consists of two components, one of them determined mainly by ionic processes inside the tip. This term determines the microelectrode resistance at low ρ_e. Superimposed on this latter relation there is a linear dependence of the microelectrode resistance on ρ_e, and this term is predominant at higher resistivities. This linear part can be obtained directly from an experimentally determined resistivity function. The tip diameter of the electrode can be calculated from a formula describing the convergence resistance proposed by Engel et al. (1972):

$$d = 8\rho_e / 3\pi^2 R_E \tag{1.2}$$

where d is the inner diameter of the microelectrode tip and R_E is the graphically determined convergence resistance. Redrawing Fig. 1.3 with an abscissa expressed in resistivity values (Schanne et al., 1968), the diameter of this electrode has been calculated to be 0.4 μm, a value in agreement with tip diameters obtained by electron microscopy (Gagné, 1970).

(*c*) *Microelectrode capacitance.* Compared to the tip potential and resistance, the capacitance of a microelectrode is a rather uncomplicated property. It results from the inside and outside electrolytes acting as electrical conductors while the glass wall of the microelectrode represents the dielectric. In fact, we are dealing with a leaky capacitor, because the tip at its smallest diameter is hydrated and has a finite resistance.

For a microelectrode with the properties mentioned on page 8, we can calculate a high-frequency limit (-3 dB) of 8 kHz using the equivalent circuit of Fig. 1.2. It has been found that the microelectrode capacitance increases with the depth of immersion in the bathing fluid, and a good estimation of its change is $1 \mu\mu F/mm$ (Nastuk and Hodgkin, 1950).

*1.2.2.2 **Double-Barreled Microelectrodes.*** To measure the input impedance of a single biological cell (see Sec. 1.3.2) with a voltage–current method, two microelectrodes have to be impaled in the cell—one for injecting the current and the other for measuring the resulting change in potential. In small cells the simultaneous impalement by two single microelectrodes is not possible; therefore, attempts have been made to use double- or multibarreled electrodes so that only a single penetration was necessary. Review articles describing the production and properties of these electrodes include Curtis (1964) and Frank and Becker (1964). In our laboratory, we prefer the method described by Coraboeuf (1969).

The equivalent electrical circuit of a double-barreled electrode is shown in Fig. 1.4. Considering that in the figure only the passive elements of a double-barreled microelectrode are shown (double-barreled microelectrodes do have tip potentials), it can be seen that basically we have two single microelectrodes in parallel, with the addition of the coupling resistance R_c and the coupling capacity C_c between the two barrels of the microelectrode. C_c can be measured using a suitable method for determining the capacitance between points 1 and 2 of the equivalent circuit. R_c is usually determined by injecting current through channel 1 and recording the resulting voltage drop through channel 2 when the electrode is immersed in the extracellular medium. As shown in Fig. 1.3, the resistance of a single microelectrode depends upon ρ_e. From this, we can assume that R_c changes upon penetration of a double-barreled microelectrode into a cell. This was confirmed by Wardell and Tomita (1967). A theoretical treatment of the electrical properties of double-barreled microelectrodes has been provided by Rush et al. (1968).

Methods exist for compensating electrically for R_c and C_c (Tomita, 1969a), but no ideal compensation, especially for C_c, has been obtained. A compensation for R_c is possible as long as R_c does not change much when the electrode tip enters the cell. When R_c varies, the measurement of the change in transmembrane potential upon current injection will include a voltage drop across a part of R_c. Similar considerations hold for concentric microelectrodes. However, it is possible to reduce or eliminate the coupling

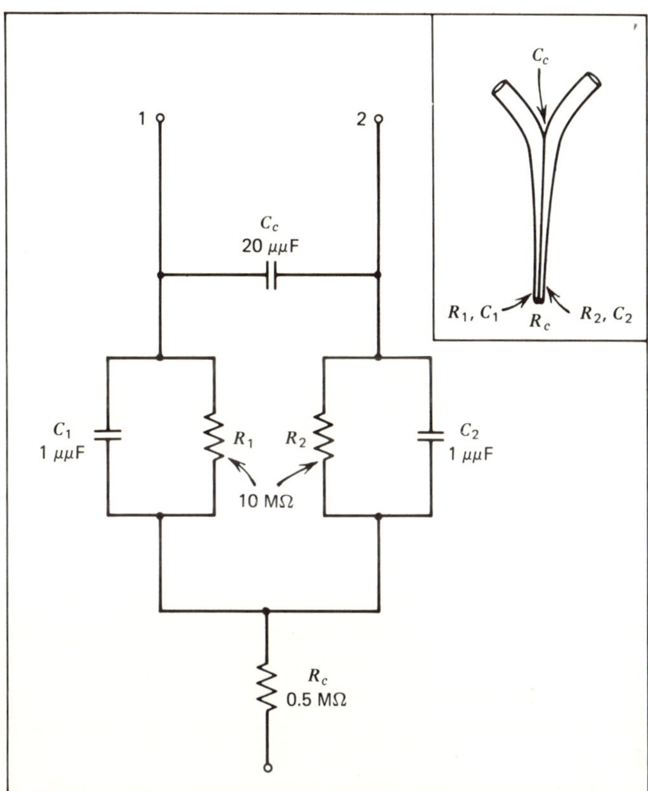

Figure 1.4. Equivalent circuit of the passive elements of a double-barreled microelectrode. R_1, C_1 and R_2, C_2 are the resistance and capacity of barrels 1 and 2, respectively; these elements are subject to the same simplifications discussed for the resistance and capacitance of a single microelectrode. R_c and C_c are the coupling resistance and coupling capacity, respectively. The numerical values are approximations found in a typical electrode. Inset: Localization of the passive electrical elements of a double-barreled microelectrode.

resistance of concentric electrodes by advancing the inner electrode relative to the tip of the outer electrode. This procedure is effective even with a diameter of 1 μm of the outer electrode (Tomita, 1969a).

According to the theory of double-barreled microelectrodes, it should be possible to reduce or eliminate C_c or R_c when an assembly of two closely spaced individual microelectrodes are used instead of a double-barreled microelectrode. For this, two individual microelectrodes are adjusted by means of a micromanipulator under microscopic control and then glued together (Nagai and Prosser, 1963b; Frank and Tauc, 1964; Tarr and

Sperelakis, 1964). The electrodes are glued together either before or after filling. It is extremely important to choose gluing material that will not shrink and cause misalignment.

Using two individual microelectrodes glued together reduces the coupling between them; but with higher currents, some electrical interaction between the electrodes still appears. The same was found when two independent microelectrodes were closely spaced, as shown by Falk and Fatt (1964).

1.2.2.3 Current-Carrying Microelectrodes. Sometimes a microelectrode must carry a substantial electrical current: for microinjection of charged molecules, intracellular stimulation, injection of current for measuring the input resistance, and in some voltage-clamp experiments. Only the last application is pertinent to our discussion here.

A current of 10^{-7} A causes a current density of 10 A/cm^2 at the tip of a microelectrode with 1 μm tip diameter. Since the electrode is filled with an electrolyte, such a high current density will certainly affect the electrical properties of the microelectrode, especially its resistance which is the most important variable affected by current flow. The change in microelectrode resistance as a function of the electrical current is not yet fully understood, but Rubio and Zubieta (1961) have shown that it increases with increasing dc current. These authors also showed that with higher current densities the current–voltage relation of a microelectrode becomes nonlinear and time dependent. The nature of this phenomenon has not yet been systematically investigated, and here only some factors influencing the current–voltage relation of microelectrodes are enumerated: current, waveform, polarity, and duration of the pulse.

It has been reported that filling the microelectrode with potassium citrate (Boistel and Fatt, 1958; Trautwein, Dudel, and Peper, 1965) or potassium sulfate (Frank and Tauc, 1964) would make its resistance less sensitive to current flow. However, a systematic investigation of the effectiveness of this procedure has not yet been published. As a general rule, the resistance of microelectrodes used for the injection of current should be kept as low as possible.

1.3 EXAMPLES OF CIRCUITS FOR BIOLOGICAL IMPEDANCE MEASUREMENTS

Since Du Bois-Reymond (1849) published a circuit for measuring resistances on nerve and muscle with a voltage–current method, numerous circuits have been employed. Some of the early studies are only of historical interest today. The circuits presented here were selected as typical of those presently used for a given experimental situation. Circuits for large electrodes (mainly platinized electrodes) are generally different from those employing glass microelectrodes. These differences are determined largely by the high impedance of microelectrodes and the low impedance of large-surface electrodes. Therefore, large-surface electrodes were used for measurements in the whole range of technically possible frequencies, whereas glass microelectrodes can be used only for impedance measurements with frequencies up to 10 kHz. On the other hand, measurements with large electrodes can yield a precision of 1%, whereas the precision for an impedance measurement with glass microelectrodes is 5%, provided conditions are favorable. A review of the construction details of measuring circuits for macroelectrodes has been published by Schwan (1963). A similar review for microelectrode circuits does not exist, and the references given in Table 1.3 should be consulted. For this book, circuits will be divided into three groups: circuits for large-surface electrodes, circuits for glass microelectrodes, and circuits for measurements using controlled voltage or current.

1.3.1 Circuits for Large-Surface Electrodes

Table 1.1 summarizes the methods most commonly used with large electrodes. In the second column the basic configuration for the method described is presented. The biological structures on which representative measurements have been performed are listed in the third column. Absolute values for the circuit elements are presented only when they are of special interest. The six methods listed in Table 1.1 all employ either some kind of reversible or platinized electrodes. Platinized electrodes are often used in conjunction with special measuring cells derived from conductivity cells used in electrochemistry. Some of the more important types are discussed in Appendix 1.

Circuit 1 illustrated in Table 1.1 is widely used for studies on uniform cable-like biological structures, such as nonmyelinated nerve fibers. This circuit employs two fixed electrodes El_1 and El_2 for injecting current, which can be monitored as a voltage drop across the known resistance R. This resistance is usually in the range of 10–100 MΩ in order to form, together with the generator G, a source of constant current. In practice, the current is monitored across a smaller resistor of about 1 kΩ in series with R. This allows the use of a conventional measuring device for v_2. The voltage-recording circuit consisting of electrodes El_3 and El_4 and voltmeter v_1 has a fixed electrode El_4, whereas El_3 can be displaced along the longitudinal axis of the specimen. With this circuit the applicability of the cable theory to a nonmyelinated nerve was proven (Hodgkin and Rushton,

Table 1.1 Circuits Used for Impedance Measurements with Large-Surface Electrodes

Number	Circuit	Structure	References
1	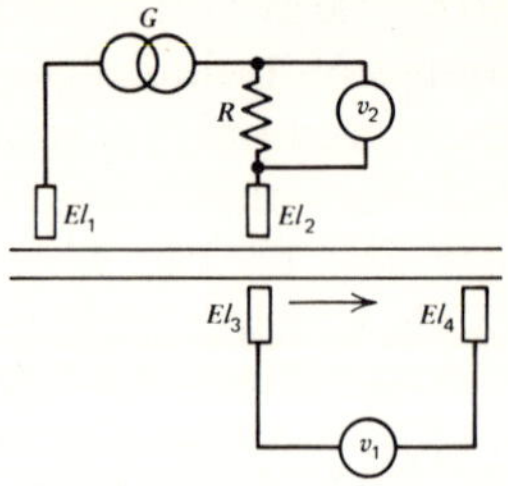	Nerve (*Homarus vulgaris*)	Hodgkin and Rushton (1946)
2	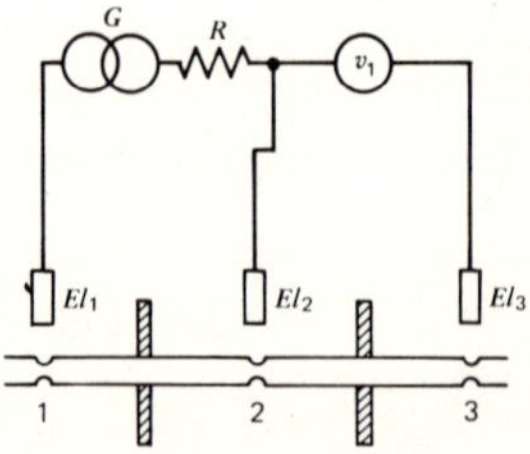	Myelinated nerve (single nerve fiber, frog)	Stämpfli (1958)
3	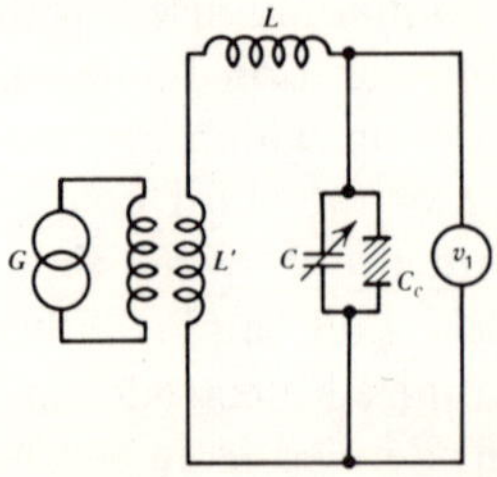	Not specified	Schwan (1963)
4	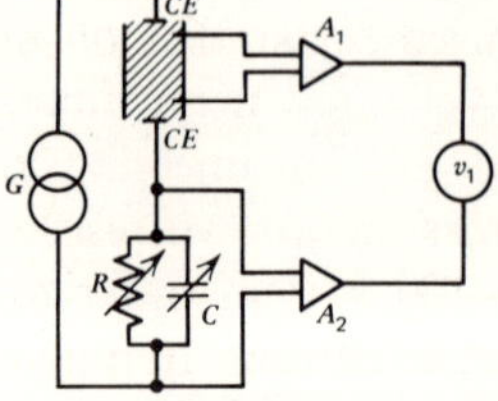	Not specified	Schwan (1963)

Table 1.1 *Continued*

Number	Circuit	Structure	References
5	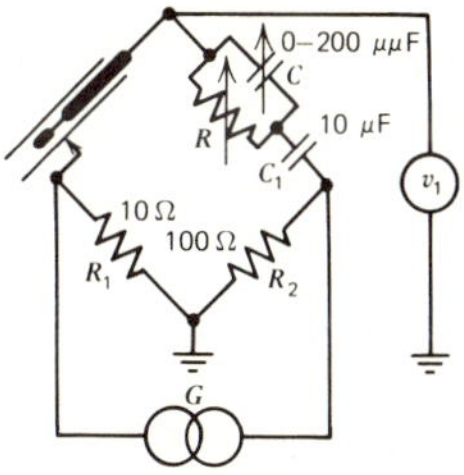	Giant axon (*Loligo*)	Tasaki and Hagiwara (1957b) Tasaki (1959a, b)
6	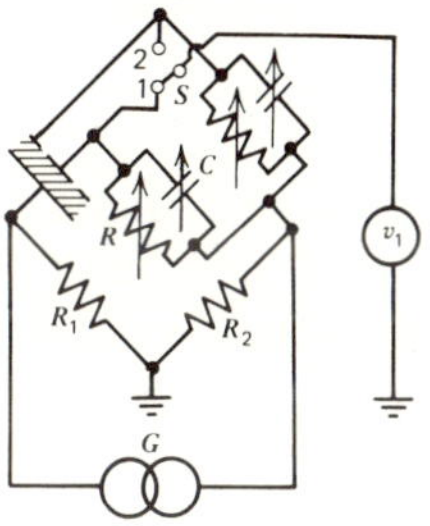	Cell suspensions	Cole and Gross (1949)

1946). A simplification of this method was reported that is less powerful but permits the determination of the membrane resistance provided certain assumptions are made (Hodgkin, 1947).

Circuit 2 was used to measure the voltage–current characteristic of a node of Ranvier in a myelinated nerve. The node under investigation is placed beneath electrode El_2. The neighboring nodes in compartments 1 and 3 are separated from compartment 2 (containing El_2) by insulating partitions. The circuit is equivalent to circuit 1, but the central electrode combines the functions of El_2 and El_3. The electrodes consist of chlorided silver wires immersed in agar made up with concentrated KCl solution. To establish a reference value for the membrane potential, the membranes of the nodes in compartments 1 and 3 are inactivated with cocaine. The principle of electrically isolating a section of a biological structure has led recently to the development of several measuring circuits for smooth and cardiac muscles.

Circuit 3 is basically a parallel-resonance circuit formed by L, L', and C. First the circuit is brought to resonance without the conductivity cell C_c; then the conductivity cell containing the preparation is added to the circuit, which is then brought back to resonance by readjusting C. By this procedure the impedance of the biological structure can be calculated

(Schwan, 1963). This method has been used to extend the usable frequency range for impedance measurements up to 100 MHz (Rajewski, 1938; Schwan, 1963).

Circuit 4 is essentially a voltage–current method used with a measuring cell that permits elimination of the polarization impedance of the current electrodes *CE* (see Appendix 1, Fig. A1.1b). Since a high sensitivity is needed, a compensation circuit was chosen. When v_1 reads null after adjusting R and C, the reading of these elements gives the impedance of the biological specimen. Usually, the measurement is made using the substitution method (Appendix 1). The amplifiers A_1 and A_2 are of equal gain, and each has a high input impedance. The values of R and C depend upon electrical characteristics of the measuring cell and the structure under study.

Circuit 5 illustrates an application of an impedance bridge to measurements on a large biological cell, the giant axon of the squid (*Loligo*). The electrodes used here are about 100 μm in diameter, which represents a limiting case for large-surface electrodes. Similar types of bridges can be adapted to measurements with classical two-terminal measuring cells (Appendix 1, Fig. A1.1). The values given for the elements in the bridge are typical for its application with a giant axon. R_1 and R_2 are kept low to place one terminal of the measuring cell close to ground potential, and the output of generator G must be isolated. The precision of a bridge used in measurements with ordinary two-terminal cells can be made much higher than that of an impedance bridge for microelectrodes. This affects the design of the balancing arm containing C and R: In most cases R consists of decade resistors and C consists of a high-quality variable capacitor.

A special application of a Wheatstone bridge is the three-terminal bridge shown as circuit 6. This bridge, which is used for measurements of the dielectric constant, has two balancing arms, each consisting of a parallel variable resistor and capacitor. The switch S permits selective balancing of either arm 1 (S in position 1) or arm 2 (S in position 2). The bridge is designed for use with guard-ring cells (Appendix 1, Fig. A1.1f). Arm 1 is connected to the central electrode of the cell and is used for determining the biological impedance and the dielectric constant. Arm 2 is connected to the guard ring, and its balance provides equal potential across the central electrodes and the guard ring. This procedure can eliminate the effect of the stray field on the electric field across the central electrodes. For a modern version of this bridge, see Bernhardt (1972). The type of specimen and the range of frequency for which such a bridge can be employed depend largely on the design of the guard-ring measuring cell. Three terminal bridges were used between 50 Hz and 5 MHz. Table 1.2

Table 1.2 Examples of Bridge Circuits Used for Determinations of Biological Impedances

Type	Frequency Range	Reference
Wheatstone	800 Hz to 4.5 MHz	Fricke and Morse (1926)
Wheatstone (universal bridge)	30 Hz to 10 MHz	Cole and Curtis (1937)
Wheatstone	10 Hz to 200 kHz	Schwan and Sittel (1953)
Schering (modern version: Hewlett-Packard RX-Meter 250 B)	500 kHz to 250 MHz	Pauly (1959)
Wheatstone (3-terminal bridge)	50 Hz to 5 MHz	Cole and Gross (1949)
Wheatstone (3-terminal bridge)	20 Hz to 500 kHz	Bernhardt (1972)

gives the frequency range of bridge circuits used successfully in the determination of biological impedances and lists the sources of technical details.

1.3.2 Circuits for Microelectrodes

The methods used for determining impedances with microelectrodes are summarized in Table 1.3. The same arrangement has been chosen as for Table 1.1. In the circuits of Table 1.3 the input impedance of the measuring instrument, v_1, has to be higher than in the circuits for large-surface electrodes. Therefore, cathode-follower or source-follower input stages are imperative.

Circuit 1 is the classical circuit for determining the electrical characteristics of longitudinal biological cells with intracellular electrodes. El_1 and El_2 are glass microelectrodes. El_1 is the potential electrode that records the shift of the transmembrane potential caused by the current injected through El_2. Usually, El_1 is positioned at different distances from El_2 in order to analyze the current distribution within the cell. The current-injecting circuit consists of electrode El_2, resistances R and R_1, and the generator G. The resistance R is usually made high (50–100 MΩ) so that the current flowing in this circuit is independent of changes in the resistance of El_2. Its second function is to prevent current driven by the membrane potential from flowing out of the cell. The value of R_1 is, in most cases, 1–10 kΩ, and a voltmeter v_2 connected across it serves to measure the current.

Circuit 2 is a modification of circuit 1. It is used to determine the input impedance of cells when the insertion of two separate microelectrodes is not possible. For this, double-barreled or concentric microelectrodes can be used in a single penetration. This circuit is also used when the site of

Table 1.3 Circuits Used for Impedance Measurements with Microelectrodes

Number	Circuit	Structure	References
1	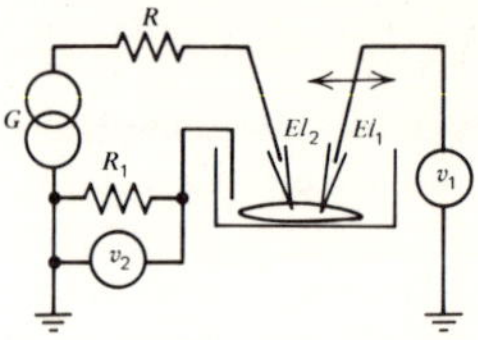	Purkinje fibers (kid)	Weidmann (1952)
2	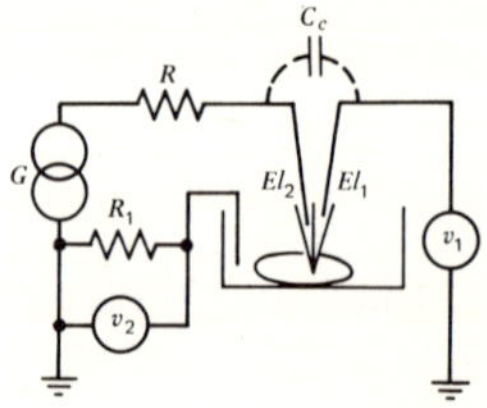	Motoneuron (cat)	Coombs, Eccles, and Fatt (1955)
3	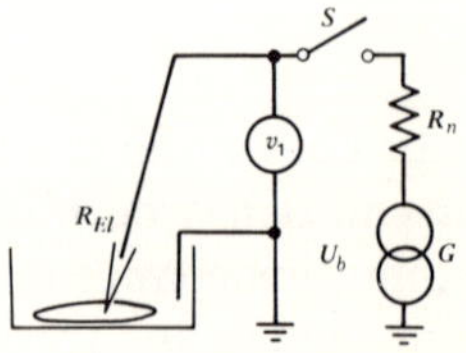	Sartorius (frog)	Schanne, Kawata, Schäfer, and Lavallée (1966)
4	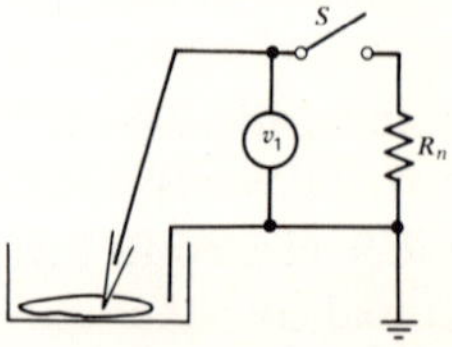	Sartorius (frog)	Kawata (1965)
5	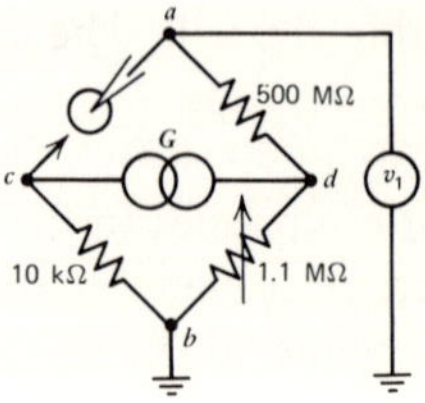	Cultured cardiac cells (chick)	Sperelakis and Lehmkuhl (1964)
6	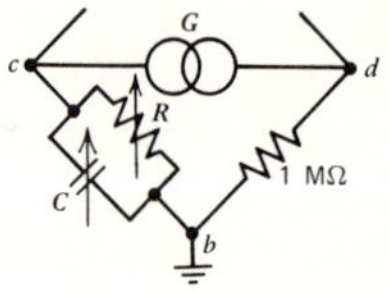	Cultured cardiac cells (chick)	Sperelakis and Lehmkuhl (1964)

penetration cannot be controlled microscopically. However, the electrical complications (coupling capacity C_c, indicated by broken line) of this type of microelectrode makes this method less precise than method 1 (see Sec. 1.2.2.2).

Circuit 3 is a simple circuit for measuring a membrane potential when switch S is open. When this switch is closed, a precision resistance of known value R_n, together with the electrode resistance R_{El}, form a voltage divider. The measuring voltage (U_b) is supplied by the generator G, and the voltage drop across R_{El} is measured. The impedance of a biological cell can then be measured by the subtraction principle, measuring electrode resistance first outside the cell and then inside the cell, provided the interaction impedance Z_{int} can be neglected (see Sec. 1.1), which is generally not the case. The method can be used with square pulses and sinusoidal signals.

Circuit 4 in Table 1.3 is basically the same as circuit 3. The essential difference between them is that in circuit 4 the voltage source to measure the apparent resistance is derived from the cell itself. Nevertheless, for this type of measurement the electrode resistance in the medium must be determined by another method. Therefore, method 4 is subject to the same restrictions as method 3.

Circuit 5 employs a typical bridge used with square-wave signals and glass microelectrodes. Such a circuit allows the measurement of the resistive component of a cell impedance. Resistance measurements with such a bridge are made by substitution; first the bridge is balanced in the medium and then, once the electrode has penetrated the cell, the bridge is rebalanced and the difference between the two readings is considered the input resistance of the cell. Since the resistance of the microelectrode enters into such a measurement as a constant, Eq. 1.1 holds, and an apparent resistance is measured. Moreover, circuit 5 is also used for stimulating and recording through the same electrode. Once the cell is penetrated, a subthreshold current pulse is injected, and the bridge is then balanced. When a suprathreshold stimulus is applied afterwards, no stimulus is recorded because the bridge is balanced, but the action potential can be measured between points a and b. This is possible because the source of the action potential is situated in the electrode arm of the bridge and not in the stimulator branch.

It is essential that the reference electrode of the system be close to ground potential. Therefore the resistance between b and c is made low. On the other hand, the resistance between a and d reduces the effective input impedance of v_1, which can cause a slight reduction of the measured value of the transmembrane potential. This can be avoided by opening branch a–d with a switch and using the bridge only for the resistance measurements.

Circuit 5 has also been used for impedance measurements where the deflection of the membrane potential as a function of an injected current was measured (Sperelakis and Lehmkuhl, 1964). The current was measured as a voltage drop across an additional resistor (about 50 MΩ) in branch *a–d*. In performing a measurement, the bridge is first balanced with the microelectrode in the extracellular medium, then the electrode is inserted into the cell, and a voltage drop ΔU is measured at the point where the current I is injected. Thus $\Delta U/I$ is equal to the input resistance R_{inp}, provided the interaction resistance of the microelectrode can be neglected. When the microelectrode resistance changes, again an apparent resistance is measured and Eq. 1.1 applies.

Circuit 6 shows an example of an impedance bridge. The circuit above points *d* and *c* is identical with the corresponding portion of circuit 5. Here the parallel circuit R and C permits balance of the impedance of the cell. The interaction resistance must be controlled as in the previous circuit to obtain the true input resistance.

1.3.3 Circuits for Measurements with Controlled Voltage or Current

The aim of the methods described here is not primarily the measurement of impedances, specifically the impedance of the membrane. Sometimes it is of interest to study cell membrane responses with either the current or the potential as independent variables (current-clamp or voltage-clamp). In the case of voltage-clamp measurements, the membrane potential is displaced from a reference value by an amount ΔV and the time-varying membrane current I_m is measured. Under experimentally manipulated conditions, the geometry of the current path through the membrane can be controlled and the current density J_m calculated:

$$J_m = I_m / A \tag{1.3}$$

where A is the surface through which the membrane current I_m passes. The membrane resistance R_m is given by

$$R_m = \Delta V / J_m \tag{1.4}$$

Table 1.4 is a summary of the basic circuits of voltage and current clamps used with various biological structures. The circuits are reduced to the essential elements to facilitate understanding of their basic functioning.

Circuit 1 is the classical voltage-clamp circuit as used by Hodgkin, Huxley, and Katz (1952) for the giant axon of *Loligo*. It will be used here to explain the principle underlying the method. The function generator F delivers the command signal U to the input of the control amplifier FA.

Consequently, the difference between the command signal and the membrane potential (V), sensed by the potential electrode P, appears at the input of FA, causing a current to be sent through the current electrode CE and across the membrane until the command signal and membrane potential are equal. It is possible to change the membrane potential V by either step or sawtooth (ramp) functions. The current flowing through the membrane delimited by two insulating partitions is measured with the meter A.

We may now describe circuit 1 provided no current is drawn through the potential electrode P by the equation:

$$\mu(U-V)=V+IR \tag{1.5}$$

where μ is the gain of the feedback amplifier FA, R represents the sum of the load resistances in the current circuit, and I is the current measured by the meter A. Solving for V, we obtain:

$$V=U\mu/(1+\mu)-IR/(1+\mu) \tag{1.6}$$

Inspection of Eq. 1.6 reveals several important properties of a voltage-clamp circuit:

1. The membrane potential becomes equal to the command signal U when the gain of FA is high, since $\mu/(1+\mu)$ approximates 1, and $IR/(1+\mu)$ becomes 0.
2. To make $IR/(1+\mu)$ negligibly small, I must be low; this is easy to achieve with small membrane surfaces like the node of Ranvier or by applying the isolated spot technique (see Table 1.4, circuit 9), but the current I cannot be kept low in the case of the giant axon.
3. In clamping the giant axon, it is necessary to reduce R to keep the second term of Eq. 1.6 negligible by using a low-resistance current electrode.
4. The characteristics of the amplifier FA are extremely important. Its gain has to be constant over a considerable bandwidth, to follow rapid changes of the command signal. The amplifier must have a low drift and a high input impedance. In recent years, it has become common practice to take advantage of operational amplifiers.
5. Since in an actual circuit R has an additional capacitive component—for instance, the capacitance of the membrane—the ratio of the time constant of the load impedance Z to the effective time constant of the amplifier input (as determined by its input resistance, its input capacitance, and the capacitance of the shielded leads to the input terminals) must be considered to prevent the system from oscillating. The ratio of the two time constants changes according to whether one uses systems having a small or a large time constant of the load impedance, (Katz and Schwartz, 1967).

Table 1.4 Circuits for Measurements with Current- and Voltage-Clamp Methods (For meaning of symbols see text.)

Number	Circuit	Structure	References
1	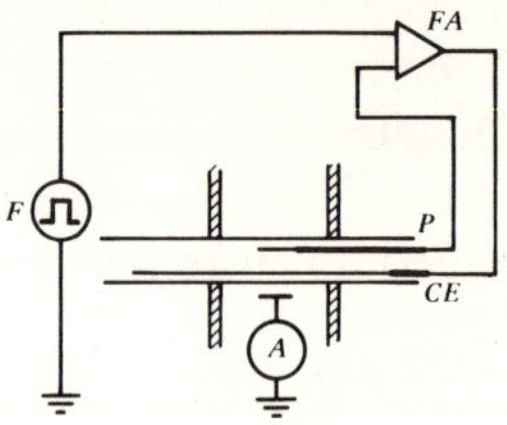	Giant axon (*Loligo*)	Hodgkin, Huxley, and Katz (1952)
2	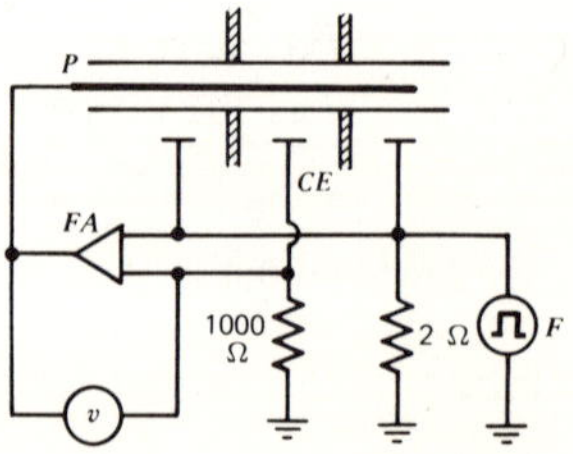	Giant axon (*Loligo*)	Marmont (1949)
3	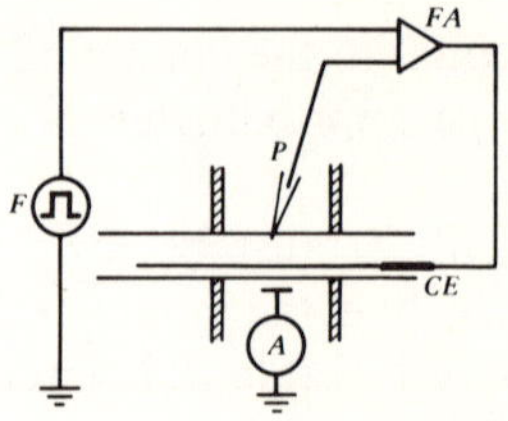	Giant axon	Cole and Moore (1960)
4	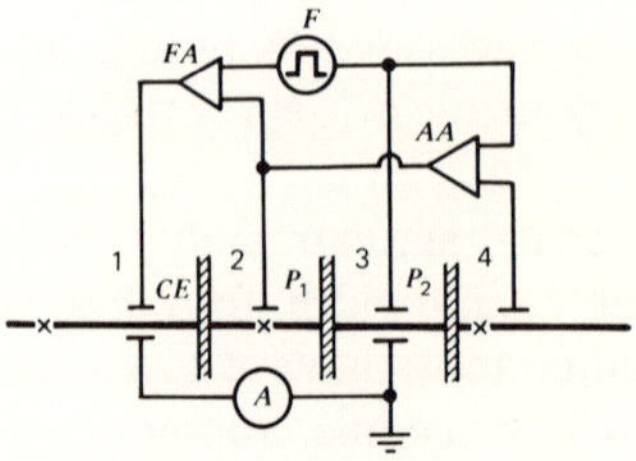	Ranvier node of single fiber from sciatic nerve (frog)	Dodge and Frankenhaeuser (1958)
5a	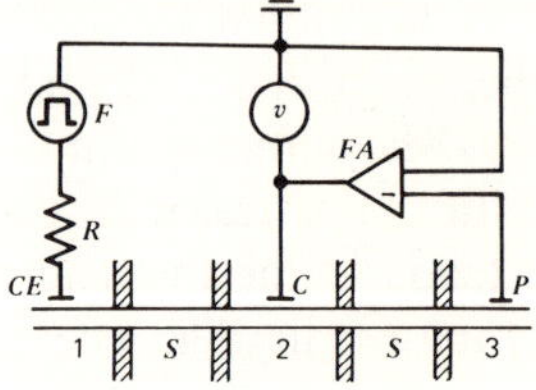	Skeletal muscle, single fibers (frog)	Stämpfli (1963)

Table 1.4 *Continued*

Number	Circuits	Structure	References
5b	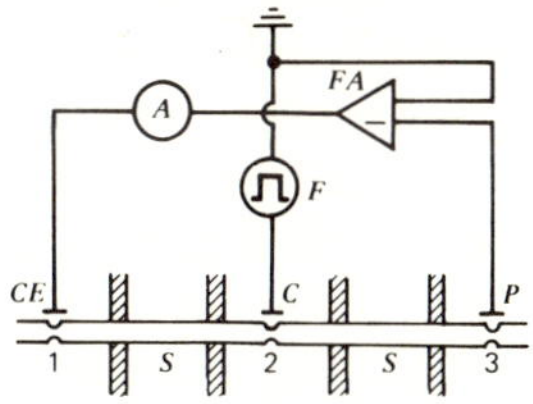	Bundles of nerve fibers (frog) Atrial trabecula (frog)	Bergman and Stämpfli (1966) Rougier, Vassort, and Stämpfli (1968)
6	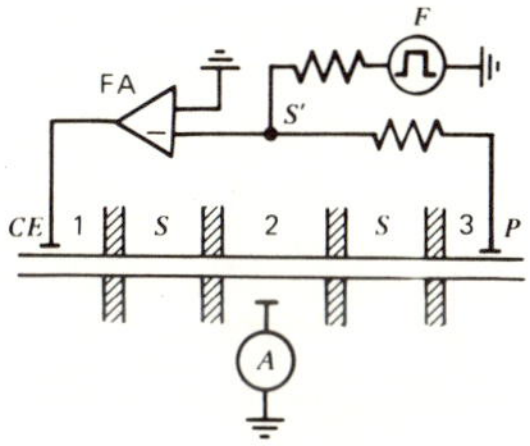	Lobster giant axon Atrial trabeculae (frog)	Julian, Moore, and Goldman (1962a, b)
7	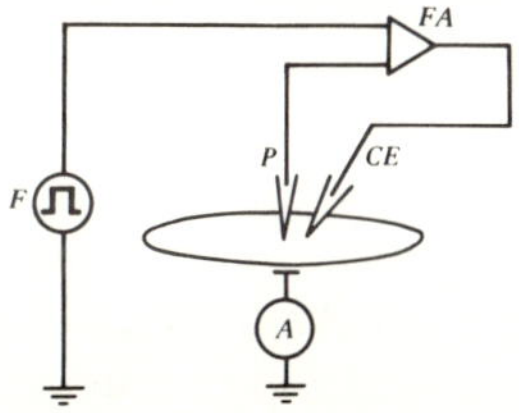	Ganglion of puffer fish	Hagiwara and Saito (1959a)
8	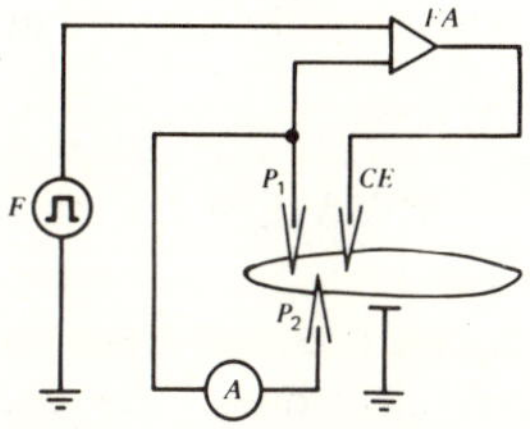	Sartorius muscle (frog)	Adrian, Chandler, and Hodgkin (1966)
9	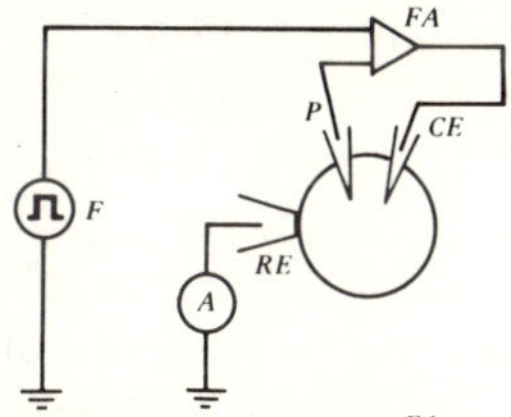	Ganglion cell (*Aplysia*)	Frank and Tauc (1964)
10	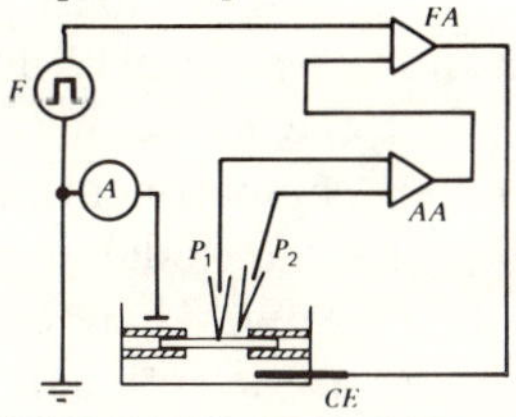	Eel electro-plaque	Nakamura, Nakajima, and Grundfest (1965a)

Such a simple analysis of a voltage-clamp circuit can indicate only some of the problems involved in its application. For anyone wishing to use the voltage-clamp technique, it is imperative to consult the literature. One useful review article is by Moore and Cole (1963). In the original circuit of Hodgkin, Huxley, and Katz potential and current electrodes were small-diameter platinum-blackened silver wires introduced longitudinally into the axon. These wires were wound around a small glass capillary and insulated with shellac, except for the parts indicated with thin lines in the circuit 1 diagram, which were electrolytically chlorided. The diameter of the electrode was kept to 120 μm, since electrodes with a diameter larger than 150 μm damage the axon (Hodgkin and Huxley, 1945). The partitions indicated in the figure were made of Plexiglas and served as guards for the solution in the lateral compartments, which were held at ground potential. The central compartment was 7 mm wide, which permitted good control of its geometry allowing current densities to be measured.

Circuit 2 of Table 1.4 employs the classical circuit of a current clamp. The input of the amplifier *FA* receives the difference between the command signal from the function generator *F* (which is in this case a current signal), and the current flowing through the membrane is picked up by the central current electrode *CE*. The current is injected through the internal electrode *P* until the current flow corresponds to the command signal. Thus the voltage required for maintaining a constant current is monitored by the meter v. The electrodes on both sides of the measuring chamber are guard electrodes. This method has been successfully applied to the giant axon of the squid.

Circuit 3 represents a refinement of circuit 1. An intracellular glass microelectrode is used for the potential control. When the microelectrode tip remains close to the inner surface of the membrane, better control of the potential is obtained than in the procedure using a central potential electrode. The arrangement of the guard electrodes, the width of the measuring chamber, and the functioning of the circuit are the same as in circuit 1.

Circuit 4 of Table 1.4 has been used for a large number of voltage-clamp experiments on myelinated single fibers of frog nerves by the group of Dodge and Frankenhaeuser. Three partitions of petroleum jelly seals, 150 μm thick, were arranged on the nerve so that the two central compartments 2 and 3 had a width of 200–300 μm. The lateral compartments were considerably wider. An x indicates the position of a Ranvier node. It can be shown that with the actual dimensions of the measuring circuit the potential change across the partition between compartments 2 and 3 is equal to the potential change produced across the nodal membrane in compartment 2 (Frankenhaeuser and Vallbo, 1969). Therefore, the displacement of the membrane potential caused by a command signal is controlled by electrodes P_1 and P_2 as it appears across the partition between compartments 2 and 3. It is characteristic of this method that the clamped potential appears between P_1 and P_2 and it does not represent the actual potential difference across the nodal membrane in compartment 2. The circuit involving the auxiliary amplifier *AA* serves to stabilize the potential between P_1 and P_2. The control amplifier *FA* supplies the current necessary to shift the potential across the nodal membrane in compartment 2. It is essential for the application of this method that the node in compartment 4 be inactivated. This is achieved by cocaine-Ringer's solution, or, preferably, isotonic KCl. The calculation of the current density through the nodal membrane depends upon the correct estimate of the nodal geometry. Dodge and Frankenhaeuser (1958) believe that their calculated current density is correct within a factor of 10.

Circuit 5a shows a current-clamp circuit that represents a further development of the gap method already discussed for circuit 2 in Table 1.1. The insulation between compartments was greatly improved by perfusing the spaces marked *S* with isotonic sucrose solution. The spaces are bound by petroleum jelly seals, as indicated by the shadowed areas. Provided the central compartment 2 is small compared to the length constant of the preparation, the space clamp is reasonably effective. Compartment 2 contains the membrane surface under investigation, which is perfused with test or control solution. Usually the membrane in compartments 1 and 3 is inactivated with isotonic KC1. Then the potential between electrodes *C* and *P* corresponds to the membrane potential, which is measured by v. This method was applied to myelinated nerves (Bergman and Stämpfli, 1966), but its most extensive use has been for potential measurements in frog auricles and smooth muscles (Rougier, Vassort, and Stämpfli, 1968; Kuriyama and Tomita, 1970).

Circuit 5b represents a voltage-clamp version of circuit 5a (Bergman and Stämpfli, 1966). The arrangement of the compartments and that of the sucrose gaps is the same as for circuit 5a. When the node in compartment 3 is inactivated, the interior of the fiber bundle is connected to the summing point of the amplifier. A signal from the function generator *F* will create a potential difference across the nodal membrane that will be counteracted by the current injected from the amplifier through the current electrode *CE*.

A disadvantage of this method is that it does not allow measurement of the membrane potential. However, one can obtain information of a change of the transmembrane potential (not its absolute value) by combining the circuits for the voltage and current clamps. In practice, this amounts to changing the configuration of the circuits during an experiment, which is technically feasible. The short-circuit factor (determined by shunt currents through imperfect insulation of the middle and lateral compartments) can be estimated by comparing the amplitude of an action potential measured with microelectrodes in compartment 2 and an action potential obtained under current clamp. On the other hand, the construction of the sucrose compartment *S*, as described for circuits 5, avoids a large hyperpolarization usually found at large interfaces of sucrose and Ringer's solutions. In the case of circuits 5, a hyperpolarization of only 7–10 mV in muscle fibers, over a 1–2 hr period, has been observed. Control over the geometry of the central compartment is difficult, and the determination of current densities is fairly uncertain.

Circuit 6 (Table 1.4) illustrates an alternative for the double-sucrose-gap method. The main difference between circuits 5b and 6 is that there are no petroleum jelly divisions between compartments 1–3 and sucrose compartments *S*. An ingenious device delimits and controls the membrane area under investigation by the flow of sucrose in compartments *S*. The method has the disadvantage that there is a strong hyperpolarization at the interface of sucrose and test solution. The control circuit is more straightforward: The summing point S' is outside the preparation, and the potential is measured between the inactivated membrane in compartment 3 and the electrode in compartment 2. This potential is compared at the summing point with the command signal from the generator *F*. The method has been successfully adapted to voltage-clamp measurements in trabeculae of frog auricles (Haas and Tarr, 1969).

The voltage-clamp circuit shown as number 7 in Table 1.4 has been applied to structures where penetration of two microelectrodes is possible. Hagiwara and Saito (1959a) applied this method to ganglion cells of the puffer fish, Kishimoto (1965) to the algae *Nitella*, and Takeuchi and Takeuchi (1959) to the end-plate membrane. The functioning of the circuit is straightforward, but the high resistance of the current-carrying microelectrodes reduces its effectiveness. With this method, no direct control over the electrotonic spread of the potential is possible, and thus the current density through the membrane cannot be determined.

Circuit 8 of Table 1.4 illustrates an elaborate voltage-clamp method used only in frog skeletal muscle. Three microelectrodes are introduced at the end of the muscle fiber, close to the tendon. Adrian and Freygang (1962a,b) showed that when the electrodes are arranged and connected as shown in the diagram the electrotonic potential can be controlled. According to Adrian, Chandler, and Hodgkin (1966), the potential distribution can be controlled to within $\pm 5\%$, provided the current electrode is impaled 250 μm from the end of the fiber. Therefore, the method permits evaluation of the current density.

Circuit 9 has been applied to the ganglion of the mollusk *Aplysia* by Frank and Tauc (1964). The functioning of the voltage-clamp circuit is straightforward. One microelectrode is used for potential control and another for current injection. A glass electrode of about 20 μm diameter, positioned directly on the outside of the cell membrane, serves as reference electrode *RE* to permit the measurement of the current over a known membrane surface (the area of the electrode opening). Provided there is no leak between the reference electrode and the outside of the membrane and the potential electrode is positioned close enough to the reference electrode to permit efficient potential control, the method permits determination of the current density. Circuit 9 is applicable to cells of unknown or complicated geometry, but the cells must be large enough to allow introduction of two microelectrodes and it must be possible to observe microscopically the positioning of all the electrodes.

Circuit 10 represents an application of the voltage-clamp method to the electroplaque of the electric eel. This organ has a nonelectrogenic low-resistance membrane (lower surface of preparation in the sketch) and an electrogenic side with a higher membrane resistance (top of preparation in sketch). The current is injected through the low-resistance side of the organ, and the voltage across the electrogenic membrane is controlled by means of two microelectrodes P_1 and P_2. The partitions supporting the electric organ also isolate the outside of its electrogenic membrane from the outside of its nonelectrogenic surface. The geometry of this partition controls the surface through which the current flows; determination of the current density is therefore possible.

1.4 MODELS RELATING INPUT IMPEDANCE TO ELECTRICAL CELL CONSTANTS

The preceding sections dealt mainly with the problem of how to separate the input impedance from the apparent impedance. Once this has been accomplished, the next step is to relate the input impedance to the electrical cell constants by a suitable model (see also Sec. 1.1). Table 1.5 lists models used to calculate electrical cell constants from the input impedance. The table shows either the equivalent electrical circuit or a topological representation of the model, lists examples of structures to which the model was applied, and cites key references.

The first circuit represents a model for the membrane impedance. It is used mainly for measurements where the surface of the membrane exposed to current flow, for example, the surface of a hole over which the membrane is extended, can be controlled by the experimenter and voltage-clamp measurements with simple structures where the geometry of the test compartment can be controlled. The relation between the input

Table 1.5 Models for Interpreting Input Impedance Measurements

Model	Schematic	Structure	Reference
1	R_{inp}, C_{inp}	Two-dimensional artificial membranes	Mueller and Rudin (1969)
		Toad skin	Whittembury (1964)
2	R_1, R_2, C	Different biological tissues	Philippson (1921)
		Frog atria	Tarr and Trank (1971)
3	R_1, R_2', $R_2'' \propto R_i$, $C \propto C_m$, $\propto Z_c$, R_2	Skeletal muscle	Bozler and Cole (1935)
4	r_e, a, r_m, c_m, b, r_i	Nonmyelinated axons	Hodgkin and Rushton (1946)
		Skeletal muscle	Katz (1948)

5	For a large reference electrode and a small intracellular potential electrode.	Skeletal muscle	Fatt and Katz (1951)
	Schematic similar to number 4.	Purkinje fibers	Weidmann (1952)
6	Same as for number 5. Modification for microelectrodes with alternating current	Skeletal muscle	Tasaki and Hagiwara (1957a)
7	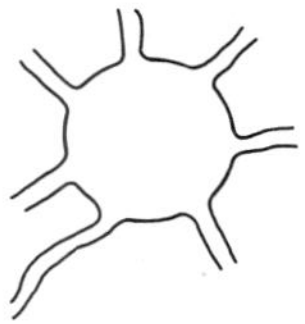	Motoneuron	Coombs, Curtis, and Eccles (1959)
8	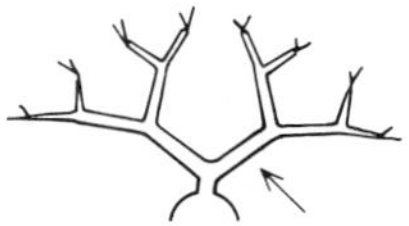	Neurons	Rall (1959, 1960, 1964, 1967)
9	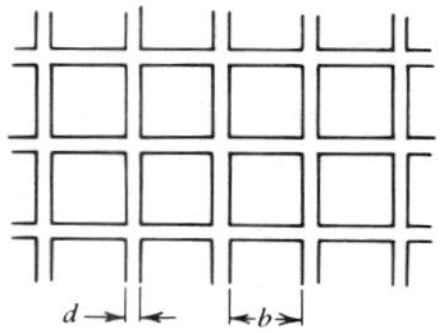	Ventricular muscle (mouse)	Tanaka and Sasaki (1966)
10	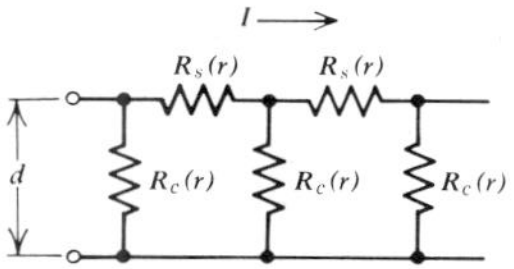	Gut epithelium	Shiba (1971)

resistance and the membrane resistance is shown by:

$$R_m = \Delta V/J = AR_{\text{inp}} \qquad (1.7)$$

Equation 1.7 defines the membrane resistance R_m as the ratio of the voltage drop ΔV across the membrane over the current density J. A is the area of the membrane surface exposed to current flow. Equation 1.7 defines the membrane resistance as a surface resistance with the dimension Ω cm^2; consequently, the respective conductance has the dimension mho/cm^2. Similarly, we obtain the membrane capacitance from the time constant of the input impedance τ_{inp}

$$C_m = \tau_{\text{inp}}/R_m \qquad (1.8)$$

This equation shows that the dimension of the membrane capacitance is F/cm^2. If the magnitude of the input impedance is measured as a function of frequency, it is related to the input resistance by:

$$Z_{\text{inp}} = R_{\text{inp}}/\left(1+\omega^2\tau^2_{\text{inp}}\right)^{1/2} \qquad (1.9)$$

Equation 1.7 shows that the area of the membrane is the proportionality factor between R_m and R_{inp}. This was used to estimate changes in membrane resistance without actually calculating R_m; in such a case, $\Delta R_m \propto \Delta R_{\text{inp}}$. This shortcut can be applied when (1) the current density over the entire membrane surface is homogeneous and (2) when the resistances of the solutions at both sides of the membrane are small compared to the input impedance. For typical applications of this procedure, see Junge and Moore (1966) and Chalazonitis et al. (1967).

Model 2 of Table 1.5 has been applied to a variety of tissues, but generally it cannot be correlated to the electrical cell constants. This model is historically important because it was used in early quantitative impedance measurements to demonstrate the fundamental electrical properties of biological bulk material (Philippson, 1921). An important electrical property of such a circuit is that it can be represented in the complex plane by a semicircle with its center on the real axis (see Appendix 2, Fig. A2.2). The derivation of this impedance locus is described in Appendix 2.

The input impedance of the model is related to the two limiting resistances R_0 and R_∞, which are obtained when Z_{inp} is measured at very low ($\omega=0$) and very high ($\omega=\infty$) frequencies according to the relation:

$$Z_{\text{inp}} = \left[\left(R_0^2+\omega^2\tau^2R_\infty^2\right)/\left(1+\omega^2\tau^2\right)\right]^{1/2} \qquad (1.10)$$

where $R_0=(R_1+R_2)$, $R_\infty=R_1$, and τ represents the time constant R_2C. Tarr and Trank (1971) used the model to interpret results of voltage-clamp measurements on frog atrium. They identified R_2 and C with the cell membrane, and R_1 was attributed to a resistance component in series with the parallel resistance-capacitance network of the membrane. Such an analysis does not yield the cell constants directly, but they can be calculated using additional information derived from the geometry of the preparation (see Sec. 3.2.3).

Model 3 has been used mainly to interpret impedance measurements of cell suspensions. The term suspension is used here in a very general sense; for example, a skeletal muscle can be considered a population of muscle fibers suspended in the extracellular medium. The equivalent circuit of the model contains two resistances R_1 and R_2 and a capacity C in series with R_2. The magnitude of its input impedance is also given by Eq. 1.10, but the definitions of R_0, R_∞, and τ are different and in this case are as follows:

$$\left.\begin{aligned} R_0 &= R_1 \\ R_\infty &= R_1R_2/(R_1+R_2) \\ \tau &= C(R_1+R_2) \end{aligned}\right\} \tag{1.11}$$

The impedance locus of the model is the same as that of model 2 (see Appendix 2, Fig. A2.2). In fact, the circuits of models 2 and 3 can be made electrically equivalent by a suitable choice of their components (Zobel, 1923; Fatt, 1964). In order to interpret a measurement on a biological structure in terms of electrical cell constants, R_1 can be ascribed to the extracellular medium. The same is true for R_2' as part of R_2 (Table 1.5). The elements in the box indicated by broken lines in model 3 are proportional to the cell impedance Z_c and they are related to the membrane capacitance C_m and the cytoplasmic resistivity R_i through Eqs. 1.18 and 1.19 (see Table 1.6). Solving these equations for R_i and C_m, we obtain:

$$R_i = R_\infty^* \frac{1-\rho'}{(1+\rho')^2}\left(\frac{R_\infty^*}{R_0^*-R_\infty^*} - \frac{1-\rho'}{1+\rho'}\right) \tag{1.20}$$

and

$$C_m = \frac{(1+\rho')^2}{4\rho'}\left(\frac{C^*}{a}\right) \tag{1.21}$$

The symbols are defined in Table 1.6. Moreover, Eqs. 1.17 and 1.19 show the relationship between the electrical cell constants and the resistivity of the extracellular medium.

Table 1.6 The Suspension Equation*

$$Z^*_{\text{inp}} = \frac{R_e[(\gamma+\rho')Z^*_c+(1-\rho')R_e]}{[\gamma(1-\rho')Z^*_c+(1+\gamma\rho')R_e]} \tag{1.12}$$

For cylindrical cells parallel to the electrical field, $\gamma=1$ (Velick and Gorin, 1940) and

$$Z^*_c = R_i-(j/a\omega C_m) \quad \text{for} \quad R_m\to\infty \tag{1.13}$$

$$Z^*_{\text{inp}} = \frac{R_e\{(1+\rho')[R_i-(j/a\omega C_m)]+(1-\rho')R_e\}}{\{(1-\rho')[R_i-j/(a\omega C_m)]+(1+\rho')\}R_e} \tag{1.14}$$

$$R^*_0 = R_e[(1+\rho')/(1-\rho')] \tag{1.15}$$

$$R^*_\infty = R_e[(1+\rho')R_i+(1-\rho')R_e]/[(1-\rho')R_i+(1+\rho')R_e] \tag{1.16}$$

$$R^*_1 = R_e(1+\rho')/(1-\rho') \tag{1.17}$$

$$R^*_2 = \underbrace{\left[R_e(1-\rho')^2/4\rho'\right]}_{R'_2\text{ (Table 1.5, model 3)}} + \underbrace{\left[R_i(1+\rho')^2/4\rho'\right]}_{R''_2\text{ (Table 1.5, model 3)}} \tag{1.18}$$

$$C^* = aC_m\left[4\rho'/(1+\rho')^2\right] \tag{1.19}$$

*Symbols: Z^*_{inp}, specific input impedance of suspension, Ω cm; Z^*_c, specific impedance of cell, Ω cm; R_e, resistivity of extracellular medium, Ω cm; R_i, resistivity of intracellular medium, Ω cm; C_m, capacity of 1 cm^2 of membrane, F/cm^2; R^*_0, resistivity of suspension for $\omega\to$zero; R^*_∞, resistivity of suspension, for $\omega\to\infty$; R^*_1, R^*_2, and C^*, specific resistances and capacity related to corresponding symbols (without asterisks) in schematic of model 3, Table 1.5 by Eq. A1.1; ρ', volume concentration; γ, form factor; a, cell radius.

Equation 1.12 in Table 1.6 relates Z^*, the resistivity of a suspension, to the electrical constants of the suspended phase. This equation, developed by Maxwell (1873) and refined by Rayleigh (1892), Fricke (1933), and Cole and Curtis (1950), is called here the suspension equation: γ is a form factor that becomes 1 in the case of cylindrical cells extended parallel to the electrical field and 2 for spherical cells. Z^*_c stands for the specific input impedance of a single cell. For instance, when Eq. 1.12 is applied to measurements on skeletal muscle (Bozler and Cole, 1935), γ becomes 1 and Z^*_c is represented by the cytoplasmic resistivity R_i in series with the specific impedance of the membrane capacity, $1/a\omega C_m$, where a is the cell radius. The shunt resistance of C_m (the membrane resistance) was assumed to be very high. Introducing the explicit expression for Z^*_c in Eq. 1.12, we obtain

Eq. 1.14; Eqs. 1.15 and 1.16 define the measured resistivities at very low and very high frequencies.

It is also possible to use more elaborate circuits for Z_c to represent the electrical properties of a single cell. This has been done successfully in investigating the electrical properties of the transverse tubular system in muscle fibers (Fatt, 1964; see also Sec. 3.1.1). Moreover, theoretical studies were undertaken to use the suspension equation for suspensions of particles with a finite membrane thickness. This problem was analyzed for a suspension of spheres (Pauly and Schwan, 1959; Schwan and Morowitz, 1962) and for a suspension of ellipsoids (Fricke, 1953b). Here is also a theoretical treatment of a suspension of structures having several layered membranes (Fricke, 1955).

The suspension equation is one of the most useful tools in interpreting biological impedance measurements. Its impact on these measurements can be compared to the importance of the linear cable theory for measurements on longitudinal cells. However, the suspension equation has the drawback that it does not generally permit evaluation of the membrane resistance.

Model 4 in Table 1.5 is the basic model for the application of the theory of a linear cable to biological cells. Equation 1.22 describes the potential distribution in a linear cable:

$$\lambda^2 \frac{\partial^2 V}{\partial x^2} - \tau_m \frac{\partial V}{\partial t} - V = 0 \tag{1.22}$$

where x is the distance between the current and the potential electrodes. The length constant λ and the membrane time constant τ_m are as defined by Eqs. 1.26 and 1.28 (page 34).

The equivalent circuit of the model represents the conductor of the core r_i (longitudinal resistance per unit length, in Ω/cm) separated from the extracellular medium r_e (external longitudinal resistance per unit length, in Ω/cm) by the impedance of the membrane r_m (membrane resistance per unit length, Ω cm) in parallel with c_m (membrane capacity per unit length, in F/cm). The relation of r_i and r_m with their respective cell constants is:

$$r_i = R_i / a^2\pi \tag{1.23}$$

and

$$r_m = R_m / 2a\pi \tag{1.24}$$

where a is the radius of the cell.

An analysis of model 4 leads to four quantities that can be determined experimentally (Hodgkin and Rushton, 1946; Katz, 1948): R_{inp},* the input resistance; λ, the length constant; m, the combined parallel resistance of external and internal media; and τ_m, the membrane time constant. The input resistance for measurements made with extracellular electrodes far from the fiber ends is given by:

$$R_{\text{inp}} = 0.5\left[r_m(r_i + r_e)\right]^{1/2} \tag{1.25}$$

and the remaining quantities are defined as:

$$\lambda = \left[r_m/(r_i + r_e)\right]^{1/2} \tag{1.26}$$

$$m = r_i r_e/(r_i + r_e) \tag{1.27}$$

$$\tau_m = r_m c_m \tag{1.28}$$

The length constant λ is a measure representing the longitudinal spread of the electrotonic potential. It is defined as the distance from a current-injecting electrode to the point where the electrotonic potential has declined to 37% ($1/e$) of its value at the point of current injection. The time constant of the membrane represents the time for a square pulse to reach 83% of its steady-state value when measured at the point of current injection (Hodgkin and Rushton, 1946). After having defined these quantities, it is now possible to relate the electrical cell constants to the input impedance:

$$R_i = \left[\pi(d/2)^2 m(1 + m\lambda)/2\right] R_{\text{inp}} \tag{1.29}$$

$$C_m = \tau_m / R_m \tag{1.30}$$

$$R_m = \left[2\pi(d/2)\lambda^2 m(2 + m\lambda)/2\right] R_{\text{inp}} + 2R_{\text{inp}}/m\lambda \tag{1.31}$$

In the discussion of model 4, it has been assumed that the measurements were made on a single fiber. However, this model has been applied also to fiber bundles, and in this case a determination of the extracellular space

*In the theory of a linear cable, the input resistance is defined as the disturbance of the membrane potential ΔV_0 over the injected current at the point of current injection (Eisenberg and Johnson, 1970). The resistance defined by the ratio of the disturbance of the membrane potential ΔV at any distance from the point of current injection to the injected current is called here effective resistance or, in accordance with the terminology used in the analysis of linear passive networks, transfer resistance (Schneider, 1970).

with electrical methods is possible; see the discussion of Katz (1948) in Sec. 3.1.2.

Model 5 represents the solution of the linear cable equation when measurements are performed with a large outside electrode and a small intracellular electrode. In this situation, the outside medium can be considered as being at isopotential and r_e becomes 0. Since r_e is eliminated from the equations, the expressions for R_{inp} and λ are simpler than the corresponding expressions for model 4:

$$R_{\text{inp}} = 0.5(r_i r_m)^{1/2} \tag{1.32}$$

$$\lambda = (r_m / r_i)^{1/2} \tag{1.33}$$

and consequently the expressions for the electrical cell constants become less complicated:

$$R_i = \pi d^2 R_{\text{inp}} / 2\lambda \tag{1.34}$$

$$R_m = 2\pi d \lambda R_{\text{inp}} \tag{1.35}$$

$$C_m = c_m / \pi d = \tau_m / R_m \tag{1.36}$$

Some simplifications of the linear cable equations were used to evaluate changes in membrane resistance when only changes in the electrotonic potential or the injected current were measured: combining Eq. 1.32 with Eqs. 1.23 and 1.24, we obtain:

$$R_m = 8 R_{\text{inp}}^2 \pi^2 a^3 / R_i \tag{1.37}$$

Considering that $R_{\text{inp}} = \Delta V_0 / I$, we obtain:

$$R_m \propto \Delta V_0^2 \tag{1.38}$$

provided that I, a, and R_i do not change under given experimental conditions. This simplification has been used to evaluate relative changes of membrane resistance during the action potential (Weidmann, 1951a) or when a cell was subjected to different treatments (Carmeliet, 1961b). On the same basis, the relation $G_m \propto I^2$ has been used to estimate changes in G_m from the currents required to shift the membrane potential by a given constant amount (Hubbard, 1963).

Model 6 in Table 1.5 represents a solution of the linear cable equation that permits the determination of the cell constants with alternating current, using microelectrodes. Here only the expression of the input resistance is necessary, because theoretically one can obtain the three equations for the determination of r_i, r_m, and τ_m by performing three measurements at three different frequencies. The relation of the constants r_i, r_m, and τ_m to the cell constants is the same as for model 5. Equation 1.39 describes the potential distribution $V(x,t)$ for frequencies >30 Hz when $\omega\tau_m \gg 1$:

$$V(x,t)=\frac{Ir_i^{1/2}}{2(\omega c_m)^{1/2}}\left\{\exp\left[-x(\omega c_m r_i/2)^{1/2}\right]\cos\left[\omega t-x(\omega r_i c_m/2)^{1/2}-\pi/4\right]\right\} \tag{1.39}$$

where I is the peak-to-peak amplitude of the injected current. Substituting the relations:

$$k=r_i^{1/2}/\left[2(2\pi c_m)^{1/2}\right] \tag{1.40}$$

and

$$h=1/\left[x(\pi r_i c_m)^{1/2}\right] \tag{1.41}$$

into Eq. 1.39 together with $V(x,t)=V$, we obtain:

$$\ln(Vf^{1/2}/I)=\ln k-f^{1/2}/h \tag{1.42}$$

The values of k and h can be measured from a plot of $\ln(Vf^{1/2}/I)$ versus $f^{1/2}$, which results in a straight line where k represents the intersection with the ordinate and $1/h$ is its slope (Tasaki and Hagiwara, 1957a). A modification of this method was used to evaluate the cable properties of taenia coli (Abe and Tomita, 1968). Using model 6, Tasaki and Hagiwara (1957a) reported a method for obtaining C_m from the foot of a propagated action potential:

$$\tau_f=1/r_i c_m v^2 \tag{1.43}$$

where τ_f is the time constant of the foot of the action potential and v is its propagation velocity.

Model 7 in Table 1.5 represents the standard neuron, widely used for the interpretation of impedance measurements on motoneurons (Coombs, Curtis, and Eccles, 1959). The soma is approximated by a sphere of 70 μm diameter, and the axon and dendrites are approximated by seven infinite linear cables with a diameter of 5 μm and a cytoplasmic resistivity of 75 Ω cm. The input resistance of this structure is considered to represent the input resistance of the soma (R_s) in parallel with the combined input resistance of the dendritic tree (R_d) so that we obtain:

$$R_{\text{inp}} = R_d R_s / (R_d + R_s) \tag{1.44}$$

Using the aforementioned numerical data, we obtain:

$$\left.\begin{aligned} R_d &= 7 \times 10^4 R_m^{1/2} \\ R_s &= 6.5 \times 10^3 R_m \end{aligned}\right\} \tag{1.45}$$

Combining Eqs. 1.44 and 1.45, we have:

$$R_{\text{inp}} = 4.55 \times 10^4 \times R_m / \left(0.65 R_m^{1/2} + 7\right) \tag{1.46}$$

This expression leads to a quadratic equation, which can be solved for R_m:

$$R_m - 0.14 \times 10^{-4} R_{\text{inp}} R_m^{1/2} - 1.54 \times 10^{-4} R_{\text{inp}} = 0 \tag{1.47}$$

In this equation, R_{inp} is expressed in multiples of $10^5\ \Omega$. Model 7 crudely approximates the real morphological structures, but its strength is that it allows the interpretation of the input resistances measured in neurons without introduction of geometrical parameters, which are difficult to quantitate.

Model 8 of Table 1.5 was developed to interpret more realistically the input impedance of a neuron. Here, the equation for a linear cable holds for a single branch, as indicated by an arrow in the diagram. However, the values for the cable elements are different for each branch. Actually, this model was conceived to describe the potential distribution in a neuron resulting from an imposed square pulse. The theory underlying it was developed for steady-state conditions (Rall, 1959) and for the analysis of transients (Rall, 1960). The model is developed to a degree that it is suitable to interpret the results of microelectrode studies, and the membrane parameters estimated on the basis of the model were different from those obtained using the standard neuron. The input resistance of model 8 is given by:

$$R_{\text{inp}} = R_m^{3/2} / \left(C D^{3/2} R_m + S R_m^{1/2}\right) \tag{1.48}$$

where S represents the area of the soma surface and the parameter C is defined as:

$$C = (\pi/2) R_i^{-1/2} \tag{1.49}$$

The quantity D represents a combined dendritic tree parameter, and under simplifying conditions it was found to be:

$$D^{3/2} \approx \sum_j d_j^{3/2} \tag{1.50}$$

With that the equation for the membrane resistance is:

$$R_m = (1+\varepsilon)C^2 D^3 R_{\text{inp}}^2 \tag{1.51}$$

where the parameter ε is defined as:

$$\varepsilon = \frac{1}{4}\left[1 + \left(1 + 4S/C^2 D^3 R_{\text{inp}}\right)^{1/2}\right]^2 - 1 \tag{1.52}$$

The definition of the parameters C and D reflects the limitations of the model since these parameters are difficult to evaluate considering the complex geometry of a neuron. This is the weak point in all of the models used for interpreting a structure with a complex geometry. Basically, the geometry of the whole group of these models, of which Rall's is one of the best, should be evaluated by morphological methods. Model 8 has been developed further by Jack and Redman (1971), who were mainly interested in computing excitatory postsynaptic potentials in the soma as a function of a brief current injection at any point of the dendritic tree. However, this model is less powerful than the one developed by Rall (1964, 1967), which became known as the compartmental model. More recently, Rall (1969a,b) used another approach to describe the electrotonic potential in neurons and showed that a combination of voltage- and current-clamp measurements can yield the respective contributions of soma membrane conductance and input conductance of the dendritic tree to the overall input conductance of a neuron.

Model 9 has been used to interpret measurements on cardiac muscle. It consists of a regular lattice, the components of which have core-conductor properties. The potential distribution for steady-state conditions is:

$$\lambda^2(d^2V/dx^2) + (\lambda^2/x)(dV/dx) - V = 0 \tag{1.53}$$

Using some approximations, we obtain for the input resistance of the lattice:

$$R_{\text{inp}} = (r_i\lambda/4)\left\{\left[\sinh(b/\lambda)\right]^{-1}\left[\exp(b/\lambda) - K_0(b/\lambda)/s\right] - 1\right\}^{-1} \tag{1.54}$$

where the meaning of b is shown in the diagram of Table 1.5, λ is defined as $(r_m/r_i)^{1/2}$, $K_0(b/\lambda)$ is a zero-order modified Bessel function, and s is defined as $K_0(\alpha/\lambda)$, α representing a constant depending on b and λ. In principle, the model is useful in interpreting the input resistance in terms of the electrical cell constants, provided a realistic estimation of b and d, the cell diameter can be obtained. A similar model was analyzed earlier by George (1961).

Model 10 represents a Heaviside's Bessel cable, proposed as the electrical equivalent for simple flat epithelial cells with low-resistive junctional membranes (Shiba, 1971). It describes the current spread in a two-dimensional cable. The following assumptions were made: (1) The junctional membranes have the same resistivity as the cytoplasm; (2) the cytoplasm constitutes a conducting fluid of uniform thickness and infinite dimension; (3) the surface membranes are a pair of thin parallel uniform sheets of leaky insulators that bound the cytoplasm on both sides; and (4) the external medium is infinitely conductive.

The electrical equivalent of the model is shown in Table 1.5. This model belongs to a type of transmission line, Heaviside's Bessel cable (Heaviside, 1899), and its elements are defined as:

$$R_s(r) = \frac{\rho}{2\pi r d}, \qquad R_c(r) = \frac{R_m}{4\pi r} \tag{1.55}$$

where r is the radius vector having the point of current injection as origin, $R_s(r)$ is the resistance of the bounding layer (surface membrane), $R_c(r)$ is the resistance of the conducting fluid, ρ is its resistivity, R_m is the membrane resistance per unit area of leaky membrane, and d is the thickness of the conducting layer. The electrotonic potential V elicited by the injection of current I satisfies the relation:

$$-dV/dr = R_s(r)I \tag{1.56}$$

for $r > a$, a being the radius of a cylindrical current-injecting electrode. The solution of Eq. 1.56, for steady-state conditions, in response to a unit current step through a very small electrode ($a \gg r$) is:

$$V(X,\infty) = (\rho/2\pi d)K_0(X) \tag{1.57}$$

where $K_0(X)$ is a zero-order modified Bessel function. X is defined as r/λ, where λ is the length constant:

$$\lambda - (dR_m/2\rho)^{1/2} \tag{1.58}$$

and appears in the expression for the membrane resistance:

$$R_m = 2\rho\lambda^2/d \tag{1.59}$$

The input resistance cannot be obtained from this formalism, since Eq. 1.58 has no solution when $x=0$. The cell constants λ and ρ are obtained by fitting the function $K_0(X)$ to the experimental data. The model has been applied to measurements performed on amphibian gastric epithelium.

Model 10 was further developed by Shiba and Kanno (1971) for the case when the electrotonic potential spreads preferentially in one direction, for instance, along a flat tissue of the rat atrial appendage (Woodbury and Crill, 1961). For this case, two length constants are needed, and the equations describe the potential distribution along the x and y axes characterized by the two length constants λ_x and λ_y. When $\lambda_x=\lambda_y$, the equations are the same as for model 10. The length constants are defined as:

$$\left.\begin{aligned}\lambda_x&=(dR_m/2\rho_{xx})^{1/2}\\ \lambda_y&=(dR_m/2\rho_{yy})^{1/2}\end{aligned}\right\} \qquad (1.60)$$

where ρ_{xx} and ρ_{yy} represent the specific resistance of the conducting fluid in the x and y directions; the other symbols are the same as in model 10. The potential distribution is described by:

$$V+dV/dT=\left[d^2V/dX^2+d^2V\right]/dY^2+4\pi\phi(X,Y)F(T) \qquad (1.61)$$

where V is the shift in membrane potential, $X=x/\lambda_x$, $Y=y/\lambda_y$, and $T=t/\tau_m$. Solutions have been derived for the transient during injection of a step function of current, steady-state conditions, and injection of a linearly rising current. For steady-state conditions, the electrotonic potential is given by:

$$V=(R_m/4\pi\lambda_x\lambda_y)K_0(r) \qquad (1.62)$$

where $K_0(r)$ is a zero-order modified Bessel function.

In addition to models 9 and 10, there are others describing the distribution of the electrotonic potential in two- or three-dimensional structures. They are all derived from the general expression:

$$\Lambda^2\nabla^2V-\tau_m\frac{\partial V}{\partial t}-V=0 \qquad (1.63)$$

which represents a special case of the Laplace equation. Equation 1.63 describes the potential distribution in a volume conductor with an internal homogeneous resistivity R_i and bounded by a thin membrane whose impedance is characterized by the time constant τ_m. Λ is a parameter with the dimension of length and related to R_i and the membrane impedance Z_m. The value of Λ varies with the frequency of the imposed signal due to the reactive properties of the membrane. However, when square pulses are employed, Λ depends in steady-state conditions only upon R_i and R_m, and consequently its value is constant. It has therefore been termed a generalized space constant (Eisenberg and Johnson, 1970).

Equation 1.63 is very general; it has to be adapted to a specific problem by a proper choice of the coordinates considering existing symmetries in the potential distribution and the geometry of the volume conductor. From this, it is clear that cylindrical and spherical volume conductors have been among the most intensely studied geometries. Moreover, the fact that the steady-state solution of Eq. 1.63 is analogous to the steady-state solution of the problem of heat flow in solids has greatly aided its theoretical analysis. There is a physical reason why, in general, a transient solution of the heat problem cannot be converted into a transient solution of the electrical problem: The material of the solid, in which heat conduction occurs, is characterized by a conductivity and a heat capacity, whereas the corresponding electrical property is only a conductivity. It is purely coincidental that the transient solution for a step of temperature imposed on a linear conductor leads to the same equations as when a voltage step is imposed on the end of a linear cable. Eisenberg and Johnson (1970) profited from the extensive collection of solutions of the heat conduction problem reported by Carslaw and Jaeger (1959) to analyze the potential distribution during impedance measurements with intracellular electrodes.

In the context of this book, we are concerned with solutions of Eq. 1.63 applied either to single cells or to small cell conglomerates consisting of electrically interconnected cells. A common characteristic of these solutions is the boundary condition that intracellular current flowing at right angles to the membrane, and immediately beneath it, is equal to the current leaving through the membrane. Another element entering the boundary conditions is the geometry of the current source, which has gained some importance for the analysis of the electrical field around a microelectrode when it serves as an intracellular source of current. It has been a problem in earlier analyses of input impedance measurements with circuit 1 of Table 1.3 that the electrotonic potential becomes infinitely large at the point of current injection in a volume conductor of complex geometry (Woodbury and Crill, 1961; Noble, 1962a; Shaw, 1964). However, as shown recently, this was only a consequence of considering the microelectrode as a point source of current. Taking into account the physical dimension of the microelectrode tip, the mathematical discontinuity of the electrotonic potential at the point of current injection can be avoided (Tarr, 1967; Minor and Maksimov, 1969; Eisenberg and Johnson, 1970).

Among the most widely studied solutions of Eq. 1.63 is its steady-state solution for the so-called linear cable model (Eq. 1.22):

$$\lambda^2 \frac{d^2V}{dx^2} - V = 0 \tag{1.64}$$

Here λ is the length constant and depends only on the resistance of the membrane and the resistivities of the intra- and extracellular media (Eq. 1.26). The basic assumption is that current spreads only in the longitudinal direction. This assumption is not strictly correct, since some current will flow across the membrane resistance. Therefore, the approximation of the linear cable model holds only as long as the longitudinal current is large compared to that flowing across the membrane. This situation exists as long as the diameter of the fiber is small compared with the length constant.

As already shown in the discussion of model 6 in Table 1.5, the linear cable equations may be solvable for measurements with sinusoidal current. In this case, r_m has to be replaced by z_m in Eqs. 1.25, 1.26, 1.32, and 1.33, and consequently the input impedance and the length constant become frequency dependent. According to Eisenberg and Johnson (1970), the potential distribution in a linear cable when intracellular electrodes are used and an alternating current is injected can be described by:

$$V(x,t) = \frac{1}{2} I_0(t)(z_m r_i)^{1/2} \exp\left[-x/(z_m/r_i)^{1/2}\right] \tag{1.65}$$

This equation resembles the potential distribution for dc conditions except that voltage and current vary with time and r_m is replaced by the membrane impedance z_m. Now, in ac analysis, z_m is represented by a complex number. Consequently, the ac length constant, λ_{ac}, is also a complex number and cannot have the same physical meaning as the dc length constant, λ_{dc}, which represents a real number. The new meaning of λ_{ac} becomes clear when we consider the exponent of Eq. 1.65 as a product $x\gamma$ where $\gamma = (a + jb)$ is a complex number called the propagation constant in transmission line theory (King, 1965). Here, a is called the attenuation constant and b is the phase constant. It has been shown by Falk and Fatt (1964) that λ^*, the reciprocal of the real part of γ, has the dimension of a length defined as the interelectrode distance when the peak value of the sinusoidal transmembrane potential is attenuated to a value $1/e$. Equation 1.65 then becomes:

$$V_{pp} = \frac{1}{2} I_{pp} (z_m r_i)^{1/2} \exp(-x/\lambda^*) \tag{1.66}$$

where I_{pp} is the peak-to-peak current at the point of current injection and λ^* can be written as:

$$\lambda^* = 1/a = 1/\left[\text{real part of } (r_i/z_m)^{1/2}\right] \tag{1.67}$$

It has been shown that when a frequency sufficiently high is used, the membrane impedance is short circuited, and λ^* may be expressed as:

$$\lambda^* = (1/\omega c_m r_i)^{1/2} \tag{1.68}$$

It is useful to know that λ_{dc} is a maximal value of the length constant. Therefore, whenever fast-changing membrane potentials are encountered, either as action potentials or as imposed square pulses, it must be taken into account that due to the high-frequency components of fast-changing voltages, λ^* and not λ will determine the electrotonic spread.

Eisenberg and Johnson (1970) also analyzed the three-dimensional electrotonic spread in a long cylindrical cell to obtain quantitative criteria for the limitation of the linear cable model. They arrived at a distribution of the electrotonic potential under steady-state conditions of the form:

$$V(x,\theta) = 0.5 I_0 r_i a\left[L(x) + S(x,\theta)\right] \tag{1.69}$$

where a is the cell radius and θ is the circumferential angle between the electrodes. The intracellular electrodes are assumed to be point sources, and consequently no solution for very small interelectrode distances was obtained. The function $L(x)$ is defined as follows:

$$L(x) \equiv (\lambda/a)\exp(-x/\lambda) \tag{1.70}$$

which represents the longitudinal decrement caused by one-dimensional spread of current. The term $S(x,\theta)$ represents the contribution of the three-dimensional current spread to the electrotonic potential. It is defined by:

$$S(x,\theta) = \sum_{n=-\infty}^{n=+\infty} \sum_{s=1}^{s=\infty} A^*_{n,s} \cos n\theta \exp\left[-j'_{n,s}(x/a)\right] \tag{1.71}$$

where:

$$A^* = A_{n,s} \frac{J_n(j'_{n,s} r/a) J_n(j_{n,s} r'/a)}{J_n^2(j'_{n,s})} \tag{1.72}$$

and:

$$A_{n,s} = \frac{j'_{n,s}}{j'^2_{n,s} - n^2} \tag{1.73}$$

In this formalism, the symbol $j'_{n,s}$ represents the sth root, in order of increasing magnitude of the function $J'(z)$. These roots and the coefficients $A_{n,s}$ were tabulated for cylindrical cells by Eisenberg and Johnson (1970). The accuracy of the series expansion (Eq. 1.71) is estimated to be quite good as long as λ is greater than two fiber radii and reasonable as long as the length constant is larger than one fiber radius.

It follows from Eq. 1.69 that if $S(s,\theta)=0$, it reduces to the well-known form of potential distribution in the linear cable:

$$V(x) = V_0 \exp(-x/\lambda) \tag{1.74}$$

The authors tabulated their results (Table 1.7), which are of considerable practical importance because they provide the correction factors for the electrotonic potential calculated with Eq. 1.74. The correction factors in the table are calculated for the case when the tips of the current-injecting and potential-recording microelectrodes are just beneath the membrane surface ($r=r'=a$). It is evident from Table 1.7 that with increasing λ the corrections become less important, and the same is true when measurements are made at interelectrode distances large compared to the fiber radius. Moreover, it is interesting to see that the correction is minimal for electrodes positioned at a circumferential angle between 45° and 90°.

Eisenberg and Johnson analyzed also the model of a thin plane cell, which is important for measurements in sheets of interconnecting cells. The geometry of the thin plane cell is shown in Fig. 1.5. Two unrestricted sheets of membrane bound a block of material with thickness d and resistivity R_i. The current is injected through electrode 1 at depth z_1 at radial position zero, whereas the electrotonic potential is measured with electrode 2 at depth z_2 at radial position r.

Using basically the same approach as in the case of a long cylindrical cell, the authors derived the following expression for the electrotonic potential:

$$V(r,z) = \frac{I_0 R_i}{2\pi d}\left\{ K_0\left[\frac{r}{d}\left(\frac{2d}{\Lambda}\right)^{1/2}\right] + 2\sum_{n=1}^{n=\infty} K_0\left(\frac{n\pi r}{d}\right) f(z_1,z_2)\right\} \tag{1.75}$$

where K_0 is a modified Bessel function of the second order (Abramowitz and Stegun, 1967), Λ is a generalized length constant defined as R_m/R_i,

Table 1.7 Correction Factors for ΔV Calculated with the Linear Cable Model to Take into Account the Effects of Three-Dimensional Current Spread in a Cylindrical Cell*

		x/a				
λ/a	θ (deg)	0.25	0.5	0.75	1	2
2.0	0	2.81	1.78	1.44	1.27	1.06
	5.6	2.67	1.76	1.43	—	—
	11.3	2.36	1.70	1.41	1.26	—
	22.5	1.79	1.53	1.35	1.23	—
	45	1.22	1.22	1.18	1.14	1.04
	90	0.83	0.89	0.94	0.96	1.00
	135	0.70	0.76	0.82	0.87	0.96
	180	0.66	0.73	0.79	0.84	0.95
4.0	0	1.85	1.34	1.18	1.11	1.02
	5.6	1.79	1.33	1.18	—	—
	11.3	1.64	1.31	1.17	1.10	—
	22.5	1.37	1.23	1.14	1.09	—
	45	1.10	1.10	1.07	1.05	1.01
	90	0.92	0.95	0.97	0.99	1.00
	135	0.86	0.90	0.93	0.95	0.99
	180	0.84	0.88	0.91	0.94	0.99
10.0	0	1.33	1.13	1.06	1.04	1.01
	5.6	1.30	1.12	1.06	—	—
	11.3	1.25	1.11	1.06	1.03	—
	22.5	1.14	1.09	1.05	1.03	—
	45	1.04	1.04	1.03	1.02	1.00
	90	0.97	0.98	0.99	1.00	1.00
	135	0.95	0.96	0.97	0.98	1.00
	180	0.94	0.96	0.97	0.98	1.00

*From Eisenberg and Johnson (1970). Reprinted by permission.

and $f(z_1,z_2)$ is given by:

$$f(z_1,z_2)=\left[\cos\frac{2z_1\beta'}{d}+\frac{d\sin(2z_1\beta'/d)}{2\Lambda\beta'}\right]\left[\cos\frac{2z_2\beta'}{d}+\frac{d\sin(2z_2\beta'/d)}{2\Lambda\beta'}\right] \tag{1.76}$$

where

$$\left.\begin{aligned}\beta'\tan\beta'&=d/2\Lambda\\ \beta'\cot\beta'&=-d/2\Lambda\end{aligned}\right\} \tag{1.77}$$

The other symbols in Eqs. 1.75–1.77 are defined in Fig. 1.5. In the case

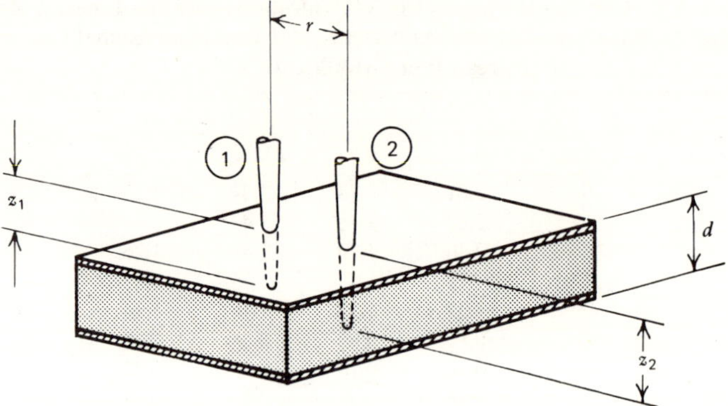

Figure 1.5. Geometry of the thin plane cell with two microelectrodes impaled. For explanation see text. Modified after Eisenberg and Johnson (1973), by permission.

when the tips of the two electrodes are just beneath one membrane, $f(z_1, z_2)$ becomes unity. Equation 1.75 can now be rewritten formally as:

$$V(r,z) = \frac{I_0 R_i}{2\pi d}[P+Q] \tag{1.78}$$

As in the case for the long cylindrical cell, the solution for the electrotonic potential consists of two terms, one dependent upon the membrane properties (P) and one independent of them (Q). Numerical values for Q have been calculated and are summarized in Table 1.8. It can be seen that the influence of Q diminishes with large interelectrode distances.

The potential distribution in a thin plane cell has been analyzed by other authors (Woodbury and Crill, 1961; Noble, 1962a; Tarr, 1967), but the formalism discussed here is more suitable for practical applications than those resulting from previous attempts.

Heppner and Plonsey (1970) derived equations for the electrical interaction through the intercalated disc region between two abutting cells. The solution had the form of a series of Bessel functions. The intercalated disc resistance was estimated using a field analysis of the junction. The closed solution for the same equivalent circuit had been originally derived by

Table 1.8 Contribution of Q to the Potential Distribution in a Thin Plane Cell*

r/d	0.1	0.15	0.2	0.25	0.3	0.4	0.5	0.75	1.00
Q	6.84	3.92	2.52	1.78	1.32	0.72	0.46	0.07	0.03

*From Eisenberg and Johnson (1970). Reprinted by permission.

Woodbury and Crill (1961) and it has been recently revised and corrected (Woodbury and Crill, 1970). Using the appropriate parameters, the disc resistance has been estimated, but the new result only slightly modified the original estimation reported in 1961.

Other models that have been applied to intercellular junctions are: (1) a delta configuration for a network representing two cells electrotonically coupled (Bennett, 1966) and (2) an analytical method to estimate the intercellular resistance in smooth muscle by measurements of injury and sucrose-gap potentials (Koide, 1967). Two models were considered: end-to-end bridges, a combination of a large number of series and parallel elements, and the lateral bridge, in which the cells were considered to be radially or laterally adjacent to each other. Although these models have been used to evaluate the electrical properties of junctional membranes, they have not been included in the table or discussed in detail because they cannot be used to interpret the actually measured input impedance in terms of cell constants.

1.5 ELEMENTS OF A BIOLOGICAL IMPEDANCE

An impedance measurement on a biological structure will result in a value for its resistance and its positive or negative reactance. One of the most important and challenging problems in electrobiology has been to relate these electrical quantities to biological structures and mechanisms. This process of identification is far from being completed. It is therefore useful to review the concepts developed to interpret resistance, capacitance, and inductance in terms of their biological correlates.

1.5.1 Resistances

In general, one can consider an ohmic resistance as a dissipative element (Cole, 1933) or a nonconservative element in which, ultimately, energy is lost as heat. In almost all biological phenomena accessible to impedance measurements, the electrical current is carried by ions. Therefore, the most plausible interpretation for a biological resistance is an ion moving under the influence of an electrochemical driving force and hindered in its way by various forces. One of these forces can be the friction in a narrow channel through which the ion is squeezed or the viscosity of the medium in which the ion is moving. It is possible to give a general formulation of a resistance between two points A and B in a biological system, for example, inside and outside a membrane (Johnson, Eyring, and Pollisar, 1963, chapter 2) when the current is carried by ion j:

$$\mu_{Aj} - \mu_{Bj} = M_j F R_j \tag{1.79}$$

where μ is the local electrochemical potential of the ion j at point A or B (see Eq. 1.85), M is the flux of the ion j, F is the Faraday, and R_j is the resistance between points A and B. Taking into account that the flux M_j is related to the current density, J_j, we obtain an equation with purely electrical terms on the right-hand side:

$$\left.\begin{aligned} M_j &= J_j/F \\ \mu_{Aj} - \mu_{Bj} &= J_j R_j \end{aligned}\right\} \qquad (1.80)$$

The meaning of the resistance term for biological measurements becomes clear if we consider that the conductivity σ of a given solution of a univalent salt is related to its equivalent ionic conductance Λ, its equivalent ionic conductivities λ_+ and λ_-, and the mobilities u of its constituent ions by a set of equations:

$$\left.\begin{aligned} \sigma &= \Lambda C/10^3 \\ \sigma &= C(\lambda_+ + \lambda_-)\times 10^3 \\ \sigma &= (u_+ + u_-)C \end{aligned}\right\} \qquad (1.81)$$

where C is the concentration expressed in equivalents/liter. These relations permit calculation of the contribution of a given ion to the measured conductivity in a solution, since the values of Λ and λ are tabulated (Landolt-Börnstein, 1960).

1.5.1.1 Classification of Resistances. The ionic mechanisms just outlined can result in different types of resistances systematized according to their voltage–current curves and their time dependence. Mauro (1961) published such a classification, distinguishing between time-invariant and time-variant resistances. In the first class he placed linear, symmetrical, and asymmetrical elements (Fig. 1.6). The linear element is the most common among the time-invariant resistances, the ohmic resistance. The symmetrical elements are also known as varistors, and the asymmetrical ones as rectifiers. Among the time-variant resistances, all three types shown in Fig. 1.6 can occur, but in general they are nonlinear. This nomenclature has been chosen according to Mauro, but it should be emphasized that the nonlinear, time-variant resistances are commonly called voltage- and time-dependent resistances in the electrobiological literature. The characteristic of this class of resistance is that their voltage–current relation changes with duration of current flow. Since these resistance elements are very important for biological systems, they are discussed in Sec. 1.5.1.2.

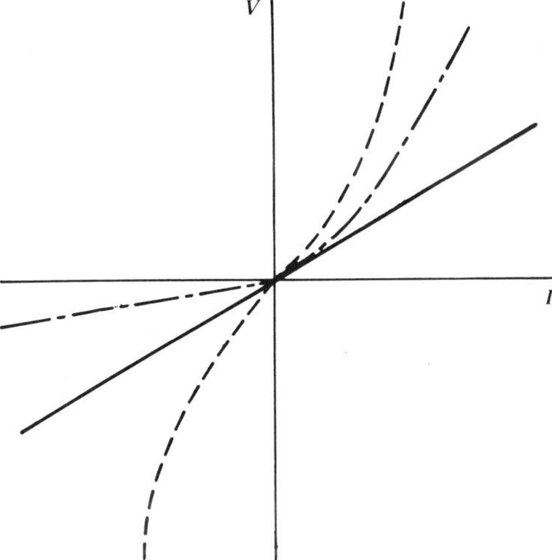

Figure 1.6. Typical voltage–current curves for different types of resistance elements: (———) linear element; (---) symmetrical element; (-·-) asymmetrical element.

The terminology used in papers dealing with rectification processes in biological membranes is quite particular to this field. Most rectifier properties of the membrane are attributed today to the potassium channel, and consequently, indications of current directions are referred to the direction in which a positive charge (potassium ion) is displaced under the influence of an electrical driving force. For example, outward current means in this context a displacement of potassium from the intracellular medium to the outside. Moreover, the rectifying property of the potassium channel, as expected from the constant field equation, is a low conductance for potential values more negative than the potassium equilibrium potential and a high conductance for more positive potentials. When, however, the rectifying property of the rectifier is reversed (a high conductance at potentials more negative than the equilibrium potential), this is called anomalous rectification. Consequently, the nonlinearity predicted by the constant-field equation is considered normal rectification. The terms outward and inward rectifier, also found in the literature, imply that the rectifier conducts preferentially outward or inward current.

Generally, the rectifier concept does not imply time-variant resistance elements, but biological rectifiers usually do. This phenomenon has been termed delayed rectification, a nonlinearity of the membrane that increases or develops with time. Such a delayed rectification can be normal or anomalous. For a recent discussion of biological rectifiers, see Zachar (1971, chapter 2).

1.5.1.2 Time-Variant Resistances. Resistance elements that vary with time, which are termed anomalous impedances, are of considerable importance for the theory of biological membranes and especially of electrically excitable ones (Cole, 1947; Teorell, 1949). Anomalous impedances will be considered here in some detail based on the report of Mauro (1961). Some more recent ideas about the nature of time-variant resistances can be found in Cole (1968, part 2).

Although a time-variant resistance is a purely dissipative element, it appears in an electrical measurement generally as an impedance with a positive or negative reactive component. To understand this, the circuit in Fig. 1.7a will be discussed. It consists of a two-terminal network containing a bias voltage V_0, a load resistance R_1, and a voltage- and time-variant resistance element $R(V,t)$. When no current is injected at terminals 1 and 2, the bias voltage V_0 drives a current I' through $R(V,t)$ and R_1. Now we can distinguish two situations: The time-variant resistance represents a thermopositive element (incandescent lamp), and a positive current pulse I is imposed on terminals 1 and 2. The current component flowing through the thermosensitive element is opposed to I'; consequently, the current flowing through $R(V,t)$ will diminish, and the voltage measured at the two terminals ($V_{1,2}$) will instantaneously increase. Then the voltage drop across $R(V,t)$ will further decrease as the value of $R(V,t)$ diminishes and $R(V,t)$ cools down. This will result in a further increase of $V_{1,2}$ as a function of time until a new steady-state level is reached. Figure 1.7b illustrates this

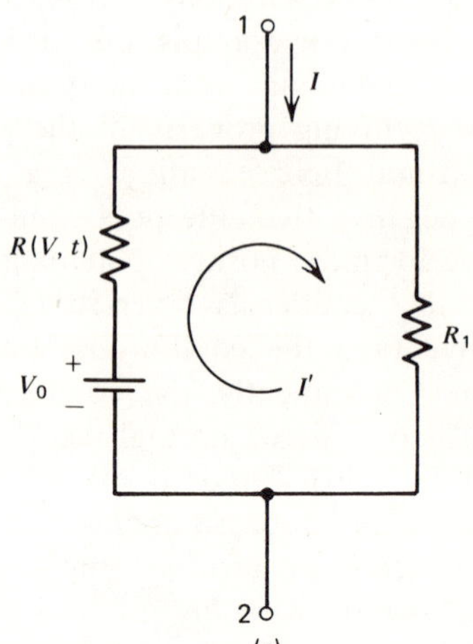

Figure 1.7. (a) Circuit containing a voltage- and time-dependent resistance $R(V,t)$ biased by a voltage V_0. (b) and (c) illustrate reactions of this circuit to small positive current pulses. (b) $R(V,t)$ increases with current and time; (c) $R(V,t)$ decreases with current and time. Insets: Equivalent linear circuits; for explanation see text. Modified from Mauro (1961). By copyright permission of The Rockefeller University Press.

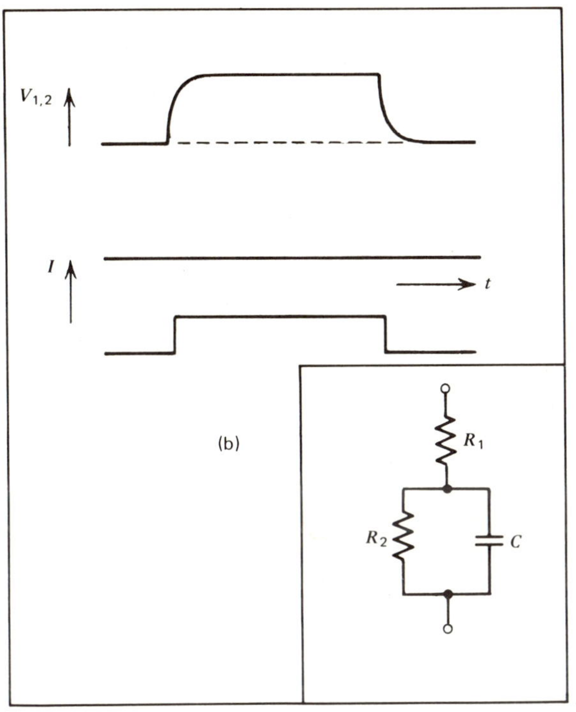

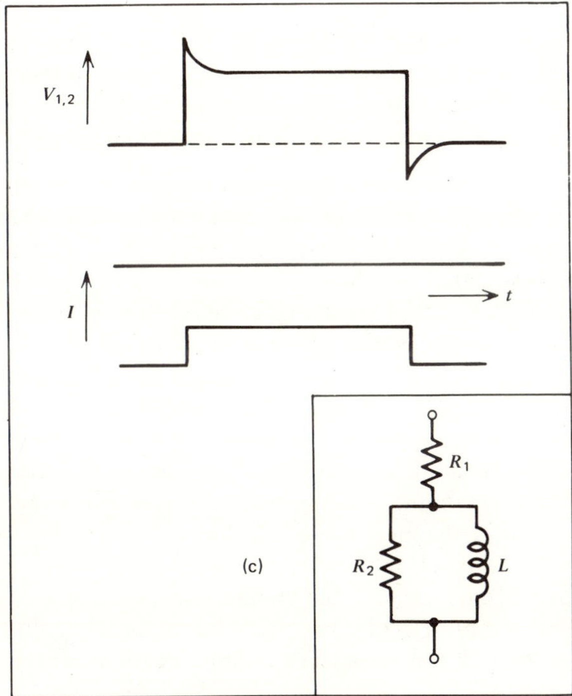

Figure 1.7 *Continued.*

sequence of events, and it can be seen that the resulting time course of $V_{1,2}$ resembles that across an RC network as shown in the inset. It is noteworthy that this network is identical to model 2 of Table 1.5.

If the time-variant element is thermonegative (thermistor) and a positive current pulse is imposed on the terminals, again the current flowing through $R(V,t)$ decreases and an instantaneous increase in $V_{1,2}$ occurs. Then, as the element $R(V,t)$ cools down, its value increases and $V_{1,2}$ decreases with time to a new steady-state level. Figure 1.7c illustrates the resulting time course of $V_{1,2}$, which resembles the transient across an RL network as shown in the inset. From this, it is plausible that a time-variant dissipative element can show apparent reactive properties. It is also evident that the bias voltage V_0 is essential for the functioning of the circuit.

1.5.2 Capacitances

Capacitive elements store energy without dissipating it (Cole, 1933). They appear in an impedance-locus plot as negative reactances, and their identification with physicochemical processes or biological structures has been a particular challenge. Important contributions were made by properly identifying negative reactances.

Here, the three most common identifications of a negative reactance will be treated: the static capacitance (1.5.2.1), the polarization capacitance (1.5.2.2), and the apparent capacitance manifest in a relaxation system of the Debye-Hückel type (1.5.2.3). Finally, the problem of interpretation when more than one time constant is present will be discussed (1.5.2.4). Another possible interpretation of the negative reactance as a voltage- and time-dependent conductance was already treated in Sec. 1.1.5.2.

1.5.2.1 Static Capacitance. Since the earliest times of impedance measurements, the cell membrane was described as a dielectric bounded by the conducting intra- and extracellular media. With such an approach, Fricke (1925a) estimated the thickness of the red cell membrane from its capacitance, 0.81 $\mu F/cm^2$. Assuming a dielectric constant of 3, Fricke calculated the thickness of the membrane to be 33 Å. This finding is remarkable because a bold estimate was made of a biological parameter based on data obtained from an indirect measurement. Moreover, the author was well aware of the speculative element contained in this early estimation of the membrane thickness: "The order of magnitude of the thickness of this membrane...is suggestive. By using a value of 3 as the dielectric constant of the membrane (a value which is rather uncertain...) we obtain..." (Fricke, 1925a).

Later measurements proved the membrane capacity to be almost a universal cell constant of 1 $\mu F/cm^2$. Nothing essential could be added to these findings until the present. The estimation of Fricke proved to be correct. However, there are still some unanswered questions: (1) How can a value of 3 for the dielectric constant be calculated for a mosaic structure consisting of lipids, proteins, and patches of aqueous ionic channels? and (2) How can a structure of about 100 Å thickness, partially shunted by ionic conductances, withstand mechanically an electric field of about 10^5 V/cm, which is also the value for the dielectric breakdown of glass? (Kohlrausch, 1953, vol. 2).

The concept of a static capacity is extremely useful for interpreting biological impedance measurements. However, its application requires a simplifying assumption: The center of the impedance locus must lie on the real axis. Actually, in most biological structures the impedance locus has a center lying in the region of the positive reactance. With the convention generally adopted, this results in an impedance locus with a center below the resistance axis (suppressed center), Fig. 1.8a. This feature of the impedance locus led to the attempt to identify the negative reactance found in biological structures with a polarization capacitance.

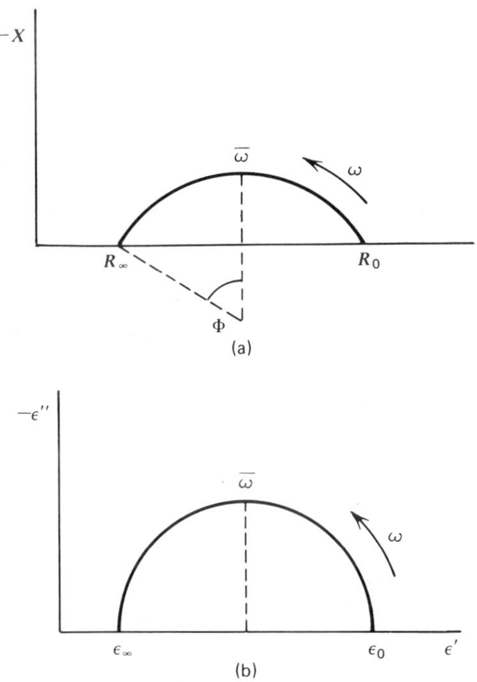

Figure 1.8. (a) Impedance locus as found in several biological tissues. R_0 and R_∞: limiting resistances; Φ: phase angle; $\bar{\omega}$: characteristic frequency. (b) Representation of the dielectric constant in the complex plane. For explanation see text.

1.5.2.2 Polarization Capacitance. In 1899, Warburg proposed a theory for nonpolarizable and selective electrodes. Electrically, these electrodes were characterized by a constant phase angle Φ of about 45°; that is, their impedance locus had a suppressed center. Although no physicochemical explanation was available for this phenomenon, it was hoped for two decades that, by analogy, the constant phase angle found in biological impedances could be related to the ion selectivity of cell membranes. The concept had been worked out theoretically by Cole (1928b, 1933) and Fricke (1932) and reviewed by Cole (1968, part 1). It postulates that a polarization element consisting of a frequency-dependent series resistance (R_{pol}) and capacitance (C_{pol}) is responsible for the constant phase angle defined as:

$$R_{\text{pol}}C_{\text{pol}}\omega = \tan\Phi = m \tag{1.82}$$

where the dependence of C_{pol} upon the frequency is defined by the empirical relation $C_{\text{pol}} = \hat{C}_{\text{pol}}\omega^{-\alpha}$, where $\hat{C}_{\text{pol}}$ is the polarization capacitance at $\omega = 1$, α and m are interdependent constants (Eq. 5.15), and Φ is the phase angle.

An impedance-locus plot, like that in Fig. 1.8a, can be created by replacing the capacitance in model 2 or 3 of Table 1.5 by a polarization element. Although phase angles between 40° and 90° were actually measured on a large number of cells (Cole and Curtis, 1950), identifying such an impedance locus with a polarization element gradually lost its attractiveness, as an early quotation illustrates (Cole, 1933): "The experimental evidence at present indicates the electrical behaviors of cell membranes are very similar to those of polarizations in solid dielectrics and at irreversible electrode surfaces—which are also unexplained. . . ." Today a powerful and general explanation for the impedance locus with a depressed center is available (see Sec. 1.5.2.4 and Chapter 5).

1.5.2.3 Negative Reactance as a Relaxation Phenomenon. Cole and Cole (1941) and R. H. Cole (1965) employed the complex dielectric constant ($\varepsilon' - j\varepsilon''$) to characterize biological systems. The idea is based on Debye's 1929 theory of the dielectric behavior of a dilute suspension of free dipoles. The real part of the dielectric constant ε' represents its storage component, and ε'' is the dissipative component. With this formalism a dielectric can be characterized in the complex plane by a semicircle with its center on the ε' axis (Fig. 1.8b). A physical interpretation of this plot is a system of diluted dipoles rotating together with their hydration shells in a fast-changing electric field. At a given frequency, the changes of the field are so fast that the dipole shells are disturbed, and the system becomes internally unstable and converts its energy into heat. This energy loss is described as heat dissipation in ε''. As can be seen from Fig. 1.8b, the heat loss is maximum at the characteristic frequency $\bar{\omega}$. From this, it is plausible that the dielectric constant could be related to experimentally accessible conductances. Defining limiting quantities by the subscripts 0 and ∞, it can be shown that:

$$\varepsilon_r = \varepsilon_\infty + (\varepsilon_0 - \varepsilon_\infty)/\left[1+(\omega\tau)^2\right]$$

and

$$\sigma = \sigma_0 + (\sigma_\infty - \sigma_0)(\omega\tau)^2/\left[1+(\omega\tau)^2\right] \tag{1.83}$$

and both are related by

$$(\varepsilon_0 - \varepsilon_\infty)\varepsilon_v/\tau = (\sigma_\infty - \sigma_0)$$

where ε_v is the dielectric constant of the free space. The time constant τ, which characterizes the relaxation time of the system, can be obtained from the characteristic frequency of the locus in Fig. 1.8b and Appendix 2, Eq. A2.13. Actually, σ represents the dissipative component ε'' in Fig. 1.8b. Figure 1.9 shows the graphical representation of the dielectric constant and the conductivity in the frequency domain. In this figure it is also indicated how the characteristic frequency of the system can be obtained. The dielectric constant* is related to the measured capacitance by Eq. A1.4 of Appendix 1. Moreover, it has been shown that the complex dielectric constant also represents a circular arc in the complex impedance or admittance plane provided σ_0 and ε_∞ can be neglected (Schwan, 1957).

1.5.2.4 Negative Reactances in Systems with More Than One Time Constant. When measurements are performed in biological tissues over the technically available range of frequencies, generally three different dispersions are obtained. These are empirically called α, β, and γ dispersions (Fig. 1.10). The characteristics of these dispersions are summarized in Table 1.9; see also Sec. 5.1. In the simplest case, each of the dispersions represents a first-order process; that is, the maximum slope for each dispersion in Fig. 1.10 is -1 for a tenfold change in frequency. Since phenomena measured as negative reactances can represent different concepts, either one or a combination of possible interpretations can underlie the experimentally observed dispersions. Any decision to identify a concept with a given dispersion must be based on additional information about the system investigated.

*The dielectric constant is identical with the capacitance of 1 cm^3 of tissue, measured in $\mu\mu$F and multiplied by 3.6 (Schwan and Kay, 1957).

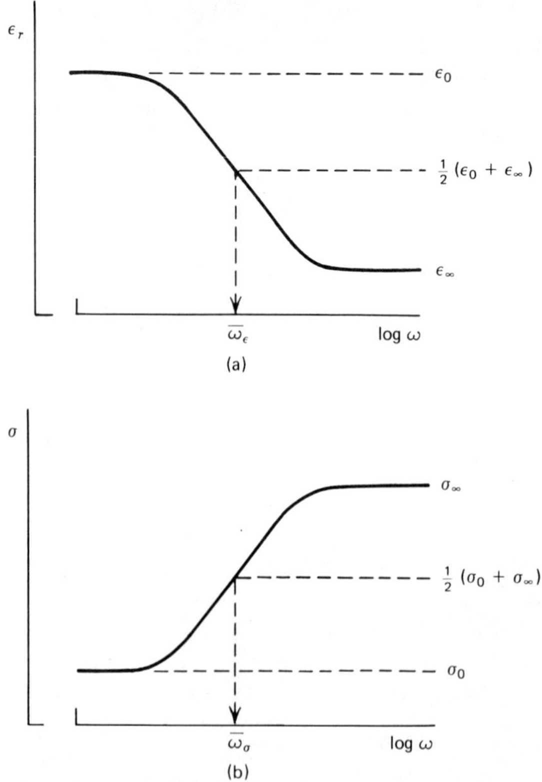

Figure 1.9. Representation of (a) the dielectric constant and (b) the conductivity of a relaxation system in the frequency domain. Modified after Schwan (1957).

Another problem is encountered when several time constants are responsible for a single dispersion. This results in circular arcs with a depressed center when the results are presented as a complex dielectric constant or a complex impedance (anomalous dispersion). As will be shown in Sec. 5.2, such a finding can be analyzed by selecting a proper distribution function for the time constants underlying the anomalous dispersion (see Eqs. 5.1). However, it was shown that a variety of different distribution functions fit the same impedance locus with the depressed center (Schwan, 1957). It is therefore important to have additional information about the parameters of the system so that a meaningful distribution function can be selected. For a practical example, see Pauly (1963). In an actual experiment, it is important to decide whether the results represent an anomalous or a normal dispersion: From the plots of ε_r and σ as a function of frequency, the characteristic frequencies $\bar{\omega}_\varepsilon$ and $\bar{\omega}_\sigma$ can be obtained (Fig. 1.9) and it can be shown that the condition $\bar{\omega}_\varepsilon < \bar{\omega}_\sigma$ indicates an anomalous dispersion.

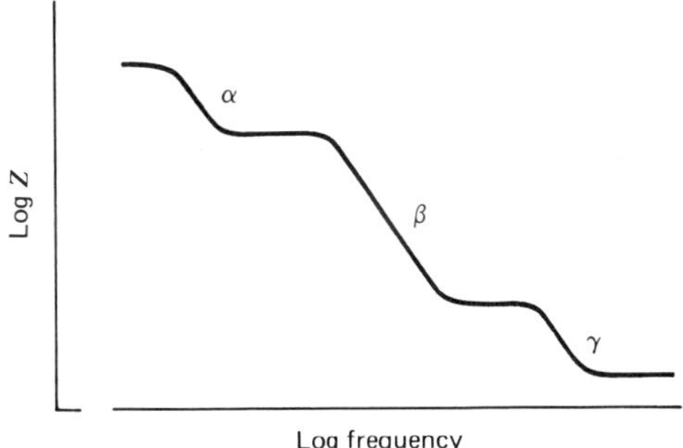

Figure 1.10. Different dispersions found in biological preparations. Modified after Schwan (1963).

1.5.3 Positive Reactance (Inductance)

For several decades, it was believed that biological impedances contained only resistances and negative reactances. Therefore, it was quite unexpected when Cole and Baker (1941) found in the squid axon an impedance locus that extended in the region of positive reactances. Further experimentation showed that this positive reactance depended upon

Table 1.9 Possible Physical Interpretation of Dispersions Found in Biological Cells (Schwan, 1957).

Dispersion	Physical Interpretation	Remarks
α (low frequency)	Gating mechanism Membrane structure	α dispersion more sensitive to age and environmental factors than β dispersion
	Surface conductance due to ionic atmosphere around cells.	α dispersion not restricted to biological cells with membrane structure.
β (radio frequency)	Structural dispersion (Maxwell-Wagner type, distribution of geometrical parameters).	
	Power law frequency dependence of membrane impedance.	
γ (high to ultrahigh frequencies)	Dipole moments of macromolecules.	Protein dispersion can be masked by structural β-dispersion.
	Surface conductance of macromolecules.	

changes in the ionic composition of the extracellular medium. An example of the influence of a change in extracellular Ca^{2+} on the impedance locus of the squid axon is shown in Fig. 1.11 (Cole and Marmont, 1942; Cole, 1968, part 1). The impedance locus in this figure was interpreted as a tuned circuit (control medium and low-Ca^{2+} medium); this circuit is shown in the inset of the figure. The quantities R, C, and L have been related to the cell membrane in a nonspecified manner. However, it was difficult to interpret the impedance locus obtained in high-Ca^{2+} medium (curve B), since there was a high capacitive component at low frequency. It took several years more until the nature of these reactances was understood as a voltage- and time-dependent conductance (see Sec. 1.5.1.2).

1.6 BIOLOGICAL SOURCES OF VOLTAGE

The nature of voltage sources found in biological systems is rather complex. There are two main groups of voltage sources: diffusion potentials and electrogenic potentials. The latter can occur only in systems where ion fluxes are coupled in a specified way to energy-yielding

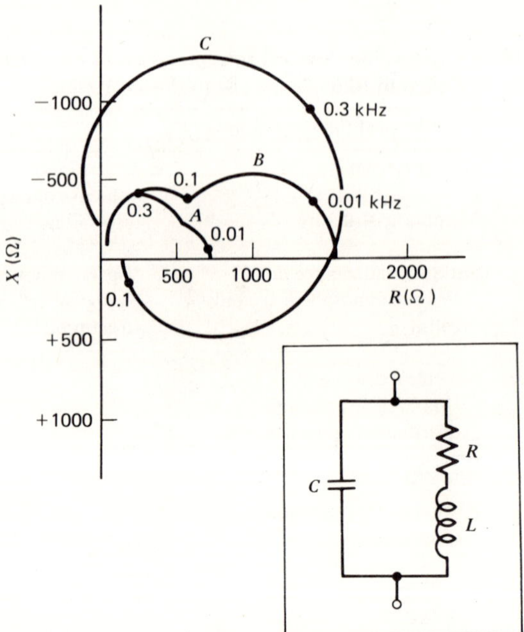

Figure 1.11. Membrane impedance of the squid axon in the complex plane. A, in seawater; B, in high calcium; C, in low calcium. Inset: Equivalent circuit. Modified from Cole (1968).

processes. In general, a quantitative treatment for diffusion potentials is available only when the system is in equilibrium or in a steady-state condition, whereas for the electrogenic potentials steady-state and non-–steady-state solutions are available.

1.6.1 Voltage Sources with Zero Net Passive Current Flow

The expression obtained for sources of potential at zero net current flow were derived by considering two separate compartments containing electrolyte solutions made up of different ions at different concentrations. The compartments are separated by a thin barrier (membrane) whose properties constrain the ion fluxes.

Derivations of the equations for the membrane potential V at zero net passive current flow are already well described (Johnson, Eyring, and Polissar, 1963; Kotyk and Janáček, 1970; Harris, 1972). Here only the different solutions of the diffusion problem will be discussed with respect to their applicability to the biophysical analysis of biological membranes. The general strategy for obtaining a solution of the problem of electrodiffusion is to derive the equations for the unidirectional fluxes normal to the diffusion barrier, taking into account the barrier's specific properties. Since the sum of all these fluxes must be zero net flux, we can calculate the membrane potential. The difficulty lies in the complexity of the diffusion equation itself (Nernst-Planck equation), a special form of which (Teorell, 1951) describes passive unidirectional flux:

$$M = \frac{\delta n}{\delta t} = Df_j C_j \frac{\delta \mu_j}{\delta x} \tag{1.84}$$

where μ_j is the electrochemical potential of the ion j given by:

$$\mu_j = \ln \mathcal{a}_j = \mu_{0j} + RT \ln f_j C_j + zFV_j \tag{1.85}$$

where n is the number of ions crossing the membrane in the x direction (normal to the membrane), D is the diffusion constant, $f_j C_j$ stands for the activity of the ion j, z is the number and sign of charges per ion, $\mathcal{a}$ is the electrochemical activity, and μ_{0j} represents the part of the partial molal free energy of the ion related neither to its concentration nor to the electrical potential. Inspection of Eqs. 1.84 and 1.85 shows that to integrate the flux equations we need to know how the concentration and the potential vary within the membrane. This is difficult to measure, and therefore the problem has been simplified by assuming either dC/dx (Henderson, 1907, 1908) or dV/dx to be constant (Goldman, 1943).

1.6.1.1 The Henderson Equation. Assuming a constant concentration gradient across a neutral membrane, Henderson obtained the following expression for the equilibrium potential:

$$V=\frac{RT}{F}\frac{(U_1-U_2)-(W_1-W_2)}{(U_1'+W_1')-(U_2'+W_2')}\ln\frac{(U_1'+W_1')}{(U_2'+W_2')} \tag{1.86}$$

where U and W are the $\Sigma\mu C$ for cations and anions, respectively, and U' and W' are the $\Sigma z\mu C$ for cations and anions, respectively, z being the charge of a given ion. The subscripts 1 and 2 denote the two compartments separated by the membrane. The assumption of a constant concentration gradient was physically justified by assuming that the solution within the membrane is efficiently mixed, thus resulting in an approximately linear concentration gradient across the membrane. It is generally believed that there is no constant concentration gradient within the membrane, and the Henderson equation is not usually used to interpret biological measurements. However, it has been found that under specified conditions it describes the membrane potential across a fixed charge membrane.

1.6.1.2 The Goldman Equation. Another way to successfully integrate the Nernst-Planck equation is to assume a constant electrical field within the membrane (Goldman, 1943). Although it is generally assumed that the constant-field hypothesis is not justified, the Goldman solution has been extensively used in the analysis of the potentials of biological membranes because it is rather manageable. In the version of Hodgkin and Katz (1949), it leads to the well-known expression for a diffusion potential caused by univalent ions:

$$V=\frac{RT}{F}\ln\frac{\Sigma P^{+}C_o^{+}+\Sigma P^{-}C_i^{-}}{\Sigma P^{+}C_i^{+}+\Sigma P^{-}C_o^{-}} \tag{1.87}$$

where the subscripts i and o stand for intra- and extracellular compartments and the indexes + and − indicate the charge of the ions. The permeability P for the different ions is defined by:

$$P^{+,-}=\mu^{+,-}\beta^{+,-}RT/dF=D^{+,-}\beta^{+,-}/d \tag{1.88}$$

where β is the partition coefficient of a given ion, d is the membrane thickness, and D represents the diffusion constant. D and β are assumed to be the same at both membrane interfaces.

The validity of the Goldman equation has been questioned because of the assumptions that underlie it: (1) a constant gradient of the electrical field within a neutral membrane and (2) that the partition coefficient β is the same at the inner and outer membrane interfaces.

Recently more and more evidence has been put forward that expressions equivalent to Eq. 1.87 can be obtained without assuming a constant field within the membrane. Based on the flux ratio relation (see Eq. A3.3 in Appendix 3), Patlak (1960) derived an expression formally equal to that of Goldman under the restriction that only univalent ions determine the potential. In this case, the expressions corresponding to the permeabilities represent rate constants of unidirectional fluxes. Moreover, it has been shown that an equation equal to Eq. 1.87 describes the phase-boundary potentials between a solution and an ion-exchange system (Eisenman, 1962). It has also been demonstrated that an equivalent of the Goldman equation can be derived rigorously from the Parlin-Eyring equation, which is based on a diffusion barrier model and is more general than, but independent of, the Nernst-Planck equation (Johnson et al., 1963, p. 759). However, the solution depends on the constant-field assumption and is restricted to univalent ions. With this evidence, one can state that Eq. 1.87 is valid under more general conditions than for those under which it was originally derived (Schwartz, 1971).

A more serious drawback of the Goldman equation (or its equivalents) seems today to be the neglect of possible contributions of surface potentials to the transmembrane potential. The assumption that the partition coefficient for a given ion is the same at the inner and outer interfaces implies a cancellation of these surface potentials. Since they cannot be measured directly, the assumption is speculative. There is a controversy on how much error the neglect of the surface potentials will introduce. One end of the spectrum has reasearchers who use the Goldman equation pragmatically as a tool as long as it yields plausible results; at the other end, one finds the critics claiming that any agreement between experimental results and those predicted by the Goldman equation "may be largely fortuitous" (Sjodin, 1961). Moreover, Johnson et al. (1963, chapter 11) reached the conclusion that the surface potentials are generally of the same order as the observed transmembrane potentials. Consequently the latter may differ considerably from the true membrane potential. On the other hand, attempts to take these criticisms into account in new formulations for the membrane potential lead in general to expressions that contain parameters that are difficult to evaluate experimentally. In view of this, it is probably wise at present to adopt the approach of Kotyk and Janáček (1970): "It may perhaps be concluded that surface potentials are likely to be of importance and that if a satisfactory fit for the membrane potential is not obtained...under various conditions, a variation of the surface potentials...may be responsible, rather than variation of permeability constants." Problems involved in the application of the Goldman equation are discussed by Johnson et al. (1963); Sjodin (1961); Lindley et al. (1967); Kotyk and Janáček (1970); and Harris (1972).

1.6.1.3 The Fixed-Charge Equation. The drawbacks of the constant-field equation were overcome by a theory developed independently by Teorell as well as Meyer and Sievers, and its final version was reported by Teorell (1951, 1953). The main assumptions are: (1) The potentials at the solution–membrane interface are Donnan potentials, (2) the potential and concentration gradients within the membrane are controlled by the Nernst-Planck equation, and (3) the membrane contains a certain amount of fixed charges.

The assumption of a Donnan equilibrium at the membrane interfaces implies that there is a nonpermeable ion in one of the compartments bordering the membrane. Assuming (1) the ions K^+ and Cl^- are exchangeable between an aqueous solution and a bulk membrane, (2) the anions R^{n-} are fixed within the membrane m, and (3) the fluid compartment o is large, a Donnan potential V_D will develop across the interface. At equilibrium there will be:

$$[K^+]_o = [Cl^-]_o = s \tag{1.89}$$

because the charges have to be balanced. Moreover, the potential V_D across the interface has to be the same for the potassium and chloride ions, and we obtain:

$$\frac{[K^+]_o}{[K^+]_m} = \frac{[Cl^-]_m}{[Cl^-]_o} = r = \exp(V_D F/RT) \tag{1.90}$$

where r is the Donnan ratio. Electroneutrality in compartment m now requires:

$$n[R^{n-}]_m + [Cl^-]_m = [K^+]_m \tag{1.91}$$

Combining Eqs. 1.89–1.91 gives:

$$[Cl^-]_m^2 + n[R^{n-}]_m[Cl^-]_m - s^2 = 0 \tag{1.92}$$

which is a quadratic equation with the solution:

$$[Cl^-]_m = -\frac{n[R^{n-}]_m}{2} + \left[\left(\frac{n[R^{n-}]_m}{2}\right)^2 + s^2\right]^{1/2} \tag{1.93}$$

With the same reasoning, an expression for $[K^+]_m$ can be obtained.

The fixed charge theory of a membrane can now be summarized as follows: The transmembrane potential is composed of two terms, the potential differences at the interfaces (Donnan potential, V_D) and the potential across the membrane proper, the diffusion potential V_{diff}

$$V_m = V_D^{i,m} + V_D^{m,o} + V_{\text{diff}}^m \tag{1.94}$$

where the indexes i and o denote the inner and outer compartments and m stands for membrane. The sum of the Donnan potentials is now:

$$V_D^{i,m} + V_D^{o,m} = \frac{RT}{F}\ln\frac{r^{m,o}}{r^{i,m}} \tag{1.95}$$

with

$$r^{i,m} = \left[1 + \left(\frac{\omega X}{2C_i}\right)^2\right]^{1/2} - \frac{\omega X}{2C_i} \tag{1.96}$$

where C_i is the concentration in the compartment i, $C_i = C_i^+ = C_i^-$; ω is $+1$ for positively charged membranes and -1 for negatively charged membranes, and X is the fixed charge density in the membrane. In analogy to Eq. 1.96, we can obtain the expression for $r^{m,o}$. In the biologically important case of a membrane separating two solutions of equal total concentration, Eq. 1.95 shows that the two Donnan potentials cancel out.

The diffusion term of Eq. 1.94 had been evaluated originally with the Henderson integration, but later the more general Planck solution (Teorell, 1951) was reported:

$$V_{\text{diff}}^m = \frac{RT}{F}\ln\xi \tag{1.97}$$

The Planck parameter ξ is a solution of the transcendental equation:

$$\frac{\xi U_o - U_i}{W_o - \xi W_i} = \frac{\ln K - \ln\xi}{\ln K + \ln\xi}\,\frac{\xi C_o^{m+} - C_i^{m+}}{C_o^{m-} - \xi C_i^{m-}} \tag{1.98}$$

where the subscripts i and o indicate the inner and the outer edge of the membrane; the indexes $+$, $-$, and m mean cation, anion, and membrane; $U = \Sigma\mu^+ C^m$, $W = \Sigma\mu^- C^m$, and C represents the sum of anions or cations and the parameter K is:

$$K = \frac{C_o^{m+} + 0.5\left[\ln(K\xi)/\ln K\right]\omega X}{C_i^{m+} + 0.5\left[\ln(K\xi)/\ln K\right]\omega X}.$$

A numerical evaluation of K and ξ is usually made by trial and error

(Plonsey and Fleming, 1969, chapter 3). It is interesting to note that ξ becomes C_o^m/C_i^m in a neutral membrane when only one electrolyte is present, and Eq. 1.97 becomes identical with the well-known Nernst equation (Eq. A3.5 in Appendix 3). Since the numerical evaluation of the membrane potential in a fixed charge membrane is difficult, the calculations reported in the literature use special assumptions such as equal total ion concentration in both compartments bounding the membrane or restriction to univalent ions. It has been shown, using Eq. 1.94, that under specified conditions a constant electrical field or a constant concentration gradient in the membrane could be obtained. Consequently, the Henderson and Goldman integrations of the Planck equation can be considered as special cases of Eq. 1.94 (Teorell, 1953). Conti and Eisenman (1965) reported solutions of the Teorell-Meyer-Sievers equation where ωX is a function of the membrane thickness. Although the Teorell-Meyer-Sievers equation is very elaborate and takes into account many more parameters of the membrane than the previously developed theories, its applicability to biological membranes is still limited. The problem is that the fixed charge concentration in a biological membrane is difficult to evaluate. Solomon (1960) attempted to calculate this quantity in erythrocyte membranes and arrived at an absurd value of 100 moles/l. It seems that there are still unknown complications in applying the fixed charge theory to biological membranes, whereas this theory has proven quite powerful when applied to artificial membranes.

1.6.1.4 The Kirkwood Equation. Kirkwood (1954) derived equations for ionic fluxes through charged membranes. The derivation is based on principles of irreversible thermodynamics. Based on Kirkwood's analysis, Caillé (1969) derived an expression for the membrane potential for zero net current:

$$V=\frac{RT}{F}\ln\left[\left(\frac{a_o^+}{a_i^+}\right)^{T_1^+}\times\cdots\times\left(\frac{a_o^+}{a_i^+}\right)^{T_m^+}\times\left(\frac{a_i^-}{a_o^-}\right)^{T_1^-}\times\cdots\times\left(\frac{a_i^-}{a_o^-}\right)^{T_n^-}\right] \tag{1.99}$$

where a stands for the activities of the respective cations and anions, and T^+, T^- are the macroscopic transport numbers for the m cations and n anions determining the potential. These transport numbers are related to the biologically important chord conductances by the relation

$$\frac{T^+}{T^-}=\frac{g^+}{g^-} \tag{1.100}$$

as shown by Hodgkin and Horowicz (1959b) and Caillé and Schanne (1972).

1.6.1.5 The Integral Resistance Equation. Finkelstein and Mauro (1963) analyzed the equivalent circuit of a membrane patch reported by Hodgkin and Huxley (1952d); see Fig. 2.12b. They assumed a linear system and steady-state conditions and obtained, for the current carried by a given ion:

$$I = E \Big/ \int_i^o \frac{dx}{uCF} \tag{1.101}$$

where the term under the integral, taken from inside to outside, is called the integral resistance of an ion. The inverse of the integral resistance is the chord conductance defined in Table 2.5. Equation 1.101 holds for cations and anions. For a membrane whose potential is determined by sodium, potassium, and chloride, Kotyk and Janáček (1970, chapter 4) showed that one obtains from Eq. 1.101 for zero net current:

$$V_m = \frac{g_K E_K + g_{Na} E_{Na} + g_{Cl} E_{Cl}}{g_K + g_{Na} + g_{Cl}} \tag{1.102}$$

1.6.2 Voltage Sources Dependent on the Activity of Pumps

By definition, the term ionic pump is used for the phenomenon of ionic fluxes directly coupled to metabolic processes (Schultz, 1969). Under specified conditions, the activities of ionic pumps can contribute to the membrane potential, and the difference between the diffusion potential and the measured membrane potential is called an electrogenic potential. Lately, considerable progress has been made in understanding electrogenic potentials, which have been found in a large number of structures. Two mechanisms can create an electrogenic potential: (1) an ionic pump transporting two different ions of the same sign in opposite directions at a ratio $\neq 1$ so that a net transport of charge occurs and (2) a charge transfer between a passive and an active transport regime within the membrane so that no net charge transfer occurs. The first mechanism has been termed a rheogenic (current-generating) system, and the latter a nonrheogenic system (Schwartz, 1971).

1.6.2.1 Rheogenic Systems. Rheogenic systems can be described by a simple expression similar to the Goldman equation (Mullins and Noda, 1963). Assuming a Na-K exchange pump, and that Na^+, K^+, and Cl^- determine the membrane potential, Cl^- being passively distributed, one obtains

$$V_m = \frac{RT}{F} \ln\left(\frac{r[\mathrm{K}]_o + b[\mathrm{Na}]_o}{r[\mathrm{K}]_i + b[\mathrm{Na}]_i} \right) \tag{1.103}$$

where b is the relative permeability $P_{\mathrm{Na}}/P_{\mathrm{K}}$ and r represents the coupling ratio (number of sodium ions pumped out per potassium ions pumped in) of the ionic pump. This equation holds under the following conditions: (1) There are no net passive movements of Cl^- ($V_m = E_{\mathrm{Cl}}$); and (2) $[\mathrm{Na}]_i$ and $[\mathrm{K}]_i$ are in a steady state. Condition 2 limits the applicability of the equation. On the other hand, it has been shown that Eq. 1.103 is not subject to the constant-field assumption. In case of $r = 1$ (neutral pump) it becomes identical with the Goldman equation (Hodgkin, 1958), which again demonstrates that the constraints originally imposed on the Goldman equation were too rigid. However, the problems arising from the neglect of the boundary potentials remain. From an operational point of view, b can be obtained by a curve-fitting procedure when the ionic pump is inactivated. Then the pump can be reactivated, V_m can be measured, and r is the only unknown, provided the intracellular ion concentrations are measured or remain constant.

Frumento (1965) derived an expression that eliminates the constraints of Eq. 1.103 and permits evaluation of V_m, $[\mathrm{Na}^+]_i$, $[\mathrm{K}^+]_i$, and $[\mathrm{Cl}^-]_i$ as functions of time. The following summary of Frumento's development follows Kotyk and Janáček (1970, chapter 4). Again a system is assumed in which Na^+, K^+ and Cl^- determine the membrane potential and which contains a sodium-potassium-coupled pump. At any time, a change in the ionic flux balance will lead to a charging process of the membrane capacity:

$$M_{\mathrm{Na}}^A + M_{\mathrm{K}}^A + M_{\mathrm{Na}}^P + M_{\mathrm{K}}^P + M_{\mathrm{Cl}}^P = \frac{C_m}{F} \frac{dV_m}{dt} \tag{1.104}$$

where M stands for net flux for a given ion species, and the indexes A and P discriminate between fluxes originating from pump activity (A) and passive diffusion mechanisms (P). The M_i^P are defined in Appendix 3 by Eq. A3.1. However, during pump activity, the intracellular ion concentrations change, and we have:

$$\left.\begin{aligned}
[\mathrm{Na}]_i(t) &= [\mathrm{Na}]_i^{t=0} + k\int_0^t M_{\mathrm{Na}}^P\,dt + k\int_0^t M_{\mathrm{Na}}^A\,dt \\
[\mathrm{K}]_i(t) &= [\mathrm{K}]_i^{t=0} + k\int_0^t M_{\mathrm{K}}^P\,dt + k\int_0^t M_{\mathrm{K}}^A\,dt \\
[\mathrm{Cl}]_i(t) &= [\mathrm{Cl}]_i^{t=0} + k\int_0^t M_{\mathrm{Cl}}^P\,dt
\end{aligned}\right\} \tag{1.105}$$

where k is the surface–volume ratio of the biological cell. Defining the coupling ratio of the pump as $M_{\mathrm{K}}^{A}=r'M_{\mathrm{Na}}^{A}$, we may combine Eqs. 1.104–1.105 to obtain a system of integral differential equations:

$$\left.\begin{aligned}
M_{\mathrm{Na}}^{P}&=P_{\mathrm{Na}}\left(\frac{-FV_m/(RT)}{\exp[-FV_m/(RT)]-1}\right)\\
&\quad\times\left\{[\mathrm{Na}]_o-[\mathrm{Na}]_i^{t=0}+k\int_0^t M_{\mathrm{Na}}^{P}\,dt+k\int_0^t M_{\mathrm{Na}}^{A}\,dt\exp[-FV_m/(RT)]\right\}\\
M_{\mathrm{K}}^{P}&=P_{\mathrm{K}}\left(\frac{-FV_m/(RT)}{\exp[-FV_m/(RT)]-1}\right)\\
&\quad\times\left\{[\mathrm{K}]_o-[\mathrm{K}]_i^{t=0}+k\int_0^t M_{\mathrm{K}}^{P}\,dt+r'k\int_0^t M_{\mathrm{K}}^{A}\,dt\exp[-FV_m/(RT)]\right\}\\
M_{\mathrm{Cl}}^{P}&=P_{\mathrm{Cl}}\left(\frac{-FV_m/(RT)}{\exp[-FV_m/(RT)]-1}\right)\\
&\quad\times\left\{[\mathrm{Cl}]_i^{t=0}+k\int_0^t M_{\mathrm{Cl}}^{P}\,dt-[\mathrm{Cl}]_o\exp[-FV_m/(RT)]\right\}\\
&(1+r')M_{\mathrm{Na}}^{A}+M_{\mathrm{Na}}^{P}+M_{\mathrm{K}}^{A}-M_{\mathrm{Cl}}^{P}-\frac{C_m}{F}\frac{dV_m}{dt}=0
\end{aligned}\right\}$$

(1.106)

There are 6 unknowns ($M_{\mathrm{Na}}^{P}, M_{\mathrm{K}}^{P}, M_{\mathrm{Cl}}^{P}, M_{\mathrm{Na}}^{A}, r', V_m$) in Eqs. 1.106, so two of these parameters must be evaluated experimentally or estimated in order to solve the system. This is demonstrated in Frumento's paper, where the method is validated with a calculation based on data from skeletal muscle for a pump working at constant rate and the case when the pump rate is proportional to the third power of the sodium concentration.

1.6.2.2 Nonrheogenic Systems. In contrast to the analysis of rheogenic systems, Schwartz (1971) showed that a pump whose activity does not result in a transfer of net charge can nevertheless be electrogenic (nonrheogenic system). The proposed mechanism for the transported ion species is shown diagrammatically in Fig. 1.12a. Here, for the same ion species an active (M^A) and two passive M_1^P, M_2^P regimes exist. However, the pump transports the ions actively only to a site x_1 within the membrane, and from there on to the outside the flux regime is passive (M_2^P). Such a

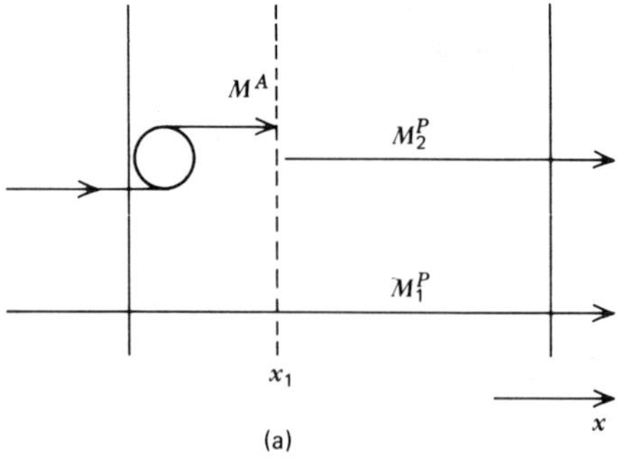

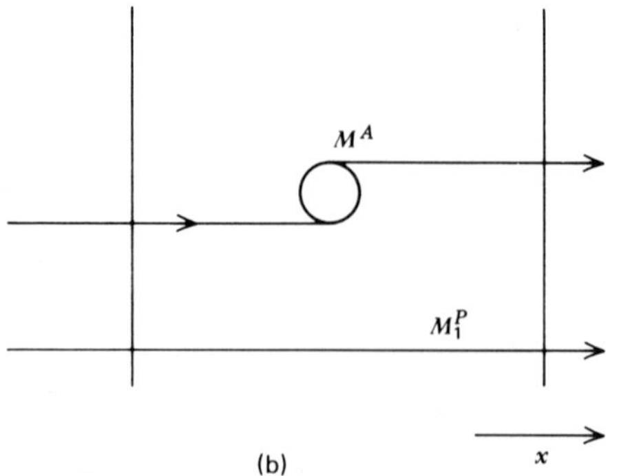

Figure 1.12. Passive (M^P) and active (M^A) fluxes of the same ion species across a membrane. (a) Passive and active regimes interacting. In the active regime, the ions are pumped to site x_1 and they leave the membrane by a diffusion mechanism (M_2^P). (b) Noninteracting passive and active regimes. For explanation, see text. Modified after Schwartz (1971). By copyright permission of The Rockefeller University Press.

transition can disturb the concentration and potential profiles within the membrane, which will in turn modify M_1^P, resulting in an interaction between the active and passive regimes. Figure 1.12b shows the traditional concept of noninteracting passive and active regimes. Inspection of Fig. 1.12a suggests that asymmetric membranes like the cells of the kidney tubules or epithelia are good candidates for interactions between the passive and active regimes, but there is, a priori, no reason to deny the existence of such a mechanism in other cell membranes.

1.7 SUMMARY AND OUTLOOK

In this introductory chapter the reason for impedance measurements on the cellular level was identified with a determination of the electrical cell constants. The cell constants can be obtained from a measurement of the input impedance of the biological structure with the aid of a suitable mathematical model. To obtain the input impedance uncontaminated by artifacts inherent in the measuring system, a thorough knowledge of the technical details of the electrical setup is mandatory; therefore, elements of electrode arrangements and measuring circuits were discussed. Once the electrical cell constants are known, their interpretation in biological terms is a crucial step that determines the significance of a biological impedance measurement. If such an interpretation cannot be made, one gathers, at best, data for a handbook, whereas the most important contributions of impedance measurements to our knowledge are directly related to a meaningful interpretation of the electrical cell constants. Therefore, the second part of this chapter contains the possible interpretations of electrical cell constants in biological or physicochemical terms.

The fact cannot be overemphasized that the results of electrical measurements must be seen in the context of knowledge obtained in other biological sciences. This is one of the areas where current research efforts are concentrated: identification of morphologically defined membrane systems with membrane capacitances and the relation of transport processes to ionic conductances and transmembrane potentials, especially electrogenic potentials. Moreover, it can be expected that the number of useful mathematical models to interpret the input impedance will substantially increase, especially the analysis of current distribution in interconnected cells arranged in two- or three-dimensional arrays.

2

Impedance Measurements on Nerves and Neurons

Because of their availability, myelinated nerves were widely investigated by electrical methods (Schaefer, 1940, 1942) until the introduction of the squid giant axon by Young (1936) as a convenient preparation. Impedance measurements on this structure led to an understanding of excitation phenomena (Hodgkin and Huxley, 1952d). Later, this theory was applied to myelinated nerves. Impedance measurements on neurons required the development of glass microelectrode technology. Introduction of the giant neurons of the sea hare (*Aplysia*) into electrophysiological research permitted analysis of the current and impedance changes accompanying synaptic transmission and spontaneous activity. This chapter discusses these phenomena as follows: Sec. 2.1, myelinated nerves; Sec. 2.2, nonmyelinated nerves; and Sec. 2.3, neurons. Within each section, the results are discussed according to the methods used in the measurements.

2.1 MYELINATED NERVES

Impedance measurements on myelinated nerves are rather scarce compared to reports on nonmyelinated fibers, because the experimental approach and the theoretical treatment are simpler for the latter. Besides, several invertebrates have large nonmyelinated axons, which are very suitable for experimentation.

2.1.1 AC Measurements on Myelinated Nerves

Alternating current measurements on myelinated nerves were centered on (1) determination of the electrical characteristics in terms of impedance loci and (2) the changes in input impedance during excitation. Some of the results are summarized in Table 2.1. Most of these measurements were

Table 2.1 Electrical Constants of Myelinated Nerves Obtained with AC Signals[a]

Conditions	Results	Remarks	Reference
Circuit 5, Table 1.1	$Z = f(\omega)$		Lullies (1930)
Model 2 Table 1.5	resistive component*: 22.2 kΩ	30 Hz to 300 kHz	
	Reactive component: 4.3 μF	1024 Hz	
Circuits 1 and 5, Table 1.1	Z_{app}: 16–17 kΩ		Danielli (1939)
	λ: 3.0 mm		
Model 4, Table 1.5	Resistivity of nerve sheath*: 12.9 kΩ cm		
Circuit 5,	Z_{inp}: 40–60 MΩ	At rest	Tasaki and Freygang (1955)
	Z_{inp}: 4–6 MΩ	During action potential	

[a]Derived quantities are indicated by *.

performed with impedance bridges connected to measuring cells with large electrodes employing frequencies from 20 Hz to 2.5 MHz (Cole and Curtis, 1936). The structure most often used was the frog sciatic nerve. Considering the input impedance as a polarization impedance, complex plane loci were obtained by plotting reactance X_s versus resistance R_s (R_1 and C in model 2 of Table 1.5; R_2 is thought to be infinite). The results are summarized in Fig. 2.1. Figure 2.1a represents a transverse measurement and shows that removal of the sheath of connective tissue dramatically reduced X_s and R_s. Schmitt (1955) reported a qualitatively similar impedance locus for the frog sciatic nerve with its sheath, where the characteristic frequency was 20 Hz and the phase angle was 79.7°. Figure 2.1b shows the result of a longitudinal resistance determination: an impedance locus that does not have a pure resistive component and seems to have three maxima. Figure 2.1c shows the result of a transverse measurement on cat sciatic nerve that corresponds qualitatively to those obtained on frog sciatic nerves. Figures 2.1a and c illustrate the large influence of the connective tissue on the measurement of the transverse impedance. The suspension equation (Table 1.6) gave a value of 91–98% for the volume concentration. For closely packed hexagonal cylinders, the value is 90.7%, suggesting that the path of current was impeded by the sheath of connective tissue. The effect of possible injury during stripping of the sheath of connective tissue was not investigated. The optic nerves of *Limulus* did not change their resistance after removing the sheath. The impedance loci for transverse measurements provided a semicircle with a depressed center and a phase angle of about 70°. Lullies (1930) found only one maximum in measurements of longitudinal impedance, because he used a frequency

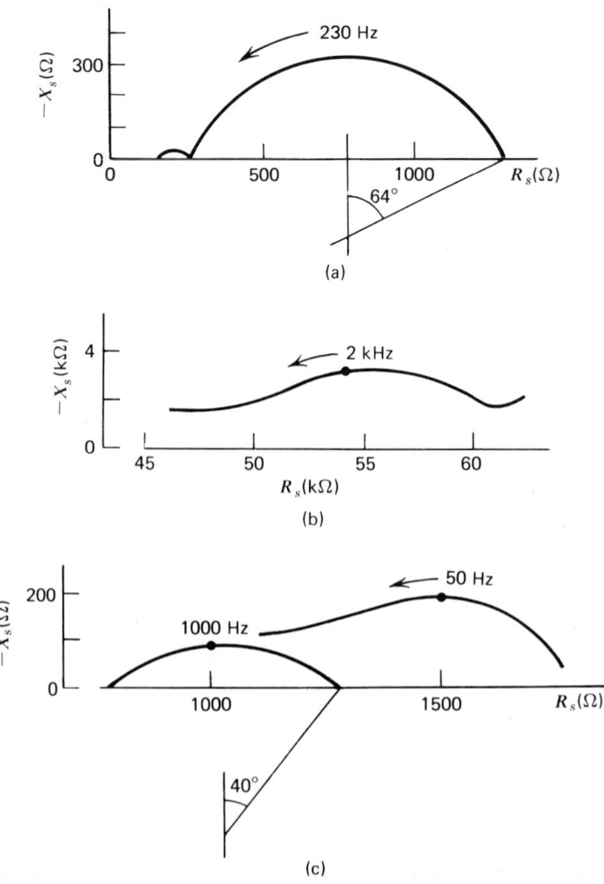

Figure 2.1. Impedance loci for series resistance and reactance of sciatic nerves. (a) Frog sciatic nerve, transverse measurement: small locus without sheath; large locus with sheath. (b) Frog sciatic nerve, longitudinal measurement. (c) Cat sciatic nerve, transverse measurement, the locus with a phase angle of 40° is the nerve without sheath, the other represents the nerve with sheath. Redrawn from Cole and Curtis (1936). By permission.

range of 30 Hz to 300 kHz, compared to 20 Hz and 2.5 MHz used by Cole and Curtis (1936). Danielli (1939) attempted to evaluate the resistivity of the myelin sheath and the length constant of a myelinated nerve, treating it as a linear cable and neglecting the influence of the nodes of Ranvier. The studies were made with a frequency of 500 Hz, and no advantage was taken of the possible ac analysis of the data using a transmission line model. Instead, the resistivity of the myelin sheath was obtained with the equation derived by Rushton (1934) for square-pulse analysis:

$$R = \left[\lambda r_e^2/(r_e + r_i)\right]\left[-\exp(-d/\lambda)\right] + \left[d r_e r_i/(r_e + r_i)\right] \tag{2.1}$$

where R is the resistance between two electrodes at a distance d, and r_i, r_e, and λ are defined as in model 4 of Table 1.5. r_m was 12.7×10^3 Ω cm in freshly excised sciatic nerves and dropped to 9.1×10^3 Ω cm after $4\frac{1}{2}$ hr of incubation. A tenfold increase in $[Ca^{2+}]_o$ and $[K^+]_o$ did not change r_m more than 30%, indicating that only an insignificant part of r_m represented the resistance of the nerve membrane. It is not known to what extent the impedance value measured at a frequency of 500 Hz was influenced by the reactive components of the system. Peiffert (1943) compared the impedance of a myelinated nerve measured at 150 kHz with its dc resistance as a function of the interelectrode distance. The high-frequency impedance was lower than the dc resistance, as predictable from Fig. 2.1. Schmitt (1955) applied a standing wave analysis to the nerve fiber and obtained the phase velocity with which a particular electric pulse proceeds in a core conductor. This phase velocity is related to the propagation of the action potential, but it can be obtained in the unexcited state of the structure. Figure 2.2 shows such a measurement obtained from the sciatic nerve of a bullfrog with and without its sheath of connective tissue.

Tasaki and Freygang (1955) measured the impedance change across a Ranvier node in frog and toad nerve fibers during the action potential using an impedance bridge operated between 3 and 4 kHz. The bridge was balanced either at rest or at minimum impedance during the action potential. The action current was also measured. The maximum impedance change was 8–10% and occurred close to the peak of the action current. A calculation showed that a vanishing membrane resistance during the action potential would cause only an 8% decrease in impedance at the measuring frequency. The drastic change in the nodal resistance during excitation was confirmed with square-pulse measurements. It was also concluded that the potassium conductance increased a maximum of 15% during repolarization; this is in contrast to the results obtained on the squid axon.

Burlakova et al. (1959) measured the time lag between the onset of the action potential and the onset of the impedance change at 35 kHz in whole sciatic nerves of frogs. Reducing the temperature from 20 to 5°C increased this time lag from 0.12 to 0.69 ms. These findings agree with those of Cole and Curtis on *Nitella* (1938a) and squid axon (1938b and 1939), but disagree with those of Tasaki and Freygang (1955).

Tasaki and Mizuguchi (1949) investigated the effects of cocaine, sinomenine, and veratrine on the time course of the impedance change

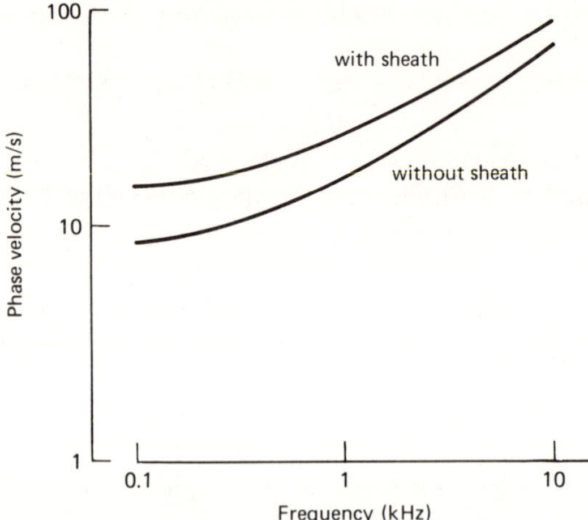

Figure 2.2. Phase velocity as a function of frequency in frog sciatic nerve with and without sheath. Modified after Schmitt (1955). By permission.

during the action potential. Measurements were performed on isolated myelinated fibers with an ac bridge operated at 2–3 kHz. Cocaine reduced the amplitude and duration of the action potential and the concomitant impedance change, whereas sinomenine and veratrine prolonged the action potential and its related impedance change. Prudnikova (1959) used an ac bridge at 35 kHz (Chailakhian and Iur'ev, 1957) to measure impedance changes in whole sciatic nerves (frogs). The impedance changed 3–5% of the initial values during the action potential. Eserine (10^{-2} M) did not produce a block after 10–15 min, but the conductance change during the action potential decreased by 30–50%.

2.1.2 Square-Pulse Measurements on Myelinated Nerves

From a study of excitation thresholds as a function of interelectrode distance, Rushton (1927) proposed the model of longitudinal current flow through a core conductor. This model was validated experimentally by measuring the longitudinal resistance of whole sciatic nerves as a function of interelectrode distance (Rushton, 1934). Short square pulses (0.8 ms) were used to simulate the current during nervous activity; Fig. 2.3 shows a typical result. The intersection of curves *A* and *C* with the ordinate (zero interpolar length) permits the separation of the two components of the measured resistance: the electrode component (R_{El}), and the effective resistance of the nerve. Curve *C* can be interpreted in terms of distribution of current across the sheath between conducting core and extracellular fluid. Despite the limitations in assuming uniform resistance for the sheath, the analysis is valid because identical curves were later obtained by Cole and Hodgkin (1939) in the squid axon. The initial part of curve *B* in Fig. 2.3 is an exponential, and the following constants can be obtained: (1) the resistance of the sheath, from the slope of curve *C* in Fig. 2.3; (2) the resistance of the core from the point where *C* intersects with the ordinate, and (3) the resistance of the interstitial fluid from the slope of a semilogarithmic plot of the exponential part of curve *B*.

Bradl (1967) also measured the longitudinal impedance in frog sciatic nerves soaked in isotonic sucrose to eliminate shunt resistances. Under one electrode the nerve was depolarized by isotonic KCl, which resulted in an injury potential between the two electrodes. Measurements were performed with a bridge consisting of resistances together with separate dc voltage sources, to compensate for asymmetry potentials at the electrodes and the injury potential. With this technique no current flowed through the preparation. The measurements were made with short-duration square pulses at a frequency of 1Hz. Using model 2 of Table 1.5, the parallel resistance and capacitance were identified with the lumped membrane impedance of

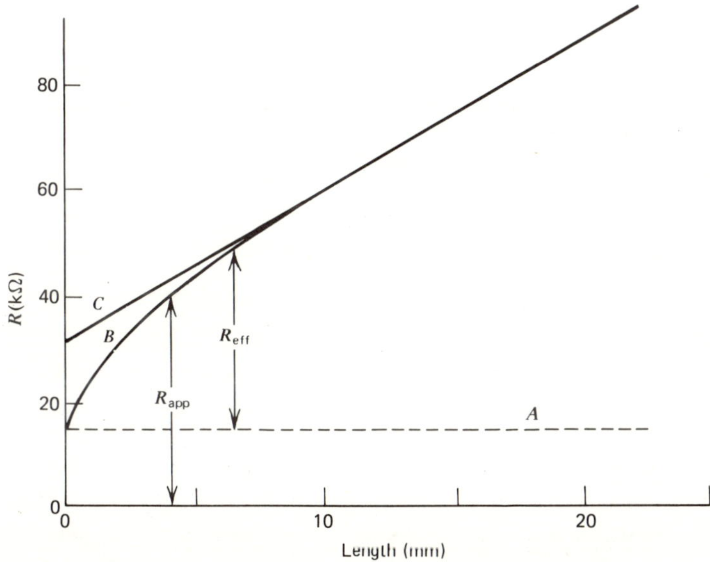

Figure 2.3. Resistance of myelinated nerve versus interpolar length. *A*, resistance of electrodes and circuit; *B*, nerve resistance at several interpolar lengths; *C*, extrapolation of linear relationship obtained for $x > 10$ mm; R_{app}, apparent resistance (measured value); R_{eff}, longitudinal resistance of nerve. Modified from Rushton (1934). By permission.

the bundle and the series resistance with the lumped longitudinal resistances. The parallel resistance was determined from the current transient after the onset of the pulse (Teorell, 1946; Schanne and Ceretti, 1971). In light of the earlier investigations on myelinated nerves (Rushton, 1934) and on squid axon (Cole and Hodgkin, 1939), the proposed identification of the components of the model seems to be an oversimplification. This could have influenced the analysis of the effects of potassium and chloride ions on the membrane resistance of the same structure (Bradl, 1970).

The action potential has been also used for estimation of the cell constants. Assuming that the myelin sheath was a perfect insulator and that current could enter or leave the fiber only at the nodes, Huxley and Stämpfli (1949) measured the external longitudinal current during the action potential, which should be equal and opposite to that inside the fiber. The potential difference across the myelin sheath (V) during the action potential was calculated according to

$$V - V_0 = r_i \int i\,dx \tag{2.2}$$

where V_0 is the resting value of V, r_i is the resistance per unit length of the core, i is the instantaneous value of the current, and x is the distance along the fiber. The current entering or leaving the myelin sheath was taken as the difference between records obtained at different positions along the fiber. The results suggested that the myelin is a passive membrane with a capacitance and resistance in parallel. These circuit elements are given by

$$G_m = \int_{\text{foot}}^{\text{end}} i_m\,dt \Big/ \int_{\text{foot}}^{\text{end}} (V - V_0)\,dt \tag{2.3}$$

$$C_m = \left[\int_{\text{foot}}^{\text{peak}} i_m\,dt - G_m \int_{\text{foot}}^{\text{peak}} (V - V_0)\,dt \right] \Big/ AP \tag{2.4}$$

where i_m is the instantaneous current density across the myelin sheath and AP, the action potential amplitude. The values of G_m and C_m thus obtained had to be corrected by a factor α because the action potential amplitude calculated with Eq. 2.2 was lower than the actual potential difference. The electrical constants of the myelin sheath of single nerve fibers of the frog are listed in Table 2.2. The results indicated that the myelin did not have electrogenic properties, which served to develop a method of recording true values of membrane potential in myelinated fibers with extracellular electrodes (Huxley and Stämpfli, 1951). Hodgkin (1951) then estimated the axoplasmic resistivity, and a value of 1.2 for α. The corrected electrical

Table 2.2 Electrical Constants of Myelinated Nerve Obtained with Square Pulses and Circuit 2 of Table 1.1

Measured Quantities	Model	Derived Quantities	References
Action current	4 of Table 1.5	R_m, 82–167 × 10^3 Ω cm^2	Huxley and Stämpfli (1949)
		C_m, 0.0023–0.0035 μF/cm^2	
		ε, 5.1–7.9	
		Resistivity of myelin, 4.2–10.5 × 10^8 Ω cm	
		R_i cytoplasm, 110 Ω cm	Hodgkin (1951)
		C_m, 0.0025 μF/cm^2	
		ε, 5.4	
		R_m, 0.16 × 10^6 Ω cm^2	
		Resistivity of myelin, 800 × 10^6 Ω cm	
Action Current charge ($A \times s$)	1 of Table 1.5	Nodal membrane:	Tasaki (1955)
		C_m: 3–7 μF/cm^2	
		R_m: 8–20 Ω cm^2	
		Myelin:	
		C_m: 0.005 μF/cm^2	
		R_m: 10^5 Ω cm^2	
		ε: 6–10	

constants are also included in Table 2.2. The calculated surface resistance and capacity for a single layer were not much different from that of a typical nerve membrane. This estimation, however, is in sharp contrast with the reported values for resistance of bilayer lipid membranes, which are of the order of 10^6 Ω cm^2 (Miyamoto and Thompson, 1967).

Tasaki (1949) developed a technique to determine the current density across the myelin sheath and the nodal membrane. Isolated fibers were placed in a three-compartment chamber so that the central compartment contained an internode (Fig. 2.4). Since the length of the internode is small compared to λ, a simultaneous stimulation of the adjacent nodes N_0 and N_1 keeps the internode isopotential. Thus, the current flowing in the central compartment reflects the current density at the internode. If a previously inactivated node occupies the central compartment, the nodal current density can be measured. Tasaki (1955) recorded simultaneously the action current and the integral of the current over time, so that the charge transferred across the myelin sheath or the node of Ranvier could

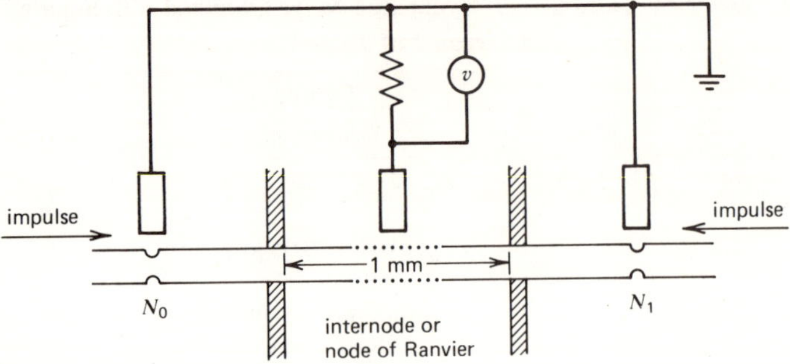

Figure 2.4. Principle of method for measuring action currents by the collision technique. For explanation, see text.

be estimated. Since the current had both capacitive and ohmic components, the total charge is given by:

$$Q(t) = CV(t) + (1/R)\int_0^t V(t)\,dt \qquad (2.5)$$

Since the action potential can be approximated by a triangular wave and the capacitative term in Eq. 2.5 cancels out because the potential in the axoplasm is the same before and after the spike, the following is obtained:

$$R = tV_{\text{peak}}/2Q_{\text{end}} \qquad (2.6)$$

From an equation derived for the charge crossing the membrane at the peak of the action potential, we obtain:

$$C = \left[\,Q_{\text{peak}} - (a/t)Q_{\text{end}}\right]/V_{\text{peak}} \qquad (2.7)$$

where a is the duration of the upstroke. For the values of V_{peak}, it is considered that the potential difference across the internode represents the total action potential amplitude and V_{peak} at an inactivated node amounts to only 70% of the action potential. The values of R and C obtained with this method are referred to 1 mm length of nerve fiber, being equivalent to c_m and r_m in the linear cable model. The constants were 1.6 $\mu\mu$F/mm and 290 MΩ mm for the myelin sheath and 1.5 $\mu\mu$F/mm and 41 MΩ mm for the Ranvier node.

Equations 2.6 and 2.7 were used to investigate changes in c_m and r_m in sodium-free Ringer's solution, hypertonic sodium chloride, and under cocaine. Cocaine (0.05%) abolished the action potential but did not change the resistance of either the node or the myelin. Higher concentrations (0.2–0.5%) raised the resistance of both in about the same proportion. The capacity remained constant. Saponin (2%) increased the leakage of current through the myelin sheath, increased c_m, and decreased r_m. The resistance of a 1 mm length of nerve containing a node of Ranvier was shown to have a higher Q_{10} than that of Ringer's solution. The capacity of the nodal membrane did not change between 15 and 25°C. The capacity and dielectric constant of the myelin sheath estimated by Huxley and Stämpfli (1949) and Tasaki (1955) are comparable (Table 2.2). The theory of saltatory conduction was confirmed because the leak current through the myelin sheath was only 10^{-4} of the current flux through the nodal membrane. The temperature dependence of the membrane resistance could satisfactorily explain the decrease in conduction velocity during temperature lowering.

Stämpfli (1958, 1959) measured current–voltage curves of Ranvier nodes under different potassium concentrations (Fig. 2.5). Increasing extracellular potassium concentration up to 10 mM produced two types of responses.

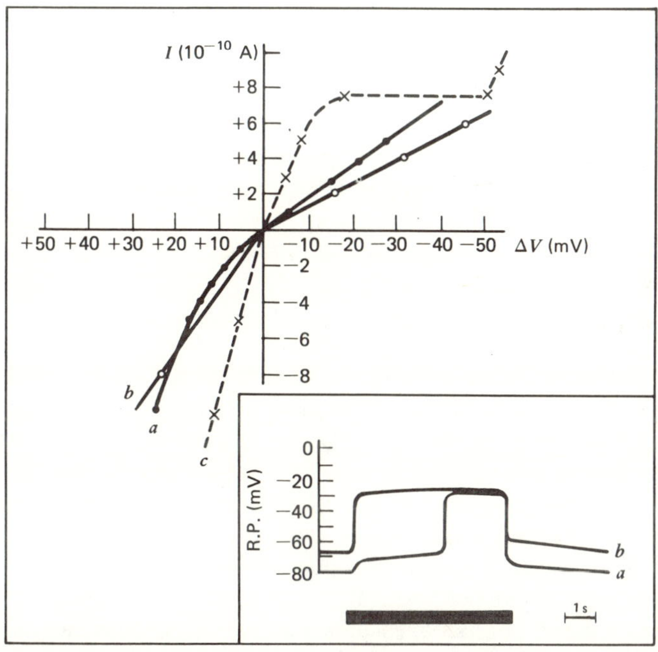

Figure 2.5. Current–voltage curves for nodes of Ranvier in state 1 (*a*), state 2 (*b*), and after equilibration with an extracellular K^+ concentration of 40 mM (*c*). The origin corresponds to a membrane potential of −70 mV for *a*, −67 mV for *b*, and −22 mV for *c*. Composed from data of Stämpfli (1959). Inset: Changes of the resting potential (RP) of a node of Ranvier to an increase in K^+ concentration to 40 mM (horizontal bar). Curve *b* is a typical response of a node in state 2 and its conversion into state 1 (*a*) by application of a hyperpolarizing pulse prior to the increase in $[K^+]_o$. Modified from Stämpfli (1959). By permission.

The first one (*a* in inset of Fig. 2.5) was characterized by a slow depolarization up to a critical level at which a sudden depolarization occurred to the potential level predicted by the Nernst equation (nodes in state 1). In the second type of response, the membrane potential immediately changed to the new steady-state level (*b* in inset of Fig. 2.5). For nodes in state 1, the current–voltage relation showed rectification for depolarizations greater than 10 mV. In contrast, the curves from nodes in state 2 bent sharply near or at the level of the resting potential. After prolonged treatment with a potassium concentration of 40 mM, the difference between nodes in state 1 and 2 disappeared. The results were interpreted as due to the existence of two equilibrium states of the membrane. The shifts from one to the other would depend upon the chloride and potassium concentration and the original value of the membrane potential. The rectification observed in the current–voltage curves would suggest that the potassium and chloride conductances change as a function of membrane potential.

Tasaki and Freygang (1955) studied the changes in membrane impedance during the action potential by superimposing short-duration square pulses on the action potential. Because of the capacitive elements in the membrane, the current reached its peak before the potential. The input resistance changed with the same time course as the change in potential. The amplitude of the electrotonic potential at the peak of the action potential was less than 10% of the value at rest. Since $r_m = 2\lambda R_{\text{inp}}$, and its value at rest was 40–60 MΩ, it was concluded that r_m dropped to less than 4–6 MΩ mm at the peak of the action potential.

2.1.3 Voltage-Clamp Measurements on Myelinated Nerves

After the voltage-clamp technique was successfully used to measure the different ionic currents in the squid axon (Hodgkin, Huxley, and Katz, 1952), attempts were made to apply this technique to other excitable structures. Several groups tried to adapt the voltage-clamp technique to the geometrical conditions required for the myelinated nerve (Del Castillo et al., 1957; Tasaki and Bak, 1958a; Dodge and Frankenhaeuser, 1958; Nonner, 1969), where the membrane currents can be measured only across the membrane of a Ranvier node. The node under study can be isolated by suitable partitions, but the measurement of the membrane potential is difficult and is usually achieved indirectly by measuring the injury potential between an inactivated node and the node under study. While this inactivation is fairly easy to obtain by bathing a node in cocaine or isotonic KCl or simply by cutting the nerve, it is difficult to obtain a sufficiently effective electrical insulation between the inactivated node and the one that is being studied. Any difference between the injury potential and the true membrane potential will affect the quality of the clamping method. In the beginning, petroleum jelly (Dodge and Frankenhaeuser, 1958) and air (Del Castillo et al., 1957) were used to obtain insulation of the partitions, until Bergman and Stämpfli (1966) proposed a sucrose-gap method. Another method using the air-gap technique was reported by Nonner (1969). The circuits used in these studies were like circuits 4 or 5 in Table 1.4, and only minor modifications were introduced to improve either the quality of the insulation or the electronic performance of the clamp.

2.1.3.1 Basic Parameters. Del Castillo et al. (1957) reported the existence of two different action potentials in the isolated sciatic nerve of the frog. The first type presented an almost exponential repolarization phase, and the inward current was not followed by an outward current. The other type of action potential had a small plateau and showed the usual outward current after the early inward current.

Tasaki and Bak (1958a), working with isolated nerves of the toad, obtained the dynamic current–voltage curve of the Ranvier node, which showed, for the early current, the well-known region of negative resistance typical of excitable biological structures (Fig. 2.6). The ratio of the low-conductance branch to the high-conductance branch of the curve was 7 : 11. In low-sodium medium (20% of normal) with 0.7% urethane and during the refractory period, a reduction in the conductance of the activated membrane was observed. Moreover, rapid decremental oscillations superimposed on the membrane current immediately after imposing a step signal on the membrane were reported. These oscillations were

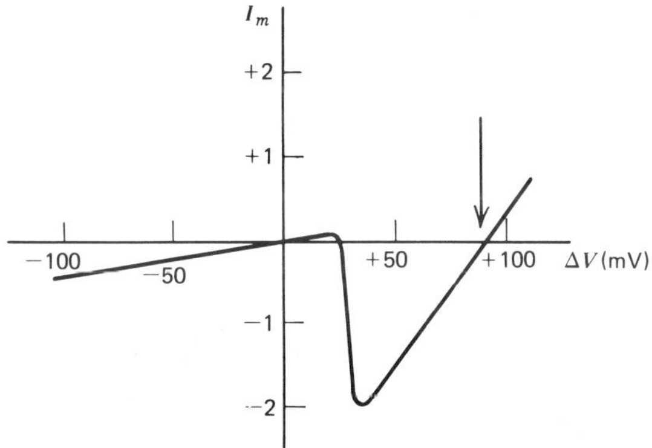

Figure 2.6. Current–voltage curve for maximum early current of nodal membrane. Membrane current in relative units. At arrow, maximum amplitude of action potential. Modified from Tasaki and Bak (1958a). By permission.

attributed to the nodal membrane and were interpreted as supporting the two-state theory of the action potential, as discussed in Sec. 2.2.3.

The method used by Dodge and Frankenhaeuser (1958) permitted better voltage control and was more stable than previous methods. Application of a voltage step on the nodal membrane gave rise to transients that interfered with the peak of early inward current at room temperature. Experiments were therefore carried out at 3°C so that the artifacts did not interfere with the recording of the ionic currents. The results confirmed that the nodal membrane behaved qualitatively like the squid axon membrane during excitation. The main difference was a higher current density than in squid axon, in agreement with Tasaki and Bak (1958a), but no oscillations attributable to the node were observed, except as instabilities of the clamping circuit, in contrast to the claim of Tasaki and Bak (1958a). Dodge and Frankenhaeuser (1958) never found a node lacking the late outward current unless the structure had previously been damaged, in contrast to the previous report of Del Castillo et al. (1957).

Dodge and Frankenhaeuser (1959), using large single fibers from the sciatic nerve of the clawed toad (*Xenopus laevis*), found that the large diameter of these fibers reduced the impedance between current pool 1 and the cytoplasm under the nodal membrane in pool 2 (method 4 of Table 1.4). This reduced the phase shift caused by this impedance and increased the stability of the system and its resolution, thus reducing the artifacts. Therefore, the experiments could be performed at room temperature. During the following years a series of papers reported a detailed analysis of the electrical behavior of the nodal membrane (Frankenhaeuser and Vallbo, 1969). These studies led to an estimation of the parameters controlling excitation in the myelinated nerve (Frankenhaeuser and Huxley, 1964). Table 2.3 lists the estimated parameters. Figure 2.7 shows

Table 2.3 Standard Data for the Ranvier Node of the Toad Myelinated Nerve

Conditions	Derived Quantities	Reference
Method 4 of Table 1.4	*RP*: −70 mV*	Frankenhaeuser and Huxley (1964)
Measured quantity, $I=f(t)$	C_m: 2 μF/cm^2	
Model 1 of Table 1.5 and Hodgkin and Huxley (1952d)	g_l: 30.3 mmho/cm^2	
	$\bar{P}_{Na}$: 8×10^{-3} cm/s†	
	$\bar{P}_K$: 1.2×10^{-3} cm/s†	
	P_p: 0.54×10^{-3} cm/s†	
	$[K]_i$: 120 mM	
	$[Na]_i$: 13.7 mM	

**RP*: resting potential.
†Permeability constants.

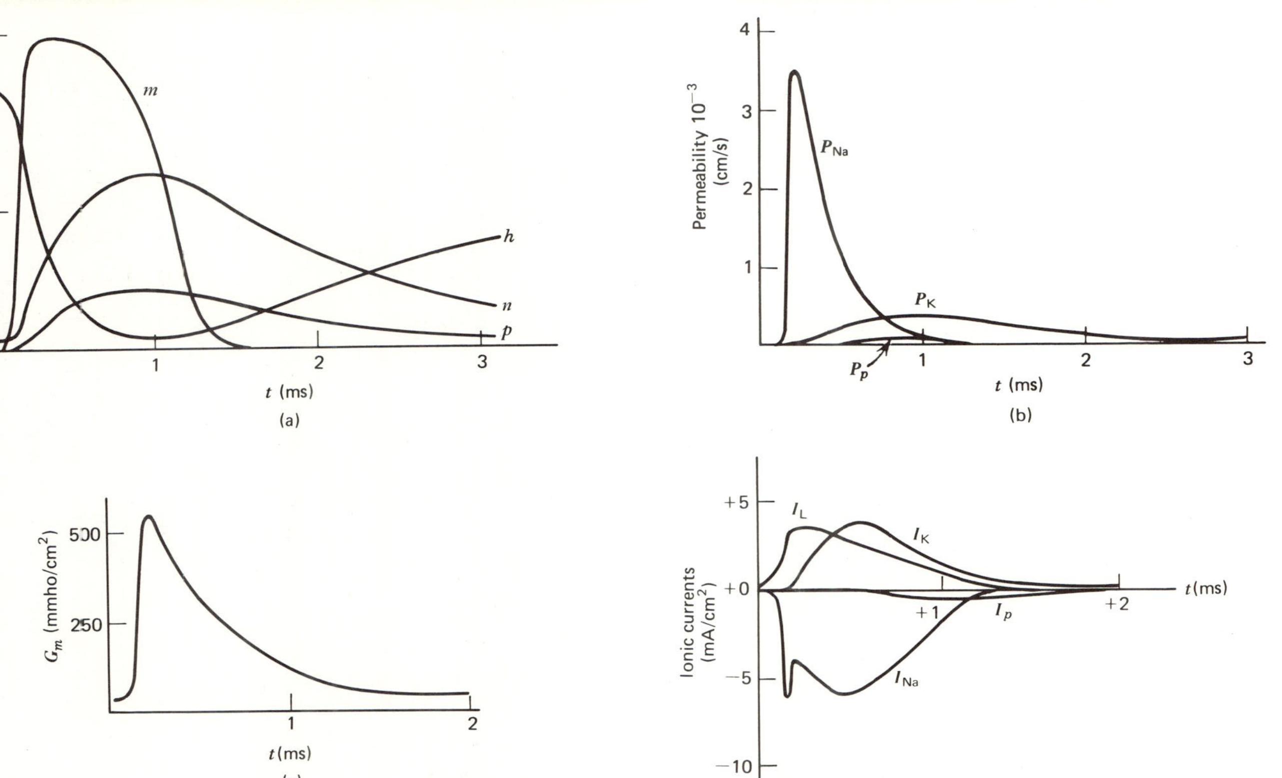

Figure 2.7. Parameters of the action potential in the nodal membrane calculated from standard data. (a) Time course of variables *m*, *h*, *n*, and *p*, the variable for the nonspecific ion permeability; (b) time course of the change in ionic permeabilities; (c) time course of the change in membrane conductance; (d) time course of ionic currents. I_L, leak current. Modified from Frankenhaeuser and Huxley (1964). By permission.

additional data obtained from these results. There is qualitative agreement between the time course of the membrane variables during the action potential in the giant axon of the squid and that found in myelinated nerves of the clawed toad. Figure 2.7 reveals the difference between the dynamic membrane parameters of the axon and the nodal membrane: (1) the permeability for the nonspecific ionic current is not constant but is governed by a variable p; (2) there are two peaks in the inward current; (3) the numerical values of the permeability constants are very high, and (4) the membrane conductances are high. There are also quantitative differences in the voltage and time dependence of the ionic conductances. The analysis by Frankenhaeuser and Huxley is based on the voltage- and time-dependent permeability constants, which are, under voltage-clamp conditions, proportional to the slope conductances (see Appendix 3, Eq. A3.9). In the case of the nodal membrane, whose parameters are much more nonlinear than those of the axonal membrane, the use of permeabilities is an advantage over the chord conductances used by Hodgkin and Huxley (1952b). By analogy with the relation found for the membrane of the squid axon, we obtain:

$$\left.\begin{aligned} P_K &= \bar{P}_K n^2 \\ P_p &= \bar{P}_p p^2 \\ P_{\text{Na}} &= \bar{P}_{\text{Na}} h m^2 \end{aligned}\right\} \qquad (2.8)$$

where P_i is the voltage- and time-dependent specific permeability for ion i, and $\bar{P}_i$ is its maximum value (permeability constant). The subscript p is the nonspecific ions. A comparison with Eqs. 2.13 shows that the variable n is only to the second power instead of to the fourth, and m is also to the second power instead of to the third; the power of the variable h remains unchanged. The relation of the variables to their respective rate constants α and β is the same as in the Hodgkin-Huxley theory. Using the same formalism, Frankenhaeuser (1965) investigated how small variations from the standard data influence the shape and time course of the recorded action potential.

Frankenhaeuser and Moore (1963) investigated the influence of temperature on the rate constants α and β for the permeability variables m, h, and n. The rate constants for h and n provided a Q_{10} value close to 3, as had been found for those of m, h, and n in the giant axon of the squid, but the Q_{10} for α_m and β_m were found to be 1.8 and 1.7, respectively. Consequently, in myelinated nerve the rising phase of the action potential is less affected by temperature than the repolarization phase.

Moore (1971) investigated the influence of temperature and Ca^{2+} on the rate constants of fibers from *Xenopus laevis* and the frog sciatic nerve.

Arrhenius plots of the rate constants $1/\tau_m$, $1/\tau_h$, and $1/\tau_n$ (Eqs. 2.18; see p. 108) yielded an activation energy of 15 kcal/mol. Moreover, it was found that (1) $1/\tau_m$ and $1/\tau_h$ increased with increasing $[Ca^{2+}]_o$, but $1/\tau_n$ was relatively independent of $[Ca^{2+}]_o$; (2) plots of τ_m and τ_h versus membrane potential showed a shift of these relations toward positive values along the voltage axis with increasing $[Ca^{2+}]_o$. The findings are in agreement with those of Hille (1968b) obtained in frog sciatic nerve.

The effects of calcium on sodium conductance were further investigated by Dubois and Bergman (1971b). Plots of g_{Na} versus V for two different calcium concentrations indicated a competition between sodium and calcium for the same membrane sites. Depolarization modified the dissociation constant for the complex Na^+-specific site, an effect similar to that of lowering the calcium concentration. The number of membrane sites appeared to be independent of the membrane potential. The affinity of the membrane sites for calcium was larger than for sodium and more significantly reduced by depolarization. These results confirm and help explain the findings of Moore (1971). The usefulness of these plots (Lineweaver-Burke) in analyzing sodium conductance in terms of affinity of the ion for a membrane site was previously shown by Dubois and Bergman (1971a).

Nonner (1969) further developed the voltage-clamp technique of Dodge and Frankenhaeuser (1958). The most important changes were the use of only one amplifier (*FA*) and the use of an air gap to improve the isolation between pools 3 and 4 (circuit 4 of Table 1.4). The air gap provided an insulation 20 times higher than with the formerly used petroleum-jelly gap. With the air gap, no current flowed along the outside of the nerve or in the cytoplasm between compartments 3 and 4. All the current from the feedback amplifier left the nerve across the nodal membrane in compartment 2. Since the extracellular fluid in compartment 3 is held at ground potential, the difference in potential between the pools in compartments 2 and 3 equals the membrane potential of the nerve and thus can be changed by a command signal from the function generator *F*. These technical improvements led to an increase in the time resolution of the clamp, which was on the order of 10–20 μs for fibers with diameters ranging between 12 and 15 μm. Therefore, voltage clamping of small fibers at room temperature was possible.

Another approach to the study of properties of the potassium system was used by groups investigating the relation of tracer fluxes to electrical parameters. Lewis and Diecke (1958) correlated the relative rubidium outflux with $[Rb^+]_o$. In myelinated nerve, the kinetics of rubidium fluxes resemble those of potassium, and the experiments were performed on the tibial nerve of the frog at room temperature. The outflux of Rb^{86} was measured at the resting potential and with the membrane clamped at zero potential. The flux at zero membrane potential showed positive coupling, which can be explained by the model of single-file diffusion. At a membrane potential of -70 mV there was negative coupling, explainable by the model of exchange diffusion (Appendix 3). It was concluded that both types of membrane transport are present in the nodal membrane but that their respective activation depends on the membrane potential. The inactivation of the delayed potassium current during strong depolarization (Frankenhaeuser and Waltman, 1959; Lüttgau, 1960), not predicted by the theory of Hodgkin and Huxley (1952d), was investigated by Lewis (1971) on the tibial nerve of the frog. During depolarization, the delayed potassium current was inactivated to 75% by a rapid process with a time constant of 15 ms followed by a slower phase of inactivation with a time constant of several seconds.

Schwarz and Vogel (1971a and b) investigated the same process in nerve fibers of *Xenopus laevis* with the method of Dodge and Frankenhaeuser (1958). At a resting potential of -30 mV, the time constant for the inactivation process was 3 s and increased with depolarization. The rate constant had a Q_{10} of 2.7 between 11 and 21°C. As in frog nerve fibers, the inactivation of the potassium mechanism was incomplete. During long-lasting depolarization it was controlled by two processes, with time constants of 0.6 s and 3.6–20 s according to the level of the depolarizing test pulse. This process could be described by the equation $P_K = \bar{P}_K n^2 k$, an extension of Eq. 2.8, where k is the sum of two variables k_1 and k_2, each controlling one phase of the inactivation process. From this, it was inferred that there might be two separate potassium channels in the tibial nerve of the clawed toad.

Bergman and Stämpfli (1966) compared the potassium permeability of motor and sensory nerve fibers, using a combination of current- and voltage-clamp techniques with two sucrose gaps. Larger depolarizations were needed to activate a potassium outward current in myelinated fibers than in nonmyelinated fibers. Current–voltage curves of steady-state current at several potassium concentrations between 2.5 and 117 mM showed a region of negative resistance for higher potassium concentrations. Motor fibers had a higher potassium permeability than sensory fibers, but there was no difference in the resting potential of the two fiber types. The reversal potentials of the current–voltage curves for different values of $[K^+]_o$ were more negative in motor fibers. Since the resting potentials of both fiber types were the same, this difference is difficult to explain.

An interesting approach to understanding the mechanism of excitation in nodal membranes was chosen by researchers who recorded the steady-state current–voltage characteristic of the nodal membrane under voltage-clamp conditions (Balk and Müller-Mohnssen, 1966). The electronic circuit was arranged so that current–voltage curves of the steady-state current could be obtained directly on an *x-y* recorder, but since the method (circuit 6 of Table 1.4) used only three compartments, leakage currents could not be neglected and the membrane current was given in relative units. Müller-Mohnssen and Balk (1965) investigated the relations between stationary and dynamic properties of the nodal membrane of single motor fibers of the frog. The basic idea behind the measurement of stationary current–voltage curves is as follows: In the traditional voltage-clamp approach, the region of negative resistance of the current–voltage curve is obtained by plotting the maximum inward current at different potential levels. A stationary measurement of this region is not possible with the nodal membrane at resting potential (-70 mV) because the current flow originating from the imposed potential is opposite to the inward current resulting from the regenerative properties of the membrane. If the membrane is depolarized beyond the region of negative resistance by increased $[K^+]_o$ or veratridine, inward potassium currents result from the application of hyperpolarizing voltages. If ramp pulses with a small rate of change (dV/dt) compared to the time-dependent processes of the membrane are imposed, the stationary currents can be measured. If the membrane conserves its properties, the resulting curve shows a well-defined region of negative resistance (Fig. 2.8). There is a high-resistance branch at more

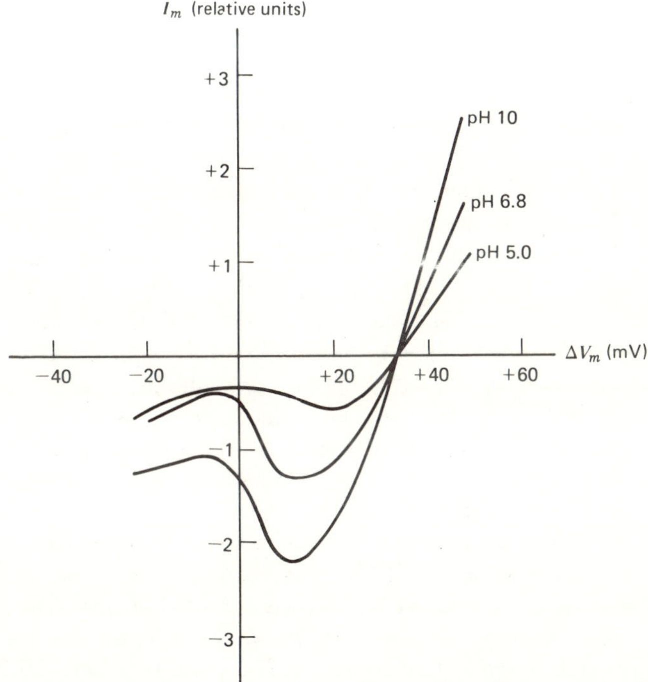

Figure 2.8. Stationary current–voltage curves from frog nerve obtained in $[K^+]_o = 40$ mM. The curves are drawn taking the resting potential in control medium as reference. Parameter: pH. Inward current negative. Modified after Müller-Mohnssen and Balk (1965). By permission.

negative potentials and a low-resistance branch in the range of less negative potentials, the transition between the two being the negative resistance. The point where the high-resistance branch intercepts the abscissa is the membrane potential in high extracellular potassium concentration. Müller-Mohnssen and Balk (1966) and Balk and Müller-Mohnssen (1968) showed that the membrane responded with action potentials when stimulated in the transitional regions or in the region of the high-resistance branch. Since the current here is carried mainly by potassium ions (the extracellular sodium concentration was only 5 mM), the action potentials were termed potassium action potentials. Decreasing the pH tended to smooth out the transitional region of the current–voltage curve until the nodal membrane had the characteristics of a simple rectifier. The occurrence of the potassium action potential was interpreted as due to a membrane whose batteries were discharged by previous depolarization but whose regenerative mechanism was unchanged (negative resistance region). In this case, K^+ ions carry the current instead of Na^+ ions. This attempt to describe an excitable membrane in terms of its stationary current–voltage characteristic is useful, since such a description allows a treatment of biological membranes in the framework of the theory developed in electronics for systems with metastable states.

2.1.3.2 Pharmacological Studies. Voltage-clamp measurements on myelinated nerves have been carried out to test the effect of neurotoxins on the different ionic channels. Ulbricht (1965, 1969) investigated the effect of veratridine, which is supposed to affect the sodium permeability because it renders the membrane more sensitive to changes in extracellular sodium. Experiments were performed on frog motor fibers with a veratridine concentration of 10^{-5} g/ml. In Ringer's solution the node showed a rectifier characteristic. Under veratridine the membrane depolarized by 60 mV and showed an N-shaped current–voltage curve similar to that obtained by Balk and Müller-Mohnssen (1966). In veratridinized nerves, the rectifier characteristic appeared in Na-free solutions, whereas in a solution containing 100% Na the N-shaped characteristic was obtained. Under veratridine, a second slow component of sodium inward current developed and the drug also decreased the amplitude of the fast inward current. Ca^{2+} counteracted some of the veratridine effects.

Hille (1968a) studied the effects of neurotoxins on large fibers of the frog sciatic nerve. Tetrodotoxin (TTX) and saxitoxin (STX) abolished the fast sodium inward current as well as the veratridine-induced slow component of the sodium current. On the other hand, these toxins did not affect the potassium and the leakage channel. At low concentrations (10^{-9} M), TTX reduced the maximum Na conductance without changing the time course of the diminished sodium currents. This means that while part of the channels are blocked, the drug has no effect on the opening of the remaining channels. A dose–response curve gave a dissociation constant of 1.2 μM for the hypothetical inhibitory complex. Identical curves were obtained in a control medium and in a medium with TEA. It was then inferred that the action occurs on the sodium channels. The effect of DDT was similar to that of veratrine. In terms of the channel hypothesis, this means that DDT causes a fraction of Na channels, which open during depolarization, to remain open for a longer time than normal.

Adam et al. (1966) reported the effects of scorpion venom on Ranvier nodes of the frog tibialis nerve. The following results were obtained: (1) Concentrations larger than 10^{-7} g/ml produced depolarization and spontaneous activity; (2) the action potential duration increased; (3) replacement of sodium by choline abolished depolarization and spontaneous activity; (4) increased $[Ca^{2+}]_o$ partially hyperpolarized the membrane; (5) N-shaped current–voltage curves were obtained in Ringer's, the range of negative resistance being between 20 and 60 mV depolarization. These results suggested that scorpion venom increases the resting sodium permeability and delays the inactivation of the sodium system.

Koppenhöfer and Schmidt (1968a and b) reported that scorpion venom from *Leiurus quinquestriatus* (Adam and Weiss, 1959) prolonged the action potential, reduced the maximum Na permeability by 40%, slowed the inactivation of the sodium permeability, reduced the maximum potassium permeability to 35% of its control value, and slowed the increase of P_K upon depolarization.

Schmitt and Schmidt (1972) investigated the antagonistic effect of Ca^{2+} on the changes induced by scorpion venom (Adam et al., 1966). Increasing $[Ca^{2+}]_o$ from 1.8 to 7.2 mM abolished the incomplete sodium inactivation typical of treated nodes, but the depression of the peak inward current was further accentuated and the peak of the current–voltage relation was shifted toward larger depolarizations. The sodium permeability reached a maximum of 4×10^{-3} cm/s at depolarizations of 100 mV and was reduced by 40% in the presence of the venom. In high calcium, the maximum P_{Na} remained unchanged despite a marked shift of the P_{Na} versus V relation. Scorpion venom reduced the maximum P_K from 0.6×10^{-3} cm/s to 0.27×10^{-3} cm/s, but it rose to 0.41×10^{-3} cm/s with increasing $[Ca^{2+}]_o$. However, no effect was found on current kinetics. The effects of calcium ions on membrane currents can explain the partial abolition of the action potential changes produced by scorpion venom.

2.2 NONMYELINATED NERVES

2.2.1 AC Measurements on Nonmyelinated Nerves

In 1936, Young described and called attention to the value of the giant axon of the squid, thus providing a rather uncomplicated preparation.

Alternating current measurements on the squid giant axon were published soon thereafter by Cole and Curtis (1938b, 1939). The impedance of the axon was measured as an equivalent parallel resistance and capacitance at frequencies between 1 kHz and 1 MHz. The impedance was resolved into series resistance R_s and reactance X_s to allow plotting the impedance locus. During the action potential, the bridge could be balanced at different time intervals, and thus the change of the impedance locus during activity was obtained (small circle in Fig. 2.9). The so-called time impedance loci can be calculated theoretically. The deviation between measured and computed time impedance loci permitted estimation that the capacity decreased by about 10% at 50 kHz, the critical frequency of the axon. The analytical treatment of the results is the same as in the paper by Cole and Curtis (1938a). Using the suspension equation, the following parameters were evaluated: $R_m = 28\ \Omega\ \text{cm}^2$, $C_m = 1.8\ \mu\text{F/cm}^2$, $R_i = 71\ \Omega$ cm, and a phase angle of 71°. It was shown that the unbalance of the bridge is proportional to the change in membrane conductance if C_m remains constant and provided that (1) the structure obeys the linear cable model and (2) the product $\beta \Delta G_m$ is small. Here ΔG_m is the change in membrane conductance and β is defined as:

$$\beta = |ds(Z - R_\infty)/(4Z)|$$

Here, d is the fiber diameter, s is a complex constant involving the resting impedance, Z is the off-balance impedance, and R_∞ is the resistance of the system at infinite frequency. Taylor (1965) measured the impedance of the squid giant axon, in the range of 10–70 kHz. The membrane capacity did not change appreciably even with long continued polarization.

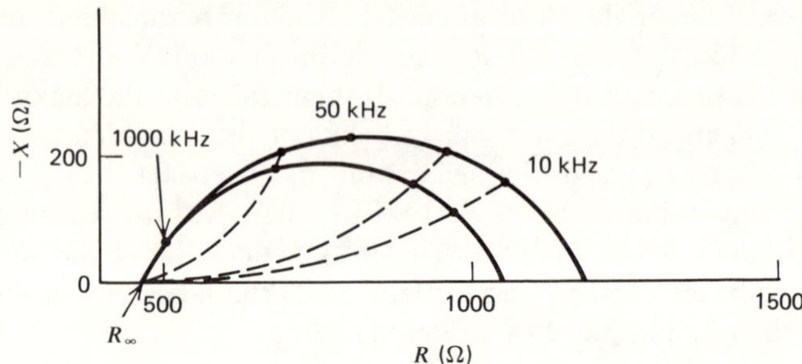

Figure 2.9. Impedance locus of the squid axon at rest (large locus) and at maximum impedance change during activity (small locus). The dashed lines (time impedance loci) connect the points of same frequency on both loci; they join at the common point R_∞ on the R axis. Modified from Cole and Curtis (1939). By copyright permission of The Rockefeller University Press.

Matsumoto et al. (1970) measured the axon impedance of the Sea of Japan squid, in the range 50 Hz to 1 MHz. The impedance loci obtained were skewed circles, as described by a system of two time constants. This result was interpreted as an indication for the mosaic structure of the axon membrane. Low sodium and high calcium did not influence the axon impedance, but in 80–100 mM KCl, the membrane capacity decreased by 10–25%. This suggested that the membrane had an electrostatic and constant capacitive component of 0.35–0.45 $\mu F/cm^2$ and a second variable capacity (0.7–1.2 $\mu F/cm^2$ in sea water) that decreased by 22–30% in high potassium. The latter was ascribed to a shielding effect exerted by monovalent cations on the electrostatic forces of the negative fixed charges in the membrane (Tasaki, 1968). Since this variable capacitance appeared only at frequencies below 2 kHz, it could not be observed in the frequency range used by Taylor (1965).

Cole and Baker (1941) measured the longitudinal impedance of the squid axon. The impedance locus was interpreted as being generated by an inductive reactance (Fig. 1.11). This finding stimulated considerable controversy and speculation (Lorente de Nó, 1947). Hodgkin and Rushton (1946) argued that the action potential would not have an overshoot if the inductive reactance represented a real inductive element. Today it is accepted that this inductive component is produced by a nonlinear conductance (see Sec. 1.5.1.2).

In 1955 Schmitt used a voltage divider method to measure the resistive component of Z_{inp} (Fig. 2.10). G_1 operates at a frequency between 20 Hz and 20 kHz. The resistance R_1 forms a voltage divider, and the voltage drop across C is measured across the Y-deflecting plates of the oscilloscope. C generates electronically a negative capacitance. When this capacitance offsets the capacitive component of Z_{inp}, only its resistive component is measured. The generator G_2 (0.01–10 Hz) imposes a time-variant current on Z_{inp}, which is simultaneously displayed at the X-deflecting plates of the oscilloscope so that the change of the input resistance as a function of the current is displayed and the current–voltage curves can be plotted. The current–voltage relation thus obtained showed a low-resistance branch on depolarization and a high-resistance branch on hyperpolarization. With low frequencies in G_2, measurements in quasi-stationary conditions are obtained. The behavior of the active membrane of a *Carcinus* nerve was also analyzed. The first derivative of the action potential was proportional to (1) the longitudinal current, and (2) the time course of net charge deposition and removal from the membrane during the impulse. The second derivative was proportional to the current density in the membrane. Displaying the action potential on the X-deflecting plates and its first derivative at the Y-deflecting plates of a CRT resulted in a loop whose

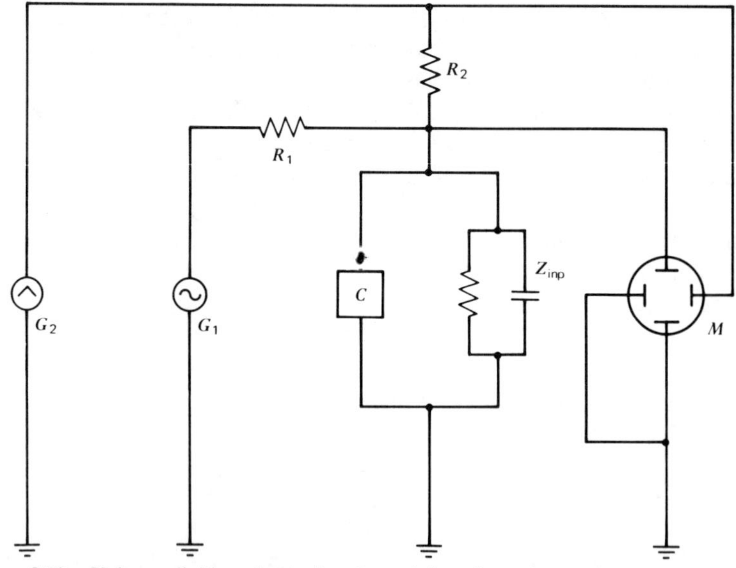

Figure 2.10. Voltage divider circuit for determining dynamic conductance in biological structures. C, compensator (generator of negative capacitance); G_1, sine wave generator; G_2, triangular wave generator; M, oscilloscope; R_1 and R_2, voltage-dividing resistances for G_1 and G_2; Z_{inp}, membrane impedance.

area was proportional to the energy released during an action potential. This loop became known as the phase-plane display (see Sec. 3.1.1).

Taylor (1965) tested several models for the squid axon membrane: the model of the constant phase angle, the model of a single relaxation time, and the so-called barrier height model, as described by Fröhlich (1958). Measurements were made in the frequency range of 10–70 kHz. The representation was as an impedance locus either in the complex plane or in the plane of the complex dielectric constant. All three models could be roughly fitted to the experimental data. However, the first was rejected because of the lack of a physical model. The single time constant model exhibited a discrepancy at the characteristic frequency in the plane of the complex dielectric constant. The barrier-height model fitted the results, assuming a finite distribution of time constants. It was concluded that there was little evidence to permit a choice between these models. More conclusive evidence was sometimes obtained by applying less sophisticated techniques to explore more biological variables, as shown by Grundfest et al. (1952, 1953), Shanes, Grundfest, and Freygang, (1953), and Amatniek et al. (1957). The measurements were performed at either 25 kHz or 31 kHz, frequencies located on the impedance locus rather far from the real axis

(Fig. 2.9). These measurements yield much less information about the input resistance than a whole impedance locus, but Cole and Curtis (1938b) showed that a small unbalance of the bridge was proportional to the membrane conductance. The results were then expressed as a percentage of change. It was reported that the increase in membrane conductance during the action potential in seawater, found by Cole and Curtis at 4°C, was also present at room temperature and that there was a decrease of conductance during the after-potential. This decrease of conductance did not occur at 10°C (Amatniek et al., 1957). Low sodium increased the threshold, decreased propagation velocity, and delayed the onset of the impedance change relative to the beginning of the action potential. The spike amplitude and the accompanying impedance change also decreased. The impedance change was affected more than the amplitude of the action potential in high potassium. The alkaloid cevadine did not affect the magnitude of the impedance change during the spike, but it prolonged the impedance change during the after-potential; it gave rise to periodic conductance increases followed by a prolonged period of elevated conductance. Veratridine caused a conductance increase during the after-potential that lasted even longer.

Another series of ac measurements on the squid axon were designed to describe impedance changes with time in a work aimed at a generalization of the Hodgkin-Huxley theory (Tasaki, 1968). In these measurements, the change in parallel resistance in the bridge represented the change in membrane resistance. Tasaki and Hagiwara (1957b) showed that the membrane impedance decreased during the rising phase of the action potential in control and TEA-treated axons. Measurements at 10 kHz revealed a slight increase in membrane impedance at the shoulder of the action potential prolonged by TEA. Tasaki (1959b) reported that in squid axon depolarized by high $[K^+]_o$, a long hyperpolarizing pulse (50 ms) evoked the sudden change in membrane potential illustrated in Fig. 2.5. Alternating current measurements revealed an increase in impedance during the hyperpolarizing response, except at their peak where the resistance approached control values. It was concluded that the voltage drop across the membrane at the peak of the hyperpolarizing response cannot account for the potential change; therefore, a change in membrane voltage must be responsible for it. It was postulated that the depolarized membrane was at a higher stable level, whereas the hyperpolarizing pulse caused a transition to a lower stable state.

This concept was further developed by the demonstration that the sodium channel is not specific for sodium. Tasaki, Singer, and Watanabe (1966) succeeded in eliciting action potentials in sodium-free media, perfusing the axon with solutions of appropriate composition. Action potentials were obtained with hydrazine, guanidine, ammonium, and rubidium replacing sodium in the outside medium. These action potentials were accompanied by a decrease in membrane resistance, measured at a frequency of 16 kHz. Tasaki, Takenaka, and Yamagishi (1968) showed that action potentials and the corresponding impedance changes could be obtained when the ionic composition differed between the intra- and extracellular media of the axon. These findings led to a new hypothesis for excitation: The macromolecules of the membrane would take either one of two stable configurations, resting or active, depending on whether the negative sites of the macromolecules were occupied by divalent cations or univalent cations. The phase transition of the macromolecules would cause an action potential (biionic action potential).

2.2.2 Square-Pulse Measurements on Nonmyelinated Nerves

Cole and Hodgkin (1939) measured the longitudinal resistance of the squid axon as a function of interelectrode distance (Fig. 2.11). Calomel electrodes (*El*) were connected to the nerve (*N*) by pools of seawater, *A* and *B*. The galvanometer *G* measured the current when a pulse of known voltage (15 s duration) was applied. The interpolar length was varied by displacement of the holder containing pool *A*. With short interelectrode distances, the current will flow in the sheath of connective tissue, but a larger proportion of current will cross the membrane and flow along the core as the interpolar distance increases. At about 8–10 mm, the relation

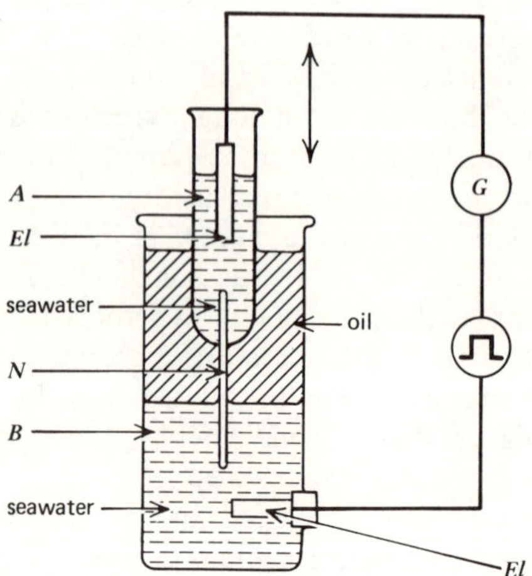

Figure 2.11. Arrangement for measuring the effective resistance as a function of interelectrode distance. For explanation, see text. Modified from Cole and Hodgkin (1939). By copyright permission of The Rockefeller University Press.

between effective resistance versus distance becomes linear, and its slope is determined by the parallel resistances of the core and the extracellular medium. The input resistance is given by:

$$R_{\text{inp}} = \frac{r_e r_i}{r_e + r_i} d + \frac{2 r_e^2 \lambda}{(r_e + r_i)\left(\dfrac{r_e + r_i}{r_i}\right)^{1/2} + \coth d/2\lambda} \tag{2.9}$$

where d is the interpolar length and the rest of the symbols are as in model 4 of Table 1.5. The electrical cell constants appear in Table 2.4. They are considered approximate because of the limitations in the applicability of Eq. 2.9 to actual experimental conditions. It was estimated that the membrane resistance could be 25% greater if the sources of error were eliminated. A corrected value of 1000 Ω cm^2 for R_m was comparable to the membrane resistance of marine algae (Blinks, 1930).

In the same year, Hodgkin and Huxley (1939) introduced the intracellular electrode technique, which made possible the direct measurement of the membrane potential in the squid giant axon. Cole and Curtis (1941) applied this technique to obtain voltage–current curves from which the value of 23 Ω cm^2 for R_m was calculated. The sharp discrepancy may be due to the estimate of r_e and r_i or the physiological condition of the axons.

The methodology and theoretical basis of the square-pulse analysis were reported by Hodgkin and Rushton (1946). The measurements were performed on axons from the walking leg of the lobster *Homarus vulgaris*. The four characteristic parameters for a linear cable model were determined in the same fiber: R_{inp}, λ, τ, and m (model 4 of Table 1.5). The current was kept small enough to avoid deviations caused by rectification or local responses. The parallel flow of current was assured because the length constant was twenty times the axon diameter. The distribution of potential as a function of distance and time followed the theoretical predictions of the model. The axoplasmic resistivity was about three times that of the seawater. Lobster axons had a larger membrane resistance than squid axons (Table 2.4).

Hodgkin (1947) determined the membrane constants in axons of the shore crab *Carcinus maenas*, using only three fixed electrodes. The length constant was given by $\lambda = l/ln(\Delta V_a/\Delta V_b)$, where l is the distance between the recording electrodes a and b and ΔV_a and ΔV_b are the electrotonic potentials at the electrodes. Table 2.4 shows that the membrane resistance of crab axons is higher than that of lobster and squid axons. Voltage–current curves were comparable to those obtained by Cole and Curtis (1941), showing a decrease in input resistance with depolarization. Departure from linearity was interpreted as a sign of subthreshold membrane activity.

Table 2.4 Electrical Constants of Nonmyelinated Nerves Obtained with Square Pulses

Method	Measured Quantities	Model	Derived Quantities	Reference
Voltage–current Fig. 2.11	$R_{eff}=f(x)$	4, Table 1.5	R_i: 29 Ω cm	Cole and Hodgkin (1939)
			R_m: 400–1100 Ω cm^2	
			λ: 1.8 to 3.8 cm (oil)	
			λ: 5 to 9.3 cm (seawater)	
1, Table 1.1	λ: 1.6 mm	4, Table 1.5	R_i: 60.5 Ω cm	Hodgkin and Rushton (1946)
	R_{inp}: 81 kΩ		R_m: 2290 (564 to 7330) Ω cm^2	
	τ_{inp}: 2.3 ms		C_m: 0.5–2 μF/cm^2	
	m: 0.83 MΩ/cm			
1, Table 1.1 (modified)	λ: 1.98 mm	4, Table 1.5	R_i: 90 Ω cm	Hodgkin (1947)
	τ_{inp}: 6.79 ms		R_m: 7653 Ω cm^2	
	r_i/r_e: 1.58		C_m: 1.1 μF/cm^2	
1, Table 1.1	λ: 1.6–1.8 mm	4, Table 1.5	R_m: 2940–3700 Ω *cm*^{2}	Katz (1948)
	τ_{inp}: 2.1–5.5 ms		R_i: 36–62 Ω *cm*	
			C_m: 0.7–1.7 μF/cm^2	
1, Table 1.1 (modified)	λ: 5.7 mm	4, Table 1.5	R_i: 63 Ω cm	Weidmann (1951b)
	R_{inp}: 67.7 kΩ		R_m: 9200 Ω cm^2	
	τ_{inp}: 14.2 ms		C_m: 1.17 μF/cm^2	
	m: 135 kΩ/cm			
1, Table 1.3	R_{inp}: 283 kΩ	5, Table 1.5	R_m: 800 Ω cm^2	Narahashi (1963a)
	τ_{inp}: 4.2 ms		C_m: 6.3 μF/cm^2	
			λ: 0.86 mm	
1, Table 1.1	λ:0.3–0.8 *mm*	4, Table 1.5	R_m: 9000 Ω cm^2	Straub (1963)
	τ_{inp}: 1.5–20 ms		R_i: 25 Ω cm	
			C_m: 0.9 μF/cm^2	
			R_m: 10,000 Ω cm^2	Straub, personal communication
			R_i: 50 Ω cm	
			C_m: 1μF/cm^2	
1, Table 1.3	Current-voltage curve	5, Table 1.5	Temperature: 20°C	Machne and Orozco (1970)
			R_m: 1.15–3.5×10^3 Ω cm^2	
			R_i: 64–197 Ω cm	
	λ: 1.6–2.3 *mm*		C_m: 1.6–1.7 *μF/cm*^{2}	
	d: 70–120 *μm*		Temperature: 10°C	
	τ_m: 4.2 ms		R_m: 1.6–3.5×10^3 Ω *cm*^{2}	
	λ: 1.8–2.4 mm		C_m: 1.6–1.7 *μF/cm*^{2}	

Table 2.4 *Continued*

Method	Measured Quantities	Model	Derived Quantities	Reference
1, Table 1.3	R_{inp}: 49 kΩ λ: 1.8 mm d: 153μm	5, Table 1.5	R_m: 870 Ω cm^2 R_i: 97 Ω cm	Yamagishi and Grundfest (1971)
1, Table 1.3	R_{inp} and R_{eff} length constant R_i: 80 Ω cm	5, Table 1.5	R_m: 8250 Ω cm^2 (electrical) R_m: 8500 Ω cm^2 (tracer) P_K: 4×10^{-7} cm/s P_{Na}: 0.048×10^{-7} cm/s P_{Cl}: 0.93×10^{-7} cm/s G_K: 86 μmho/cm^2 G_{Na}: 6 μmho/cm^2 G_{Cl}: 26 μmho/cm^2 G_m: 201 μmho/cm^2 G_K: 169 μmho/cm^2	Brinley (1965)

Weidmann (1951b) measured the electrical characteristics of *Sepia officinalis* axons and estimated the potassium leakage at rest and during activity. The values of the cell constants are comparable to those found in other invertebrate species (Table 2.4). A low membrane resistance indicated a preparation in poor condition, because the membrane resistance increased if the membrane was hyperpolarized. Since the resistance of the extracellular fluid enters the measurements when extracellular electrodes are used, the volume of fluid per unit length (volume/cm) was calculated with the equation:

$$\mathrm{Vol}_{ext} = (R_e/m)(1 + r_e/r_i) \tag{2.10}$$

and estimated as a layer of 9 μm thickness. In Eq. 2.10, $m = r_e r_i / r_e + r_i$.

Narahashi (1963a) studied the electrophysiological characteristics of insect axons in order to investigate the mechanism of action of different insecticides. He used the giant axons in the nerve chord of the cockroach and applied the linear cable model. The voltage–current relations were similar to those obtained by Cole and Curtis (1941) and Hodgkin (1947). The membrane capacity was large compared to other nerves, whereas the values for λ and R_m were within the order of magnitude reported for other nerve fibers.

Mammalian nerves are generally composed of different fiber types, but the vagus nerve in the rabbit contains 99.7% nonmyelinated fibers. It was used by Straub (1963) to measure the input resistance and the length and time constants with extracellular electrodes. The cable theory can also be applied to a bundle of identical cables in parallel, and the cell constants can be estimated if the number and size of nerve fibers are determined histologically. The values thus obtained are the constants for an average fiber. The second set of values in Table 2.4 were calculated according to more recent histological measurements (Straub, personal communication).

Changes in membrane resistance of the squid axon during activity as reported by Hodgkin, Huxley, and Katz (1952) were compared with the relative changes in conductance accompanying the action potential in crab axons by Mozhayev et al. (1963). The mean conductance at the peak of the action potential was 12 times greater than at rest. There was also a delay for the conductance to return to resting levels. These changes in conductances closely resembled those occurring in the squid axon.

Dierolf and McDonald (1969) investigated the effect of temperature acclimation on the electrical properties of giant axons from earthworms. A decrease of temperature produced an increase in input resistance. Machne and Orozco (1970) investigated the effects of calcium concentrations on the membrane potential and membrane passive properties as a function of temperature on axons from the crab *Callinectes sapidus*. The measurements were performed between 10 and 20°C. The current–voltage relation was linear for hyperpolarizing currents up to a shift in membrane potential of about 50 mV. The membrane resistance and the cytoplasmic resistivity increased when the temperature was lowered from 20 to 10°C, and the capacity did not change. Therefore, the time constant increased proportionally to the change in R_m. Lowering the calcium concentration reduced the membrane resistance to 34 and 44% of the control at 10 and 20°C, respectively. The cytoplasmic resistivity was reduced to 82–92% of normal at 10°C, but no consistent change was found at 20°C. The membrane capacity increased twofold at 10°C and 1.4 times at 20°C. In contrast to the results on squid axons (Steinbach, Spiegelman, and Kawata, 1944) calcium removal did not reduce membrane rectification in this species.

Yamagishi and Grundfest (1971) investigated the contribution of various ions to the membrane potential at rest and during activity in crayfish medial giant axons. The cell constants obtained in standard saline are listed in Table 2.4. In potassium-free media the resting potential increased by about 15 mV and the effective resistance rose by 30%. A further increase of 30–50% occurred in 20 mM sodium. R_{eff} also increased in chloride-free medium without change in potential. Calcium also increased R_{eff} and depolarized the membrane. This suggested that the changes in R_{eff} were not the consequence of the shifts in membrane potential. The current–voltage relation during the after-depolarization was steeper than at rest and linear over the whole range of applied currents. The input resistance was 36 kΩ at rest and 12 kΩ during the after-depolarization in control saline. TTX blocked the sodium spike but did not affect the hyperpolarization produced by sodium removal. It was concluded that the resting potential is determined by four ionic conductances: potassium, sodium, calcium, and chloride. The quantitative contribution of each ion depends on the presence of potassium. The experimental evidence indicated that the sodium channels for the generation of the spike differ from those involved in the resting potential.

Brinley (1965) compared electrical measurements of membrane conductance with the ionic conductances estimated from flux measurements in giant axons of the lobster *Homarus americanus*. The extracellular space was 8–12% of the axon volume, equivalent to a rim of about 2–3 μm thickness around the axon. The values obtained from both types of measurement and the ionic conductances are listed in Table 2.4. No net fluxes existed in the resting state. Stimulation of the axon at 5–20 impulses/s increased sodium influx and outflux by 6.5 and 1.3 pmol/cm^2 per impulse, while potassium influx and outflux were increased by 1.2 and 5.3 pmol/cm^2 per impulse. Assuming that the permeabilities are voltage independent, a value of 80 μmho/cm^2 was obtained for G_m. However, corrections were introduced for the variation of the resting potassium conductance with membrane potential because the latter differed from E_K by 5 mV in the lobster axon. This made a substantial contribution to the total calculated conductance, which amounted then to 118 μmho/cm^2. Application of the flux ratio equation to these results yielded 90 μmho/cm^2 for G_K, in agreement with the value found for the constant field case. The good correlation between these calculations permitted the exclusion of single-file diffusion as a mechanism to explain the measured ionic fluxes. This type of analytical treatment shows the value of the alternative approach in establishing the properties of a biological membrane.

2.2.3 Voltage-Clamp Measurements on Nonmyelinated Nerves

Some of the most remarkable results obtained in modern electrobiology will now be discussed. Even if these measurements are not primarily aimed at determining biological impedances, the results have a strong bearing on the topics treated in this volume Sections 2.2.3.1 and 2.2.3.2 describe studies performed on the squid axon in seawater or internally perfused; 2.2.3.3 presents results obtained on other axons; 2.2.3.4 illustrates special approaches to voltage-clamp measurements; and 2.2.3.5 discusses the influence of pharmacological agents on ionic currents.

2.2.3.1 Experiments Performed on Nonperfused Squid Axons. Marmont (1949) proposed a method (number 2 in Table 1.4) to measure voltage changes across the membrane of the giant axon at constant current and obtained an action potential that resembled a propagated potential. A strong current was required to change the time course of the action potential, and therefore the membrane behaved like a low-impedance voltage source, like during a conducted action potential.

In the same year, Cole (1949) published the results of voltage clamp measurements on the dynamic electrical characteristics of the axonal membrane. Since it was supposed that the electrical instability of the membrane was due to an N-shaped current–voltage characteristic, it was expected that the membrane would be stable under constant voltage conditions. In addition, it was known that the current density J_m across the membrane is:

$$J_m = C_m(dV/dt) + J_{ti} \tag{2.11}$$

where $C_m(dV/dt)$ is the capacitive current density (J_c), V is the membrane potential, and J_{ti} is the total ionic current density. When the membrane potential is clamped to a constant value ($dV/dt = 0$), J_{ti} can be measured. The full value of this method became apparent when Hodgkin and Huxley published their ionic theory of excitation. The following discussion is concerned only with the aspects of their pioneering work that are directly relevant to this book: (a) the results obtainable with this method and (b) a summary of the mathematical description of the excitation process in squid axon, since this model is used as a reference for current research.

(*a*) *Results obtainable with the voltage-clamp method.* The voltage-clamp method developed by Hodgkin, Huxley, and Katz (1952, circuit 1 of Table 1.4) was used to measure the cell constants and the dynamic electrical characteristics of the membrane. Table 2.5 summarizes these data. The membrane capacity C_m can be determined, since the charge across the membrane is proportional to the voltage change across it. With the first method, a short pulse was applied to the unclamped membrane of an axon over a defined membrane area. Q is the time integral of the current density, and $Q/\Delta V$ gives the value of C_m. An alternative method for estimating C_m is to determine the current density during the capacitive artifact and divide it by the difference of resting potential and clamped potential.

The slope conductance of the membrane can be obtained as a function of time and membrane potential from current–voltage curves constructed from recordings of the time course of the membrane current at different potentials (inset in Fig. 2.12). A different curve can be constructed for any given time. Curve I in Fig. 2.12a is constructed from the early current at 0.63 ms and curve II from the steady-state outward current. Curve II has the typical shape of a rectifier characteristic. Since this rectifying property appears late in the current versus time record, it is called delayed rectification. The tangent at any point of the curves gives the slope conductance, and curve I in Fig. 2.12 shows that G_m can be negative. Current–voltage curves provide information only about the net membrane current, not about the different current components and the ion species involved. If current–voltage curves are obtained for each ion, the chord conductance, $g_i = I_i/(V - E_i)$ for that ion can be calculated (Table 2.5). This definition of chord conductance is very general and is independent of specific assumptions about the membrane (Kornacker, 1969).

The total ionic current density J_{ti} during the action potential may be estimated by measuring the change in potential resulting from the application of a short electrical shock, if the membrane area is known. In the partitioned chamber used for voltage-clamp measurements, the action potential is limited to the membrane area in the central compartment. Such a potential has been termed a membrane action potential, in contrast to the propagated action potential. Under these conditions, the net current flowing through the membrane 300 μs after a short pulse is negligible. Then the left side of Eq. 2.11 becomes zero, and the ionic current density through the membrane is $J_{ti} = -C_m(dV/dt)$. Since $C_m(dV/dt)$ and V can be measured, J_{ti} is calculated and current–voltage curves like curve I in Fig. 2.12a can be constructed. The capacitive artifact resulting from an imposed rectangular command signal was consistent with the presence of a resistive element R_s in series with C_m as in model 2 of Table 1.5. The capacitive surge current had a time constant of 6 μs, and a value of $R_s =$ 7.3 Ω cm^2 was calculated. This series resistance was identified with a layer of connective tissue (5–20 μm thick). Moore and Cole (1963) later reported a value of 3–5 Ω cm^2 for the series resistance. The morphological correlate of this resistance has been called the Frankenhaeuser-Hodgkin space (Frankenhaeuser and Hodgkin, 1956).

The last three quantities listed in Table 2.5 are used to analyze the dynamic characteristics of the ionic channels through which charge is carried during the action potential. These measurements consist of a combination of two pulses; the first (C) is used to condition the membrane to a certain state; then the test pulse (T) is imposed on the membrane and the transient current response is recorded.

The activation of the sodium inward current can be tested with a pulse combination so that (1) the amplitude and polarity of the conditioning pulse varies but its duration is constant and (2) the membrane is clamped during the test pulse at a constant potential 44 mV more positive than the resting potential. Then the maximum inward current density ($J^T_{\max}$) can be compared with the maximum inward current obtained when no conditioning pulse was applied ($J^0_{\max}$). The ratio $J^T_{\max}/J^0_{\max}$ plotted against the

Table 2.5 Determination of Static and Dynamic Impedance Components

Derived Quantity	Technique		Results	References
C_m	$C_m = Q/\Delta V$ $Q = \int J\,dt$	Short shock, no voltage clamp	0.91 $\mu F/cm^2$	Hodgkin, Huxley, and Katz (1952)
C_m	$C_m = Q/\Delta V$ $Q = \int J_c\,dt$ $J_c = C_m dV/dt$	Step command signal, voltage clamp	0.89 $\mu F/cm^2$	Hodgkin, Huxley, and Katz (1952)
G_m	$G_m = dJ/dV$	Step command signal, voltage clamp	Voltage dependence of G_m at a given time	Hodgkin, Huxley, and Katz (1952)
g_i (chord conductance for ion i)	$g_i = J_i/(V - E_i)$ V: from command signal E_i: calc. or meas. J_i: measured	Step command signal, voltage clamp	$g_i(t)$	Hodgkin, Huxley, and Katz (1952)
J_{ti} (total ionic current density)	$J_{ti} = -C_m dV/dt$ for $J_m = 0$, $t > 300\ \mu s$ after short shock	Short shock, no voltage clamp	N-shaped I/V curve	Hodgkin, Huxley, and Katz (1952)
R_s	Obtained by analyzing the time constant of capacitive artifact	Step command signal, voltage clamp	7.3 Ω cm^2	Hodgkin, Huxley, and Katz (1952)

J^T_{max}/J^0_{max} vs ΔV	Duration of conditioning step constant, but amplitude variable; test potential at 44 mV positive with respect to resting potential	44 mV, C, T, ΔV; 44 mV, C, T, ΔV	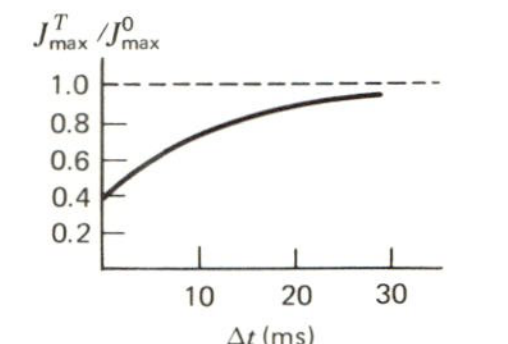	Hodgkin and Huxley (1952c)
J^T_{max}/J^0_{max} vs t	Duration of conditioning step variable, but amplitude constant: test potential at 44 mV positive with respect to resting potential	44 mV, C, T, ΔV; 44 mV, C, T, ΔV	J^T_{max}/J^0_{max}; 1.5, 1.0, 0.5; ΔV = − 16 mV; ΔV = + 8 mV; 5, 10, 15; t (ms)	Hodgkin and Huxley (1952c)
J^T_{max}/J^0_{max} vs Δt	Duration and amplitude of pulses constant, interval between pulses variable.	Δt, C, T	J^T_{max}/J^0_{max}; 1.0, 0.8, 0.6, 0.4, 0.2; 10, 20, 30; Δt (ms)	Hodgkin and Huxley (1952c)

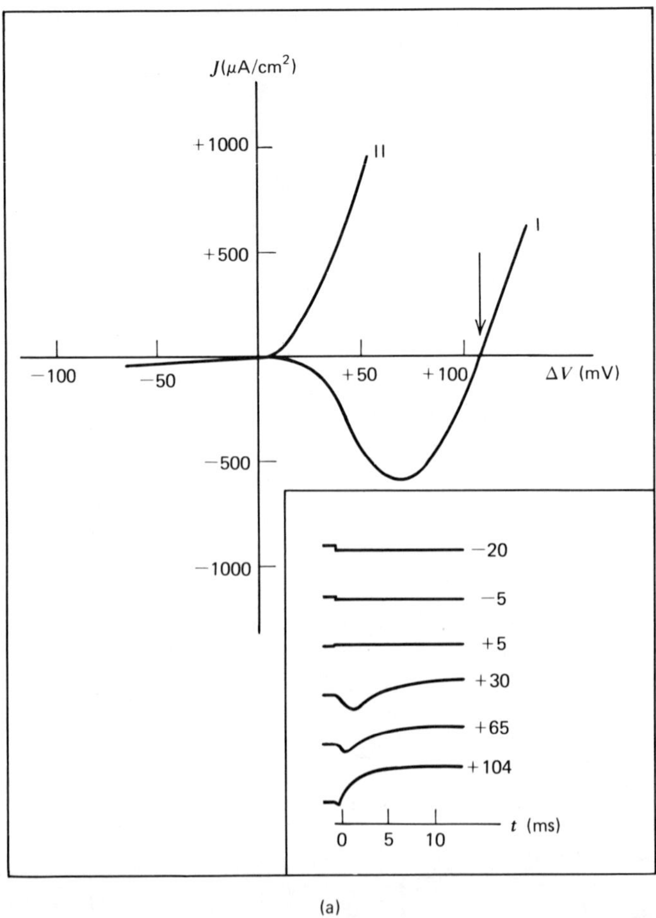

(a)

Figure 2.12. (a) Current–voltage curves obtained from records of early current, 0.63 ms after onset of command pulse (I), and steady-state outward current (II). Arrow: sodium equilibrium potential. Inset: Membrane current as a function of time at different membrane potentials. Numbers at side of traces: clamped potential relative to resting potential. Note that the maximum inward current and the steady-state outward current are reached at different times in each record. (b) Electrical circuit for a membrane under space-clamp conditions as developed by Hodgkin and Huxley. The arrows through G_K^{-1} and G_{Na}^{-1} indicate voltage- and time-dependent resistances. For explanation, see text. From Hodgkin et al. (1952) and Hodgkin and Huxley (1952d). By permission.

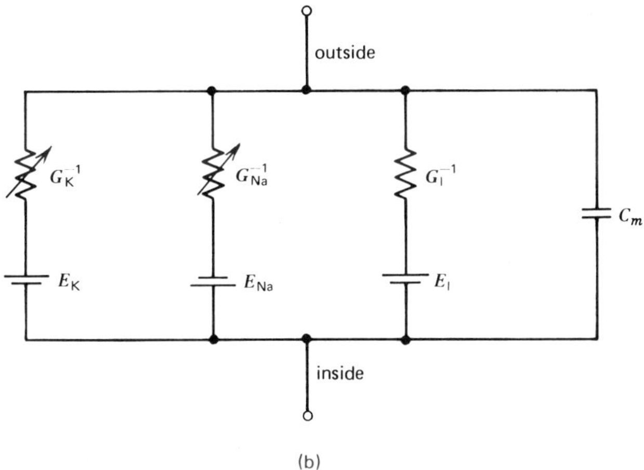

Figure 2.12. *Continued.*

displacement of the membrane potential during the conditioning pulse gives an S-shaped curve. If its maximal ordinate is assigned the amplitude 1.0, this curve represents the change of the variable h_∞ as a function of membrane potential. This variable describes the availability of the sodium system. To determine the time course of inactivation of the sodium channel, the same potential level was chosen for the test pulse. The potential of the conditioning pulse was kept constant, but its duration was varied. The ratio $J^T_{\max}/J^0_{\max}$ was plotted against the duration of C. This resulted in a family of curves where the potential during C is the parameter. The diagram in Table 2.5 represents the time course of inactivation of the sodium channel for two levels of the conditioning pulse.

To analyze the recovery from inactivation, a characteristic related to the phenomenon of refractoriness, a pair of pulses are used with identical amplitudes (displacement +44 mV from resting level) and duration (1.8 ms) but with a variable interval between pulses (Δt). The ratio $J^T_{\max}/J^0_{\max}$ is plotted against the interval between pulses. At large intervals, the ratio becomes 1.0; at 0 time it is about 0.38. The time course of the resulting curve is exponential, with a time constant τ_h of 12 ms. This time constant is characteristic for the removal of inactivation.

The methods contained in Table 2.5 were often used with minor modifications, and their great value has since been proven. The application of these techniques to the giant axon of the squid *Loligo* permitted estimation of its cell constants, at rest and upon excitation (Table 2.6), and reconstruction of the action potential.

Table 2.6 Electrical Characteristics of Squid Giant Axon (Method 1 of Table 1.4)

Measured Quantities	Model	Derived Quantities	Reference
I/V curves	Model 1 of Table 1.5	R_m: 2300 Ω cm^2 (at $I=0$)	Hodgkin et al. (1952)
	Fig. 2.12b	R_m: 35 Ω cm^2 ($\Delta V=+110$ mV)	
I_{Na}^{max} from $I(t)$		g_{Na}: 30 mmho/cm^2 ($\Delta V=+60$ mV)	
		g_{Na}: 20 mmho/cm^2 ($\Delta V=+100$ mV)	
		$[Na]_i$: 60–70 mmol/kg H_2O	
		E_{Na}: 45–50 mV	
		g_K: 28 mmho/cm^2 ($\Delta V=+100$ mV)	
		g_K: 0.23 mmho/cm^2 (at rest)	
I_K reversal potential		E_K: -80 to -85 mV	Hodgkin and Huxley (1952b)
From I_m at $V=E_K$ in choline–seawater		g_l: 0.26 mmho/cm^2	Hodgkin and Huxley (1952d)
		$(V-E_l)$: -11 mV	
		$\bar{g}_{Na}$: 160 mmho/cm^2	
		$\bar{g}_K$: 34 mmho/cm^2	
		$\bar{g}_l$: 0.26 mmho/cm^2	

(*b*) *Description of the excitation process in squid axon.* The mathematical description of the action potential in terms of voltage and time-dependent conductances is contained in a system of expressions known as the Hodgkin-Huxley equations. In the Hodgkin-Huxley model for a patch of membrane (Fig. 2.12b), the membrane current density is given by

$$J_m = C_m(dV/dt) + g_{Na}(V-E_{Na}) + g_K(V-E_K) + g_l(V-E_l) \quad (2.12)$$

where E_K and E_{Na} are Nernst potentials and E_l is defined as the potential at which the leakage current is zero. This expression follows from Eq. 2.11 and the definition of chord conductance (Table 2.5). As shown in Table 2.6, the conductance of the leakage current (g_l) is constant so $g_l=\bar{g}_l$, the maximum value of the leakage conductance. However, g_K and g_{Na} depend on voltage and time. A further complication is that g_K increases with a delay during depolarization but falls without inflection when the membrane is repolarized. g_{Na} also increases with a delay but inactivates spontaneously. To describe the rising phase of the potassium and sodium conductance, one may choose between a 3rd- or 4th-order equation or the use of a variable that in itself is a function of time and voltage. Hodgkin and Huxley (1952d) used the second approach because it made numerical treatment of the voltage-clamp data easier and provided a formal physical model to explain the observed conductance changes. They obtained the following set of equations:

$$\left.\begin{aligned} g_K &= \bar{g}_K n^4 \\ g_{Na} &= \bar{g}_{Na} m^3 h \end{aligned}\right\} \tag{2.13}$$

where $\bar{g}_i$ is a maximal constant value of the chord conductance of the ion i (limiting conductance); n and m are variables governing the activation of the potassium and sodium conductances, respectively; and h controls the inactivation of the sodium channel. n, m, and h are dimensionless and may vary between 0 and 1. Their time derivatives can be written as follows:

$$\left.\begin{aligned} dn/dt &= \alpha_n(1-n) - \beta_n n \\ dm/dt &= \alpha_m(1-m) - \beta_m m \\ dh/dt &= \alpha_h(1-h) - \beta_h h \end{aligned}\right\} \tag{2.14}$$

where α and β are rate constants with the dimension t^{-1} that depend only on the membrane potential, provided $[Ca^{2+}]_o$ and temperature remain constant. Under voltage-clamp conditions, $(dV/dt=0)$ so Eqs. 2.14 are ordinary differential equations that can be solved easily:

$$\left.\begin{aligned} n(t) &= n_\infty - (n_\infty - n_0)\exp(-t/\tau_n) \\ m(t) &= m_\infty - (m_\infty - m_0)\exp(-t/\tau_m) \\ h(t) &= h_\infty - (h_\infty - h_0)\exp(-t/\tau_h) \end{aligned}\right\} \tag{2.15}$$

with the following set of boundary conditions for $t=0$:

$$\left.\begin{aligned} n(0) &= n_0 \\ m(0) &= m_0 \\ h(0) &= h_0 \end{aligned}\right\} \tag{2.16}$$

For $t=\infty$:

$$\left.\begin{aligned} n_\infty &= \alpha_n/(\alpha_n+\beta_n) \\ m_\infty &= \alpha_m/(\alpha_m+\beta_m) \\ h_\infty &= \alpha_h/(\alpha_h+\beta_h) \end{aligned}\right\} \tag{2.17}$$

Moreover, the following definitions are proposed:

$$\left.\begin{aligned} \tau_n &= 1/(\alpha_n+\beta_n) \\ \tau_m &= 1/(\alpha_m+\beta_m) \\ \tau_h &= 1/(\alpha_h+\beta_h) \end{aligned}\right\} \tag{2.18}$$

Making several assumptions, it is possible to obtain n, m, and h (Eq. 2.15) from voltage-clamp data, and the time course of the change in chord conductances with Eqs. 2.13. The theoretical calculation of the action potential is possible combining Eq. 2.12 with Eqs. 2.13.

Hodgkin and Huxley (1952d) obtained a good agreement between calculated and recorded potentials. They also calculated the propagated action potential, the effect of temperature, propagation velocity, impedance changes during the action potential, ionic currents during the action potential, refractory period, and threshold phenomena.

A possible physical interpretation of Eqs. 2.13 and 2.14 is the following: n and m represent relative concentrations of particles at the inside surface of the membrane, and the power of the variables indicates how many of these activating particles must be present to open the potassium and sodium channels, respectively. By analogy, h is the relative concentration of inactivating molecules at the inside surface of the membrane. The sodium channel would open when three activating particles are present and are not blocked by an inactivating molecule.

Frankenhaeuser and Hodgkin (1957) studied the effect of calcium on the electrical properties of the squid giant axon. A fivefold decrease of $[Ca^{2+}]_o$ reduced the threshold, increased I_K, and decreased the fraction of the sodium-carrying system available for activation at a given membrane potential. These effects are comparable to a depolarization of 10–15 mV. Removal of Ca^{2+} led to loss of excitability with a high g_K and inactivation of the Na^+ system. Mg^{2+} had similar but weaker effects.

At the same time, a series of papers questioned the validity of the basic assumption of the Hodgkin-Huxley theory: that voltage and time-dependent conductances can explain the action potential. Tasaki and Hagiwara (1957b) claimed that the effects of TEA on the action potential could be explained by assuming that, apart from voltage- and time-dependent conductances, changes in the ionic batteries were responsible for the action potential. The result was the theory of two steady states of the membrane. Oscillations of the clamped potential in reaction to a step command, under TEA, and at different temperatures were reported (Tasaki and Bak, 1958b; Tasaki and Spyropoulos, 1958a), but it is accepted today that these oscillations were due to imperfections of the clamping method. Tasaki and Spyropoulos (1958b) superimposed a high-frequency ac signal (8 kHz for experiments at 4–6°C, 16 kHz for experiments at 22°C) on the step command of their voltage clamp to monitor the membrane current at a given membrane potential. This procedure permitted measurement of the membrane impedance as a function of voltage and time, and allowed determination of the time course of the membrane conductance and the effective membrane potential E_{eff} (the potential at which $J_m = 0$) for the combined current density through the membrane, according to:

$$J_m = (g_m + j\omega C_m)A\exp(j\omega t) + g_m(V - E_{\text{eff}}) \tag{2.19}$$

where V is the displacement from the effective membrane potential E_{eff} and A the maximum amplitude of the sinusoidal current. The second term is the current component contributed by the step command signal. When the sinusoidal signal is on, and just before the square wave command signal is imposed, the resting current density (J_m^0) is measured and g_m^0, the resting chord conductance of the membrane, can be determined. J_m^0 can be electronically subtracted from J_m. When a large increase in the membrane conductance is considered, $g_m \gg g_m^0$, and we obtain:

$$J_m^r = g_m A\exp(j\omega t) + g_m(V - E_{\text{eff}}) \tag{2.20}$$

The $J_m(t)$ record obtained from a voltage-clamp experiment with a sine wave superimposed on the step command signal can be decomposed in the current resulting from the sine wave with the maximum amplitude J_a and the slowly changing current resulting from the voltage step with the instantaneous amplitude i_s:

$$J_m^r = J_a \sin\omega t + i_s \tag{2.21}$$

From Eqs. 2.20 and 2.21, an expression for the effective emf of the membrane can be obtained:

$$E_{\text{eff}} = V - A(i_s / J_a) \tag{2.22}$$

Since i_s and J_a can be measured, and V and A are known, the time course of E_{eff} can be obtained from such an experiment, as shown in Fig. 2.13.

Conti and Palmieri (1968) investigated the effect of replacing up to 99.7% of external water by D_2O. At this concentration of D_2O, the maximum amplitude of the initial transient current and the steady-state current were depressed by about 30%. In terms of the Hodgkin-Huxley theory, this effect corresponds to a reduction of g_{Na}, g_K, and g_{Cl} by a factor 1.4 as well as that of the rate constants for the variables m, n, and h. To obtain sufficient current density without polarization effects, platinized internal electrodes were used, but this prevented determination of the resting potential (Sec. 1.2.1).

Further work with voltage clamp of the squid axon was mainly concerned with improving the method or facilitating its manipulation. Besides this aspect, interest was chiefly concentrated on obtaining more accurate information on the nature of the two main ionic channels and their selectivity. As to the refinements of the voltage-clamp method, Cole and Moore (1960) improved the effectiveness of the voltage clamp by using a microelectrode for the voltage control in the clamp circuit. Taylor, Moore, and Cole (1960) analyzed sources of error in this method, investigating the factors responsible for the appearance of notches in records of the transient inward current. Study of the intracellular potential distribution during a clamping pulse with an intracellular electrode showed that the notches arose from an incomplete space clamp, but these deficiencies must be considerable before the notches occur. Based on these measurements, they developed a model of a membrane having two excitable patches. Applying the Hodgkin-Huxley formalism to both patches, the observed notches in the current-versus-time records were simulated. Rojas,

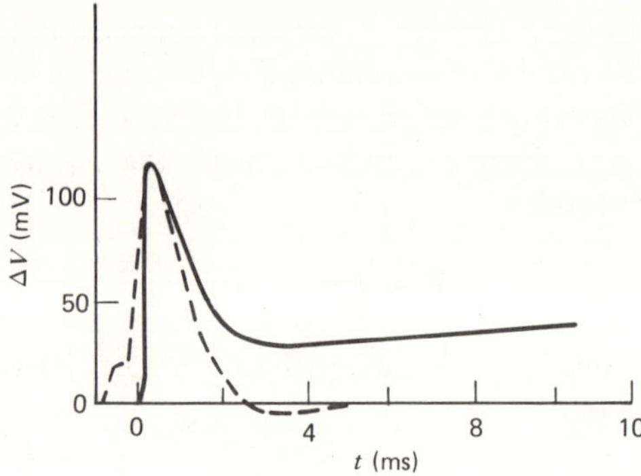

Figure 2.13. Time course of the effective membrane potential (full line) calculated according to Eq. 2.22, compared with an action potential measured under space clamp conditions (broken line). Modified after Tasaki and Spyropoulos (1958b). By permission.

Bezanilla, and Taylor (1970) modified circuit 1 of Table 1.4 to impose voltage steps at predetermined instants during the action potential. When the imposed voltage was set at the value of E_{Na} or E_K, the time course of either I_K or I_{Na} was measured. The results obtained confirmed those of Cole and Curtis (1939) and Hodgkin and Huxley (1952d).

Fishman and Cole (1969) used a time-variant command signal instead of the usual step command pulse, and the theoretical analysis of the so-called varying potential-control voltage clamp was made by Palti and Adelman (1969), based on the Hodgkin-Huxley model. Using the values of Fitzhugh (1960), the behavior of the variables m, n, and h was analyzed for triangular and sinusoidal command signals: (1) Slowly varying potentials generate pure I_K; (2) rapidly changing potentials yield I_{Na}; (3) leakage currents and capacitive currents can be obtained from properly chosen triangular command signals; (4) current–voltage characteristics and equilibrium potentials can be obtained rapidly, and (5) C_m can be determined accurately. It was found that C_m changed 1.4% per °C between 3 and 21°C, remained unchanged between 20 and 40°C, and increased considerably above 40°C. Moreover, C_m was constant between 200 and 2000 Hz, in contrast to Cole's results, which were obtained with extracellular electrodes. It was hypothesized that the α dispersion might be due to a surface admittance relaxation and not related to the dielectric constant of the membrane proper.

Fishman (1969, 1970) developed a voltage-clamp method suitable for ramp (sawtooth) command signals, which resulted in an on-line display in real time of current–voltage curves for I_K and I_{Na} obtained with slow and fast ramps, respectively.

2.2.3.2 Voltage- and Current-Clamp Experiments on Perfused Axons. Baker, Hodgkin, and Shaw (1961) and Oikawa et al. (1961) developed a technique for extruding the axoplasm and perfusing the axon with solutions of known composition. In the following years, several investigators studied the response of the membrane potential to changes in the cation concentration of the medium perfusing the axon internally (Tasaki and Shimamura, 1962; Baker et al., 1962; Narahashi, 1963b; Baker, Hodgkin, and Meves, 1964). It was found that when an artificial axoplasm was diluted with an isotonic sucrose solution, action potentials could be obtained in the presence of a very low resting potential. This led to the hypothesis that the inactivation curve (see Table 2.5) and the relation between sodium conductance and membrane potential were shifted toward more positive transmembrane potentials. This hypothesis was tested by Moore, Narahashi, and Ulbricht (1964) with voltage-clamp experiments (circuit 6 of Table 1.4). Internal perfusion with solutions containing

530 mM potassium and 446 mM sucrose and a medium containing 10.6 mM potassium and 902.1 mM glucose were used. The external medium was seawater at 7°C. The current voltage curve was shifted to more positive membrane potentials when the axon was perfused with low-potassium medium. This was found for the maximum inward current as well as for the steady-state outward current. A similar shift was found for the inactivation curve (h_∞ versus V, Table 2.5). Chandler et al. (1965) investigated this problem further and found that this shift was due to the low ionic strength of the perfusion medium. It did not occur when KCl was diluted with NaCl instead of sucrose.

It was known that axons perfused with potassium-free solutions gave action potentials lasting several seconds (Narahashi, 1963b). Baker et al. (1964) also found action potentials lasting 1–5 s, but replacing sodium with potassium in the perfusion fluid shortened the duration of the action potential to 30 ms. To explain this finding, Chandler and Meves (1970a–d) analyzed the effect of perfusion with fluoride solutions on the sodium and potassium currents (circuit 1 of Table 1.4). When the axon was perfused* with 300 mM NaF and potassium-free artificial seawater was employed outside the membrane, the membrane depolarized spontaneously to 0 mV and could then be hyperpolarized to levels of −70 to −100 mV. Depolarization from this level elicited two components of inward current that reversed at the sodium equilibrium potential and were blocked by tetrodotoxin. It was concluded that the sodium conductance was not fully inactivated by long-lasting depolarizations and that its remaining component might be responsible for the plateau of the long-lasting action potential observed under these conditions. To account for the incomplete sodium inactivation, h was assumed to be the sum of two inactivation variables, h_1 and h_2, which are related by the following scheme, where x is the inactive state:

$$h_1 \underset{\alpha_{h1}}{\overset{\beta_{h1}}{\rightleftharpoons}} x \underset{\beta_{h2}}{\overset{\alpha_{h2}}{\rightleftharpoons}} h_2 \tag{2.23}$$

In this scheme, $x = 1-(h_1+h_2)$, and α_{h1} and β_{h1} correspond to the α_h and β_h of Eq. 2.14. Figure 2.14 shows the voltage dependence of h_1 and h_2. Its molecular interpretation is consistent with a hypothetical charged particle that can move from a resting position h_1 into a blocking position x on depolarization. However, with strong depolarizations, this particle would be pulled away from the blocking position. According to this analysis, the peak sodium conductance is given by $g_{Na} = \bar{g}_{Na} m^3 h_1$, and the maintained

*Fluoride was chosen as anion because axons perfused with fluoride-containing solutions remained excitable for a longer time than those perfused with solutions containing chloride or sulfate (Tasaki et al. 1965; Adelman et al., 1966).

sodium conductance is $g_{Na} = \bar{g}_{Na} m^3 h_2$. It was shown that the latter is inactivated by long-lasting depolarizations. The readily available sodium conductance ($\bar{g}_{Na} m^3 h_1$) also inactivated slowly, as already described by Narahashi (1964) for the lobster axon. Tasaki, Singer, and Watanabe (1967) studied the interdiffusion of cations in perfused axons. Fluxes were studied at rest and during activity (trains of pulses), and the product of current and resistance at rest and during activity was determined from both tracer and electrical measurements. It was found that the electrically determined value was 30 mV, close to that obtained from tracer measurements in rest and activity in sodium-containing media. It was concluded that the squid axon membrane behaves like a cation exchanger because the relation $R_m J_m = RT/F^2$ is satisfied (Helfferich, 1962).

Atwater et al. (1969) measured ionic influx in perfused axons under controlled voltage. The sodium influx was calculated from the measured inward current under voltage clamp, and this value was then compared with the measured sodium influx. The ratio between tracer influx and electrically determined influx was 0.92. This was interpreted as a confirmation of the Hodgkin-Huxley theory, which identifies the early inward current with the sodium current. With a similar approach, Rojas and Taylor (1970) confirmed a suggestion made earlier by Baker, Hodgkin, and

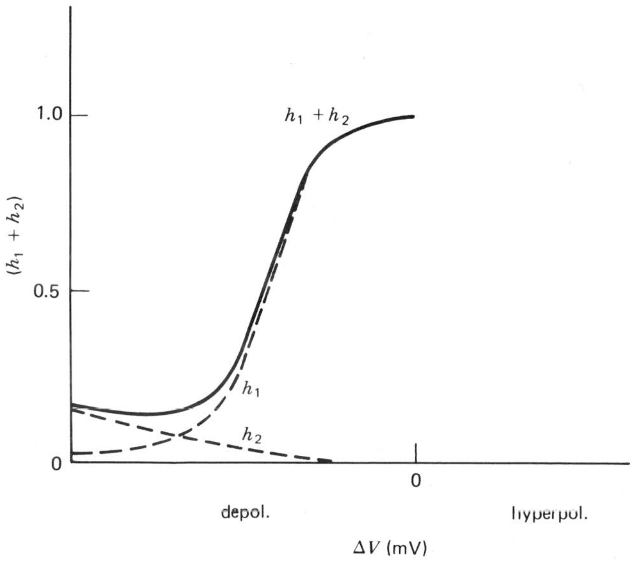

Figure 2.14. Steady-state inactivation of axons perfused with K-free solutions. Compare with inactivation of nonperfused axons in Table 2.5. For explanation, see text. Modified from Chandler and Meves (1970c). By permission.

Ridgway (1970) that there is an early calcium influx in a squid axon perfused with a fluoride medium and that this calcium influx occurs before the sodium-gating mechanism is turned on.

2.2.3.3 ***Studies on Axons Other Than That of the Squid.*** Experiments with controlled voltage on axons other than that of the squid are less numerous but reveal interesting differences from the electrical behavior of the squid giant axon. Julian, Moore, and Goldman (1962a, b) developed circuit 6 of Table 1.4 to measure membrane potentials and perform voltage-clamp measurements in the lobster giant axon. Current–voltage relations showed that the steady-state conductance in the lobster axon is smaller than in the squid axon. In the lobster axon, the steady-state current declines during long pulses, as in the frog myelinated nerve (Sec. 2.1.3). Reduction of Ca^{2+} caused depolarization, reduced E_{Na}, and shifted the current–voltage curves by 10 mV toward more negative potentials. Takata, Pickard, Lettvin, and Moore (1966) substituted lanthanum for calcium in the lobster giant axon. Lanthanum shifted the current–voltage curve for the maximum early current to more positive values (increase in threshold) and prolonged the time course of the conductance change while reducing the amplitude of the inward current. The effect of lanthanum was considered to be equal to that of a very high calcium concentration.

Binstock and Goldman (1967) described an axon with a diameter of 500–900 μm from *Myxicola infundibulum*. The current–voltage characteristics of the *Myxicola* axon were comparable to those of the axons of squid and lobster. Goldman and Binstock (1969a, b) identified the early transient current as a sodium current and found that the peak of the action potential varied with $[Na^+]_o$ with some substantial potassium permeability remaining. After the fast transient current was removed by TTX, a rectification of the leakage current was observed. This could be explained by the sum of a constant field rectification and a constant conductance. It is speculated that the latter represents a sum of ionic currents with constant field characteristics that is simply well approximated by a constant conductance over the range of measured voltages. Pichon and Boistel (1967) and Pichon (1968, 1969a, b, c) performed voltage-clamp and current-clamp measurements on the giant axon of a cockroach (*Periplaneta americana*). The membrane currents resembled those of squid axon, myelinated nerve fiber, and lobster giant axon. Inward and outward currents carried by sodium and potassium were identified. They were inhibited by TTX and TEA.

2.2.3.4 ***Special Approaches to Voltage-Clamp Measurements in Axons.*** In an attempt to clarify the molecular events underlying the excitation process, interest has focused in recent years on the study of optical phenomena such as light scattering and birefringence. Cohen, Keynes, and Hille (1968) reported that birefringence occurred in squid axon together with the onset of the action potential and lasted somewhat longer than the repolarization phase. The birefringence depended on the membrane potential. In voltage-clamp experiments, it followed closely the time course of the test pulse. Cohen and Keynes (1969) reported that the amplitude of the birefringence varied with the square of the membrane potential. In another approach, Conti and Tasaki (1970) studied the secondary fluorescence of rhodamine B, pyronin B, and ANS (8-anilinonaphthalene-1-sulfonate). Changes in the fluorescence associated with either hyperpolarization or depolarization of the membrane were observed.

Poussart (1969, 1971) used voltage-clamp methods together with measurements of membrane noise in the lobster giant axon. Most of the analysis was performed on the potassium current. Application of TTX indicated that the noise was linked to the potassium current. Figure 2.15 shows that the power density S is related to the frequency by the power function, $S(f) = Nf^{-\alpha}$, where N is a function of the membrane potential and α is a dimensionless quantity independent of the membrane potential.

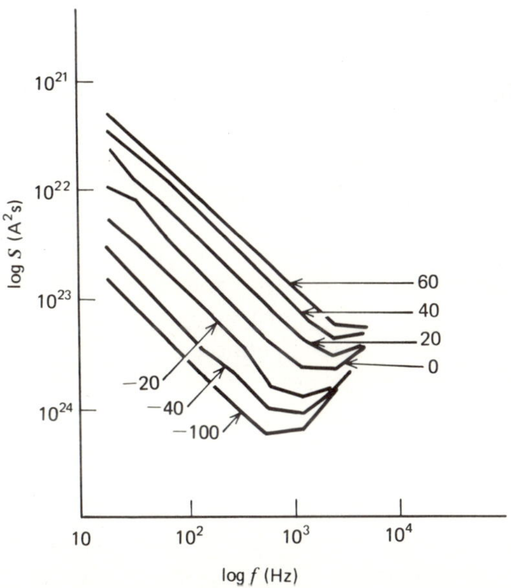

Figure 2.15. Power density spectrum of membrane noise observed in lobster axon. Parameter: holding potential. Modified after Poussart (1971). By copyright permission of The Rockefeller University Press.

Analysis of the double logarithmic plot of N versus I_K led to the following expression for the power density spectrum:

$$S(f)=(A+K'I_K^q)/f^r \tag{2.24}$$

where A, K', q, and r are parameters independent of the membrane potential. These studies confirm the results of noise measurements obtained on myelinated nerves by Verveen, Derksen, and Schick (1967) and Verveen and Derksen (1969). Although studies of membrane fluctuations might contribute to a better understanding of the nature of the excitation process (Cole, 1968), not enough data are yet available to assess the biological relevance of these measurements.

2.2.3.5 Voltage Clamp Measurements and the Action of Pharmacological Agents. Most pharmacological studies on nerve fibers deal with two groups of agents: local anesthetics and neurotoxins. Taylor (1959) reported that procaine (0.25–0.1%) reduced the rate of development of the early transient sodium current and that of the steady-state potassium current, but it did not influence the inactivation of the sodium system, in contrast to observations in Purkinje fibers (Weidmann, 1955a) and in frog nerves (Shanes, 1958). Neither was there any effect on the resting potential or on E_{Na}. The blocking action of procaine resulted from an inhibition of the sodium system. Shanes et al. (1959) compared the effects of procaine and cocaine with that of high $[Ca^{2+}]_o$. The results are summarized in Table 2.7. Local anesthetics reduced all voltage-clamp parameters, whereas the effects on E_{Na} and G_m were not very pronounced. Opposite results were obtained in high $[Ca^{2+}]_o$. The findings suggest that local anesthetics and high Ca^{2+} reduce the sensitivity of G_{Na} to depolarization. This effect persists at all levels of depolarization with the anesthetics, while in high $[Ca^{2+}]_o$ it disappears with higher depolarization and G_{Na} even increases so that inactivation is reduced. Blaustein and Goldman (1966) confirmed these results and postulated competition of calcium and procaine for a binding site in the membrane. An increase in $[Ca^{2+}]_o$ reduced the effectiveness of procaine in decreasing G_{Na}.

Armstrong and Binstock (1964) reported that alcohols with two to five carbons depolarized the nerve membrane and depressed G_{Na} much more than G_K. Simultaneously, Moore, Narahashi, and Ulbricht (1964) reported that ethanol affected $\bar{g}_{Na}$ and $\bar{g}_K$ in the giant axon. The effect on $\bar{g}_K$ contradicted the results obtained by Armstrong and Binstock (1964), but this effect was attributed to the hyperpolarization caused by the junction of sucrose with artificial seawater. Blaustein (1968a, b) reported that the effect of tropine esters and barbiturates was similar to that of procaine.

Table 2.7 Influence of Procaine, Cocaine, and Ca^{2+} on Electrical Membrane Parameters of Squid Giant Axon (Shanes et al., 1959)*

Current-Clamp		Voltage-Clamp		Remarks
33.9 45.9 41.0	threshold (mV)	2.61 0.54 4.02	I_{Na}(mA/cm^2)	For $\Delta V = +55$ mV
108 80 105	action potential amplitude (mV)	121 112 117	E_{Na} (mV)	Expressed as depolarization from resting potential
488 308 447	dV/dt (V/s)	50.0 14.4 83.6	G_{Na} (mmho/cm^2)	Linear part of I_{Na}/V curve at large depolarizations
388 197 420	dV/dt repol. (V/s)			
14.4 17.9 18.7	after-potential amplitude (mV)	29.1 4.6 31.8	dI_{Na}/dt (A/s cm^2)	For $\Delta V = +55$ mV
3.3 4.5 3.4	After-potential (half-time) (ms)	3.63 0.91 7.76	I_K (mA/cm^2)	At depolarization to $V = E_{Na}$
		34.1 8.9 76.2	G_K (mmho/cm^2)	Linear part of outward current in I_K/V curve
		10.4 3.5 18.9	dI_K/dt (A/s cm^2)	At depolarization to $V = E_{Na}$
		0.87 0.70 0.77	G_m (mmho/cm^2)	Slope of I/V curve between 50 and 100 mV hyperpolarization

*Within each type of measurement, the data are listed in the following sequence: control (seawater); 0.1% cocaine, and 0.1% procaine (pooled values); fivefold normal $[Ca^{2+}]_o$. I, dV/dt, and dI/dt measured at their maxima.

Much research has been carried out on the so-called neurotoxins. Narahashi, Moore, and Scott (1964) and Takata, Moore, Kao, and Fuhrman (1966) reported that tetrodotoxin (TTX) did not change the resting potential of lobster axons but selectively blocked the fast inward current. Internal perfusion of the axon with TTX did not have any effect (Nakamura et al., 1965b). Narahashi, Moore, and Frazier (1969) reported that TTX has a much stronger effect at pH 7 than at pH 9. It was concluded that TTX is active as a cation. The blocking effect of TTX on the sodium system (10^{-7} M) was reported by Binstock and Goldman (1969) for the axon of *Myxicola infundibulum*. Narahashi and Moore (1968) compared the effect of different neuroactive agents on axons of squid and lobster. TTX and saxitoxin (STX) blocked selectively the site for the sodium mechanism. In contrast, local anesthetics and pentobarbital reversibly blocked the conductance increases occurring during peak transient and steady-state currents. Narahashi, Deguchi, and Albuquerque (1971) reported that batrachotoxin (BTX) acted by depolarizing the membrane but the transient Na current was not markedly affected by this depolarization.

2.3 NEURONS

The size of neurons made it necessary to develop microelectrode techniques for measuring their impedance.

2.3.1 AC Measurements on Neurons

Fessard and Tauc (1956) measured the decrease in input impedance during the action potential in ganglion cells of *Aplysia*. The impedance change was only 1%, and this small impedance drop was attributed to a large capacity shunting the membrane, but it may be that an apparent resistance was measured and that ΔR_{El} was large compared to R_{inp}. Hagiwara and Saito (1959a) measured the resting passive properties of the cell membrane in the supramedullary ganglion of the puffer fish. The input capacity was 8×10^{-3} μF and the membrane capacity was 5–15 μF/cm^2. Hagiwara and Tasaki (1958) reported that a decrease in input impedance of the postsynaptic axon occurred with the synaptic potential while, as expected, a large decrease was observed when presynaptic stimulation gave rise to a postsynaptic action potential. Motoneurons have also been investigated with ac signals to decide between the neuron model of Coombs et al. (1959) and that of Rall (models 7 and 8 of Table 1.5). Smith, Wuerken, and Frank (1967) tried to determine experimentally which model was closer to the actual experimental situation in cat spinal motoneurons. It was observed that inhibitory postsynaptic potentials (ipsp's) were accompanied by an increase in impedance, but less than half of the excitatory postsynaptic potentials (epsp's) showed a detectable impedance decrease. Consequently, model 7 of Table 1.5 effectively explained the inhibitory synapses, located close to the soma, but only Rall's model took into account the influence of part of the excitatory synapses that appear to be located on the dendrites. The input resistance was measured with a bridge circuit (circuit 6 of Table 1.3), and it should be remarked that the use of double-barreled microelectrodes in the circuit eliminated the danger of confounding the input impedance with the quantity $\Delta R_{\mathrm{El}} + Z_{\mathrm{inp}}$. Nelson and Lux (1970) used phase angle determinations (2 Hz to 1 kHz) to determine the effective electrotonic length of the dendrites and the dendritic to soma conductance ratio in cat motoneurons. An analysis of long voltage transients showed the existence of two time constants. The late part of the semilogarithmic plot of the rising phase of the transient versus time represented a straight line. This indicated, according to Lux (1967), that the dendritic length is limited. The relation of the phase angle to frequency showed that the phase lag increased continuously with frequency, and no maxima or minima were observed. A range of 5–10 for the ratio of dendritic input conductance to somatic surface conductance was estimated assuming that the phase angles at high frequencies were overestimated due to instrumental errors.

2.3.2 Square-Pulse Measurements on Neurons

Microelectrodes have been used to investigate electrical changes in motoneurons in response to orthodromic or antidromic stimulation (Brock et al., 1952; Araki et al., 1953). The passive properties of the cell have been studied using bridge circuits with a single microelectrode and double-barreled microelectrodes. Araki and Otani (1955) used a bridge to measure the input resistance of toad spinal motoneurons; most probably an apparent resistance was measured (Eq. 1.1). The high values obtained (Table 2.8) can thus be explained, since Frank and Fuortes (1956) showed that the change in microelectrode resistance inside the cell (on the order of a few MΩ) contributes substantially to the measured quantity when the measurement is performed with a single microelectrode.

Coombs et al. (1955) studied the properties of motoneuron membranes with double-barreled microelectrodes. Current–voltage curves were linear throughout most of the range of current applied ($\pm 3 \times 10^{-8}$ A). The cell constants (Table 2.8) were estimated with model 1 of Table 1.5, assuming a cytoplasmic resistivity of 50 Ω cm. The time constant was measured from the decaying phase of a postsynaptic potential, which would correspond to τ_m if synaptic activity were uniformly distributed over the surface of the soma and the dendrites. The resulting membrane capacity was higher than the value found in the axon membrane, probably because of the assumptions underlying its estimation. More accurate estimates of these parameters were reported by Coombs et al. (1959) and Curtis and Eccles (1959). Cross-compensation circuits were used to eliminate the coupling resistance of double-barreled microelectrodes and model 7 of Table 1.5 was developed. The cell constants were calculated assuming $R_i = 75$ Ω cm. The

Table 2.8 Electrical Constants of Neurons Obtained with the Square-Pulse Analysis

Circuit*	Measured Quantities	Model†	Derived Quantities	Reference
5	R_{app}: 3–6 MΩ τ_{inp}: 3–6 ms	1	R_m: 269 Ω cm^2 C_m: 17.5 μF/cm^2	Araki and Otani (1955)
2	R_{inp}: 0.4–1.3 MΩ (I/V curve) τ_m: 4 ms	1 5	R_m: 150 Ω cm^2 (soma) C_m: 8 μF/cm^2 R_m: 500 Ω cm^2 (dendrites) λ: 350 μm	Coombs, Eccles, and Fatt (1955)
2	R_{inp}: 0.5–2 MΩ (I/V curve)	7	R_m: 172–1390 Ω cm^2 (soma and dendrites) R_s: 1.06–9 MΩ (soma) R_d: 0.92–2.6 MΩ (dendrites) λ: 0.17–0.485 mm (dendrites) C_m: 5 μF/cm^2	Coombs, Curtis, and Eccles (1959)
5 (modified)	ΔR_{inp}: 1.65 MΩ τ_{inp}: 1–1.4 ms	1	R_m: 1000 Ω cm^2 C_m: 1–1.5 μF/cm^2	Frank and Fuortes (1956)
1	R_{inp}: 2.2 MΩ τ_{inp}: 50 ms	1	R_m: 2200 Ω cm^2 C_m: 23 μF/cm^2	Fessard and Tauc (1956)
1	R_{inp}: 1–3.4 MΩ τ_{inp}: 30 ms	8	R_m: 987 Ω cm^2 C_m: 29.8 μF/cm^2	Maiskii (1964)
1	R_{inp}: 12.1 MΩ (I/V curve) τ_{inp}: 54 ms	1	R_m: 8300 Ω cm^2 C_m: 6.5 μF/cm^2	Meves (1968)
1	R_{inp}: 2.3 MΩ (I/V curve) R_{inp}: 21 MΩ (I/V curve)	1	R_m: ≅1400 Ω cm^2 ($\Delta V = -20$ mV) R_m: ≅14,000 Ω cm^2 ($\Delta V = +10$ mV)	Tauc and Kandel (1964) Kandel and Tauc (1966)
1	R_{inp}: 80 kΩ (depolarization) (I/V curve) R_{inp}: 270 kΩ (hyperpolarization) τ_{inp}: 15 ms	1	R_m: 500 Ω cm^2 C_m: 30 μF/cm^2	Bennett, Crain, and Grundfest (1959)
1	R_{inp}: 0.6–2.5 MΩ (I/V curve) τ_{inp}: 4-6 ms	1	R_m: 500–1000 Ω cm^2 C_m:5–15 μF/cm^2	Hagiwara and Saito (1957, 1959a)

Table 2.8 *Continued*

Circuit*	Measured Quantities	Model†	Derived Quantities	Reference
1	R_{inp}: I/V curves τ_m: 200–1000 ms	1	R_m: 0.1–1.5×10^6 Ω cm^2 C_m: 0.1–1 μF/cm^2 P_K: 0.1–2.5×10^{-8} cm/s	Marmor (1971a)
1	R_{inp}: 0.4 MΩ C_{inp}: 0.2 μF	1	At pH 8.0: G_m: 40×10^{-6} mho/cm^2 g_K: 22.8×10^{-6} mho/cm^2 g_{Cl}: 8.8×10^{-6} mho/cm^2 g_{Na}: 8.4×10^{-6} mho/cm^2 P_K: 0.21×10^{-6} cm/s P_{Cl}: 0.08×10^{-6} cm/s P_{Na}: 0.004×10^{-6} cm/s At pH 5.0 G_m: 75×10^{-6} mho/cm^2 g_K: 35×10^{-6} mho/cm^2 g_{Cl}: 30×10^{-6} mho/cm^2 g_{Na}: 10×10^{-6} mho/cm^2 P_K: 0.32×10^{-6} cm/s P_{Cl}: 0.27×10^{-6} cm/s P_{Na}: 0.005×10^{-6} cm/s	Brown, Walker, and Sutton (1970)

*All circuit numbers refer to those of Table 1.3.
†All models are those of Table 1.5.

geometry of the dendritic tree was verified with histological methods to justify the assumed distribution of current. The time course of the electrotonic potential was more abrupt than expected for uniform distribution of current. Rall (1957) showed that this deviation depends upon the number and size of dendrites. This structural feature explains the discrepancy between membrane time constants measured from the decay of the postsynaptic potential (Coombs, Eccles, and Fatt, 1955) and those measured from the electrotonic potential produced by current injection (Frank and Fuortes, 1956). The membrane time constant, calculated for a current distribution ratio between dendrites and soma equal to 2.3, was 3.1 ms (Coombs, Curtis, and Eccles, 1959).

Frank and Fuortes (1956) applied a modification of a bridge method to measure the input resistance in cat spinal motoneurons. A long pulse was applied through a microelectrode during antidromic or orthodromic stimulation. Action potentials were then recorded with and without injection of current. The input resistance was obtained from the difference between them. Assuming that the current flow does not affect the mechanism generating the spike, the input resistance can be estimated from the shifts of the membrane potential at the resting level and at the summit of the action potential, according to

$$\Delta V_{ap}/I = R^r_{inp} - R^p_{inp} \tag{2.25}$$

where ΔV_{ap} is the change in action potential amplitude, and R^r_{inp}, R^p_{inp} are the input resistances at rest and at the peak of the action potential, respectively. Plots of action potential amplitude versus current gave a linear relationship, the slope of which corresponds to the right-hand term of Eq. 2.25. The value thus obtained underestimates the input resistance if the membrane resistance is not negligible at the peak of the action potential. The good agreement between the values thus obtained (Table 2.8) and those of Coombs, Curtis, and Eccles (1959) indicates that the assumptions involved in these measurements are applicable to cat motoneurons. The measured values were not apparent resistances because the bridge was balanced for all the voltage drops occurring across the various resistances in the circuit, including the change in electrode resistance. The cell constants (Table 2.8) are within the same range as those reported for the axon membrane. Rall (1959) analyzed the results of Frank and Fuortes (1956) and Coombs et al. (1956, 1959) according to model 8 of Table 1.5. Either the geometrical factor or the membrane resistance had to be fitted to the R_{inp} value. A range of 1000–8000 Ω cm^2 for R_m was suggested. The lower value corresponds to the lowest R_{inp} measured, 0.5 MΩ. That would be the true input resistance of a large cell with a rather high value for the combined dendritic tree parameter. Recalculation of reported time constants showed that the decay of a postsynaptic potential gives a better measure of the soma membrane time constant than the estimation from the electrotonic potential, because the current distribution is more uniform. Kernell (1966) showed that the dendrite to soma conductance ratio was of the same order of magnitude for small and large motoneurons and therefore, according to Rall (1959), the input resistance of the cell is a function of the reciprocal of its soma surface and model 1 of Table 1.5 is applicable.

According to the model of Eccles (1966), hyperpolarization should increase the magnitude of the epsp, but in cat motoneurons this usually does not occur. This can be explained by nonlinearities in membrane properties. Nelson and Frank (1967) obtained current–voltage curves by superimposing action potentials on long-duration square pulses. The best fit was obtained by a high-resistance branch (7.3 MΩ) in the depolarizing side and a low-resistance branch (2.95 MΩ) in the hyperpolarizing side. The ratio between the two was called the anomalous rectification or AR ratio. In nine cells the mean AR ratio was 1.8, with a range from 1.0 to 2.2. Comparing the amplitude of electrotonic potentials for equal current pulses of opposite polarity, the ratio obtained also showed a 1.5- to 2-fold higher input resistance on depolarization than on hyperpolarization. Anomalous rectification appeared in curves measured soon after the onset of the pulse and in the ones measured at the end of it. When short square pulses were used, the input resistance also decreased with hyperpolarization. Although most of the cells showed anomalous rectification in current–voltage curves, the response of the epsp's to hyperpolarizing current varied. This was explained by the complex geometry of the neuron, because the current distribution is not the same for epsp's arising in the dendritic tree as for those elicited in the soma. The effect of the polarizing current will be greater on the latter. The response of an epsp to hyperpolarizing current can give indications about the localization of the synapsis with respect to the cell body.

The discovery of large ganglionic cells in some species of invertebrates allowed the performance of more accurate determinations of cell electrical constants, as reported by Fessard and Tauc (1956) for *Aplysia* ganglion cells (Table 2.8). Maiskii (1964) studied the membrane properties of ganglionic cells of the snail *Helix pomatia*. The neuron is composed of the cellular body and a single axon; the recorded input resistance represents the combined input resistance of the soma and the axon in parallel. Assuming a homogeneous membrane, the cell constants can be obtained with the following equations (Rall, 1959):

$$1/R_{\text{inp}} = (A/R_m) + 1/\left(4.9 \times 10^5 \times R_m^{1/2}\right)$$

$$1/R_{\text{inp}} = (A/R_m) + \left[(\pi/2)\left(R_i^{-1/2}\right)xd^{3/2}\right]/R_m^{1/2} \qquad (2.26)$$

Ionic requirements for electrical activity of ganglionic cells in *Helix aspersa* and *Helix pomatia* were studied by Meves (1968). Voltage–current curves in normal medium showed rectification on the depolarizing side of the curve. It was found that the contribution of the axon to the measured input resistance was negligible. Cocaine (0.4%) reduced the rectification, the resting potential, the overshoot, and the rate of rise of the action potential. The same effects were obtained in low sodium, but electrical activity persisted in sodium-free medium. Calcium concentration affected the overshoot and the rate of rise.

Tauc and Kandel (1964) reported a strong rectification in voltage–current curves of metacerebral giant cells in *Helix aspersa*, between +10 and −20 mV. The corresponding membrane resistances were 14 kΩ cm^2 and 1.4 kΩ cm^2, respectively (Table 2.8). This rectification may play an important role because it occurs within the physiological range of membrane polarization (epsp's of ±5 mV) and it can serve to regulate the output of the axon. Further evidence of the voltage dependence of R_m was reported by Kandel and Tauc (1966). Acetylcholine produced depolarization but did not change R_{inp}, but if the membrane potential was kept at resting level during application of acetylcholine, a marked decrease in R_{inp} occurred. The degree of rectification varies according to the species as shown by Ochs (1967) in ganglionic cells of *Helix memoralis*. R_{inp} decreased between the resting potential and 25–30 mV of hyperpolarization. At higher degrees of hyperpolarization, the slope of the curve did not change, but a time-dependent fall in R_{inp} was observed. Chalazonitis et al. (1967) reported that rectification persisted at different temperatures. R_{inp} decreased with increasing temperature. Lowering the temperature from 23°C to 11.4°C produced depolarization and an increase in R_m. The change in R_m might have resulted from the changes in membrane potential because of the rectifying properties of the membrane. Gola (1967) reported that the input resistance of neurons from the snail *Helix pomatia* increased an average of 30% during hypoxia. Depolarization was also observed. Keeping the membrane potential at resting level by current injection did not abolish the change in R_{inp} in neurons showing anomalous rectification and enhanced it in those with normal rectification.

Junge and Moore (1966) recorded the electrical activity of pacemaker neurons and applied short square pulses at different moments of the spontaneous depolarization. The size of the electrotonic potentials did not vary, and it was concluded that R_m reached a rather constant level early in the activation cycle. In contrast, Watanabe et al. (1967) reported an increase in R_{inp} during the spontaneous depolarization between periodic burst discharges in the pacemaker neurons of the heart ganglion of a crustacean. Bennett et al. (1959) measured current–voltage curves on the neurons of the Atlantic puffer, *Spheroides maculatus*. The input resistance was higher for hyperpolarizing currents than for depolarizing currents (Table 2.8). The membrane resistance during the hyperpolarization that follows the spike was lower than at rest. Strong hyperpolarizing currents elicited a slow decay of the electrotonic potentials, indicating a time-dependent change in R_m. Depolarizations above threshold triggered spikes, and the slope of the current–voltage curve changed drastically, suggesting a fall in R_{inp} during activity. Hagiwara and Saito (1959a) investigated the supramedullary ganglion cells of the puffer and reported linear current–

voltage relationships, with slopes ranging between 0.6 and 2.5 MΩ. The lack of rectification upon depolarization was attributed to the low current strength, always below threshold.

Ito and Oshima (1965) analyzed the marked overshoots and undershoots of the membrane potential appearing at the onset and cessation of current steps in cat spinal motoneurons. Current–voltage curves showed the nonlinearities described by Nelson and Frank (1967). The time course of the potential changes during and after depolarizing and hyperpolarizing current steps could not be explained by applying model 8 of Table 1.5 or with the kinetics of g_K of the squid axon. Voltage-clamp experiments (Sec. 2.3.3) have shown the existence of more than one component for I_K, which could explain the findings of Ito and Oshima (1965).

Some reports (Marmor, 1970; Marmor and Gorman, 1970; Marmor and Salmoiraghi, 1970) indicated that the membrane properties of a molluscan neuron, the *Anisodoris* G cell, depended upon temperature and ionic environment. Gorman and Marmor (1970) reported that the membrane potential of the G cell could be predicted by the constant field equation at temperatures less than 5°C but not at higher temperatures, where an electrogenic sodium pump became active. Marmor (1971a) measured the current–voltage curves resulting from injection of square pulses and ramp signals. Action potentials and delayed rectification appeared during the depolarizing phase of ramp signals, regardless of temperature. Inward-going rectification disappeared at 5°C and in the absence of K^+. When the cell was hyperpolarized with ramp pulses, the potential followed the current transiently but then sagged back to a lower value despite the continuous change in the current. This high conductance state occurred between −90 and −140 mV, and hysteresis appeared in the current–voltage relation. The development of the high conductance state depended upon $[Ca^{2+}]_o$, but not upon $[K^+]$, $[Na^+]$, and $[Cl^-]$. The ratio of axonal to somatic conductance was larger than 5.0, and therefore the cell constants in Table 2.8 were estimated according to model 8 of Table 1.5. The time constant, R_m, and the axon to soma conductance ratio varied with temperature and potential, but the temperature dependence disappeared in the absence of K^+. C_m was independent of membrane potential at all temperatures. The absolute P_K was calculated using the constant field equation and electrical measurements. Inward-going rectification depended on the absolute value of the membrane potential, but it was not influenced by the activity of the electrogenic sodium pump (Marmor, 1971b). Inhibition of the pump by ouabain caused a large depolarization, but G_m and the pattern of rectification were the same at any given value of membrane potential. Therefore, the ionic pump was considered a constant current source, and its contribution to the resting potential was calculated with an equation similar to Eq. 1.103, where the constant current contribution is introduced into the sum of ionic currents contained in the Goldman equation (Sec. 1.6.2). The total current generated by the pump was 10–20 nA, and the uncoupled sodium efflux was 0.2–4.0 pmol/cm^2 s. Brodwick and Junge (1972) studied the post-stimulus hyperpolarization (PSH) appearing after a burst of action potentials or passage of outward current in *Aplysia* giant neurons. The input conductance increased during the PSH. These responses were not abolished by a calcium-free medium containing TTX or by inhibition of the electrogenic pump. The reversal potential for the PSH varied with $[K^+]_o$ but not with $[Cl^-]_o$ or $[Na^+]_o$. The PSH was then attributed to slow changes in g_K. A model consisting of two parallel channels was proposed. The first contained the resting potential in series with the resting conductance, and the second was composed of the equilibrium potential for the PSH, in series with the PSH conductance.

Brown and Berman (1970) investigated the mechanism of the excitatory effect of CO_2 on *Aplysia* neurons, which was mediated by the concomitant changes in pH. At pH 8.0, the input slope conductance (G_{inp}) had its lowest value (1.2 μmho) around the resting potential (-55 mV), and it increased to 2.9 μmho upon depolarization to -40 mV (onset of delayed rectification). Anomalous rectification appeared upon hyperpolarization to -60 mV where G_{inp} increased to 3.3 μmho. At pH 6.5 the current–voltage relation became linear because of an increase of G_{inp} to 2.4 μmho for zero current and a decrease from 3.3 to 2.4 μmho at the hyperpolarized end of the curve. This response occurred in neurons of the abdominal ganglion. The pacemaker neurons were much less sensitive to pH variations, although they showed the same type of changes. These effects were attributed to a direct effect of $[H^+]$ on G_m and not to changes in resting potential, because half the cells responded with depolarization and the rest of them with hyperpolarization. To explain this apparently contradictory effect of the increase in G_m on the resting potential, Brown, Walker, and Sutton (1970) estimated the individual ionic conductances and their modifications upon lowering of the pH. Hyperpolarization in low pH reversed to depolarization upon reduction of $[Cl^-]_o$ to one-half. The depolarization observed in other cells could be reversed to hyperpolarization if the cells were previously depleted of chloride. The values of E_{Cl} and E_K were calculated from measurements of intracellular activities of these ions. Two groups of cells were found. In the first one, the cells had a mean membrane potential of -56.4 mV and responded with hyperpolarization to a fall in pH from 8.0 to 5.0. Their E_{Cl} was -63.6 mV. In the second group, the cells had an E_{Cl} of -53.3 mV and the mean resting potential was -58 mV. These neurons responded to acidification with depolarization. The mean E_K was -80 mV. Changes in g_K and g_{Cl} with pH were estimated from plots of resting potential versus $[Cl^-]_o$ or $[K^+]_o$ at two pH values: 8.0 and 5.0. From the transport numbers for potassium and chloride calculated for both pH values, the individual ionic conductances were obtained (Table 2.8). The table shows that at pH 5.0, g_K increased to 150% of control, g_{Cl} increased to 340%, and G_{Na} to 110%. The ionic permeabilities were calculated with the constant field equation (Eq. 1.87). It was concluded that the depolarizing or hyperpolarizing effect of a fall in pH depends mainly on the increase in g_{Cl} because, although g_K also increases, this can lead only to hyperpolarization because the membrane potential is always lower than E_K. The variability in the response depends, therefore, upon the relation of E_{Cl} to the membrane potential. These results offered an explanation for the opposite effects of CO_2, which produces excitation or inhibition according to the cell type: The result of an increase in CO_2 or a lowering in pH will depend on the relation between the membrane potential and the equilibrium potential for the ion whose conductance is more affected by $[H^+]$.

2.3.3 Voltage-Clamp Measurements on Neurons

Voltage-clamp measurements on neurons have been used mainly to study spike genesis and to identify the charge-carrying ions during the transient inward current. Few investigations of drug action were performed on these cells.

Hagiwara and Saito (1957, 1959a, b) reported that current–voltage curves measured in giant neurons of the puffer fish showed the typical N-shape for the early current and the diode characteristic for the steady-state current. The N-shape characteristic indicated that the soma membrane is excitable. The sub- and supraoesophageal ganglion of the mollusk *Onchidium verruculatum* also presented an early inward current and a late outward current. The fast inward current was blocked by 2% urethane. TEA did not influence the resting G_m but decreased the degree of late rectification. The soma membrane of these ganglia was excitable. Hagiwara, Watanabe, and Saito (1959) applied circuit 7 of Table 1.4 to the cardiac ganglion of a lobster. There was a weak inward current and a late steady-state current, but the current–voltage curve of the early current did not show the typical N-shape. This indicated that the soma membrane was unexcitable. The observed inward current was due to an action potential originating in the axon, in a region where the space clamp is becoming noneffective.

Terzuolo and Araki (1959) and Araki and Terzuolo (1962) performed voltage-clamp measurements on motoneurons of the cat. When the soma membrane was clamped to a value close to the resting potential (ΔV: ± 10

mV) only one peak of inward current was observed upon antidromic stimulation. At higher depolarizations, a second transient of inward current appeared. This result suggested that the initial part of the action potential results from the activity of the axonal cone while the soma membrane gives rise to the second component. Frank, Fuortes, and Nelson (1959) reported a two- to threefold reduction of R_m during activity. From this, it was concluded that only part of the soma-dendritic complex was activated at a given time. A later report by the same group (Nelson and Frank, 1964) contains a detailed analysis of the method (circuit 7 in Table 1.4). Frank and Tauc (1964) studied ganglia of the mollusks *Aplysia depilans* and *Helix pomatia* (circuit 9, Table 1.4). The electrically isolated membrane patches behaved qualitatively like other excitable membranes: There was a variable transient inward current, but in every instance a large late outward current was found. Acetylcholine was applied electrophoretically to the ganglia cells that are classified as D cells (depolarized and excited by Ach) and H cells (hyperpolarized and inhibited by Ach; Tauc and Gerschenfeld, 1960). Figure 2.16 shows the effects of acetylcholine on the late steady-state current and the reversal potential.

Chamberlain and Kerkut (1967, 1969) reported that the inward current found in neurons of the snail *Helix aspersa* decreased in sodium-free solutions. Removal of Ca^{2+} abolished the remaining inward current. TTX (10^{-8}–10^{-5} g/ml) did not affect it, whereas injection of EDTA into the cell made the inward current more sensitive to changes of $[Na^+]_o$ compared to cells with normal $[Ca^{2+}]_i$. This finding indicated an influence of Ca^{2+} on the fast inward current, but the role of Na^+ and Ca^{2+} in its production could not be elucidated. In contrast, Geduldig and Junge (1968) found that the soma action potential had two components, one dependent upon Na^+ and the other upon Ca^{2+}. In calcium-free solutions, the sodium-component could be blocked by TTX. These two components of the action potential were further investigated by Geduldig and Gruener (1970). Figure 2.17 shows current–voltage curves obtained for the early transient current and the steady-state current under different experimental conditions. The peak inward current under TTX was independent of $[Na^+]_o$ and was ascribed to Ca^{2+}. Different channels for Na^+ and Ca^{2+} were postulated, because g_{Na} and g_{Ca} showed different inactivation kinetics. In calcium-free medium, the inactivation curve was similar to that of the sodium inactivation in the squid axon (Table 2.5), but inactivation in sodium-free medium was different. Hyperpolarizing pulses inactivated the calcium system. A slight depolarization removed inactivation, and only strong depolarizing pulses inactivated this system (Fig. 2.18). These results were corroborated by Krishtal and Magura (1970) on neurons of different mollusks. A decrease of $[Na^+]_o$ had little influence on the overshoot of the action potential. Increase in $[Ca^{2+}]_o$ did not influence the inward current but increased the amplitude of the overshoot. This effect cannot be

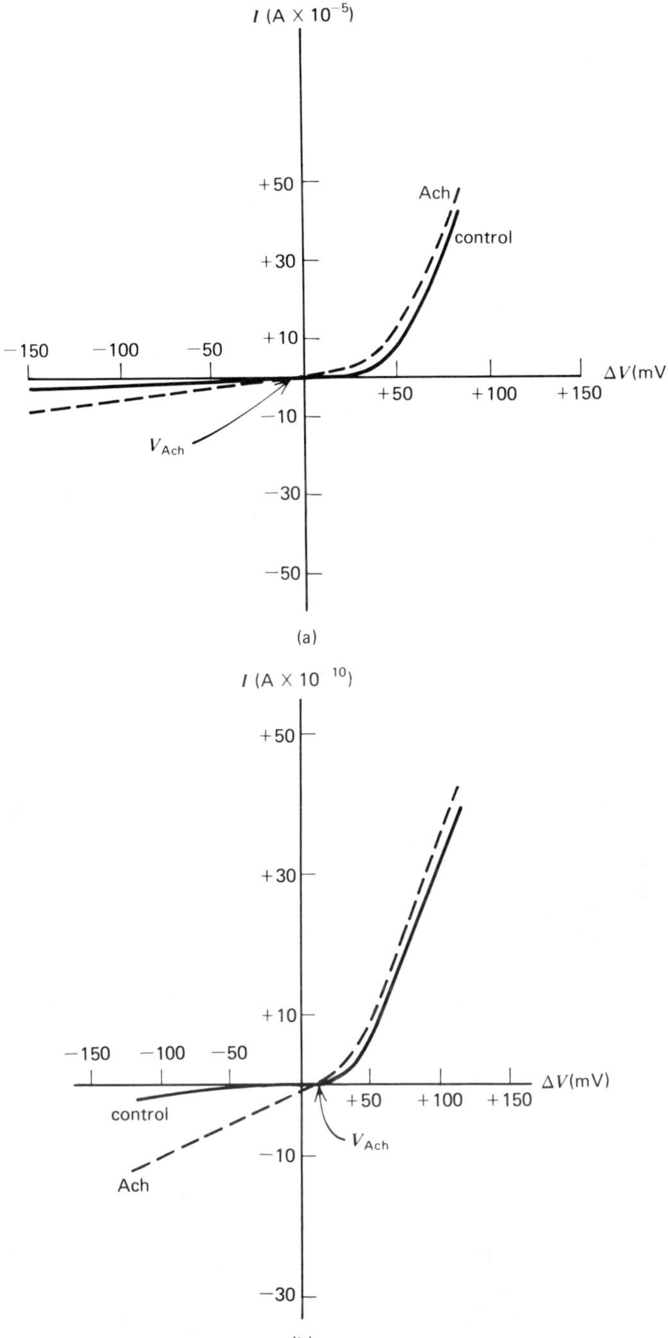

Figure 2.16. Effect of acetylcholine on H (a) and D (b) cells in molluskan ganglia. Abscissa: displacement from resting potential. Ordinate: current. V_{Ach} is the reversal potential under acetylcholine. Modified from Frank and Tauc (1964). By permission.

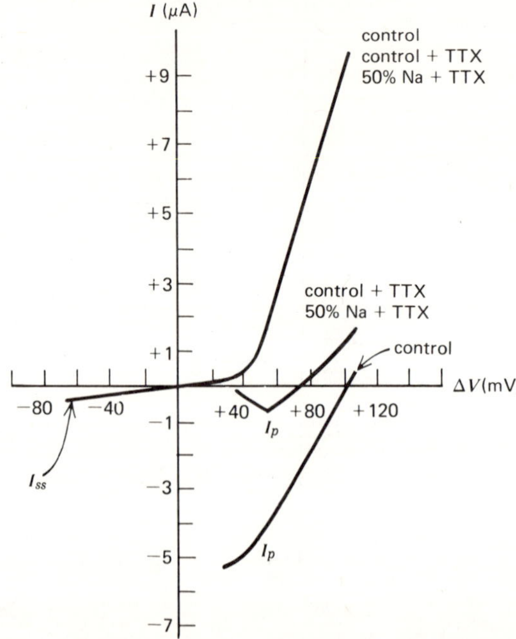

Figure 2.17. Current–voltage curves of the steady-state current I_{ss} and the peak transient current I_p of *Aplysia* neurons in different media. Modified after Geduldig and Gruener (1970). By permission.

explained if Na^+ is the only charge carrier across the membrane. It seems that Ca^{2+} starts to carry inward current when the dV/dt begins to decline during the upstroke of the action potential.

In a completely different approach, Thomas (1969) used a slow voltage clamp to measure the current produced by the electrogenic pump in giant neurons of the common snail, *Helix aspersa*. Iontophoretic injection of Na^+ caused a hyperpolarization of 14 mV. When ouabain was applied prior to injection of Na^+, the hyperpolarization was reduced to 3 mV. Ouabain alone depolarized the membrane by 2 mV without changing R_m. The current recorded with the membrane potential clamped to the resting level reflects the current of the electrogenic pump. During injection of Na^+ the current rose at a constant rate. It was found that two-thirds of the extruded Na^+ was coupled to the active transport of other ions and that the uncoupled remaining third produced the electrogenic effect.

Neher and Lux (1969) developed a modification of method 9 of Table 1.4 to clamp an isolated spot of the membrane of the suboesophageal ganglion of the snail *Helix pomatia*. Using this method, Neher and Lux (1971) and Neher (1971) reported that the currents in isolated patches of

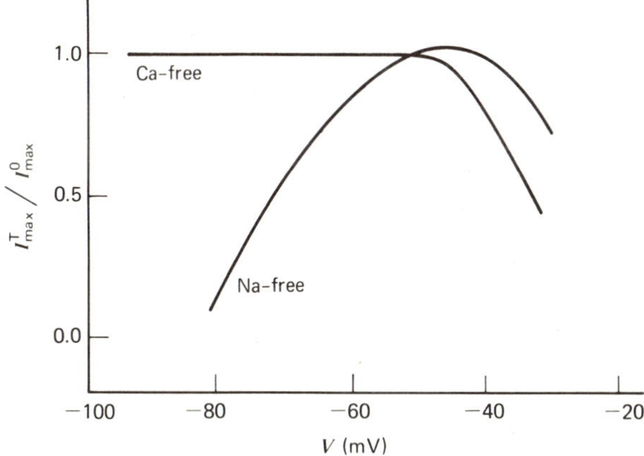

Figure 2.18. Inactivation of the fast inward current in Na-free and Ca-free solutions. Ordinate: relative transient current (see Table 2.5). Abscissa: V during conditioning step. From Geduldig and Gruener (1970). By permission.

the soma membrane could be compared to those of the squid giant axon but the ratio $\bar{g}_{Na}/\bar{g}_K$ was 0.5 in the neuron as compared to 3.3 in the squid axon. Two distinct features were found: (1) a slow inactivation of the late I_K, which was too complex to be described by a combination of first-order variables as is done in the Hodgkin-Huxley formalism; (2) a fast transient subthreshold outward current upon hyperpolarization. This current was inactivated at the resting potential, had the same reversal potential as the late outward current, and was insensitive to ouabain and $[Cl^-]_o$. A similar current was reported by Stevens (1969) in *Archidoris* neurons, and it was considered part of the pacemaker mechanism in these neurons. Neher (1971) investigated the relation of the transient fast outward current to the transient inward current. It was confirmed that the fast outward current is carried by K^+. Its kinetics were similar to those of the transient inward current, but inactivation was slower. The inward current inactivated with two time constants. It was proposed that the fast outward current keeps the membrane hyperpolarized after the action potential. Upon inactivation, the membrane is slowly depolarized by a weak constant outward current until threshold is reached, resulting in the firing of the pacemaker.

Connor and Stevens (1971a) studied the neural somata of some Nudibranchia. Since the currents in the soma were found to be more complex than those in the squid axon, three different methods were used to determine the equilibrium potentials for the respective currents. These methods defined the equilibrium potential with respect to a current flowing through a specified channel and not with respect to a single ion species. They were based on the equations for the chord conductances (Table 2.5).

Under particular experimental conditions, the instantaneous value of the chord conductance for a given current can be measured, and it is constant for short time intervals. Using the instantaneous current–voltage relation for g, one can compare the current obtained just before applying the clamp potential (I_0) with the current obtained a short time after the clamp potential has been applied (I'). Since, under these conditions, g is constant, one obtains:

$$I_0/I' = (V - E)/(V' - E) \tag{2.27}$$

where E is the only unknown and can be calculated. This method has the advantage that the equilibrium potential can be obtained from a single measurement. The soma membrane showed a rapid inward current followed by a delayed outward current. The inward current was carried by Na^+ and Ca^{2+} because the spikes persisted in sodium-free solutions and under TTX, but they were abolished in solutions containing neither Na^+ nor Ca^{2+}. The outward current activated and inactivated with a more complex mechanism than in the squid axon. It was carried by K^+. Connor and Stevens (1971b) also studied a fast transient outward current that appeared during hyperpolarization and inactivated at the level of the resting potential. The predominant current-carrying ion was K^+, and the activation and inactivation processes had exponential time courses. This fast outward current was not influenced by TEA, and it was concluded that the transport mechanisms for the two outward currents were operationally different.

Connor and Stevens (1971c) formulated a theory of repetitive firing in isolated neuron somata based mainly on the fast outward current component. Equations for the three voltage- and time-dependent conductances $g_{\mathrm{I}}(V,t)$, $g_A(V,t)$, and $g_{\mathrm{K}}(V,t)$ were derived. The suscripts I, A, and K indicate the rapid inward current, the rapid outward current, and the late outward current. The result of the formal analysis for the inward current is:

$$\left.\begin{aligned} g_{\mathrm{I}}(V,t) &= \bar{g}_{\mathrm{I}} A_{\mathrm{I}}^3(V,t) B_{\mathrm{I}}(V,t) \\ \tau_{A\mathrm{I}}(V)\, dA_{\mathrm{I}}(V,t)/dt + A_{\mathrm{I}}(V,t) &= A_{\mathrm{I}}(V,\infty) \\ \tau_{B\mathrm{I}}(V)\, dB_{\mathrm{I}}(V,t)/dt + B_{\mathrm{I}}(V,t) &= B_{\mathrm{I}}(V,\infty) \end{aligned}\right\} \tag{2.28}$$

The symbols A and B denote activation and inactivation and $\bar{g}$ is the maximum conductance. Although this formulation is similar to the Hodgkin-Huxley equations, the difference is that Hodgkin and Huxley obtained their description by a curve-fitting process based on empirical relations describing the voltage dependence of their rate constants α and β (Eq. 2.14), whereas Connor and Stevens obtained the same information for their processes A and B by a piecewise linear approximation to experimental data. The procedure is the following: the description of g_{I} requires the estimation of g_{I}, $\tau_{A\mathrm{I}}$, $\tau_{B\mathrm{I}}$, $A_{\mathrm{I}}(V,\infty)$, and $B_{\mathrm{I}}(V,\infty)$. $\bar{g}_{\mathrm{I}}$ is estimated from current versus time records as the maximum inward-going input conductance. In order to obtain $\tau_{B\mathrm{I}}(V)$, semilogarithmic plots for the approach of experimentally determined g_{I} toward its steady-state values were prepared for a range of clamping voltages. Fitting the experimentally obtained points by a straight line gives $\tau_{B\mathrm{I}}(V)$. The steady-state inactivation, $B_{\mathrm{I}}(V,\infty)$, was obtained as in Table 2.5. The activation $A_{\mathrm{I}}^3(V,\infty)$ can be estimated from a series of clamps from a fixed holding potential to various levels of the clamped potential. Provided activation occurs rapidly compared to inactivation, $\bar{g}_{\mathrm{I}}A_{\mathrm{I}}^3(V,\infty)$ may be found by extrapolating the decaying conductance curves back to zero time, where V equals the holding potential. Since $g_{\mathrm{I}}(V,t)$ is measured, $A_{\mathrm{I}}^3(V,\infty)$ can be evaluated. The value of $\tau_{A\mathrm{I}}$ was estimated assuming an exponential time course from the onset of the command signal to the peak inward current. The remaining membrane conductances were similarly described, and the following expression for $I_m(t)$ was obtained:

$$I_m(t) = C(dV/dt) + \Sigma\, \bar{g}_i A_i^n(V,t) B_i(V,t)(V - E_i) \qquad (2.29)$$

where i denotes a given ionic channel and n an integer.

Good agreement was found between recorded and computed action potentials. The time course of the different ionic currents during a series of action potentials is shown in Fig. 2.19. These computations revealed that inactivation of I_A is largely responsible for the depolarization of the membrane between two action potentials. They also predicted that the intervals between the action potentials elicited by a depolarizing pulse are longer at the beginning of the train of impulses than when the frequency has reached a steady-state level. Gola and Romey (1971) also found a rapid K^+ outward current in the giant neurons of *Helix pomatia*. Current–voltage curves obtained with triangular pulses revealed anomalous rectification and hysteresis at $dI/dt > 8$ mA/s.

Meves and Wellhöner (1969) and Leicht et al. (1970) reported that veratridine depolarized the giant neurons of *Helix pomatia* by 20–30 mV. The action potential showed a hyperpolarizing after-potential of long duration. Replacing NaCl by Tris-HCl eliminated the depolarization, but TTX did not abolish the effects. It was concluded that veratridine induces a slow sodium permeability that is not sensitive to TTX. Aconitine did not show any effect.

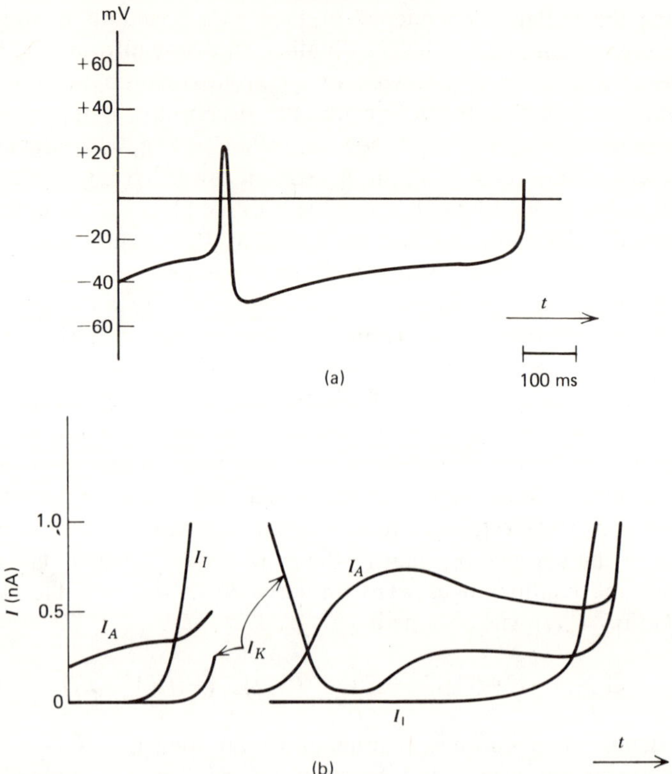

Figure 2.19. Computed action potential (a) and ionic currents (b) of a neuron cell resulting from a stimulus (1–6 mA). Ionic currents during the action potential not shown for sake of clarity. Ordinate: membrane potential in (a) and input current in (b). Abscissa: time. Modified after Connor and Stevens (1971c). By permission..

2.4 SUMMARY AND OUTLOOK

Throughout this chapter it has become evident that impedance measurements in nerves and neurons have contributed to our understanding of nervous excitation and synaptic transmission. Even before the ionic theory of excitation was formulated, the analysis of the cable properties of nerve fibers by Rushton and by Hodgkin and Rushton supported the local circuit theory as the mechanism of impulse propagation. The study of dynamic membrane conductances led to the Hodgkin-Huxley formalism, which was rapidly extended to other excitable structures. The Hodgkin-Huxley equations were then applied to the myelinated nerve and the different types of muscle. They were modified and adapted to fit the particular characteristics of the excitation process in these cells. The description of the active currents in terms of voltage- and time-dependent conductances proved to be a very useful tool in the study of excitation phenomena. However, as stated in Chapter 1, the nature of these conductances remains to be disclosed. The molecular mechanism responsible for the increase in permeability of a given ionic channel is not known. Recent advances were made studying the effects of inhibitors such as TTX, STX, TEA, etc. on ionic currents and their kinetics. The analysis of these results suggested that the gating mechanism lies in the inner face of the membrane and that the inhibitory effect of TTX occurs through a block of the entrance of the sodium channel. The mechanism of the voltage dependence of the ionic conductances remains not only one of the most challenging questions but also one of the most difficult problems to attack from an experimental point of view. Results from electrical measurements will have to be complemented by information obtained with molecular probes or methods supplying evidence at the molecular level to disclose the nature of this mechanism. The study of the electrical properties of neurons developed as a logical followup of the investigations in nerve fibers. Analysis of the electrotonic spread elicited by injection of current, by afferent stimulation, or by local application of the transmitter has supplied information on membrane properties and localization of synapses. Here, the application of an appropriate electrical equivalent was an essential step in the elaboration of the conclusions. Membrane mechanisms for the excitatory and inhibitory postsynaptic potentials were then proposed. Studies on large neurons from invertebrates contributed to our understanding of some apparently contradictory effects of CO_2, one of the most potent regulators of nervous function in mammals. However, despite the good deal of present knowledge on the basic mechanism of synaptic transmission, much remains to be explored in the field of integration of functions, elaboration of inputs, and storage of information at the cellular level. Here again, electrical measurements will complement the knowledge obtained in other scientific domains.

3

Impedance Measurements in Muscular Tissue

Impedance measurements in muscular tissue are characterized by a great variety of approaches according to the property under investigation and the particular features of each type of muscle. Throughout this chapter the main differences between the three types of muscle are emphasized, especially with respect to the interpretation of the input impedance and the limitations encountered in cardiac and smooth muscle when the electrical cell constants are inferred from measurements of input or apparent resistances. The recent development of voltage-clamp methods for these structures has permitted the study of membrane currents and their kinetics.

3.1 SKELETAL MUSCLE

In skeletal muscle, much of the evidence obtained with voltage-clamp methods has confirmed earlier results or theoretical predictions based on measurements performed with ac signals or long-duration current pulses, thus illustrating the relevance of an adequate electrical model. The lack of an adequate model and technique for smooth and cardiac muscle has prevented a parallel theoretical development.

3.1.1 AC Measurements on Skeletal Muscle

Skeletal muscle is one of the oldest structures used in impedance measurements, and the evolution of the different measuring methods is reflected in these studies. The four different trends that underlie ac measurements on this structure will be treated as follows: 3.1.1.1, systematic studies of the muscle impedance; 3.1.1.2, relation between impedance and morphological structure; 3.1.1.3, influence of biological variables on the impedance; and 3.1.1.4, methodological studies, where the skeletal muscle was chosen as a convenient preparation.

3.1.1.1 Systematic Studies of the Impedance of Skeletal Muscle. Hartree and Hill (1921) measured the resistivity of frog sartorius muscle (killed by chloroform vapor and resting in air or oxygen) at temperatures between 0 and 20°C. At 18°C, the conductivity of the muscle was 5.8 mmho/cm. This value is the same as that of a 0.36% NaCl solution. The low muscle conductivity was ascribed to membranes and other structures that hinder the diffusion of ions. The values reported (Table 3.1) are rather low compared with conductivities measured under comparable conditions in skeletal muscles (Geddes and Baker, 1967), but Höber (1913) found similar values for the cytoplasmic resistivity (165–330 Ω cm) of the frog gastrocnemius. In contrast to this, Buchthal (1934) found a resistivity of 3000 Ω cm. Reviews of systematic studies of muscle impedance can be found in Gildemeister (1928), Cole and Curtis (1950), and Cole (1968). Philippson (1921) measured the impedance of guinea pig muscle and proposed the model shown as 2 in Table 1.5. The series resistance was ascribed to the conductivity of the cytoplasm, and the parallel circuit to the properties of the membrane. The capacity changed with frequency according to Eq. 5.14. There was no change in the power factor ($\alpha = 0.41$) or in the series resistance, but C'_{pol} increased, and the parallel resistance decreased 1 hr after death. Fricke (1931) found that results from a rabbit muscle could be interpreted as a membrane impedance consisting of a frequency-dependent resistance and capacity with a constant phase angle of 65° (Sec. 1.5.2.2). At the same time, Cole (1932, 1933) reported that the phase angle was 71° in cat diaphragm, much lower than that found in suspensions of blood cells (See Chapter 5).

Bozler and Cole (1935) extended these investigations to the frog sartorius muscle in rigor. Phase angle and infinite frequency resistance (determined at frequencies between 1 kHz and 1 MHz) remained unchanged, indicating that the cytoplasmic resistivity does not change during rigor. The suspension equation (Table 1.6) was used to estimate the cytoplasmic resistivity. The volume concentration S' was estimated from measurements of R^* performed in normal Ringer's solution, 50% Ringer's-50% isosmotic sucrose, and 25% Ringer's-75% isosmotic sucrose. A value of 553–564 Ω cm was found for the cytoplasmic resistivity.

Cole and Curtis (1936) published an impedance locus for the frog sartorius with a phase angle of 47°, variable between 45 and 63°. Measurements of the longitudinal impedance locus at varying electrode distances showed that R_0 and R_∞ increased linearly with distance, whereas the maximum reactance leveled off at an interelectrode distance of about

Table 3.1 Results of AC Measurements in Skeletal Muscle

Method	Measured Quantities	Model	Derived Quantities	Reference
Voltage–current	Transverse impedance	Eq. A1.1	Resistivity of muscle: 0°C 290 Ω cm 5°C 233 Ω cm 10°C 210 Ω cm 15°C 175 Ω cm 20°C 147 Ω cm	Hartree and Hill (1921)
5 of Table 1.1	Impedance loci	3 of Table 1.5	R_i: 253–264 Ω cm ρ': 76.3–77.3%	Bozler and Cole (1935)
Voltage–current	Impedance loci	2 and 5 of Table 1.5	At 5 kHz: C_m: 3.5 μF/cm^2 ϕ: 71°	Strickholm (1961)
Voltage–current	Impedance loci	1 of Table 1.5	At 400 Hz: R_m: 1000 Ω cm^2 C_m: 3 μF/cm^2	Strickholm (1962)
5 of Table 1.1	Impedance loci	3 of Table 1.5	C_m: 2.58 μF/cm^2 R_i: 202 Ω cm R_x: 24 kΩ cm (Fig. 3.1) C_x: 53.9 μF/cm^2 (Fig. 3.1)	Fatt (1964)
1 of Table 1.3	Impedance loci	6 of Table 1.5	Frog muscle: R_s: 3100 Ω cm^2 C_s: 2.6 μF/cm^2 R_e: 330 Ω cm^2 C_e: 4.1 μF/cm^2 Crayfish muscle: R_s: 680 Ω cm^2 C_s: 2.6 μF/cm^2 R_e: 35 Ω cm^2 C_e: 1.7 μF/cm^2	Falk and Fatt (1964)
5 of Table 1.1	Impedance loci	3 of Table 1.5	R_i: 176–310 Ω cm ϕ: 71° ρ': 75–81%	Guttman (1939)

10 mm. The phase angle remained constant, and the change in reactance was attributed to electrode interaction. However, no distinction could be made between an interpretation of these results as polarization impedance or as a statistical distribution of static capacities (Wagner, 1913). Schwan (1954) extended ac measurements down to 10 Hz and found an increase of the capacity and the specific resistance at frequencies below 1 kHz. The low-frequency limiting value for the dielectric constant was more than 10^6,

higher than the values reported for any other material at that time. The results showed that the theory of the constant phase angle did not apply to this low-frequency dispersion. Expressing the results as series resistance and reactance, a semicircle with a suppressed center was obtained for frequencies above 1 kHz with a phase angle of 75°, and another semicircle was obtained for the lower frequencies with a characteristic frequency of 70 Hz, compatible with the assumption of a dispersion having its origin in the structure of the membrane.

Another approach was chosen by Strickholm (1961, 1962) to clarify the nature of the depressed center in the impedance locus observed in frog sartorius muscle. The impedance loci from single fibers had a constant phase angle of 71° at 5 kHz. A change in the impedance locus of sartorius muscle during excitation was observed. These results permitted the exclusion of a statistical distribution of fibers with different diameters as a cause of the depressed center of the impedance locus. It was proposed that the membrane consisted of a multitude of microscopic membrane sites, each having frequency-independent resistances and capacitances.

3.1.1.2 Studies of the Relation Between Impedance and Structure. Two fundamental papers (Fatt, 1964; Falk and Fatt, 1964) stimulated a new interest in the interrelation between electrical impedance and the membrane systems of skeletal muscle. Fatt (1964) measured the parallel resistance and capacity of frog sartorius muscles between 1.5 Hz and 130 kHz. The results were plotted in the complex impedance plane and fitted to circles with their centers lying on the real axis (Fig. 3.1a). The values of the characteristic frequencies and the limiting resistances R_1, R_2, and R_3 were obtained from these loci. The analysis of the impedance of a single fiber was based on the suspension equation, where Z_m was interpreted as a two-time-constant system with the electrical configuration shown in the inset of Fig. 3.1. Here C_s and C_x are surface capacities, R_i is the cytoplasmic resistivity, R_x is a bulk resistivity in series with C_x, and a is the radius of the fiber. The aforementioned quantities and also the volume concentration ρ' are related to the basic quantities of Fig. 3.1 by the following set of equations:

$$\left.\begin{aligned}
&\rho'=(R_1-R_e)/(R_1+R_e)\\
&C_x=(1/\bar{\omega}_1 a)\left[(R_1-R_2)/(R_1^2-R_e^2)\right]\\
&R_i=\frac{(R_1R_2-R_e^2)(R_1R_3-R_e^2)}{(R_2-R_3)(R_1^2-R_e^2)}\\
&R_x=(R_1R_2-R_e^2)/(R_1-R_2)\\
&C_s=(1/\bar{\omega}_2 a)\left\{\left[(R_2-R_3)(R_1^2-R_e^2)\right]/(R_1R_2-R_e^2)^2\right\}
\end{aligned}\right\}\qquad(3.1)$$

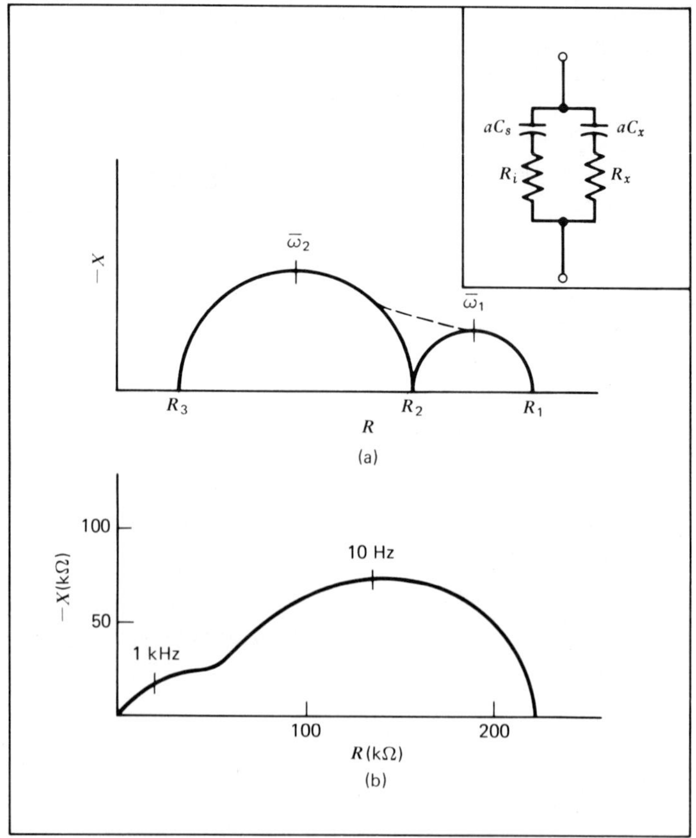

Figure 3.1. (a) Parallel resistance and reactance of frog sartorius. The characteristic frequencies $\bar{\omega}_1$ and $\bar{\omega}_2$ as well as R_1 and R_3 were easily obtained, but the extrapolation for R_2 was more difficult because the actual impedance loci followed the dashed line. Inset: interpretation of the fiber specific impedance Z_f. Modified from Fatt (1964). (b) Parallel resistance and reactance of a frog sartorius muscle in normal Ringer's solution measured with intracellular electrodes. Modified from Falk and Fatt (1964). See text for explanation. By permission.

Increasing the resistivity of the extracellular medium (R_e) enhanced the semicircle characteristic for the low-frequency dispersion. The only quantity that changed was R_x. The increase of R_x with increasing R_e depended on total ion concentration and not on the ionic species. Hypertonic solutions reduced the low-frequency dispersion and caused a decrease in volume concentration and an increase in cytoplasmic resistivity. This was explained by an increase in intracellular ion concentration when the fibers are shrinking. Hypertonic solutions with a constant $[K^+]_o \times [Cl^-]_o$ product caused a gradual decrease of C_x, indicating that the structure responsible for C_x undergoes a gradual change in hypertonic solutions. After being immersed in solutions of an ionic strength less than 20 mM and then returned to an isotonic solution, the muscle showed an irreversible loss of its low-frequency dispersion. This is called a burst muscle. Electrically, an increase of R_i and a decrease in C_s were observed together with the vanishing of R_x and C_x. The increase in R_i was explained by an increase in viscosity of the cytoplasm. The elements responsible for the low-frequency dispersion, R_x and C_x were ascribed to the T-system of the muscle. These results gave a lower value of C_s than Fatt and Katz (1951) obtained for the membrane capacity (Tables 3.1 and 3.2). This problem was further investigated by Falk and Fatt (1964). Measurements were made on frog sartorius and crayfish (*Astacus fluvialis*) muscle over a frequency range from 1 Hz to 10 kHz. Figure 3.1b shows a typical impedance locus obtained for a frog sartorius fiber. In contrast to the impedance loci obtained with extracellular electrodes (Fig. 3.1a), this locus emphasizes the low-frequency dispersion. Its analysis is based on model 5 of Table 1.5, where the transverse elements were replaced by the two-time-constant network shown in Fig. 3.2a. Figure 3.2b shows the relation between the parameters R_x, C_x, R_e, and C_e. Pugsley (1966) confirmed the results of Falk and Fatt for the frog sartorius and found for the toad sartorius only a one-time-constant equivalent circuit. The series resistance r_e was attributed to a membrane across the openings of the tubular system. In the case of the toad sartorius, this resistance was very small ($<4\ \Omega\ \text{cm}^2$). The lack of r_e in toad muscles was used to propose a hypothesis to explain the absence of the late after-potential in the toad sartorius.

This hypothesis was contested by Freygang, Rapoport, and Peachey (1967), who did not find the ultrastructure of the tubular system in the toad different from that of the frog and reported a late after-potential in the toad muscle. Their results confirmed again the results of Falk and Fatt (1964), but in hypertonic Ringer's solution the surface of the transverse tubular system increased, as well as the capacity c_e attributed to it. The resistance r_e did not fall, and it was concluded that it is probably not located in the transverse tubular system. In addition, the phase angle versus frequency diagram showed systematic deviations when it was fitted to the circuit of Fig. 3.2a (Fig. 3.3), suggesting that the actual equivalent circuit is more complex than the one proposed by Falk and Fatt (1964). Based on these findings, Falk (1968) analyzed theoretically the circuit of Fig. 3.2b. He derived three different models with the following properties: (1) The resistivity of the interior of the tubules was negligible, $R_x \to 0$; (2) the component R_e was negligible, and (3) the circuit of Fig. 3.2b was valid and none of its elements could be neglected. Considering that the

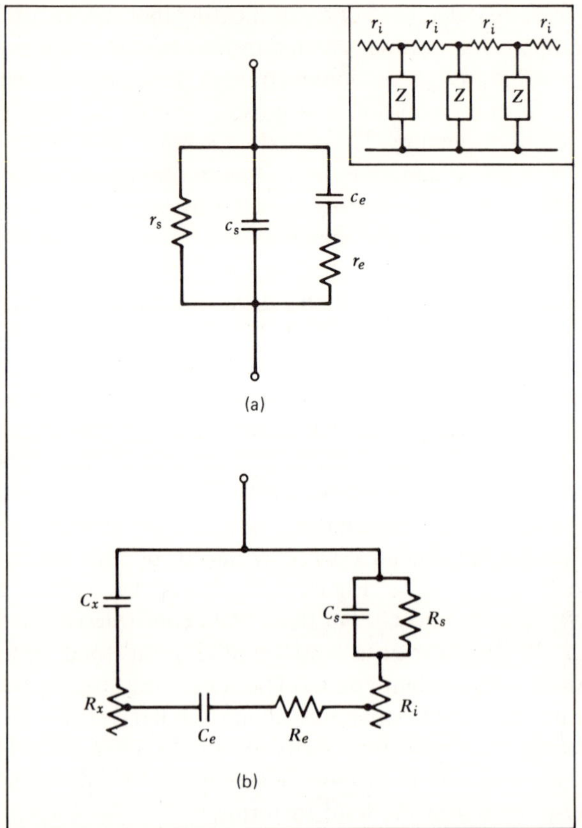

Figure 3.2. (a) Inside-outside membrane impedance of a muscle fiber as used to explain the impedance loci obtained with intracellular electrodes. Inset: generalized linear cable model where the inside-outside impedance Z was replaced by the two-time constant network shown in (a). (b) Relation between the parameters R_x, C_x obtained with extracellular electrodes and R_e, C_e measured with intracellular electrodes. Modified after Falk and Fatt (1964). By permission.

propagated action potential triggers a contraction by shifting the potential across C_e to a critical value, the relation between a disturbance (sinusoidal, exponential, and step function) of the membrane potential V_m and the time course of the voltage (V_c) across C_e was investigated. Except for model 1, a time delay between V_m and V_e across the distributed elements of C_e can be expected, which would delay the activation of the contractile system from the outermost filaments to those at the center of the muscle fiber. Gonzalez-Serratos (1966, 1967) had shown a delay of 1 ms. This finding ruled out model 1 and also model 2, for which a theoretical delay of 2.6 ms was predicted It was then assumed that model 3 most closely approxi-

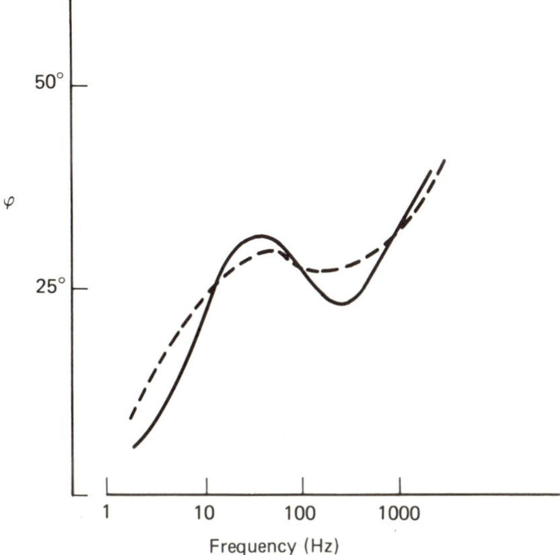

Figure 3.3. Theoretical (full line) and measured (dashed line) change of the phase angle of the admittance versus frequency plot in frog sartorius immersed in hypertonic sucrose solution. Modified after Freygang et al. (1967). By copyright permission of The Rockefeller University Press.

mated reality, with a predicted delay of 0.25 ms. Schneider (1970) performed an ac analysis of the transfer impedance of single frog sartorius muscle fibers. The results were presented as plots of impedance or phase angle as a function of frequency. Fitting the results to the lumped circuit of Fig. 3.2a resulted in a misfit. A better fit was obtained with a distributed circuit (lattice model of Adrian, Chandler, and Hodgkin, 1969) than with the lumped circuit (Table 3.4).

Eisenberg (1967) used an ac method to refine the equivalent circuit of crab muscle. The results were presented as impedance loci in the complex plane. It was already known that this muscle showed a very high capacity, about 40 $\mu F/cm^2$ (Fatt and Katz, 1953a). Measurements on nine fibers were reported. The locus of the input impedance of seven fibers could be represented by the circuit shown in Fig. 3.4: The left branch is related to the surface membrane (sarcolemma), and the right is ascribed to the tubules. This circuit is more complex than the circuits proposed for the frog sartorius. It contains one element more than the simplest possible circuit (canonical circuit*) that can describe the result obtained. This

*The simplest possible network to describe a given set of measurements is called a canonical form of circuit.

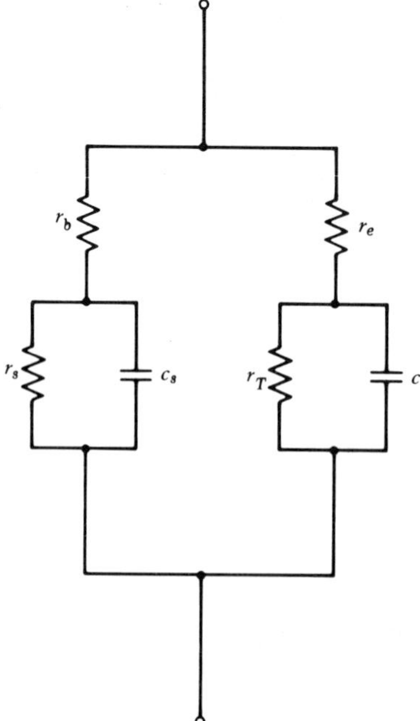

Figure 3.4. Equivalent circuit proposed for inside-outside impedance of a crab muscle fiber. This circuit replaces Z in the generalized linear cable model for microelectrodes; see inset of Fig. 3.2a. For explanation, see text. Modified from Eisenberg (1967). By copyright permission of The Rockefeller University Press.

circuit was chosen because of additional information on the ultrastructure of the muscle, furnished by the work of Peachey and Huxley (1964) and Peachey (1965a, 1967).

3.1.1.3 Influence of Biological Variables on Muscle Impedance. Most papers in this group are of only historical interest today, but many interesting findings in the older literature, not analyzable in the light of today's standards, await reinvestigation. An example is the work of Höber (1912, 1913), who claimed, based on impedance measurements, that about one-third of the intracellular potassium is bound. Although many results were obtained later by alternative approaches, this discussion has not been settled, and today the subject is still highly controversial (Troshin, 1961; Hazlewood, 1973). In the early 1930s, impedance measurements in muscle were performed in Europe and in North America, but there seems to have been an almost complete lack of communication between these groups of researchers. The following illustrates well the situation. Achelis (1932) reported an increase in the series resistance and a decrease of the capacitance in frog muscle during excitation, and these results were not considered contradictory to the hypothesized increase in membrane permeability during activity. Quensel (1932) investigated the same preparation under different conditions, and the results were ascribed to changes in membrane permeability. At that time, the suspension equation had already been successfully applied to blood cells, and it was known that it was difficult to measure R_m with Quensel's approach. So it seems that the implications of the suspension equation were unknown in Europe. This lack of communication is underlined by the fact that Bozler and Cole (1935) virtually repeated one of Quensel's experiments with monoiodoacetate without quoting the experiment of 1932.

Guttman (1939) measured the transverse impedance of frog sartorius muscles between 200 Hz and 1 MHz. The control values (Table 3.1), obtained as loci of series resistance and capacitance, were comparable to those of Bozler and Cole (1935). After 2 hours, isotonic solutions of $MgCl_2$ and KCl did not have any significant effect, whereas isotonic solutions of NaCl decreased R_p slightly and $CaCl_2$ and $BaCl_2$ decreased it drastically. These results can probably be explained by a combination of change in cytoplasmic resistivity due to the Donnan effect and a direct effect on the membrane. A group of anesthetics and surface-active substances were also investigated. R_p was found to increase and then reverse into a decrease at higher concentrations. In Ringer's solution saturated with chloroform, the series resistance in the complex impedance plane coincided with R_∞ after 16 min. This was interpreted as a destruction of low-conductance barriers, the cell membranes.

Barbashova and Moskalenko (1961) reported a shift of the β dispersion to higher frequencies in hypoxic rat muscle. An interesting result was reported by Mashima and Washio (1968), who compared voltage–current curves obtained with ac and long-duration square pulses on the frog semitendinosus muscle. There was a lower imput impedance at 800 Hz, which was explained by a decrease of the reactive component of the input impedance at this frequency.

Bromm and Simon (1971) chose the same method for a comparative study of the effect of long-duration square pulses and alternating current up to 10 kHz in the frog iliofibularis. A small depolarization of the cell membrane occurred at frequencies above 500 Hz, which favored spontaneous activity. This depolarization was related to the sodium channel because it was abolished by TTX, and it was explained as caused by a preferential activation of the sodium system by alternating current.

3.1.1.4 Methodological Studies with Skeletal Muscle as a Model. Tasaki and Hagiwara (1957a) adapted the cable theory for ac measurements. They found that the time constant of the foot of the action potential was related to the membrane capacity independent of the geometry of the structure (Eq. 1.43). They also found that the membrane capacity was constant up to 700 Hz.

Jenerick (1961, 1963) used phase plane trajectories to analyze the action potential of a frog sartorius muscle. The advantage of recording an action potential in the phase plane is that regions that appear very steep in the conventional potential-versus-time record appear expanded in the phase plane, as shown in Fig. 3.5. The trajectory resembles a compressed ellipse with three linear parts. The region of the rate of rise of the action potential is expanded. The cable equation for the propagated action potential can be expressed as:

$$\frac{1}{k_1}\frac{d^2V}{dt^2} - \frac{dV}{dt} = \frac{J_i}{C_m} \tag{3.2}$$

where $k_1 = 2R_iC_mv^2/a$ is the inverse of the time constant of the foot of the action potential τ_f (Tasaki and Hagiwara, 1957a). Equation 3.2 can be used to calculate the sodium current during the action potential assuming: (1) g_{Na} reaches and maintains a constant level soon after dV/dt has reached its maximum and (2) during this time, $g_{\mathrm{K}} < g_{\mathrm{Na}}$. Combining Eq. 3.2 with the expression for the chord conductance (Table 2.5), $\bar{g}_{\mathrm{Na}}$ can be calculated from:

$$\bar{g}_{\mathrm{Na}} = C_m \frac{k}{k_1}(k + k_1) \tag{3.3}$$

provided C_m is known. In Eq. 3.3, k is the slope of the second linear part of the phase plane trajectory. The intercept of the extrapolation of this part gives the sodium equilibrium potential, E_{Na}. With this procedure, Jenerick (1961) calculated the time course of the total ionic current during the action potential in frog sartorius muscle. The largest ionic current occurred during the upstroke and lasted less than 0.5 ms at room temperature. The net ionic influx was 7 pmol/cm^2. The inward current decreased when $[\mathrm{Na}^+]_o$ was reduced from 110 to 55 mM. This shifted E_{Na} from +37.5 to +22.5 mV, while the firing level (at −45 mV) was not affected. The value of E_{Na} in normal Ringer's solution corresponded to an intracellular sodium concentration of 20–25 mM. The effect of different buffers and changes in $[\mathrm{K}^+]_o$ were analyzed. In Tris buffer, all parameters were reduced by 20–50%, except the negative after-potential. All the changes in different $[\mathrm{K}^+]_o$ were explained as consequences of the reduced height of the action potential amplitude and not as a specific effect of potassium concentration.

Jenerick (1964) further developed the analysis of the phase plane and derived the following equation:

$$\frac{J_i}{C_m} = \frac{dV}{dt}\left(\frac{m}{k_1} - 1\right) \tag{3.4}$$

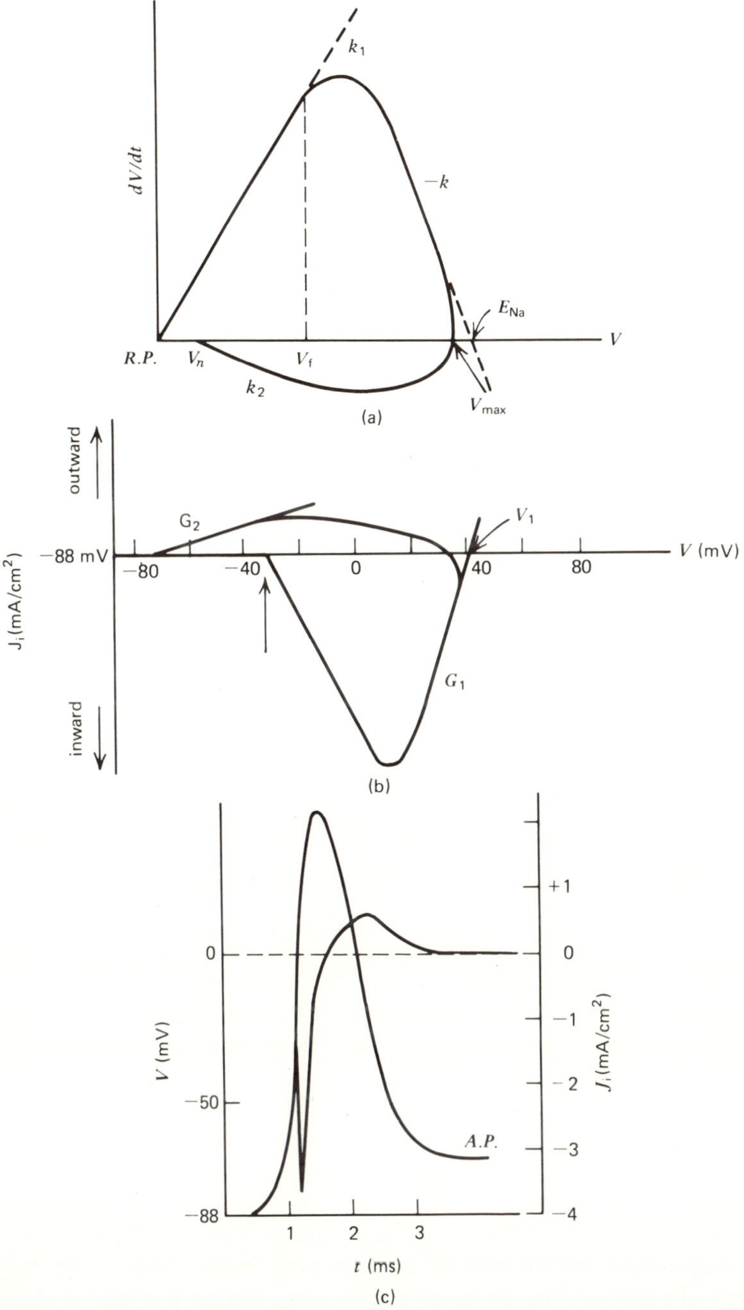

Figure 3.5. (a) Phase plane trajectory: R.P., resting potential; V_{max}, spike potential; V_n, negative after potential; V_f, firing level of the action potential. Modified after Jenerick (1963). (b) Current–voltage curve of the net ionic current during the action potential. (c) Ionic current as a function of time in relation to conventional recording of the action potential (A.P.). For explanation, see text. Modified after Jenerick (1964). By copyright permission of The Rockefeller University Press.

where $m=[k_1(dV/dt+J_i/C_m)]/(dV/dt)$. This equation permitted a point-to-point evaluation of the current density J_i from a phase plane display. From it, the current–voltage relation and the time course of the ionic current during the action potential were obtained (Fig. 3.5b, c). This plot showed that (1) the peak inward current was correlated with dV/dt max; (2) the peak outward current was closely correlated with the maximum rate of fall of the action potential; (3) the inflection point indicated by a vertical arrow in Fig. 3.5b (excitation potential) is roughly correlated with the firing level obtainable from the phase plane display; (4) V_1, which corresponds to E_{Na}, varied logarithmically with $[Na^+]_o$; (5) G_1 (maximum membrane conductance) was independent of $[Na^+]_o$; and (6) peak J_i changed linearly with $[Na^+]_o$. The quantitative results are listed in Table 3.10.

Inoue (1971) published a method to facilitate the extraction of the current–voltage curve from the phase plane trajectory, and the method was validated on the frog sartorius muscle. Assuming that C_m remains constant during the action potential, the current–voltage curve can be obtained as a difference of d^2V/dt^2 versus (V) and the phase plane trajectory, provided that a proper scale factor can be found for the actual recording conditions. This scale factor can be taken into account by the following reasoning: During the foot of the action potential the ionic current is zero. Therefore, the amplification of the channel recording the second derivative is chosen so that the initial trajectories of the first and second derivative are equal. The method of phase plane trajectories has also been used by Morlock, Benady, and Grundfest (1968) in eel electroplaques and by Paes de Carvalho et al. (1969) in cardiac muscle.

3.1.2 Square-Pulse Measurements on Skeletal Muscle

Square-pulse measurements of skeletal muscle were performed mainly by researchers whose prime interest was the study of bioelectric phenomena. The discussion in this section has been divided according to the different approaches used: 3.1.2.1, measurements of the electrical cell constants; 3.1.2.2, the rectifier properties of the muscle membrane; 3.1.2.3, estimations of the ionic permeabilities, and 3.1.2.4, the effects of biological variables. The frog sartorius muscle was the most widely used structure, and circuits 1 of Tables 1.1 and 1.3 were generally applied. The electrical cell constants were estimated, applying models 4 and 5 of Table 1.5. For this analysis, four measurements must be made on the same fiber: R_{inp}, λ, d, and τ. In general, no more than a few penetrations can be made in the same cell, and these four basic quantities are obtained from plots like those of Fig. 3.6a and b. However, it will be seen that several assumptions, short-cuts, or indirect methods of estimating a variable have been used when R_{inp}, λ, d, and τ could not all be determined in the same cell.

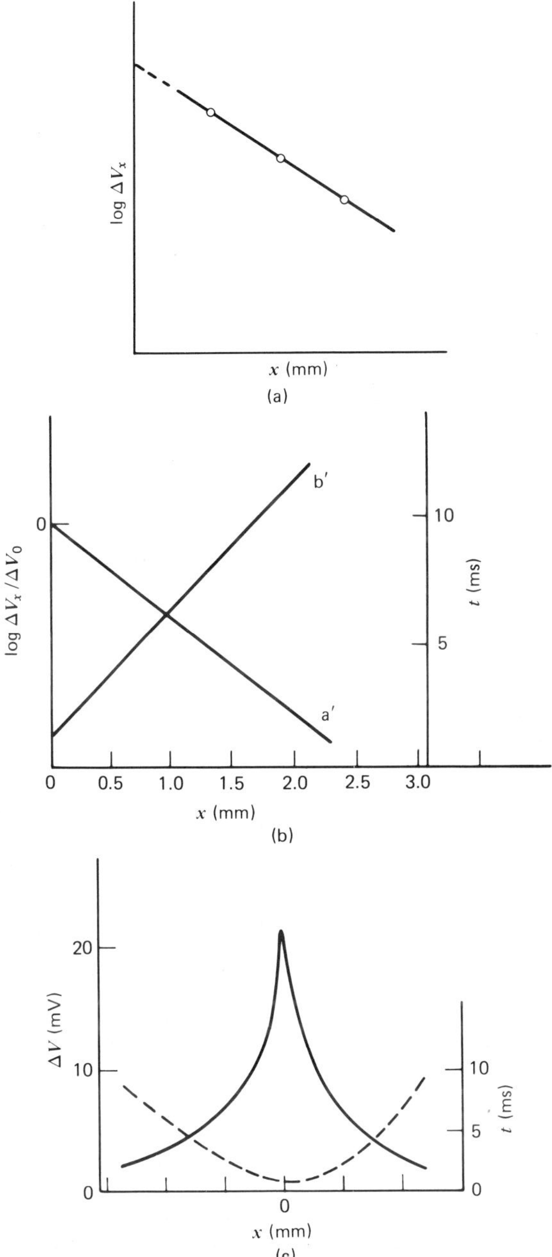

Figure 3.6. Plots currently used to determine the input resistance and the length constant from measurements of the electrotonic decay as a function of interelectrode distance (x). (a) Logarithm of the electrotonic potential versus x. (b) Determination of λ and τ_{inp} for a bundle of four muscle fibers. Curve a', spatial decay of the electrotonic potential $\log(\Delta V_x / \Delta V_0)$ versus x, slope λ. Curve b', half-time of the electrotonic potential versus x, slope τ_{inp}. (c) Spread of epp along a muscle fiber. Full line, peak amplitude; dashed line, time of rise of epp. At $x = 0$, region of endplate.

3.1.2.1 Electrical Cell Constants of Skeletal Muscle. Katz (1948) measured the cell constants in small bundles and whole muscles of the frog *Rana temporaria*. Voltage–current curves were linear for small depolarizing currents and in a wider range for hyperpolarizing currents, from which the electrical cell constants for the resting membrane were estimated. The specific resistance of the myoplasm was 156 Ω cm in 9 small bundles and 157 Ω cm in 14 extensor muscles at 22°C. Table 3.2 lists the results from all the experiments. Fatt and Katz (1951) determined the cell constants of frog sartorius muscle with square-pulse analysis and applied the cable theory to endplate potentials (epp) measured in curarized nerve-muscle preparations. Figure 3.6c shows the spatial decline of the epp along the fiber. This potential also decayed exponentially with a time constant equal to the membrane time constant determined by the square-pulse analysis. It was calculated that the endplate charge reached a maximum within 1.5–3 ms and then decayed exponentially with a time constant of 20–27 ms. This finding indicated that the rising phase of the epp was caused by the interaction of the transmitter with the membrane, and that the decay of the epp was determined by the cable properties of the fiber. The cell constants calculated from the decay of the epp ($R_m = 3300$–4100 Ω cm^2, and $C_m =$ 6–7 μF/cm^2) agree with those determined with the direct method (Table 3.2). Changes in membrane resistance during activity were estimated by applying short square pulses during the action potential. Two minima in membrane resistance were detected, one during the rise and the other during the fall of the action potential. The input resistance measured during the falling phase of the spike was 21 kΩ.

The cell constants of crab muscle (*Portunus depurator* and *Carcinus maenas*) were determined by Fatt and Katz (1953a), who found that crustacean muscle fibers have a large membrane capacity and a low membrane resistance (Table 3.3). Increasing the temperature from 2 to 17°C increased the resting potential and decreased the membrane resistance and myoplasmic resistivity, giving Q_{10} values of 2.5 and 1.2 for R_m and R_i, respectively. The Q_{10} for myoplasmic resistivity did not differ markedly from that of ordinary electrolytes. Following large depolarizations, two types of responses were observed: local spikes of few millivolts and action potentials that were large enough to propagate along the fiber. Conduction velocity was 29 cm/s; a crab axon 30 μm in diameter conducts at 450 cm/s (Hodgkin, 1947). The cytoplasmic resistivity is about the same for crab muscle and the axon, so that the low conduction velocity in the former is consistent with its large membrane capacity. The electrical activity and membrane properties of crayfish muscle (*Astacus fluvialis*) were further investigated by Fatt and Ginsborg (1958). Voltage–current curves were linear within a ΔV of ± 20 mV. A strong depolarizing pulse produced a local depolarization superimposed on the electrotonic poten-

Table 3.2 Electrical Cell Constants of Frog Muscle

Method	Measured Quantities	Model	Derived Quantities	Reference
1 of Table 1.1	V/I curve λ: 0.65 mm τ_{inp}: 9 ms	4 of Table 1.5	Bundles of 2–4 fibers: R_m: 1500 Ω cm^2 R_i: 188 Ω cm C_m: 6 μF/cm^2	Katz (1948)
	V/I curve λ: 1.1 mm τ_{inp}: 18.5 ms		Muscle extensor longus: R_m: 4300 Ω cm^2 R_i: 255 Ω cm C_m: 4.4 μF/cm^2	
1 of Table 1.3	R_{inp}: 204 kΩ λ: 24 mm τ_{inp}: 34.5 ms	5 of Table 1.5	R_m: 4100 Ω cm^2 C_m: 8 μF/cm^2 d: 137 μm R_i: 250 Ω cm (assumed)	Fatt and Katz (1951)
Fig. 2.11	R_{eff} vs x	Eq. 1.37	T: 18°C R_m: 9400 Ω cm^2 R_i: 130 Ω cm λ: 1.1 mm (in air) λ: 4.4 mm (in Ringer's)	Tamasige (1950)
1 of Table 1.3	R_{eff}: 294 Ω λ: 1.18 mm τ_{inp}: 15.4 ms d: 80 μm	5 of Table 1.5	R_m: 1740 Ω cm^2 C_m: 8.8 μF/cm^2	Portela et al. (1970)
1 of Table 1.3	R_{inp}: 149–192 kΩ λ: 1.4–1.8 mm τ_{inp}: 19–28 ms	5 of Table 1.5	At T: 19–24°C: d: 116–139 μm R_i: 220–247 Ω cm (assumed) R_m: 1628–2415 Ω cm^2 C_m: 8.7–12 μF/cm^2	DelCastillo and Machne (1953)

Table 3.2 *Continued*

Method	Measured Quantities	Model	Derived Quantities	Reference
	R_{inp}: 201–291 kΩ λ: 1.6–2 mm τ_{inp}: 31–35ms		At T: 1.2–4.2°C: d: 123–153 μm R_i: 370–390 Ω cm R_m: 3410–4193 Ω cm^2 C_m: 7.6–11.3 μF/cm^2	
1 of Table 1.3	R_{inp}: 304 kΩ	5 of Table 1.5	$[K]_o$: 2.5 mM R_m: 2500 Ω cm^2 λ: 1.54 mm	Jenerick (1953)
	R_{inp}: 80 kΩ		$[K]_o$: 110 mM R_m: 170 Ω cm^2 λ: 0.4 mm	
1 of Table 1.3	λ from log $\Delta V_x/\Delta V_0$ vs x R_{inp}: 167–395 kΩ λ: 2.2–5.2 mm	5 of Table 1.5	$[K]_o$: 2.5 mM: R_m: 1880–7200 Ω cm^2 R_i: 41–139 Ω cm C_m: 2.2–7.6 μF/cm^2	Maiskii (1963)
	R_{inp}: 81–250 kΩ λ: 1.2–2.9 mm		$[K]_o$: 20 mM: R_m: 700–2100 Ω cm^2 R_i: 46.5–152 Ω cm C_m: 3.7–9.5 μF/cm^2	
	R_{inp}: 47–101 kΩ λ: 1.2–2.1 mm		$[K]_o$: 100mM: R_m: 358–775 Ω cm^2 R_i: 22.5–67.9 Ω cm	
4 of Table 1.3	R_{app}: 3.04 MΩ R_{inp}: 437 kΩ		$[K]_o$: 2.5 mM: R_m: 3850 Ω cm^2	Schanne, Kawata, Schäfer, and Lavallée (1966)

	R_{inp}: 79.4 kΩ	5 of Table 1.5	$[K]_o$: 100 mM: R_m: 313 Ω cm² R_i: 100.6 Ω cm	
2 of Table 1.3	R_{inp}: not reported R_i: 246 Ω cm	5 of Table 1.5	R_m: 4430 Ω cm²	Schanne and Ceretti (1965)
1 of Table 1.3	R_{inp}: 430–730 kΩ λ: 1.74–2.42 mm R_{inp}: 370–910 kΩ λ: 1.35–2.2 mm	5 of Table 1.5	Fibers with intact tubules R_m: 3858 Ω cm² C_m: 6.1 μF/cm² R_i: 133 Ω cm Detubulated fibers: R_m: 3759 Ω cm² C_m: 2.24 μF/cm² R_i: 192 Ω cm²	Eisenberg and Gage (1967)
1 of Table 1.3	R_{inp}, λ, τ_{inp}: not reported	5 of Table 1.5	R_i: 169 Ω cm (20°C) R_i: 299 Ω cm (2°C) C_m: 4.6 μF/cm², R_m: 4760 Ω cm² — d: 50 μm C_m: 6.1 μF/cm², R_m: 3030 Ω cm² — d: 80 μm C_m: 8.5 μF/cm², R_m: 2700 Ω cm² — d: 130 μm	Hodgkin and Nakajima (1972a)

tial, with rapid changes of potential like a damped oscillation. The first peak, which developed with an increasing rate of rise, was taken as evidence of a regenerative response. The very large length constant of some fibers made it impossible to detect the decay of the electrotonic potential over the length of the fiber, so the membrane constants were estimated for fibers with a small λ (Table 3.3). Increasing $[K^+]_o$ and removing Ca^{2+} reduced the membrane resistance. This indicated a strong participation of K^+ and Ca^{2+} in the resting membrane conductance. No signs of regenerative responses were observed in citrate solutions or calcium-free media, but removal of Na^+ did not change the voltage–current characteristics or abolish regenerative responses. It was concluded that entry of Ca^{2+} during depolarization was responsible for the production of the action potentials, which could also be elicited in the presence of Sr^{2+} or Ba^{2+}.

The effect of temperature on crustacean muscle (Fatt and Katz, 1953a) showed clearly that the myoplasmic resistivity was not a constant. This observation was extended to frog muscle by Tamasige (1950), applying the method of Cole and Hodgkin (1939) to single fibers dissected from the frog semitendinosus. The value of R_i (Table 3.2) appears reliable because the geometry of the preparation was carefully controlled and the method used makes it possible to obtain this value directly. Figure 3.7 shows that R_i and R_m decreased with increasing temperature, and the change in myoplasmic resistivity was not that of an electrolyte solution (curve *c* of Fig. 3.7). This finding is in disagreement with the reports of Hartree and Hill (1921) for frog muscle and Fatt and Katz (1953a) for crustacean muscle. Del Castillo and Machne (1953) estimated a Q_{10} of 1.35 for R_m, assuming that R_i changed with temperature as a saline solution of the same ionic concentration as the extracellular medium. However, it seems that R_m changed more than estimated, because there was an increase in λ with low temperatures that would not have occurred from proportionally equal changes in R_m and R_i.

Jenerick (1953) estimated changes in R_m upon variations of $[K^+]_o$ assuming that R_i remained constant, but Maiskii (1963) reported that R_m and R_i changed with the potassium concentration in frog sartorius (Fig. 3.8a). Voltage–current curves were linear from half threshold current strength (less than $0.5–1\times10^{-7}$ A) up to 2×10^{-7} A in the hyperpolarizing side. This is a characteristic of phasic fibers, which, in contrast to tonic fibers, present a rather wide range of linearity in the voltage–current curve (Portela et al., 1970). Table 3.2 shows that the change in input resistance is due to changes in membrane resistance and cytoplasmic resistivity. Statistical analysis showed that R_{inp} and λ decreased in 20 mM potassium as a consequence of the change in membrane resistance, whereas at 100 mM $[K^+]_o$ a concomitant decrease in cytoplasmic resistivity contributed to lower R_{inp} further and λ did not change as compared to its value in 20 mM.

Table 3.3 Electrical Constants of Muscle Fibers of Species Other than the Frog (Method: Circuit 1 of Table 1.3, Model 5 of Table 1.5)

Measured Quantities	Derived Quantities	Reference
R_{inp}: 10 kΩ λ: 0.9 mm τ_{inp}: 4.6 ms	R_m: 116 Ω cm² R_i: 69 Ω cm C_m: 42 μF/cm²	Fatt and Katz (1953a); crab
R_{inp}: 150 kΩ λ: not reported R_{inp}: 11–20 kΩ	R_m: 1000 Ω cm², R_i: 125 Ω cm, C_m: 20 μF/cm² (short λ) R_m: up to 5000 Ω cm² (long λ)	Fatt and Ginsborg (1958); crayfish
R_{inp}: 0.9–1.3 MΩ τ_{inp}: 33.6 ms	λ: 1.19–1.85 mm R_m: 3850–6140 Ω cm² C_m: 4.9–8.5 μF/cm² d: 20.2–28.2 μm	Levine (1966); tortoise
R_{inp}: 320–1440 kΩ λ: 0.5–0.75 mm τ_{inp}: 1.2–2.1 ms	At T: 22°C: d: 36–52 μm R_m: 506–872 Ω cm² C_m: 2.1–3.6 μF/cm²	Boyd and Martin (1959); cat:
R_{inp}: 650–700 kΩ λ: 1.0–1.3 mm τ_{inp}: 4.8–5.6 ms	At T: 37°C: d: 34–52 μm R_m: 1220–1620 Ω cm² C_m: 3–3.9 μF/cm²	
R_{inp}: 357 kΩ λ: 0.57 mm τ_{inp}: 3.1 ms	R_i: 176 Ω cm: radius: 21 μm R_m: 542 Ω cm² C_m: 5.9 μF/cm²	Kiyohara and Sato (1967); rat
R_{inp}: 621 kΩ λ: 0.46 mm τ_{inp}: 2 ms	R_i: 194 Ω cm: radius: 15 μm R_m: 540 Ω cm² C_m: 3.8 μF/cm²	
R_{eff}: 310–640 kΩ λ: 0.42–0.44 mm τ_{inp}: 1.1–1.7 ms	d: 20–50 μm: R_m: 351–464 Ω cm² C_m: 2.54–3.82 μF/cm²	Zolovick et al. (1970); rat
R_{eff}: 210–430 kΩ λ: 0.5–0.61 mm τ_{inp}: 1.2–1.9 ms	d: 34–60 μm R_m: 360–569 Ω cm² C_m: 3.1–3.85 μF/cm²	
R_{inp}: 310–640 kΩ λ: 1.75–2.65 mm τ_{inp}: 14–23 ms	R_i: 125 Ω cm: d: 48–82 μm R_m: 3200–5100 Ω cm² C_m: 3.3–6.4 μF/cm²	Elmquist et al. (1960); human muscle

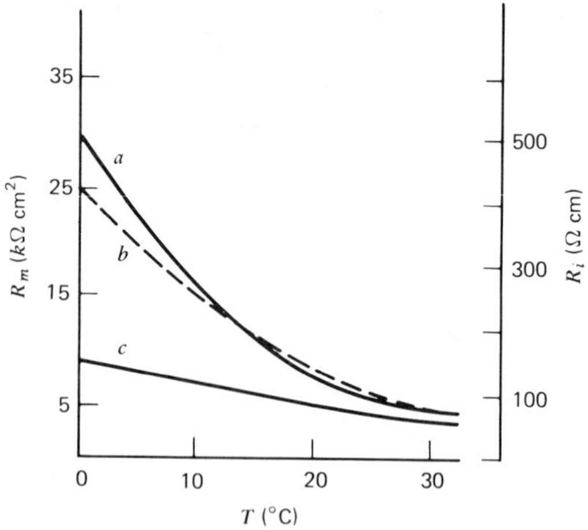

Figure 3.7. Changes in R_i (curve *a*) and R_m (curve *b*) with temperature. Curve *c*, resistivity of Ringer's solution. Modified from Tamasige (1950). By permission.

Schanne, Kawata, Schäfer, and Lavallée (1966) confirmed these findings by measuring the input resistance of frog sartorius fibers together with the apparent resistance of the cell. As discussed in Chapter 1, the apparent resistance in this experimental situation is formed by two components, the input resistance of the cell in series with the resistance of the electrode that changes upon penetration. This property of the microelectrode has been used to estimate directly the cytoplasmic resistivity (Schanne, 1969). Since the difference between the input resistance and the apparent resistance is a function of the myoplasmic resistivity, it is evident from Fig. 3.8b that R_i decreased at high potassium concentrations. The value of R_i for $[K^+]_o =$ 100 mM (Table 3.2) was estimated with an equation developed by Zablow (in Amatniek, 1958). The findings of Schanne, Kawata, Schäfer, and Lavallée (1966) indicated that high values for R_{inp} of frog sartorius reported in the literature (Sperelakis, Hoshiko, and Berne, 1960) were actually apparent resistances. The property of a microelectrode to change its resistance as a function of the resistivity of the medium was used by Schanne and Ceretti (1965) to measure the input resistance and cytoplasmic resistivity with a double-barreled microelectrode. The value for R_i thus obtained (246 Ω cm) agreed with the classical value reported by Katz (1948).

Falk and Fatt (1964) also used long-duration square pulses to test their two-time-constant model for the muscle membrane. For times longer than 1.3 ms, the theoretical curve for the two-time-constant model fit very well with the experimental curve. A good fit was also found for the one-time-constant model for times longer than 15 ms, because the rise of the

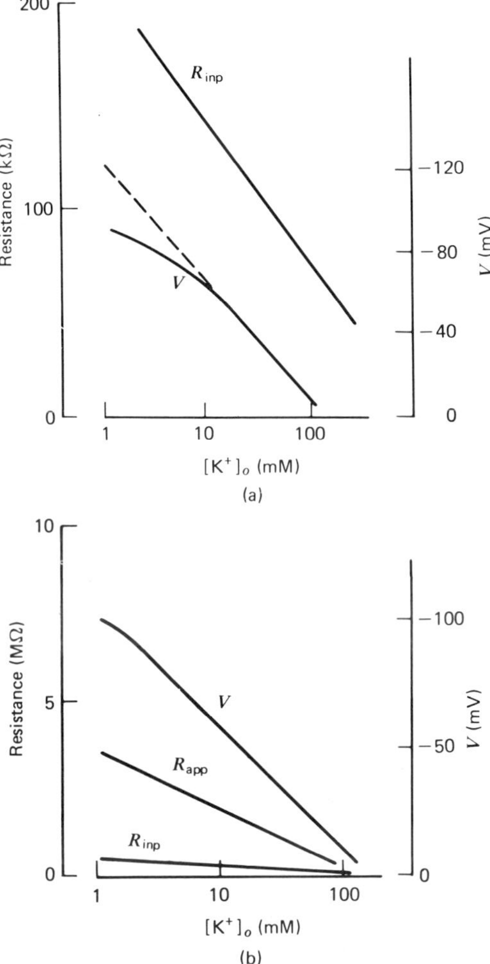

Figure 3.8. (a) Changes in input resistance R_{inp} and membrane potential (V) with potassium concentration in a muscle fiber. Modified from Maiskii (1963), by permission. (b) Changes in input resistante (R_{inp}), apparent resistance (R_{app}), and membrane potential (V) with potassium concentration. Means of several fibers. Modified from Schanne, Kawata, Schäfer, and Lavallée (1966), by copyright permission of The Rockefeller University Press.

electrotonic potential beyond its half-time approaches that of the single-time-constant model. Therefore, the values of τ_{inp} obtained by Katz (1948) and Fatt and Katz (1951) would correspond to the combined input time constant: $\tau_{\text{inp}} = (C_e + C_s)R_s$ (Fig. 3.2a). Adrian and Freygang (1962a) had already suggested that the high membrane capacity in muscle represented the capacity of the surface membrane in parallel with the capacity of the transverse tubular membranes. Moreover, the ratio of the tubular membrane area to the surface membrane area agreed well with the ratio of the two capacity components (Peachey, 1965b; Peachey and Schild, 1968). Howell and Jenden (1967) showed that the transverse tubules were absent from muscles that had been soaked in Ringer's solution with 400 mM glycerol for 1 hour and then returned to normal Ringer's medium. Eisenberg and Eisenberg (1968) applied this method and used horseradish peroxidase as an extracellular marker. They obtained a good estimation of the effective tubular membrane area that would account for the second capacitive component found by Falk and Fatt (1964). Eisenberg and Gage (1967) and Gage and Eisenberg (1969) measured the electrical cell constants in normal muscles and in muscles where the sarcotubular system had been disrupted by the glycerol treatment, and separated the capacitances of the surface and tubular membranes. Table 3.2 shows that the membrane resistance and cytoplasmic resistivity of glycerol-treated muscles are within the range of values reported for normal muscles. The membrane capacity was reduced by one-third in the muscle with disrupted tubular system, while it had normal values in the muscle immersed in glycerol-Ringer's solution. Since glycerol alone does not change the membrane properties, the structural localization of the two capacitive components of the muscle fiber was confirmed. Moreover, the agreement between the capacity values reported by Fatt and Katz (1964) and Gage and Eisenberg (1969) is striking. If the value for the surface capacitance is corrected for the 2% of tubules remaining intact after glycerol treatment, one obtains 2.1 $\mu F/cm^2$.

Nakajima and Hodgkin (1970) and Hodgkin and Nakajima (1972a, b) evaluated the contribution of the transverse tubular system to the capacity and resistance measured at the surface membrane. Provided that the current spreads into the middle of the fibers, the tubular system contributes to the measured resistance and capacity, and these quantities should then be related to the fiber diameter. The experiments were performed on isolated twitch fibers of sartorius and semitendinosus muscle of *Rana temporaria* at room temperature and at 2–4°C. The Q_{10} values for G_i, G_m, and C_m were 1.35, 1.39, and 1.05, respectively.

The model developed by Adrian et al. (1969), described in Sec. 3.1.3, predicts that the combined membrane conductance and capacity per unit area of fiber surface (G_m and C_m) in fibers of diameter less than 100 μm should vary linearly with the fiber diameter d, according to:

$$\left.\begin{aligned} G_m &= G_s + \tfrac{1}{2}a\overline{G}_w \\ C_m &= C_s + \tfrac{1}{2}a\overline{C}_w \end{aligned}\right\} \qquad (3.5)$$

where a is the radius; G_s and C_s are the conductance and capacity, respectively, of the surface membrane; and $\bar{G}_w, \bar{C}_w$ are the mean conductance and capacity, respectively, of the tubular wall per unit volume (see Eqs. 3.19). Extrapolating the relation C_m versus d to the origin resulted in C_s equal to 1.15 μF/cm^2. The values for G_s and G_m were 112 μmho/cm^2 and 213 μmho/cm^2, respectively, for fibers less than 120 μm in diameter. The effect of fiber diameter and detubulation on the capacity determined from the foot of the action potential (Eq. 1.43) was also evaluated, considering the ratio of current to voltage during the foot of the action potential as an admittance (Y_f) arising from an effective capacity (C_f) and conductance (G_f) in parallel, so that:

$$C_f = Y_f\tau_f - \tau_f G_f \tag{3.6}$$

In detubulated fibers, $G_f = G_m$, so Eq. 3.6 yields C_s. The values of τ_f and C_f (2.59 μF/cm^2) were independent of fiber diameter. Disruption of the transverse tubular system did not modify τ_f. C_f^s (see below) calculated from Eq. 3.6 averaged 0.9 μF/cm^2, and C_s measured with a current step was 1.87 μF/cm^2. The results are shown in Fig. 3.9. The combined membrane capacity C_m was calculated from:

$$C_m = \left[C_s + \left(a\bar{C}_w/2\right)\right]\left\{1 - \left[I_1(a/\lambda_T)/I_0(a/\lambda_T)\right]^2\right\} \tag{3.7}$$

where the symbols have the same meaning as in Eqs. 3.5 and 3.19 (see page 193). In detubulated fibers, the admittance Y_f for an exponentially rising voltage is given by $Y_f = C_f^s/\tau_f + G_s$. This relation was used to obtain C_f^s, which, as expected, is independent of fiber diameter. The curve C_f was calculated from the relation $C_f = C_f^s + Y_t\tau_f$, where Y_t is the admittance of the tubular system during the foot of the action potential and is obtained from Eq. 3.19. It was found that only the outer zones of the tubular network contribute to the effective capacity, because the length constant of the tubules for an exponentially rising voltage was 6.5 μm. That explains why C_f is fairly independent of the fiber diameter. The relationship between G_m and d was calculated from the following expression, also obtained from Eq. 3.19:

$$G_m = G_s + \bar{G}_w(a/2)(2\lambda_T/2)\left[I_1(a/\lambda_T)/I_0(a/\lambda_T)\right] \tag{3.8}$$

with values of $G_s = 0.11$ mmho/cm^2, $G_w = 0.03$ mmho/cm^2, and $\lambda_T =$ 100 μm. The conclusion that G_w is lower than G_s agrees with the conclusion of Eisenberg and Gage (1969) that a large fraction of the membrane permeability (mainly chloride permeability) is located in the surface membrane.

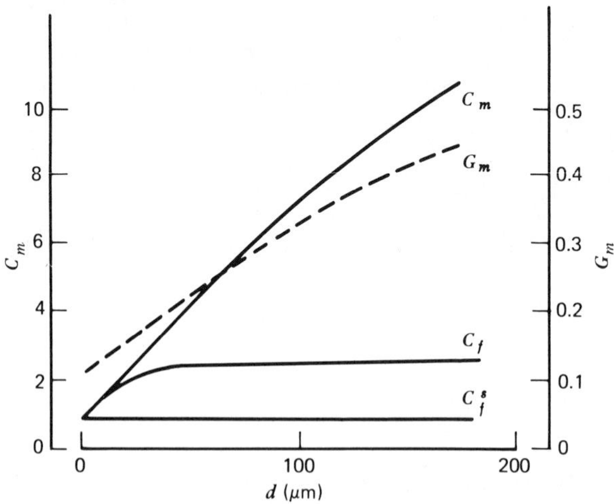

Figure 3.9. Membrane conductance and capacity as a function of fiber diameter. C_m, combined membrance capacity; G_m, combined membrane conductance; C_f, capacity measured from the foot of the action potential in normal fibers; C_f^s, capacity measured from the foot of the action potential in detubulated fibers. Scale at left ordinate, capacity in μF/cm²; scale at right, conductance in mmho/cm². Modified from Hodgkin and Nakajima (1972b), by permission.

The contribution of the tubular system to the capacity and conductance measured at the surface membrane was confirmed by this study. Table 3.4 lists the electrical constants for the surface and tubular membranes as reported by several investigators using different methods. To facilitate comparison, the symbols used by Adrian, Chandler, and Hodgkin (1969) were adopted for this table. The subscripts T (tubular) and s (surface) are equivalent to the subscripts e and m used by Falk and Fatt (1964). Some of the original data have been recalculated and converted to other units. The combined membrane resistance and capacity are given by

$$\left.\begin{aligned} C_m &= C_s + C_T \\ R_m &= R_s R_T/(R_s + R_T) \end{aligned}\right\} \tag{3.9}$$

where all the values are expressed per unit area of fiber surface. The subscript w is used for constants of the tubular wall expressed per unit area of tubular membrane.

Differences in membrane capacity between various types of muscle fibers can reflect a difference in structure of the surface membrane or in the development of the sarcotubular system. This system is poorly developed in the slow or tonic fibers of frog muscle. Adrian and Peachey (1965)

Table 3.4 Electrical Constants of Surface and Tubular Membranes of Frog Twitch Fibers*

C_m	C_s	C_T	C_w	C_f	G_m	G_s	G_T	G_w	Ref.†
6.70	2.60	4.10			0.32		3.03		1
	1.61	2.73				0.50	3.82		2a
	0.92		0.96			0.25		0.06	2b
6.10	2.24	3.86	1.0		0.30	0.25	0.04		3
6.10	1.15	4.90	0.9	0.9	0.33	0.11	0.22	0.03	4
6.76	1.0	5.76			0.34	0.50	0.29	0.05	5

*Values of capacities in $\mu F/cm^2$, and conductances in $mmho/cm^2$.
†References: 1, Falk and Fatt (1964); 2, Schneider (1970), values estimated with lumped model in 2a and with distributed model in 2b; 3, Eisenberg and Gage (1969), Gage and Eisenberg (1969); 4, Hodgkin and Nakajima (1972a, b); 5, Adrian, Chandler, and Hodgkin (1969).

measured the electrical properties in the tonus bundle from the frog iliofibularis muscle. The membrane resistance of slow fibers was about nine times as large as the membrane resistance of the twitch fibers (Table 3.5). Since the ratio of the time constants was only 2, it was concluded that the membrane capacity of a slow fiber was comparable to that reported for the nerve membrane. Because of the pronounced nonlinearity of the current–voltage curve close to the resting potential, it is preferable to calculate C_m from a plot of ΔV versus $t^{1/2}$, which gives an estimation of C_m independent of the contribution of the membrane resistance to the time constant (Hodgkin and Rushton, 1946). The low values of the lumped membrane capacity are consistent with a poorly developed tubular system.

These properties of slow muscle fibers were confirmed by Stefani and Steinbach (1969). The values of the cell constants for twitch and slow fibers are listed in Table 3.5. The mean diameter did not differ significantly from the diameter calculated assuming 250 Ω cm for R_i. Measurements of resting potential as a function of $[K^+]_o$ fitted the Goldman equation for a permeability ratio P_{Na}/P_K of 0.02. The resting potassium permeability was estimated as 1×10^{-7} cm/s, five to ten times lower than in twitch fibers. Anomalous rectification was not observed in high external potassium or normal medium, as shown in Fig. 3.10, and in agreement with previous reports of Oomura and Tomita (1960). A large increase in effective resistance was observed when the calcium concentration was raised.

Burke and Ginsborg (1956a) showed that the complications of determining the time constant from the electrotonic potential in fibers with strong rectifying properties can be avoided if measurements are made from the decay of the nonpropagated depolarization following nerve stimulation.

Table 3.5 Cell Constants of Twitch and Slow Muscle Fibers (Method: Circuit 1 of Table 1.3, Model 5 of Table 1.5)

Measured Quantities	Derived Quantities	Reference
	Twitch fibers:	Adrian and Peachey (1965)
R_{eff}: 800 kΩ	R_m: 3140 Ω cm²	
τ_{inp}: 21.5 ms	C_m: 6.8 μF/cm²	
	Slow fibers:	
R_{eff}: 2.43 MΩ	R_m: 29,000 Ω cm²	
τ_{inp}: 46 ms	C_m: 1.6 μF/cm²	
	Twitch fibers:	Stefani and Steinbach (1969)
R_{eff}: 575 kΩ	R_m: 5430 Ω cm²	
λ: 2.03 mm	C_m: 4.14 μF/cm²	
τ_{inp}: 22.2 ms		
	Slow fibers:	
R_{eff}: 4.44 MΩ	R_m: 11,200 Ω cm²	
λ: 9.6 mm	C_m: 3.24 μF/cm²	
τ_{inp}: 351 ms		
	Twitch fibers:	Proske and Vaughan (1968)
R_{inp}: 300–800 kΩ	R_m: 800–4800 Ω cm²	
τ_{inp}: 6.2–40 ms	C_m: 6–9.6 μF/cm²	
	d: 55 μm	
	Slow fibers:	
R_{inp}: 0.6–2.4 MΩ	R_m: 3900–51,200 Ω cm²	
τ_{inp}: 19–54 ms	C_m: 1–3 μF/cm²	
	d: 55 μm	
R_{inp}: 13.9 kΩ	Type A:	Atwood (1963)
λ: 0.4 mm	R_m: 88.5 Ω cm²	
τ_{inp}: 4 ms	R_i: 650 Ω cm	
d: 300–800 μm	C_m: 45 μF/cm²	
	Type B:	
R_{inp}: 145 kΩ	R_m: 1117 Ω cm²	
λ: 2.39 mm	R_i: 53 Ω cm	
τ_{inp}: 41.4 ms	C_m: 36 μF/cm²	
d: 80–120 μm	Type C:	
R_{inp}: 29.8 kΩ	R_m: 282 Ω cm²	
λ: 1.55 mm	R_i: 63 Ω cm	
τ_{inp}: 15 ms	C_m: 54.5 μF/cm²	
d: 150–260 μm	Red muscle (slow):	Hidaka and Toida (1969)
R_{inp}: 1.43 MΩ	R_m: 10,900 Ω cm²	
λ: 2.1 mm	R_i: 389 Ω cm	
τ_{inp}: 26.6 ms	C_m: 2.55 μF/cm²	
	White Muscle:	
R_{inp}: 1.0 MΩ	R_m: 7000 Ω cm²	
λ: 1.9 mm	R_i: 294 Ω cm	
τ_{inp}: 48.4 ms	C_m: 7.23 μF/cm²	

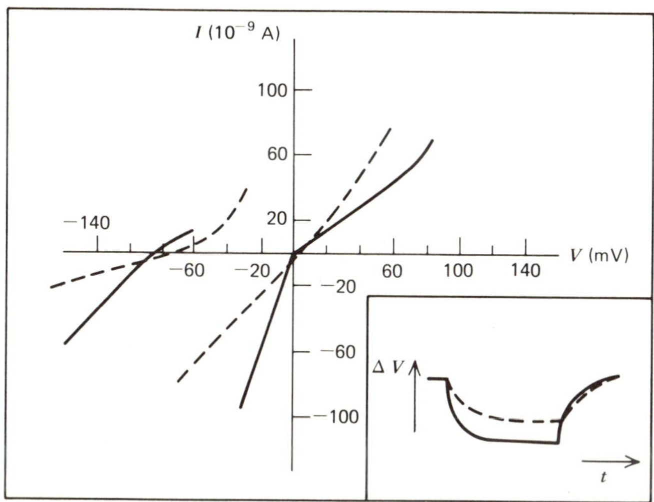

Figure 3.10. Current–voltage curves of twitch (full lines) and slow (dashed lines) muscle fibers of the frog in Ringer's (resting potential around −80 mV) and in isotonic K_2SO_4 solution (resting potential close to 0 mV). Note lack of anomalous rectification in slow fibers. Inset: time course of electrotonic potentials in both fiber types. Modified from Stefani and Steinbach (1969), by permission.

This potential is called the small junctional potential (sjp) and differs from the endplate potential in that it does not give rise to a propagated action potential and decays with a terminal phase of hyperpolarization. In addition, the innervation of the slow muscle fiber is multiple, so that nervous stimulation depolarizes the fiber rather uniformly and the sjp does not decay with distance, in contrast to an electrotonic potential. At a time equal to τ_m, the sjp declines to 37% of its maximum, while the electrotonic potential, which falls off with distance, declines to 16%. Reptile muscles present two types of innervation: a very distinctive endplate, and multiple innervation with grape-like terminations, called "en grappe." Proske and Vaughan (1968) measured the passive electrical properties of both fiber types in the scalenus muscle of the blue tongue lizard. The membrane constant for fibers that generated action potentials are comparable to those of twitch fibers of the frog. Fibers with slow-decaying junction potentials showed signs of delayed rectification; their membrane resistance was higher and their membrane capacity lower. These two histological and functional types of muscle are therefore comparable to the fast and slow fibers of the frog. The same analogy exists with red (slow) and white (fast) muscles from fishes (Hidaka and Toida, 1969), as shown in Table 3.5. The electrical properties of tortoise muscle and its sensitivity to acetylcholine were tested by Levine (1966). The electrical cell constants were of the same order of magnitude as for frog muscle (Table 3.3). Although the fibers responded to acetylcholine along the whole surface, only the endplate region was highly sensitive. It was then concluded that this muscle is composed mainly of twitch fibers.

Atwood (1963) reported the existence of three main types of fibers in closer muscles of *Carcinus maenas*. The three fiber types were distinguished on the basis of their electrical responses to indirect stimulation and were called A, B, and C. Square-pulse analysis of these fibers showed that the electrotonic potential decayed exponentially with distance in A fibers and in many type C fibers. However, type B and the remaining portion of type C fibers decayed much more slowly and did not follow an exponential function. Table 3.5 shows that the differences in input resistance are not solely attributable to the different diameters but reflect actual differences in membrane resistances. It was concluded that the various patterns of electrical responses were due to the particular electrical properties of the different fiber types.

Mammalian skeletal muscles have not been so extensively studied as amphibian muscles because of the experimental complications arising from temperature and oxygen requirements of mammalian tissue. Measurements of fiber diameter are usually not possible by direct microscopic examination, and fiber size is estimated by histological methods. Boyd and Martin (1959) measured the membrane constants in the isolated tenuissimus muscle of the cat at 37 and 22°C. Voltage–current curves were linear from a hyperpolarization of 30 mV to near threshold depolarization. The myoplasmic resistivity was taken as 250 Ω cm (as in frog sartorius) and corrected for higher ionic concentrations in mammalian muscle and for temperature. The value thus found is in agreement with that obtained by Weidmann (1952) for Purkinje fibers. The value of R_m obtained at 22°C is of the same order of magnitude as that found for rat muscle at 27°C (Kiyohara and Sato, 1967).

Mammalian muscles also present two types of fibers, which differ in the magnitudes of their resting and action potentials. Kiyohara and Sato (1967) measured comparatively the membrane properties of predominantly red (soleus) and white (extensor digitorum longus) muscles in the rat (Table 3.3). The myoplasmic resistivity was estimated as that of an electrolyte solution of the same composition as the cytoplasm. The intracellular ionic concentrations for both muscles had been determined by Yonemura (1967). The values thus obtained were 176 Ω cm for the extensor digitorum longus muscle and 194 Ω cm for the soleus. The membrane resistance was the same in both types of muscle, and the difference observed in the values of R_{inp} in rat muscle were due to different fiber diameters. The membrane capacity of white muscle was larger than that of red muscle, probably because of a better development of the transverse tubular system.

Muscle fibers from rat diaphragm also differ in their diameters, but the membrane resistance was the same in two groups of fibers with diameters ranging between 24–31 μm and 34–41 μm as shown in Table 3.3 (Zolovick et al., 1970).

Elmquist et al. (1960) determined the cell constants in bundles of human intercostal muscle obtained during surgery. The resting potential ranged from -70 to -90 mV. The values of R_m were higher than the ones reported for rat and cat muscle and comparable to the frog sartorius membrane (Table 3.3). McComas et al. (1968) determined the cell constants in limb muscles of three volunteer subjects, with circuit 5 of Table 1.3. The mean resting potential of 134 fibers was -83 mV, the action potential amplitude in 30 fibers was 104 mV, the mean value of the apparent resistance in 19 fibers was 270 kΩ, and the time constant was 6.3 ms. The input resistance of limb muscle would be much lower than that of intercostal muscle if apparent resistances were corrected for ΔR_{El}. If human muscle is homogeneous in its membrane properties, such a difference in input resistance can be explained only by a much larger fiber diameter in limb muscles.

3.1.2.2 Rectifier Properties of the Muscle Membrane. Katz (1949) reported a higher resistance for outward currents in muscles immersed in isotonic solutions of K_2SO_4, a finding that agrees with the low value of P_K for outward movement of potassium found by Hodgkin and Horowicz (1959b). The contribution of the potassium conductance to rectification was confirmed by Hutter and Noble (1960a), who reported an increase in anomalous rectification in methylsulfate solutions. Adrian and Freygang (1962a) developed a method to estimate time- and voltage-dependent conductances, using three microelectrodes impaled near the end of the fiber, as in the experimental arrangement of method 8 of Table 1.4, but P_1 and P_2 are voltage-recording electrodes and CE is connected to a square-pulse generator. The voltage drop (ΔV) at P_1 and the difference of ΔV between P_1 and P_2 are recorded. It was shown that the ratio of these two quantities is practically proportional to the chord conductance of the membrane (g_m). The membrane conductance changed with time, membrane potential, and interval between pulses. If measured 100 ms after the onset of the pulse, the conductance increased with hyperpolarization and decreased as a function of time (Fig. 3.11b). For any given value of membrane potential, the conductance decreased with time. The membrane conductance decreased with depolarizing currents, reaching a minimum at depolarizations of 20 mV. With larger depolarizations, the conductance increased again, but frequently with a delay after the onset of current. These conductance changes did not occur or were opposite in potassium-free solutions. A two-membrane model was then proposed, with a single channel for chloride and two parallel channels for potassium, rectifying in opposite directions and located in the surface membrane and the tubular membrane. The latter would be the site where anomalous rectification

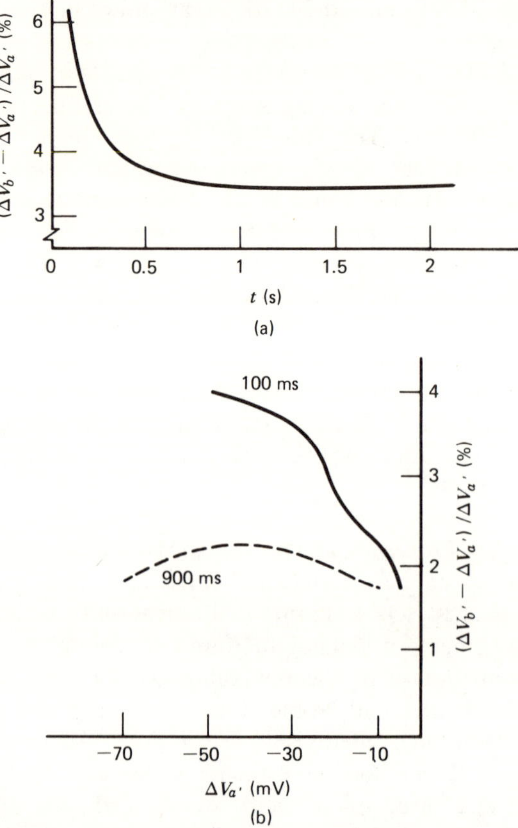

Figure 3.11. (a) Variation of membrane chord conductance during a current pulse of 2 s duration in Ringer's solution. (b) Changes in chord conductance with membrane potential and time measured in chloride-free Ringer's at 100 and 900 ms after the onset of the current pulse. Modified from Adrian and Freygang (1962a), by permission.

originates. The potassium permeability of the tubular membrane would depend on the potential across the membrane, and time-dependent changes of potassium conductance would depend on alterations of the potassium concentration in the space between the two membranes. This is still a controversial point, but some results from voltage-clamp measurements seem to contradict this explanation (see Sec. 3.1.3).

Nakajima, Iwasaki, and Obata (1962) obtained delayed rectification in sartorius muscles in Ringer's with TTX, in choline Ringer's, in 40 mM Na_2SO_4 with TTX, and in 40 mM K_2SO_4. Anomalous rectification developed at the end of long depolarizing pulses, indicating an inactivation of the transient increase in G_K elicited by depolarization (Fig. 3.12a). The normal resting potential was −20 mV. The membrane was hyperpolarized

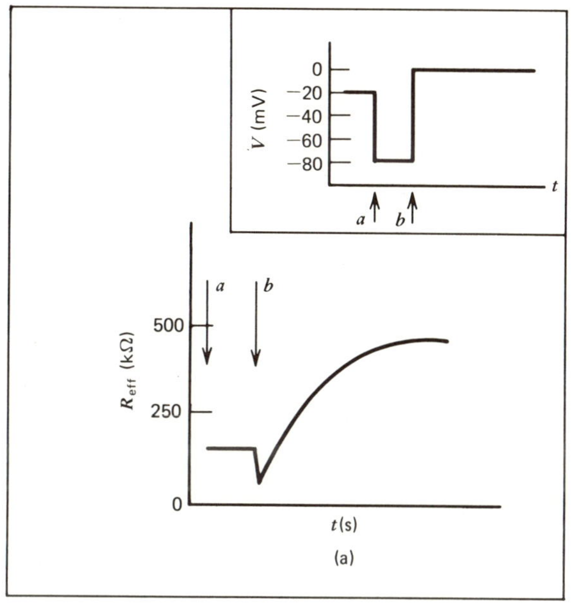

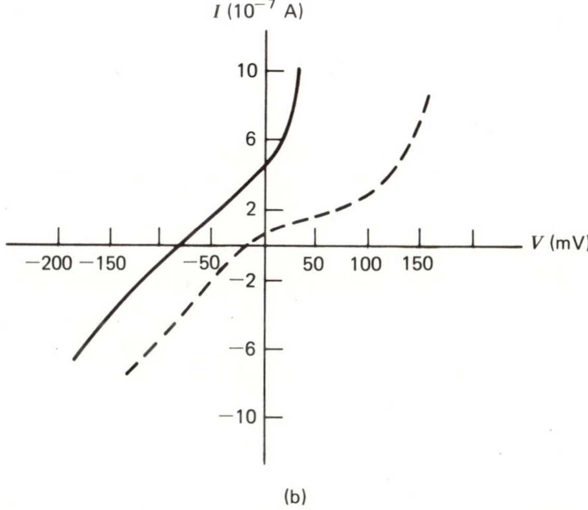

Figure 3.12. Delayed and anomalous rectification in muscle fiber immersed in 40 mM K_2SO_4 solution. (a) Time course of the change in effective resistance upon depolarization from −80 to 0 mV. Inset: pulse program. (b) Current–voltage curves obtained with (full line) and without (dashed line) conditioning hyperpolarization. Modified from Nakajima, Iwasaki, and Obata (1962), by copyright permission of The Rockefeller University Press.

to -80 mV by a conditioning pulse from *a* to *b*. At *b*, a current pulse was injected such as to bring the potential to zero. Upon depolarization, the effective resistance decreased (delayed rectification), but this was followed by a progressive increase due to the appearance of anomalous rectification. Figure 3.12b shows a marked anomalous rectification up to $+70$ mV. A much lower effective resistance at $V=0$ was the characteristic of the membrane that had not undergone a maintained depolarization. The ratio of g_K at -32 mV to g_K at $+70$ mV was 6 : 1.

Eisenberg and Gage (1969) showed that voltage–current curves in intact fibers as well as in glycerol-treated fibers were characterized by anomalous rectification, indicating that this property is not confined exclusively to tubular membranes, in contrast to what had been proposed by Hodgkin and Horowicz (1959b) and Adrian and Freygang (1962a). This was further investigated by Takeda and Oomura (1969) in intact and detubulated fibers of frog muscle at pH 5.6 and pH 7.5. It was postulated that the depolarizing anomalous rectification and the early increase of conductance upon hyperpolarization involve the membranes of the tubular system. The delayed conductance increase upon hyperpolarization seemed to depend on properties of the surface membrane, and the insensitivity of this delayed mechanism to changes in extracellular pH suggests its dependence upon G_K. In addition, delayed rectification seemed to be associated with the conductance of the surface membrane, because it persisted in detubulated fibers. However, a residual inwardly rectifying component was reported to remain in detubulated fibers immersed in isotonic solutions of potassium methylsulfate (Hutter, 1969). This component was abolished by formaldehyde. In intact fibers, formaldehyde also abolished the nonlinear characteristic of the current–voltage relation.

Adrian and Peachey (1965) reported a large increase in membrane resistance upon strong hyperpolarization in frog slow fibers (Table 3.5). These nonlinearities were confirmed by Burke and Ginsborg (1956a) in the iliofibularis muscle of frogs, where the small junctional potentials (sjp) elicited by nerve stimulation end with a hyperpolarizing phase. The value of R_{inp} at 30 mV of hyperpolarization was about four times as large as at rest and it fell to one-tenth of the resting value upon strong depolarization. The hyperpolarizing phase of the sjp was attributed to this rectification, caused by changes in g_K.

3.1.2.3 Estimation of Ionic Permeabilities. Hodgkin and Horowicz (1959b) estimated P_K, P_{Cl}, g_K, and g_{Cl} in frog muscle, from measurements of resting potential in different ionic media. The method was based on the assumption that the membrane potential is described by the Goldman equation (Eq. 1.87) and that only K^+ and Cl^- carry current through the membrane at rest. In the absence of external current, $I_K + I_{Cl} = 0$, and the membrane potential V can be described by the expression:

$$V = E_K T_K + E_{Cl} T_{Cl} \tag{3.10}$$

where $T_K = g_K/(g_K + g_{Cl})$, and $T_{Cl} = 1 - T_K$. If $V \cong E_K \cong E_{Cl}$ and the shift in potential produced by a given ionic change is small, the transport numbers are obtained as follows:

$$\left.\begin{aligned} T_K &= (dV/dE_K)E_{Cl} \\ T_{Cl} &= (dV/dE_{Cl})E_K \end{aligned}\right\} \tag{3.11}$$

where the macroscopic transport numbers T_K and T_{Cl} represent the fraction of the membrane current carried by potassium and chloride when E_K and E_{Cl} are kept constant and V varies. When the extracellular concentrations are changed, the following relations hold:

$$\left.\begin{aligned} 58T_K &= (dV/d\log[\mathrm{K}]_o)[\mathrm{Cl}]_o \\ 58T_{Cl} &= (dV/d\log[\mathrm{Cl}]_o)[\mathrm{K}]_o \end{aligned}\right\} \tag{3.12}$$

Under certain assumptions, the ionic permeabilities and conductances can be estimated from the time course of the potential shift that occurs upon changes in ionic concentrations. The results are listed in Table 3.6. The total membrane conductance corresponds to R_m of 3900 Ω cm^2. The permeability for outward movement of potassium was about twentyfold lower than for inward movement, whereas P_{Cl} was the same for both directions. At the resting potential, the chloride-to-potassium conductance ratio was 2 : 1.

The same method was applied by Hinkle et al. (1971) to estimate the ionic conductances of single fibers dissected from an abdominal muscle of the crayfish. The transport numbers T_K, T_{Cl}, and T_{Na} were calculated (Table 3.6). The average resting potential in van Harreveld solution was −74.5 mV. The sum of T_K and T_{Cl} reached a value close to 1 only when the sodium permeability was taken into account: $P_{Na}/P_K = 0.01$. In depolarized fibers, the sum of T_K and T_{Cl} was only 0.3, indicating a marked deviation of the membrane from a system permeable to only these two ions. The estimated values of P_K and P_{Cl} were considered to represent only the order of magnitude, because a critical analysis of the results showed that the assumptions underlying the theoretical treatment were not strictly applicable to this muscle. Despite these restrictions, the results showed a striking difference between the membrane properties of frog and crayfish muscle. In the latter, P_K is high compared to P_{Cl}.

Table 3.6 Ionic Permeabilities and Conductances of Surface and Tubular Membranes

Measured Quantities	Derived Quantities	Remarks	Reference
ΔV upon sudden shift in ionic concentrations	$G_K + G_{Cl}$: 255 μmho/cm^2	Frog sartorius	Hodgkin and Horowicz (1959b)
	P_K: 0.05×10^{-6} cm/s	Outward direction	
	P_K: 8×10^{-6} cm/s	Inward direction	
	P_K: 1–2×10^{-6} cm/s	At $V = E_K$	
	g_K: 80 μmho/cm^2	Resting, $V - E_K \cong 7$ mV	
	g_K: 15 μmho/cm^2	Outward I_K, at $V = +70$ mV	
	P_{Cl}: 4×10^{-6} cm/s	Both directions	
	g_{Cl}: 190 μmho/cm^2		
ΔV upon sudden shift in ionic concentrations	T_K: 0.67	Crayfish muscle, assuming negligible P_{Na}	Hinkle et al. (1971)
	T_{Cl}: 0.19		
	T_K: 0.78	Assuming $P_{Na}/P_K = 0.01$	
dV/dt upon $\Delta[\mathrm{Cl}]_o$	P_K: 114×10^{-6} cm/s	Inward direction	
	P_{Cl}: 46×10^{-6} cm/s		
$\Delta[\mathrm{Cl}]_i$ upon $\Delta[\mathrm{Cl}]_o$	P_K: 81×10^{-6} cm/s	Inward direction	
	P_{Cl}: 20×10^{-6} cm/s		
dV/dt upon $\Delta[\mathrm{K}]_o$	P_K: 27×10^{-6} cm/s	Outward direction	
	P_{Cl}: 22×10^{-6} cm/s		
$\Delta[\mathrm{Cl}]_i$ upon $\Delta[\mathrm{K}]_o$	P_K: 24×10^{-6} cm/s	Outward direction	
	P_{Cl}: 15×10^{-6} cm/s		
	g_K: 7.4 mmho/cm^2		
	g_{Cl}: 3.2 mmho/cm^2		
$d[\mathrm{Cl}]_i/dt$	R_m: 290 Ω cm^2		
dV/dt	g_K: 8.3 mmho/cm^2	GABA: 20 μg/ml	
	g_{Cl}: 41 mmho/cm^2		
Circuit 5, Table 1.5			Hutter and Padsha (1959)
R_{inp}: 625 kΩ		Chloride Ringer's	
$t_{1/2}$: 5.4 ms			
R_{inp}: 870 kΩ		Nitrate Ringer's	
$t_{1/2}$: 10.5 ms			
Circuit 1 of Table 1.3			
R_{inp}: 330–590 kΩ		Chloride Ringer's	
λ: 1.2–1.4 mm			
R_{inp}: 0.54–1.02 MΩ		Nitrate Ringer's;	
λ: 1.8–2.3 mm		frog sartorius	
	Model 5 of Table 1.5		Adrian and Freygang (1962a)
$(\Delta V_{b'} - \Delta V_{a'})/\Delta V_{a'}$ (method: Fig. 3.9)	P_{Cl}; 2.2×10^{-6} cm/s		
	P_K: 18×10^{-6} cm/s	Inward going current	
	P_{Na}: 0.9×10^{-6} cm/s		

Table 3.6 *Continued*

Measured Quantities	Derived Quantities	Remarks	Reference
Circuit 1 of Table 1.3	Model 5 of Table 1.5		Eisenberg and Gage (1969)
R_{inp}: 660 kΩ	G_m: 295 μmho/cm^2	Whole muscle, pH 7.2	
R_{inp}: 1.6 MΩ	g_K: 92 μmho/cm^2	Whole muscle, pH 5.6	
R_{inp}: 626 kΩ	$g_{Cl}+g_K^s$: 252 μmho/cm^2	Disrupted T system, pH 7.2	
R_{inp}: 1.89 MΩ	g_K^s: 42 μmho/cm^2	Disrupted T system, pH 5.6	
	g_{Cl}^s: 210 μmho/cm^2	Corrected values: $G_K^s=28$,	
	g_K^t: 52 μmho/cm^2	$G_K^t=55$, $G_{Cl}^s=219$ (μmho/cm^2)	
	g_{Cl}^t: −7 μmho/cm^2		
ΔV and τ upon ionic changes	g_{Cl}/g_K: 0.70	pH 7.4	Orentlicher and Reuben (1971)
	τ_1: 40 s		
	τ_2: 2.6 s		
	T_K: 0.60		
	g_{Cl}/g_K: 1.04	pH 6.8;	
	τ_1: 40 s	crayfish muscle fibers	
	τ_2: 1.9 s		
R_{inp}. I/V curve (Circuit 1 of Table 1.3)	Model 5 of Table 1.5	Chloride-free Ringer's	Harris and Ochs (1966)
	R_m: 20,000 Ω cm^2	$[K]_o$: 5 mM, 4°C	
	R_m: 1300 Ω cm^2	$[K]_o$: 5 mM, 20°C	
		Chloride-Ringer's	
	R_m: 2500 Ω cm^2	$[K]_o$: 5 mM, 4°C	
	R_m: 2210 Ω cm^2	$[K]_o$: 5 mM, 20°C	
	R_m: 27,000–58,000 Ω cm^2	Cocaine, other anesthetics and antihistaminics	
	R_m: 35,000 Ω cm^2	10 mM Rb, Cl^--free, 4°C	
	R_m: 21,600 Ω cm^2	10 mM Rb, Cl^--free, 20°C	
	P_K: 0.6×10^{-6} cm/s		
	P_{Cl}: $5\text{–}9\times10^{-6}$ cm/s		

The high contribution of chloride to the membrane conductance of frog muscle was further confirmed by Hutter and Padsha (1959), who reported that replacement of chloride by nitrate increased the input resistance and the time constant of sartorius muscles. The relative membrane resistances for different anions were $Cl^-:Br^-:NO_3^-:I^-=1:1.5:2.0:2.3$. The contribution of chloride to the total resting conductance was investigated by Hutter and Noble (1960a), who replaced chloride by the nonpermeant anions methylsulfate or pyroglutamate. The membrane conductance decreased in chloride-free media. The contribution of chloride to the total membrane conductance was estimated as 68%. This figure is close to the chloride conductance calculated by Hodgkin and Horowicz (1959b).

The evidence is consistent with a membrane potential mainly determined by potassium and chloride, but Hodgkin and Horowicz (1960) reported that the time course of the potential shift during changes in $[Cl^-]_o$ and $[K^+]_o$ was more complicated than expected from a uniform distribution of P_{Cl} and P_K over the membrane surface. It was then postulated that the membrane permeable to potassium was mainly that of the tubular system, and that the high chloride permeability would be a property of the surface membrane. This was confirmed by Eisenberg and Gage (1969) almost ten years later. In the meantime, Adrian and Freygang (1962a) had estimated that the chloride permeability contributes to two-thirds of the total resting conductance.

Eisenberg and Gage (1969) separated the surface and tubular components of the ionic conductances by manipulating the extracellular pH of intact and detubulated fibers. Neglecting the contribution of sodium ions, and assuming that potassium and chloride conductances are distributed over the surface and tubular membranes, the total resting membrane conductance is given by:

$$G_m = G^s_{Cl} + G^t_{Cl} + G^s_K + G^t_K \tag{3.13}$$

where the indexes s and t denote surface and tubular, respectively. If the chloride conductance is abolished by lowering the pH (Hutter and Warner, 1967), the remaining conductance is $G_K = G^s_K + G^t_K$. In glycerol-treated fibers, only the surface membrane conductance remains, so that $G^s = G^s_{Cl} + G^s_K$. Finally, the measured conductance is reduced to G^s_K if glycerol-treated muscles are put in a medium at pH 5.6. The values of the different conductance components are listed in Table 3.4; the individual ionic conductances are in Table 3.6. The negative value for G^t_{Cl} is not significantly different from zero. The results indicated that the potassium conductance is evenly distributed between the surface and tubular membranes. On the other hand, chloride conductance is exclusively restricted to the surface membrane. The results agreed with those of Adrian and Freygang (1962a) for the estimation of the proportional magnitude of the chloride conductance in relation to the total membrane conductance and the existence of two membrane systems for potassium.

Orentlicher and Reuben (1971) confirmed the hypothesis of Girardier et al. (1963), who had suggested that the morphological distribution of conductances in crayfish muscle is the reverse of that found in frog. Their findings partly agreed with the report of Hinkle et al. (1971). However, a disagreement was found regarding the existence of a diffusion barrier separating the site of the chloride conductance from the bathing solution, as indicated by the slowness of responses to changes in $[Cl^-]_o$, which in the isolated fibers from the walking-leg muscles developed with time constants of up to 15 min. Anatomical studies of fibers isolated from the walking-leg muscles (Brandt et al., 1965; Brandt et al., 1968) revealed that the sarcolemma is densely invaginated. The core of these invaginations constitutes a barrier to diffusion between the bath and a large fraction of the membrane area. Depolarizations induced by an increase of $[K^+]_o$ developed with a short half-time, suggesting a superficial localization of g_K. This was confirmed by addition of $BaCl_2$ (inhibitor of K^+ permeability), which resulted in a rapid decrease of membrane conductance. Rapid withdrawal of chloride depolarized the membrane. In contrast to frog muscle, a reduction of pH from 7.4 to 6.8 enhanced chloride permeability and caused a larger depolarization and a more rapid decay of the voltage transient upon chloride removal.

Interpretation of these data required a precise analysis of the transient phase of the potential shift upon changes in $[Cl^-]_o$. This analysis was complicated by the presence of the diffusional barrier represented by the membrane invaginations. The differential equation describing the flux in an invagination is:

$$w(dC/dt) = wD\left(d^2C/dx^2\right) + jp \tag{3.14}$$

where w is the cross-sectional area of the invagination; C, concentration; D, the diffusion constant in the solution; and j is defined as $j = H([Cl^-]_i - r[Cl^-]_o)$, where H is the flux rate constant and r is given by $r = \exp(VF/RT)$. Equation 3.14 is the common diffusion equation in which the term jp takes into account the flux into the invagination. Multiplying by the invagination area per unit area of fiber surface gives the total flux, from which the expressions for P_{Cl} and I_{Cl} are derived:

$$\begin{aligned} P_{Cl} &= HNLp/\pi dl \\ I_{Cl} &= FP_{Cl}\left([Cl]_i - r[Cl]_o\right) \end{aligned} \tag{3.15}$$

where N, L, p, d, and l are geometrical parameters. Differentiation of Eq. 3.15 for I_{Cl} with respect to V gives the slope conductance G_{Cl}, which for the resting level is expressed as:

$$G_{Cl} = P_{Cl}\left(F^2/RT\right)[Cl]_i \tag{3.16}$$

For a 200-μm fiber equilibrated in control saline, less than half the total conductance corresponded to chloride (Table 3.6).

The selectivity characteristics of surface and tubular membranes of crayfish muscle were confirmed by Law and Atwood (1971) through comparative measurements of effective resistance (R_{eff}) in short and long sarcomere fibers. The number of tubules is much greater in short sarcomere fibers (SSF) than in long sarcomere fibers (LSF). The mean effective resistance of SSF (100 kΩ) was lower than in LSF (168 kΩ), as expected if the tubular membranes are the site of an important fraction of the total conductance. A similar morphological distribution of ionic permeabilities was described by Mounier and Guilbault (1970) for muscles of the crab *Carcinus maenas*. However, in contrast to other crustacean muscles, a reduction of extracellular pH from 7.8 to 4.25 increased the membrane resistance to 180% its control value (Mounier, Vilain, and Guilbault, 1970). This change was accompanied by depolarization and an increase in the relative permeability P_{Cl}/P_K, due to a decrease in P_K.

Harris and Ochs (1966) measured current–voltage curves and the cellular ionic content of frog sartorius to investigate the mechanism of K^+ influx. After 16–40 hr of storage at 4°C, the $[K^+]_i$ was 100 mM, and the resting potential was -65 mV. The latter rose rapidly to -70 to -80 mV upon rewarming. The potassium equilibrium potential was lower than the membrane potential for periods up to 1 hr or longer. Measurements in chloride-free medium permitted the calculation of the potassium conductance (Table 3.6) and the potassium influx resulting from the difference between E_K and V. The potassium conductance increased upon rewarming. Estimations of the potassium influx from the analysis of ionic concentrations agreed with the values reported by Harris and Sjodin (1961) from measurements of the uptake of labeled potassium in depleted fibers. It was estimated that a potential difference ($E_K - V$) of 15 mV across the measured membrane resistance could account for the calculated potassium uptake. Since this estimation agreed with the measured value, an active process for K^+ influx need not be postulated. The calculated ionic permeabilities are within the same order of magnitude as those published by others (Table 3.6).

Using the same experimental approach, Adrian and Slayman (1966) investigated the mechanism involved in potassium and rubidium transport occurring during active extrusion of sodium. Sodium-rich, potassium-depleted muscles extruded sodium and replaced it by potassium or rubidium upon rewarming in solutions containing 10 mM of K^+ or Rb^+. Figure 3.13 shows that upon rewarming in the recovery solutions, the membrane potential increased steadily during the first half hour and became 20–40

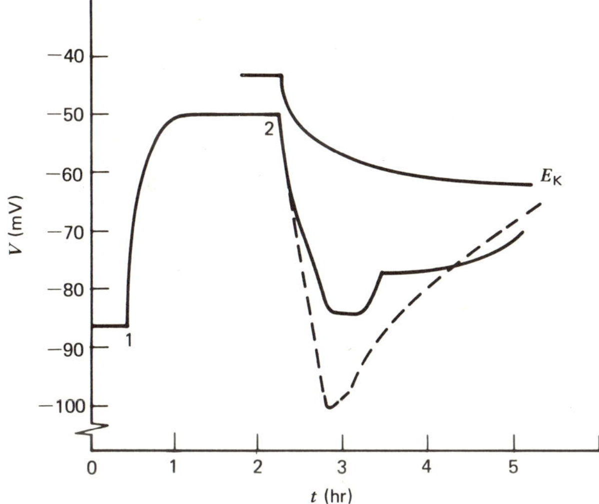

Figure 3.13. Membrane potential of frog muscle measured under the following conditions: from 1 to 2, at 1°C, $[K^+]_o = 10$ mM; at 2, rewarming to 21°C; full line, muscle in 10 mM K^+; dashed line, muscle in 10 mM Rb^+. Upper curve: potassium equilibrium potential as calculated from parallel measurements of ionic concentrations. Modified from Adrian and Slayman (1966), by permission.

mV higher than the calculated E_K. Cooling a muscle during the hyperpolarization caused a return of the potential to values close to E_K. It was estimated that 50–100% of the potassium influx during active sodium extrusion could be driven by the difference between V and E_K. The findings of Harris and Ochs (1966) and Adrian and Slayman (1966) show how a combination of analytical methods with the determination of the electrical constants can lead to the characterization of membrane properties such as permeabilities, ionic fluxes, and transport mechanisms.

3.1.2.4 The Effects of Some Biological Variables on Muscle Electrical Properties. Ishiko and Sato (1960) determined the membrane constants at normal length and at 30% stretch in several toad muscles. The results are listed in Table 3.7. No significant change of R_m and C_m occurred with stretch, although a definite tendency of R_m to decrease and C_m to increase was observed. The magnitude of the changes was of the order of 14%. The increase in propagation velocity observed in muscle during stretch could not be totally attributed to changes in membrane capacity, but a small depolarization occurred, which could reduce the threshold current enough to account for an increase of up to 50% in propagation velocity.

Table 3.7 Effect of Various Agents on Muscle Electrical Cell Constants

Method	Measured Quantities	Model	Derived Quantities	Reference
1 of Table 1.3		5 of Table 1.5	Length, 100%:	Ishiko and Sato (1960)
	R_{inp}: 127–301 kΩ λ: 1.4–1.7 mm τ_{inp}: 6.7–11 ms		R_m: 1100–2990 Ω cm^2 d: 66–98 μm C_m: 4.7–7.1 μF/cm^2	
			Length, 130%:	
	R_{inp}: 132–420 kΩ λ: 1.2–1.5 mm τ_{inp}: 6.5–12 ms		R_m: 870–2650 Ω cm^2 d: 52–86 μm C_m: 4.5–8.9 μF/cm^2	
1 of Table 1.1		4 of Table 1.5	Control:	Nicholls (1965)
	λ: 1.06 mm τ_{inp}: 17.2ms		R_m: 4970 Ω cm^2 R_i: 328 Ω cm C_m: 4.1μF/cm^2	
			Denervated:	
	λ: 1.46 mm τ_{inp}: 29.3 ms		R_m: 10,300 Ω cm^2 R_i: 350 Ω cm C_m: 3.1 μF/cm^2	
1 of Table 1.3	R_{inp}: I/V curve λ: 1.04 mm τ_{inp}: 8.1 ms	5 of Table 1.5	Control: R_m: 1680 Ω cm^2 C_m: 4.9 μF/cm^2	Hubbard (1963)
			Denervated, 28 days:	
	λ: 1.38 mm τ_{inp}: 15 ms		R_m: 2840 Ω cm^2 C_m 5.4 μF/cm^2	
			Control, 57 days:	
	λ: 1.68 mm τ_{inp}: 15.6 ms		R_m: 2650 Ω cm^2 C_m: 5.9 μF/cm^2	
			Denervated, 57 days	
	λ: 1.71 mm τ_{inp}: 18 ms		R_m: 3970 Ω cm^2 C_m: 4.6 μF/cm^2	
1 and 3 of Table 1.3		5 of Table 1.5		Law and Atwoo (1972)
	R_{eff}: 535 kΩ R_i: 50 Ω cm d: 30 μm		R_m: 1516 Ω cm^2 (control)	
	R_{eff}: 234 kΩ R_i: 50 Ω cm d: 39 μm		R_m: 636 Ω cm^2 (dystrophic)	
	R_{eff}: 209 kΩ R_i: 51 Ω cm		R_m: 515 Ω cm^2 (innervated, dystrophic)	

Table 3.7 *Continued*

Method	Measured Quantities	Model	Derived Quantities	Reference
	d: 39 μm			
	R_{eff}: 254 kΩ R_i: 48 Ω cm d: 39 μm		R_m: 754 Ω cm^2 (functionally denervated, dystrophic)	
	R_{eff}: 688 kΩ R_i: 59 Ω cm d: 26 μm		R_m: 1436 Ω cm^2 surgically denervated, normal)	
	R_{eff}: 660 kΩ R_i: 56 Ω cm d: 29 μm		R_m: 1934 Ω cm^2 (surgically denervated, dystrophic)	
1 of Table 1.3	R_{inp}: V/I curves τ_{inp}: 26.0 ms	5, Table 1.5	Control: r_m: 293 kΩ cm r_i: 6.41 MΩ/cm	Freygang et al. (1967)
	τ_{inp}: 28.4 ms		Hypertonic Sucrose: r_m: 240 kΩ cm r_i: 8.97 MΩ/cm	
	τ_{inp}: 20.7 ms		Hypertonic NaCl: r_m: 218 kΩ cm r_i: 8.90 MΩ/cm	

The modifications of the muscle membrane following denervation were first investigated by Nicholls (1956) in the foot muscles of the frog (Table 3.7). The measurements were performed one month after denervation. Atrophy occurred in a few muscles, but the mean values of total membrane area were not significantly reduced. Denervated muscles had an R_m twice that of normal muscles, but R_i and C_m did not change. Accordingly, propagation velocity was not altered. The increase in R_m with denervation was attributed to a decrease in potassium permeability. This interpretation was supported by Hubbard (1963), who investigated the changes in chloride and potassium conductance after 28, 40, and 57 days of denervation. The length constant and the membrane resistance increased significantly with denervation. Potassium conductance increased and chloride conductance decreased with hyperpolarization. In denervated fibers the resting chloride conductance remained at the same level as the controls, and potassium conductance decreased by about 40%. Law and Atwood (1972) compared the effects of surgical and natural denervation on the membrane properties of normal and dystrophic fibers in mice of the Bar Harbor 129 strain, which present spontaneous muscular dystrophy. The results listed in Table 3.7 show that dystrophic fibers have lower R_{eff} and R_m than muscle fibers from normal mice. Within the group of dystrophic fibers, the effective resistance of functionally denervated fibers was higher than that of innervated ones. R_m was not much altered with denervation in normal muscles, because an increase in R_{eff} was offset by an increase in R_i and a decrease in fiber diameter. However, the much larger increase of R_{eff} in dystrophic muscle reflected a substantial increase in R_m. These results seem to exclude the possibility that the changes of the membrane electrical properties in dystrophic fibers are secondary to the lack of the trophic influence of innervation.

Knutsson (1961) showed that ethanol depolarizes the muscle membrane in frogs and decreases the membrane resistance. This decrease in membrane resistance suggests that the ionic permeabilities are modified. Knutsson and Katz (1967) reported that the slight depolarization (2–7 mV) occurring in 50 mM ethanol was suppressed in sodium-free Ringer's and enhanced in sodium-rich medium. No effect of ethanol on voltage–current curves of fibers kept in chloride-free, high-$[K^+]_o$ (80 mM) solutions could be detected. It was concluded that ethanol does not affect P_K. However, since the membrane potential was very low in high-$[K^+]_o$, this result does not necessarily imply that ethanol did not alter P_K, because its effect may have been masked by the influence of $[K^+]_o$ on P_K and the voltage dependence of g_K.

Freygang et al. (1967) reported the effects of hypertonic sucrose and NaCl on the time constant, r_m, and r_i of sartorius muscle fibers: r_m declined 18 and 26% in muscles immersed in hypertonic sucrose or hypertonic NaCl, respectively, and r_i increased (Table 3.7). Since the intracellular potassium concentration does not change by immersion of the muscle in hypertonic sucrose (Freygang et al., 1964), it was estimated that the increase in r_i corresponds to a fall of 30% of the conductivity of the myoplasm per unit potassium concentration of the myoplasm. Changes in fiber diameter did not influence this result, because r_i was calculated from the relation $r_i = 2R_{\text{inp}}/\lambda$. The changes in the time constant were ascribed to an increase in the area of the T system in hypertonic sucrose. This was confirmed by electronmicroscopy.

The effects of barium, tonicity, and CO_2 on membrane resistance and ionic conductance of frog sartorius muscles were reported by Sperelakis, Schneider, and Harris (1967), Sperelakis and Schneider (1968), and Sperelakis (1969b). Table 3.8 summarizes these results, which were obtained in Ringer's and chloride-free Ringer's. The table lists two values of

R_m for solutions of different tonicity. They were obtained assuming a constant R_i (214 Ω cm) for the upper value, and that R_i changed with the osmotic strength of the medium (lower value). The individual ionic conductances were separated on the assumption that g_K was the same for any given tonicity independent of the presence of chloride, a questionable assumption considering the profound structural changes of the T system that accompany variations of tonicity (Freygang et al., 1967). In chloride-free Ringer's, Ba^{2+} increased the resting R_m. In hypertonic sucrose, the calculated fiber diameter decreased from 84 μm to 61 μm, on the assumption that R_i remained constant. Addition of barium to this solution caused a depolarization from a mean value of −76 mV to −32 mV and an increase of R_m. The fact that Ba^{2+} did not change the membrane resistance in normal Ringer's and increased it in chloride-free Ringer's supports the view that it acts mainly on g_K. In chloride-free Ringer's, R_m increased in the range from hypotonic to isotonic solutions but fell considerably in

Table 3.8 Effects of Tonicity, Barium, and CO_2 on Electrical Cell Constants and Ionic Conductances of Frog Sartorius Muscle Fibers*

Experimental Condition	R_{inp} (MΩ)	λ (mm)	R_m (Ω cm²)	G_m (μmho/cm²)	g_K (μmho/cm²)	g_{Cl} (μmho/cm²)
Ringer's	0.52	1.6	3800			
Ringer's, 9 mM Ba	0.66	1.7	5100			
Chloride-free Ringer's	1.0	4.9	27,000			
Cl^--free, hypertonic	1.0	2.4	9700			
0.5 mM Ba^{2+}	1.0	4.6	29,500			
Cl^--free, high K^+	0.24	1.2	1500			
0.5 mM Ba^{2+}	0.45	2.9	6700			
Ringer's, hypotonic	0.36	1.1	1700			
			2100	476	46	430
Ringer's, isotonic	0.65	1.8	4700			
			5000	200	37	163
Ringer's, hypertonic	1.0	1.6	5500			
			6200	160	140	22
Cl^--free, hypotonic	1.1	3.1	14,500			
			21,900	46	46	
Cl^--free, isotonic	1.1	4.9	27,900	37	37	
Cl^--free, hypertonic	0.99	2.1	7000			
			7500	139	139	
Ringer's	I/V		3800–5000		42	185
Ringer's, 5% CO_2			4600–5800		46	162
Ringer's, 100% CO_2			21,000–27,000			

*Data from Sperelakis, Schneider, and Harris, 1967; Sperelakis and Schneider, 1968; Sperelakis, 1969.

hypertonic medium. This was explained by (1) a marked tubular swelling in hypertonic solutions and a slight shrinking in hypotonic medium; (2) the impermeability of the tubular membrane to Cl^-, and (3) the permeability of the surface membrane to Cl^- and K^+.

The effect of increasing CO_2 concentration (a rise in R_m) disappeared in chloride-free medium, confirming the conclusions of Hutter and Warner (1967). At 100% CO_2, g_{Cl} was totally abolished. These results cannot be attributed completely to the shift in pH produced by the high concentration of CO_2, because Meves and Völkner (1958) found an increase in R_m of 1.7 to 3.6 times with a CO_2 concentration of 32% at normal pH. Moreover, the total inactivation of g_{Cl} reported by Sperelakis (1969b) occurred at pH 6.1, whereas Hutter and Warner (1967) obtained it only at pH 5.1 and Eisenberg and Gage (1969) estimated a residual chloride conductance of 4% at pH 5.6.

3.1.3 Voltage-Clamp Measurements in Skeletal Muscle

3.1.3.1 The Ionic Currents in Skeletal Muscle. This paragraph discusses (a) the results obtained in frog muscle and (b) the ionic currents in muscle from invertebrates.

(*a*) *Ionic currents in frog muscle..* One of the first attempts to control the potential of a restricted area of the membrane was carried out by Adrian and Freygang (1962b), using a circuit based on one they had employed earlier (Adrian and Freygang, 1962a). Although the method did not permit control of the membrane potential, the ionic current could be estimated within a time as short as 5 ms after application of the pulse. Current–voltage curves showed anomalous rectification. The slope of a linear part of the curve found at values between +30 and +60 mV was interpreted as representing G_K. Table 3.9 shows that the P_K value agrees well with the estimate of Hodgkin and Horowicz (1959b).

Another indirect approach to studying the ionic current during excitation was that of Strickholm (1962). A small membrane area was electrically isolated by placing a smooth tipped electrode (20 μm diameter) against the cell surface. With large depolarizations, the active current flowing through the membrane contributed to the recorded voltage U_a' and was given by $I_m = -(U_a' - U_a)/R_s$, where U_a is the voltage drop for a negligible current flow and R_s is the leak resistance between the electrode tip and the underlying cell surface. This technique did not allow effective control of the membrane potential because no feedback circuit was used, but the method yielded curves similar to the current versus time relations obtained under voltage-clamp conditions. Progressive depolarization elicited transient inward currents, but outward currents were not observed even with

Table 3.9 Electrical Properties of Skeletal Muscle Measured with Voltage-Clamp Methods

Method	Measured Quantities	Model	Derived Quantities	Reference
8 of Table 1.4	I/V curve: mA/cm^2 mV	5 of Table 1.5	$[K]_o = 100$ mM; $V = -10$ mV: R_m: 2800 Ω cm^2 P_K: 0.76×10^{-6} cm/s $V = 30$–60 mV: R_m: 11,570 Ω cm^2 P_K: 0.19×10^{-6} cm/s	Adrian and Freygang (1962b)
8 of Table 1.4	r_i: 6.3 MΩ/cm τ_{inp}: 1.28 ms $i_m = f(V_2 - V_1)$	5 of Table 1.5	λ: 1.8 mm R_m: 4500 Ω cm^2 C_m: 5.2 $\mu F/cm^2$ d: 64 μm τ_f: 0.127 ms C_f: 2.76 $\mu F/cm^2$	Adrian, Chandler, and Hodgkin (1970a)
5b of Table 1.4	$I_c = f(t)$	1 of Table 1.5	R_m: 1000 Ω cm^2 C_s: 0.9 $\mu F/cm^2$ C_T: 2.1 $\mu F/cm^2$ R_s: 17.5 Ω cm^2 R_T: 175 Ω cm^2	Ildefonse and Rougier (1972)
8 of Table 1.4	Voltage current R_{inp}: V/I curves	5 of Table 1.5	Ringer's: R_m: 3530 Ω cm^2 P_K: 4×10^{-6} cm/s (inward) P_K: 0.5×10^{-6} cm/s (outward) SO_4^{2-}-Ringer's: R_m: 5580 Ω cm^2 P_K: 1×10^{-6} cm/s (inward) P_K: 0.5×10^{-6} cm/s (outward)	Stanfield (1970)
7 of Table 1.4	R_{inp}: I/V curves λ_T: 40–91 μm	1 of Table 1.5	R_m: 800–1300 Ω cm^2 R_T: 1000–1800 Ω cm^2	Adrian et al. (1969)

strong depolarizations. The total charge transferred was 8.6×10^{-6} coulombs/cm^2, a value within the order of magnitude estimated by other methods.

Frankenhaeuser et al. (1966) applied the voltage-clamp method developed by Dodge and Frankenhaeuser (4 in Table 1.4) to isolated fibers from iliofibularis and semitendinosus muscles of the frog *Rana rudibunda*. Poor sealing of the partitions resulted in either current electrode polarization (defective sealing between 2 and 3) or a defective space clamp (loose sealing between 1 and 2). Control measurements with microelectrodes showed that the method was reliable for current-clamp measurements, but it was not possible to obtain effective voltage control. These difficulties were attributed to the nature of the outside-inside impedance in pool 1, the low membrane impedance in pool 3 (Z_{m3}), and the regenerative membrane currents occurring at the partitions 1-2 and 2-3 (circuit 4, Table 1.4). In a myelinated nerve, Z_{m3} is very high (myelin sheath) and there is no active current flowing through it, but in the muscle Z_m is uniform along the fiber, and the activity of the membrane at partitions 1-2 and 2-3 makes it very difficult to obtain an effective space clamp. Moore (1972) improved this technique so that effective voltage control was possible. The membrane current was calculated as $I_m = V_1 / Z_{1-b} A_m$, where V_1 is the potential in pool 1, Z_{1-b} is the impedance from pool 1 to the inside of the fiber in pool 2, and A_m is the membrane area in pool 2. Isotonic RbCl in pool 1 removed the rectifying properties of the membrane, and more accurate measurements of the steady-state outward current were obtained because the influence of a variable impedance in the current path was eliminated. In agreement with the results of others (see below), two time constants were required to describe the turning off of the inward current and the turning on of the late outward currents. The oil-gap technique of Cole and Hodgkin (1939) was used to estimate the resistance of the seals and to calculate the cell constants. The seal resistance per unit length was 10 MΩ/mm and the internal longitudinal resistance was 0.5 MΩ/mm, which resulted in an internal resistivity of about 300 Ω cm. The mean R_{eff} was 700 kΩ; λ was 0.25 mm, and R_m was 670–7400 Ω cm^2.

Adapting the method of Adrian and Freygang (1962a, b), Adrian et al. (1966, 1968) developed a technique to control the voltage over a short length at the end of a muscle fiber. The method (8 of Table 1.4) was later analyzed and discussed in detail (Adrian et al., 1970a). The membrane current per unit length (i_m) at electrode P_1 is proportional to $2(\Delta V_2 - \Delta V_1) 3 l^2 r_i$, where the subscripts indicate the voltage at P_1 and P_2, l is the distance from the fiber end ($x=0$) to P_1 ($x=l$), and r_i is the resistance per unit length of cytoplasm. It was found that a distance of 125 μm between electrodes 1 and 2 was small enough to give reliable results, but the equation for i_m is a less accurate approximation if pronounced nonlinearities or time variations occur, so that the method is less satisfactory for studying the regenerative inward current. For a cylindrical fiber, the membrane current density J_m is given by:

$$J_m = a(\Delta V_2 - \Delta V_1)/3R_i l^2 \tag{3.17}$$

where a is the fiber radius. The method was tested by determining the passive electrical properties of the fiber (Table 3.9). The following equation was derived to calculate the membrane capacity.

$$C_m = \frac{2}{3l^2 r_i} \frac{\Delta V(2l)}{\Delta V^2(l,\infty)} \int_0^\infty \Delta V' dt \tag{3.18}$$

This equation can be applied to any of the equivalent networks currently employed for the combined membrane capacity (Falk and Fatt, 1964; Adrian et al., 1969).

The experiments were carried out in hypertonic Ringer's solution to minimize muscle contraction. The time course and components of the membrane current were comparable to those found in nerve. The inadequacy of the method to give precise data on early currents is shown by the discrepancy found between the mean value of E_{Na} (+20 mV) and the mean overshoot of the action potential (+29 mV). The time course of inactivation was approximately exponential with a time constant of 30 ms, and the time constant for removal of inactivation was 4–7 ms, both at 2°C. The delayed current was measured in TTX-Ringer's. Depolarizations beyond −40 mV gave rise to an outward current very similar to that seen in squid axon, but, in contrast to it, this current inactivated with a time constant of 2 s. The value of E_K was −85 mV in Ringer's with 2.5 mM K^+ and −52 mV at a $[K^+]_o$ of 20 mM. The values of E_K and resting potential could be fitted to a P_{Na}/P_K of 0.01 in the resting fiber and 0.03 during delayed rectification. During prolonged depolarization, the delayed current was rapidly and completely inactivated following an exponential time course with a time constant of 0.6 s at −24 mV and 1 s at −40 mV. Table 3.10 lists some of the derived values obtained by applying the Hodgkin-Huxley theory. The limiting conductances in muscle are about half the values for the squid axon (Table 2.6), but the absolute permeabilities are roughly the same for K^+ or a little greater for Na^+ because the ionic gradients are lower in the frog muscle immersed in hypertonic sucrose Ringer's than in the squid axon in seawater. The theoretical action potential was computed according to the Hodgkin-Huxley theory and the electrical equivalent of Fig. 3.2a, assuming that all the ionic current was

Table 3.10 Determinants of the Action Potential in Frog Sartorius Muscle*

	Phase-Plane Trajectory	Voltage-Clamp
V_{max} (mV)	+30 to +35	+16 to +35
E_{Na} (mV)	+42	+15 to +22
$\bar{g}_{Na}$ (mmho/cm^2)	190	55–70
$\bar{g}_K$ (mmho/cm^2)	21	8.5–20
$[Na]_i$ (mM)	17	—
Threshold (mV)	−65	−70
Q_{Na} (pmol/cm^2)	7.4	7.6
Q_K (pmol/cm^2)	5.6	5.2

*Data from Jenerick (1964) for the phase-plane trajectory and Adrian et al. (1970a) for the voltage-clamp.

carried across the surface membrane. Agreement was found between measured and calculated action potentials, the latter having a negative after-potential of the correct duration. Experimental values obtained in glycerol-treated muscles were not different from the controls. This supported the assumption that ionic current flow was restricted to the surface membrane.

The double sucrose-gap technique developed by Stämpfli (1954, 1963) and applied by Bergman and Stämpfli (1966) to voltage-clamp measurements in myelinated nerves was adapted for measurements in skeletal muscle by Rougier, Vassort, and Ildefonse (1968) and Ildefonse and Rougier (1968, 1969). The main drawback of this method (circuit 5b, Table 1.4) is that the true membrane potential of the fiber is not known under voltage-clamp conditions, but it gives more accurate measurements of the early membrane currents. This feature permitted the analysis of the capacitive surge at the onset of a step potential change and the study of the kinetics of activation and inactivation of the early inward current (Ildefonse and Rougier, 1972). The capacitive current decayed in two phases with time constants of 0.015 and 0.37 ms (Fig. 3.14). In detubulated fibers, only the initial fast component of decay remained. The values for the elements of the circuit in Fig. 3.14 are listed in Table 3.9. The resistance R_s concerns the surface membrane, and a small fraction of it would correspond to the resistance of the extracellular solution in compartment 2. The resistance R_T was attributed to the core of the tubular system, although it is larger than the value estimated by Adrian et al. (1969) for the radial resistance of the tubular fluid, and R_m is the membrane resistance.

Figure 3.15a shows the current–voltage curve for the maximum inward current and the effect of reducing $[Na^+]_o$. The reversal potential was close

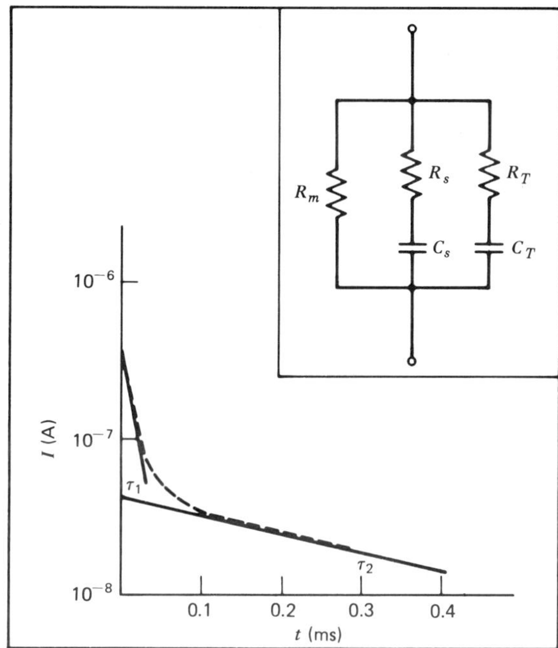

Figure 3.14. Semilogarithmic plot of the decay of capacitive current. The time constants of the fast and slow components of decay are represented as τ_1 and τ_2. Inset: equivalent circuit proposed to account for the time course of I_c. The subscripts refer to tubular system (T), surface of the fiber (s), and membrane (m). Modified from Ildefonse and Rougier (1972), by permission.

to a ΔV of $+150$ mV in Ringer's, and decreased to $+130$ mV in 37% of normal $[Na^+]_o$. The inward current disappeared in choline-Ringer's and Ringer's-TTX. The time course of removal of inactivation of the sodium current and its inactivation curve are qualitatively comparable to those of the squid axon, but the availability of the sodium system in muscle is 90% at the level of the resting potential. The voltage dependence of the h and m systems agreed closely with the results of Adrian et al. (1970a). Intracellular measurements of the potential distribution showed a homogeneous polarization when the fiber length in the central compartment was less than 0.4 mm. Isenberg (1971) used an intracellular electrode to control the membrane potential while the current was applied as in method 5b of Table 1.4. The current–voltage curve for the early current resembled those reported by others. However, the reversal potential ($+50$ mV) was higher than the value estimated by Adrian et al. (1970a).

Mandrino (1971) reported a decrease of the maximum sodium current elicited by a test pulse applied after long hyperpolarizing conditioning

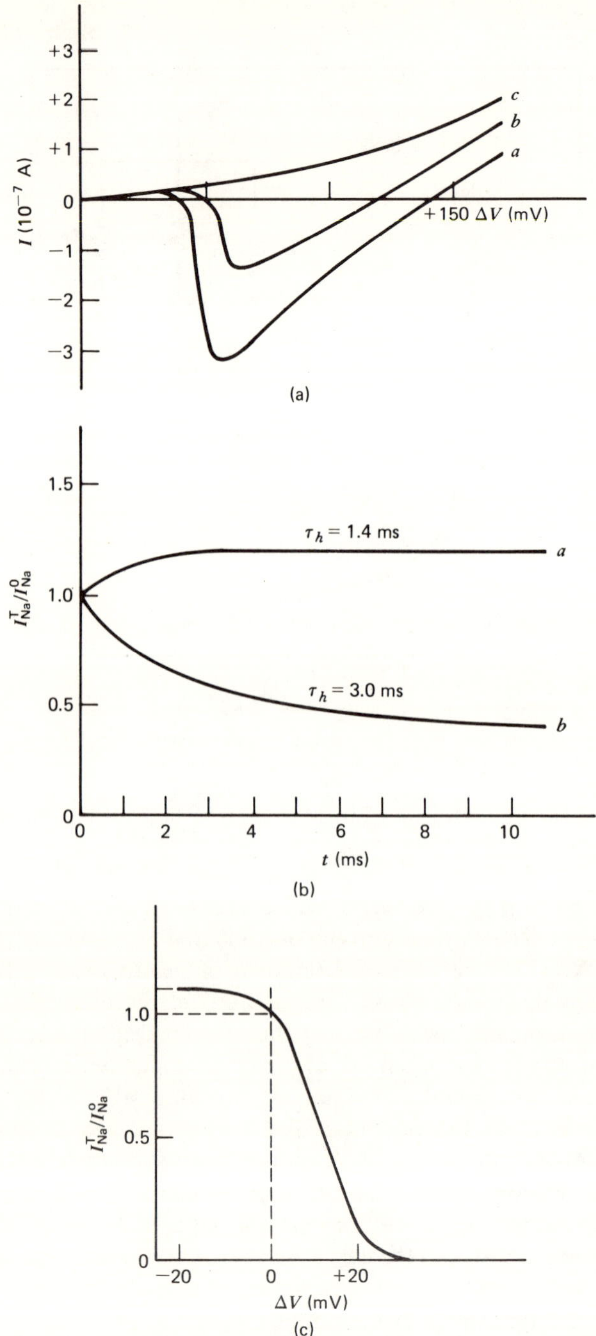

Figure 3.15. (a) Current–voltage curves for the maximum early current in skeletal muscle: *a*, in Ringer's solution; *b*, in 37% of the normal $[Na^+]_o$; *c*, in Ringer's-TTX. (b) Removal of inactivation of I_{Na}: test pulse, −58 mV; conditioning pulses, −10 mV in *a* and +10 mV in *b*. (c) Inactivation curve. For the principle of the methods and interpretation, see discussion of Table 2.5, Sec. 2.2.3. Modified from Ildefonse and Rougier (1972), by permission.

pulses. This type of response disappeared when TEA was added to the Ringer's solution. It was interpreted that potassium movement across the tubular wall during the conditioning pulse could influence the sodium current.

(*b*) *Ionic currents in striated muscle from invertebrates.* Strickholm (1963) recorded the active currents in a closer muscle of the crab *Cancer magister*. The membrane of these fibers does not generate a propagated action potential. Contraction is initiated or abolished by local junctional potentials (Hoyle and Wiersma, 1958a, b), but upon depolarization the membrane reacts qualitatively as the squid axon membrane. The reversal potential for the inward current was +41 mV. In addition to the early current, there is a delayed outward current whose current–voltage characteristic was linear over the whole potential range studied (−40 to +68 mV). The maximum inward current was 0.15 mA/cm^2, much lower than in frog sartorius muscle (5–10 mA/cm^2). This weak membrane current explained the absence of propagated action potentials in crab muscle. The charge transferred, estimated as 3×10^{-7} $coulomb/cm^2$, is one order of magnitude smaller than for other muscle fibers that generate action potentials.

Hagiwara, Hayashi, and Takahashi (1969) measured the membrane currents in barnacle muscle. These fibers are large enough to permit the insertion of two longitudinal electrodes in the cytoplasm and also to exchange the cytoplasm with solutions of known compositions. After intracellular injection of EGTA, the membrane developed all-or-none responses, due to an increase in g_{Ca} (Hagiwara and Naka, 1964). Depolarization of the membrane elicited an early inward current with a reversal potential close to the peak value of the action potential. Current–voltage curves were similar to those described for the squid axon. The plateau for minimum inactivation of the early current was reached at a potential of −45 mV. The potential level at the peak of the spike was determined by external Ca^{2+} and internal K^+ concentrations (Hagiwara and Naka, 1964; Hagiwara, Chichibu, and Naka, 1964), so the early current should be transported by Ca^{2+} and K^+. Under the experimental conditions used, $[K^+]_i$ was very high (450 mM) compared to $[K^+]_o$ (8 mM), whereas $[Ca^{2+}]_i$ was negligible compared to $[Ca^{2+}]_o$ (20 mM). Therefore, the inward current was carried by Ca^{2+} and the outward current by K^+ ions. Only the calcium channel showed spontaneous inactivation. Procaine did not alter the amplitude of the calcium spike, but delayed repolarization, and induced the development of a long plateau after the depolarization phase (Hagiwara and Nakajima, 1966). In voltage-clamped fibers, procaine depressed the potassium outward current without changing the calcium current. Cobalt markedly suppressed the inward calcium current. Increases in $[Ca^{2+}]_o$ resulted in a higher spike amplitude, and the threshold was shifted to more positive potentials.

Similar results were reported by Takeda (1967) for crayfish muscle fibers. Procaine treatment induced all-or-none responses. The conducted action potential consisted of a fast rising phase, a plateau, and a steep repolarization phase. Records of action current showed an initial outward component immediately followed by a large inward surge that lasted during the rest of the rising phase. A maintained outward current coincided with the plateau, and a fall in this current preceded the fast repolarization in the potential recording. A linear relationship existed between $\log[Ca^{2+}]_o$ and the amplitude of the overshoot. The action potential was abolished by Mn^{2+}, but it was insensitive to TTX. No anomalous rectification was seen in current–voltage curves for small depolarizations. Dudel, Morad, and Rüdel (1968) studied the contraction threshold and the current–voltage relation of crayfish muscle fibers upon substitution of chloride. Propionate did not penetrate the membrane, whereas nitrate and methylsulfate did, the permeability for nitrate being as high as for chloride, or even higher than for chloride upon depolarization. GABA increased chloride permeability in van Harreveld solution and also increased nitrate permeability. The chloride conductance represented as much as 60% of the total membrane conductance.

3.1.3.2 The Potassium Current: Rectification Phenomena. The time course of activation and inactivation of the potassium current were studied by Ildefonse and Rougier (1968, 1969). Spontaneous inactivation was not present with pulses up to 1.2 s duration. The inactivation occurred earlier if the membrane was previously hyperpolarized or the frequency of the pulses increased. The threshold for delayed rectification was found at depolarization steps of 40–50 mV, and the instantaneous current–voltage curve confirmed the linearity reported by Adrian et al. (1968). No delayed rectification was found in muscles previously treated with glycerol. The outward current and the degree of rectification decreased in TEA-Ringer's. Adrian et al. (1970b) reported that the outward current appearing during the depolarization consisted of a fast component corresponding to the delayed rectifying system with a mean equilibrium potential of -75 mV and a slow component with an equilibrium potential 10 mV more negative than that of the fast system. Figure 3.16a shows that the slow system contributes very little to the total current during the first 50 ms and can then be neglected. Both are inactivated: the fast one completely and the slow one to one-third. The results were consistent with the hypothesis that the after-potentials could be produced by the slow component.

The characteristics of the inward rectifier described by Adrian and Freygang (1962a) were investigated during strong hyperpolarization of the membrane. Figure 3.16b shows that the rectification for the initial current was stronger than that for the steady-state current, and a region of

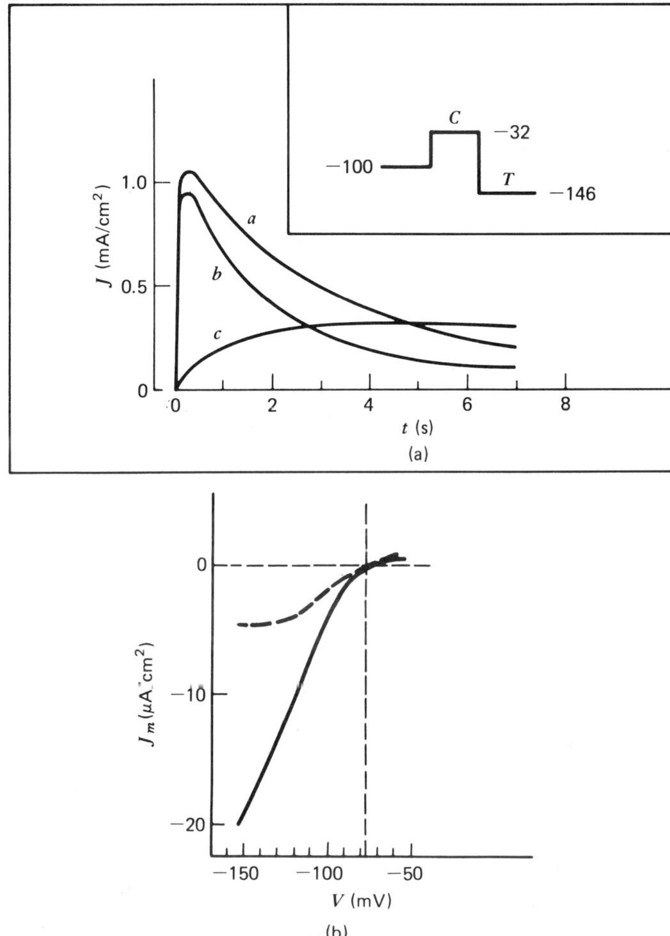

Figure 3.16. (a) Membrane current measured at large hyperpolarizations: *a*, total current; *b*, fast component; *c*, slow component. Inset: pulse pattern. (b) Current–voltage relations for potassium current upon hyperpolarization, obtained in sulfate-Ringer's. Full line: instantaneous current–voltage curve with the holding potential at −88 mV. Dashed line: steady-state current–voltage relation 2 s after onset of the test pulse. Modified from Adrian et al. (1970b), by permission.

negative slope appeared for potentials more negative than −150 mV. An S-shaped characteristic for K^+ inward current was also described by Isenberg and Küchler (1970b) in voltage–current curves measured in high-$[K^+]$ Ringer's (83 mM). This inward-going rectification had been attributed to potassium depletion in the tubular system by Adrian and Freygang (1962a). The negative slope appearing in the current–voltage curve was not consistent with the hypothesis of tubular K^+-depletion (Adrian and Freygang, 1962a) but with the current being a function of a variable (P) that decreases with hyperpolarization. Replacing K^+ by Rb^+ eliminated the current through the inward rectifier, confirming previous results of Adrian (1964).

These observations led to the identification of the different pathways of the potassium current: (1) a delayed rectifying channel, showing a spontaneous and complete inactivation; (2) a slow channel where the conductance increases with depolarization but at a rate one or two orders of magnitude slower than in the fast one; (3) an inward rectifying channel, activated with hyperpolarization but presenting also a process of inactivation. The first channel corresponds to the repolarization phase of the action potential and the early after-potentials; the second can be identified with the late after-potentials; and the third gives the slow hyperpolarization reported by Adrian and Freygang (1962a). Indirect evidence suggested that the delayed rectifier is in the surface membrane and the other two channels would correspond to the membranes of the tubular system.

Almers (1972a, b) showed that a large fraction of the decline in g_K was due to tubular potassium depletion. At potentials less negative than -125 mV, recovery occurred at a slow rate, whereas at more negative potentials recovery showed a slow and a rapid phase, indicating that a rapidly recovering process contributed to the fall in conductance. These two components had different temperature dependence: a $Q_{10}<2$ for the slow process and a Q_{10} between 2.8 and 3.0 for the fast one. The low Q_{10} of the slow component indicated a diffusion-limited process. It was suggested that K^+ would carry the inward current through the tubular membrane during hyperpolarization. Upon returning to the resting potential, K^+ could not move outward through the tubular membrane (due to its rectifier properties), and refilling of the tubules with potassium would occur predominantly by diffusion from the outside medium. This hypothesis was tested experimentally, and it was observed that the conditioning pulse shifted the membrane potential for zero current by -4 mV. It was shown theoretically that E_K across the tubular membrane can change substantially without major shifts of E_K and V across the surface membrane. It was concluded that at potentials less negative than -120 mV, tubular depletion of K^+ could account for the decline of potassium conductance, and the outward current through the tubular walls contributed only 3% to the refilling of the T-system with K^+ during the recovery phase. At more negative potentials, the conductance decline was attributed to K^+ depletion and the permeability change postulated by Adrian et al. (1970b). The current through the inward rectifier (I_r) could then be described by $I_r = P(V,t)f(V,E_K^s)$, where $P(V,t)$ represents a time- and voltage-dependent permeability and $f(V, E_K^s)$ is the instantaneous current–voltage relation. The change of the permeability variable would contribute to reduce the membrane conductance to 30% its initial value at -200 mV. At membrane potentials of about -150 mV, up to 80% of the conductance fall could be accounted for by tubular potassium depletion. It was estimated that the tubular fraction of the potassium conductance amounted to 78% and that hyperpolarization to about -150 mV could reduce the average tubular $[K^+]$ by more than 50%. The space enclosed by the T system was calculated as less than 0.8% of the fiber volume. These estimations agree with figures previously reported.

Kao and Stanfield (1968, 1970) investigated the effects of chloride replacement and addition of some cations on input resistance, outward current, and mechanical threshold in frog sartorius fibers. The input resistance in iodide and sulfate Ringer's was significantly higher than in chloride Ringer's (370, 440, and 260 kΩ, respectively). The threshold for the inward current was not modified by the different anions except by iodide, which shifted it from -58.5 mV to -63.1 mV. The threshold for rectification was -52 mV in chloride Ringer's, it was more negative with all the other anions except in sulfate, where it raised to -44.7 mV. Neither the time course of potassium inactivation nor the potential dependence of the potassium inactivation was changed by anion replacement. Tetraethylammonium (TEA) chloride increased the input resistance and reduced the potassium inward-going rectification. The increase of the input resistance was larger in TEA-iodide than in TEA-chloride or NaI; this effect was probably due to a reduction in g_K by TEA. Several quaternary ammonium compounds reduced the potassium conductance and altered its kinetics, but they did not have any appreciable effects on mechanical or rectification thresholds. These findings were interpreted as evidence for a blocking effect of tetraethylammonium ions on membrane potassium conductance, since they decreased the anomalous rectification as well as the normal rectification. The selective effect of anions on rectification threshold, without affecting the spike threshold, suggested that anions are adsorbed more readily on the T-system than on the surface membrane.

The effects of TEA and Zn^{2+} on the muscle membrane were further investigated by Stanfield (1970). TEA decreased the amplitude of the inward current elicited by hyperpolarization. The membrane resistance increased in 115 mM TEA. These effects were reversible. Current–voltage curves measured in sulfate solutions containing 100 mM K^+ and either 150 mM TEA or 150 mM Na^+ showed inward-going rectification, but the inward currents were reduced by 88% in the presence of TEA. Assuming that all the current was carried by potassium, this finding indicates the existence of a constant P_K of 0.5×10^{-6} cm/s (Table 3.9) that is not altered by TEA. Measurements of membrane potential at two potassium concentrations suggested that in the presence of TEA the membrane potential is more dependent on chloride and less dependent on potassium than in normal Ringer's. Zinc ions had very little effect on the potassium conductance. The membrane resistance increased in Zn^{2+}-Ringer's (Table 3.9), and this effect was attributed to a decrease of g_{Cl}. These results supported Adrian's view (1960) that the muscle membrane has two potassium channels in parallel that rectify in opposite directions, the channel responsible for the delayed potassium current being different from that involved in the resting potassium conductance.

3.1.3.3 Voltage-Clamp Measurements Related to Excitation-Contraction Coupling. Although many reports exist on the role of depolarization in the triggering of muscle contraction, only the work relating the ionic conductances with the contractile response is discussed here.

(*a*) *Excitation-contraction coupling in frog muscle.* Kao and Stanfield (1968, 1970) determined the threshold for a mechanical response as the membrane potential at which a change in muscle birefringence was observed. This occurred at a membrane potential 1 mV more negative than that producing a visible local contraction and was not modified by TTX. The voltage dependence of contraction was investigated by Hagiwara, Takahashi, and Junge (1968) in barnacle muscle fibers. A step change in membrane potential produced an increase in muscle tension, whose final level was related to the membrane potential by an S-shaped curve. The slope of it depended on $[Ca^{2+}]_o$.

These results are in agreement with the findings of Costantin (1968), who studied the effect of calcium and magnesium on the contraction threshold and on the threshold for the increase in ionic conductances with depolarization. Threshold was taken as the minimum depolarization to produce a brisk local contraction. In plots of membrane current versus membrane potential, the potassium conductance threshold appeared very close to the contraction threshold. Varying $[Ca^{2+}]_o$ from 0.2 to 10 mM shifted the sodium threshold by more than 30 mV, while the shifts in the potassium and contraction thresholds were 12 and 15 mV, respectively.

The kinetics of the activation of muscle contraction were studied by Adrian, Chandler, and Hodgkin (1969). Strength-duration curves for contraction and delayed rectification were obtained by measuring the threshold for contraction and for a given small increase in conductance with pulses of variable amplitude and duration. The resulting curves had a different shape, suggesting that the underlying mechanisms have a different time dependence and that the occurrence of contraction does not depend on a given, fixed increase in conductance. The effectiveness of the action potential as a stimulus for contraction was assessed by evaluating the potential distribution in the T-system upon a sudden shift in potential at the surface membrane. Under some assumptions on the geometry of the transverse tubular system, a set of equations were derived to estimate its electrical properties and the potential distribution under voltage-clamp conditions. The mean capacitance ($\bar{C}_w$) and the mean conductance ($\bar{G}_w$) of the tubular membrane per unit volume of muscle fiber, and the effective radial conductivity of the lumen ($\bar{G}_L$) in mho/cm are given by:

$$\left.\begin{aligned} \bar{G}_L &= G_L \rho' \gamma \\ \bar{C}_w &= C_w \rho' / \xi \\ \bar{G}_w &= G_w \rho' / \xi \end{aligned}\right\} \tag{3.19}$$

where G_L is the specific conductivity of the lumen, C_w and G_w are the capacitance and conductance per unit area of tubular membrane, ρ' is the fraction of the total muscle volume occupied by the tubules, ξ is the volume-to-surface ratio of the tubules, and γ is a form factor whose value depends on the arrangement of the tubular network. The radial current per unit length (i_r) is:

$$i_r = 2\pi r \bar{G}_L \frac{d\Delta V}{dr} \tag{3.20}$$

where ΔV is the potential shift across the tubular membrane and r is the distance in the radial direction. The current through the tubular membrane at the steady state is given by:

$$\Delta V / \Delta V_a = I_0(r/\lambda_T) / I_0(a/\lambda_T) \tag{3.21}$$

where a is the fiber radius, ΔV_a the potential at $r = a$, and λ_T the length constant for the tubule. The tubular current density at the surface of the fiber (J_a), for $a/\lambda_T = 0$, will be

$$J_a / \Delta V_a = a\bar{G}_w / 2 \tag{3.22}$$

This expression describes the dc component of the current density in the tubular region at the surface of the fiber. The effective capacity of the tubular system, referred to unit area of fiber surface, can be estimated with little error with the simple expression:

$$C_T = a\bar{C}_w / 2 \tag{3.23}$$

for the condition $a/\lambda_T < 1$.

Table 3.4 lists the practical constants and parameters describing the electrical properties of the tubular system, calculated from basic constants as reported in the current literature: $a=4\times10^{-3}$ cm; $\rho'=3\times10^{-3}$; $\xi=10^{-6}$ cm; $C_w=1\ \mu F/cm^2$; $G_w=0.5\times10^{-4}$ mho/cm^2; $G_L=10^{-2}$ mho/cm; $\gamma=\frac{1}{2}$. The quantity $R_sR_T/(R_s+R_T)$ (Eq. 3.9) is the combined membrane resistance (R_m in Table 3.2).

A plot of $\Delta V/\Delta V_a$ versus r/a with time as a parameter showed that the potential in the center of a fiber of 40 μm radius reaches half its final value in 0.65 ms, while only 0.1 ms is required to reach the half-time at a distance of 30 μm from the axis. This result suggested that the normal action potential would be adequate to trigger the contractile response.

However, measurements of the passive electrical properties of the T-system and the radial spread of contraction in single fibers of frog semitendinosus muscle performed by Adrian, Costantin, and Peachey (1969) indicated that the safety factor of the action potential to activate the myofibrils at the center of a muscle fiber is not larger than unity, if this activation is accomplished by electrotonic spread along the T-system. The length constant λ_T increased with the fiber radius and the value of λ_T fell within the predicted range of 80–105 μm only for the largest fibers. Table 3.9 shows that the values of R_m and R_T are very close, so that at the potential level of contraction threshold the tubular resistance appears to be as large as the combined resistance measured at rest. The existence of a mechanism for a regenerative potential change in the tubular membrane during normal excitation-contraction coupling could be neither rejected nor postulated from these experiments, because they were performed on fibers exposed to TTX and therefore any possible increase in sodium conductance in the tubules had been eliminated. Costantin (1970) used the same experimental approach on single fibers of the frog semitendinosus muscle. When long depolarizing pulses (200 ms) were applied in the absence of TTX, the depolarization of the T-tubules at the axis of the fiber seemed to remain within 3% of the surface depolarization. The radial spread of contraction was reduced by lowering $[Na^+]_o$ or addition of TTX. Short depolarizing pulses (3 ms) resulted in either simultaneous contraction of superficial and axial myofibrils or even stronger contractions of the latter. The opposite was found in TTX-50% Na-Ringer's, where contraction spread radially as surface depolarization increased. A net inward current was consistently observed with brief depolarizations, very seldom with long pulses, and never in the presence of TTX. These results implied the existence of a net inward current in the region involved in the contractile response. The geometry of the tubular system, and the differences in the ionic conductances of the surface and tubular membranes, would explain how a net inward current could flow through the latter without generating an action potential along the whole fiber. Additional evidence in support of this interpretation was supplied by Bezanilla et al. (1972). Moreover, Takeda and Oomura (1968, 1970, 1971, 1972) reported that the sarcotubular system of frog muscle fibers exhibited a regenerative response when immersed in fluoride-rich solutions and in propionate-containing EDTA. This response was blocked by picrotoxin and disappeared in glycerol-treated fibers. Indirect evidence for a regenerative response in the tubular system of cardiac muscle has also been reported (Hermsmeyer et al., 1972). Eisenberg and Costantin (1971) provided the formal basis for Costantin's interpretation of the potential gradient found in the *T*-system. It was shown that when the depolarization of the axially located tubules is greater than the depolarization of the tubules at the surface, the radial current flowing at $r=\xi$ is the total i_w crossing the tubular membranes from the axis of the fiber to a radius $r=\xi$. Since the radial derivative of potential, dV/dr, at this point was negative, an inward current resulted.

(*b*) *Excitation-contraction coupling in muscles from invertebrates.* Using a voltage–current method (1 in Table 1.3), Atwood (1963) measured the mechanical thresholds in muscle fibers of the crab *Carcinus maenas*. Depolarization induced by high $[K^+]_o$ resulted in tension development when the membrane potential reached a value of −55 mV. Dudel et al. (1968) reported that mechanical tension developed almost immediately when crayfish muscle fibers were depolarized to levels close to the threshold for inward current. The force of contraction rose with pulse length, attaining a plateau for pulses with 1 ms duration. Matsumura (1972a, b) analyzed the voltage dependence of contractile activity in crayfish muscle fibers and investigated the effect of divalent ions on this process. The mechanical threshold for potassium contracture was −53 mV, and −50 to −58 mV for depolarizations longer than 0.5 s. Peak tension and the rate of rise of tension increased with membrane depolarization. The time course of tension development could be described by a sum of two exponentials such that the tension at a given time after the onset of the pulse, $T(t)$, is expressed by

$$T(t)=T_\infty-\{A\exp(-t/\tau_1)+B\exp(-t/\tau_2)\} \tag{3.24}$$

where T_∞ is the tension reached at the end of the pulse. A and B are constant in time but voltage-dependent; τ_1 and τ_2 are the time constants. During relaxation, tension fell according to

$$T^r(t)=C\exp(-t/\tau_1')-D\exp(-t/\tau_2')+E\exp(-t/\tau_3') \tag{3.25}$$

indicating that the time course of relaxation consists of three exponentials.

$T^r(t)$ is the tension existing at time t after the end of the pulse and $[C-(D+E)]$ is the value of $T^r(t)$ at $t=0$, in other words, the tension at the end of the pulse. Therefore, C, D, and E have the same voltage and time dependence as T_∞. The time constants τ_1 and τ_1' were related to the elastic and viscous properties of the muscle, τ_2 was related to the development of the active state, and the last two terms of Eq. 3.25 would represent the decay of the active state, determined by the uptake of Ca^{2+} by the sarcoplasmic reticulum. Divalent cations shifted the mechanical threshold toward more positive values and decreased the maximum tension developed at a given membrane potential. The order of this inhibitory effect was

$$\mathrm{Cd} > \mathrm{Co} \gg \mathrm{Mn} > \mathrm{Ni} > \mathrm{Ca} \gg \mathrm{Mg} > \mathrm{Sr} > \mathrm{Ba}$$

Suarez-Kurtz et al. (1972) ruled out the possibility that a potassium current contributes to electromechanical coupling in crayfish muscle fibers. Their findings indicated that a membrane calcium current was required to trigger contractile activity, but the effect of Ca^{2+} on mechanical thresholds depended on the presence or absence of other divalent cations. This result supported the hypothesis of Hagiwara and Nakajima (1966) that calcium ions are adsorbed to sites at the surface membrane and that divalent cations may competitively occupy the calcium binding sites.

3.1.4 Square-Pulse and Voltage-Clamp Measurements on Endplates

The membrane of the endplate region of a muscle fiber lacks the dynamic properties inherent to the surface membrane. The endplate potential (epp) can be compared to an excitatory postsynaptic potential resulting from the interaction of acetylcholine and the membrane sites sensitive to it. Acetylcholine produces a depolarization of about 20 mV in 1.5–3 ms whose decay depends on the passive electrical properties of the fiber. Fatt and Katz (1951) showed that during the epp, the endplate membrane suffers a transient insulation breakdown and becomes a nonselective ion sink, equivalent to placing a leakage resistance across the muscle membrane. This resistance was estimated to be of the order of 20–30 kΩ, one order of magnitude less than the input resistance of the resting membrane. During normal transmission, the endplate leakage resistance is in parallel with the resistance of the active membrane, and it short-circuits, to some degree, the region of the cell membrane close to the endplate, thus reducing the overshoot of the action potential recorded at the endplate or close to it. Del Castillo and Katz (1954) reported that the depolarizing action of the transmitter was reversed to hyperpolarization when the membrane potential was about −15 mV. This value represents the reversal potential of the endplate membrane during the transmitter action. This model and its application to many other biological structures were extensively discussed by Ginsborg (1967).

Maeno (1966) estimated the changes in sodium and potassium conductances during the endplate potential in frog-nerve muscle preparations under the effect of procaine. Since at the resting potential the response of the endplate is a depolarization, it was assumed that the time course of the potential change represented the time course of the change in sodium conductance (Δg_{Na}). The ratio $\Delta g_{Na}/\Delta g_K$ was estimated from the equation

$$\Delta g_{Na}/\Delta g_K = (E_K - E_{ep})/(E_{ep} - E_{Na}) \tag{3.26}$$

where E_{ep} is the endplate equilibrium potential and the other symbols have the usual meaning. Procaine decreased the amplitude of the epp, shortened its rise time, and changed the ratio $\Delta g_{Na}/\Delta g_K$ from 2.75 to 2.11. From this and other results, it was concluded that procaine decreased Δg_{Na} to 64% of its control value, while Δg_K decreased to 82%. This indicates that procaine acts on independent ionic conductances, g_K and g_{Na}, in contrast to the classical model for the endplate membrane, which assumed a single, nonspecific variable conductance.

Nastuk and Parsons (1970) explored the role of calcium and other cations on receptor inactivation in frog muscle. Among the ions studied, calcium was most effective in accelerating post-junctional membrane inactivation. Magnesium inhibited inactivation and antagonized Ca^{2+}.

In contrast to amphibian and mammalian muscles, crustacean muscles are doubly innervated, by motor and inhibitory nerve fibers. Fatt and Katz (1953b) reported that stimulation of inhibitory fibers decreased the amplitude of the endplate potentials elicited by stimulation of the motor nerve in nerve-muscle preparations from the crabs *Eupagurus bernhardus* and *Carcinus maenas*. As in frog muscle, the decay of the epp depended on the passive electrical properties of the fiber, and this decay became faster when the inhibitory nerve was stimulated during the epp. The equilibrium potential for the inhibitory effect was at the level of the normal resting potential, so that the inhibitory action took place without changes in the membrane potential. The results were explained with an equivalent circuit consisting of two parallel conductances representing two independent ionic fluxes: One serves to maintain the resting potential at the level of the equilibrium potential, and the other will tend to depolarize the membrane. The resulting membrane potential will depend on the relative magnitude of the two conductances. This model for the mechanism of excitation and inhibition in crustacean muscle resembles that of Maeno (1966) for the action of procaine on amphibian endplate.

Boistel and Fatt (1958) studied the effect of chloride and potassium concentrations on the reversal potential for inhibition in crayfish muscles. Plots of the amplitude of the inhibitory junctional potential versus the membrane potential followed an S-shaped curve. This nonlinearity was attributed to rectification of the nonjunctional membrane, whose voltage–current curve had a slope of 100 kΩ at zero current, but decreased when the potential was displaced more than 10 mV in either direction. Potassium concentrations of 0–20 mM did not modify the equilibrium potential for inhibition. Rectification disappeared in 20 mM potassium, and the input resistance at zero current fell to one-third of its resting value. Replacement of chloride by nonpermeant anions caused a fall in resting potential of 10–20 mV without altering the voltage–current curve. Despite membrane depolarization, the inhibitory junctional potentials appeared also as depolarizations and the reversal potential was 15 mV more positive than the resting potential. This is consistent with an increase of chloride conductance of the junctional membrane during inhibitory action. The results also indicated that the potassium conductance of the junctional membranè is lower than the potassium conductance of the nonjunctional membrane.

Takeuchi and Takeuchi (1959) measured the membrane current at the endplate region during the active phase of the epp, on a nerve-muscle preparation of frog sartorius. Stimulation of the nerve under voltage-clamp conditions resulted in an inward current reaching a maximum at 0.77 ms and then falling exponentially with a half-time of 1.08 ms. Current–voltage curves were linear for membrane potentials between −120 and −50 mV and the reversal potential was −10 to −20 mV, in agreement with the value reported by Del Castillo and Katz (1954). A value of 380 kΩ was obtained for the shunting resistance. This value is larger than other estimates for the endplate in normal medium because the experiments were performed in curarized muscles, where the effect of the transmitter was partly blocked.

The voltage-clamp method was improved by Takeuchi and Takeuchi (1960a), who showed that sartorius and M. extensor longus digitorum IV had endplate currents of the same amplitude, but the amplitude of the epp in M. extensor longus digitorum IV was twice that found in sartorius muscle. The time course of the current was the same in both muscles, but the input resistance was 440 kΩ in sartorius and 810 kΩ in the toe muscle. This showed that the amplitude of the epp is a function of potential and input resistance. Similar conclusions were drawn by Oomura and Tomita (1960) from voltage-clamp measurements of endplate current (epc) in normal medium and nitrate Ringer's. The latter shifted the curve epc versus epp, so that a much larger epp appeared for the same value of endplate current. Sodium-free medium caused depolarization and shifted the reversal potential to values closer to the normal resting potential. Therefore, stimulation of the nerve produced a hyperpolarizing endplate potential.

Takeuchi and Takeuchi (1960b) reported that tubocurarine changed the slope of the current–voltage curve without altering the reversal potential. Lowering the sodium concentration to 33.6 mM shifted the equilibrium potential toward the resting potential by 17 mV. Electrophoretical injection of Na^+ had the same effect. An increase in $[K^+]_o$ from 0.5 to 4.5 mM made the reversal potential more positive by 28 mV. A linear relationship between the equilibrium potential and $\log[K^+]_o$ was found. Changes in chloride concentration indicated a much smaller contribution of chloride ions to the endplate potential than to the resting potential of the conducting membrane. It was postulated that the epp resulted from changes of the permeability to potassium and sodium ions, but not to chloride. From Eq. 2.12, the following expression was obtained for the endplate current:

$$I_{ep} = (\Delta g_{Na} + \Delta g_K)\left(V - \frac{E_K + (\Delta g_{Na}/\Delta g_K)E_{Na}}{1 + \Delta g_{Na}/\Delta g_K} \right) \tag{3.27}$$

where I_{ep} is the endplate current, Δg is the corresponding conductance change, and the second factor in the equation represents the difference between the membrane potential and the endplate equilibrium potential. A value of 1.29 was obtained for $\Delta g_{Na}/\Delta g_K$. The endplate current will then be composed of inward sodium current or outward potassium current according to the sign of the difference between the membrane potential and the equilibrium potential. It was concluded that the transmitter increases the endplate conductance to sodium and potassium with a constant ratio of $\Delta g_{Na}/\Delta g_K$.

Takeuchi (1963) reported that increasing $[Ca^{2+}]_o$ shifted the reversal potential to more negative values and increased the conductance change at the peak of the transmitter action. This was attributed to the larger transmitter release occurring in high-calcium media (Del Castillo and Stark, 1952). The effect of calcium ions on the current–voltage curves was consistent with a reduction of the $\Delta g_{Na}/\Delta g_K$ ratio to 70% of the value found in 2 mM Ca^{2+}. A probable increase in membrane calcium conductance during transmitter action could not be excluded in interpreting the results. Therefore, the terms E_{Ca} and Δg_{Ca} were included in Eq. 3.27 to account for the contribution of calcium ions to the endplate current. The best fit to the experimental curves was obtained with the following values: $\Delta g_{Ca}/\Delta g_K$ accounting for 11% of $\Delta g_{Na}/\Delta g_K$ in 30 mM Ca^{2+} and 7.7% in 2 mM Ca^{2+}. The contribution of calcium ions to the endplate current is small compared to sodium and potassium in normal medium, but it becomes important when extracellular sodium is greatly reduced. This suggests a competition between Ca^{2+} and Na^+ for membrane sites or a Ca^{2+} channel in the endplate, as postulated for other structures.

Deguchi and Narahashi (1971) showed that the sodium and potassium components of the epc are differently affected by procaine, thus supporting the endplate model proposed by Maeno (1966). Procaine (6.5×10^{-5} M, and 2×10^{-4} M) decreased the peak amplitude of I_{Na} and I_K, its effect being larger on I_{Na}. The half-time of decay for I_{Na} was shortened, while that for I_K was prolonged. All these effects were concentration dependent. Current–voltage curves for the peak current were linear over the range -100 to -50 mV in control conditions, and a value of 1.79 was obtained for g_{Na}/g_K. A late inward current appeared under procaine. The current–voltage curves for this current showed strong rectification, in contrast to the slope conductance for the peak current. These observations suggested the existence of a third component in the endplate current. This supported the hypothesis of Takeuchi (1963) and favored the view that separate channels exist for sodium and potassium in the endplate membrane.

On the basis of a study of miniature endplate currents, Gage and Armstrong (1968) suggested that the time course of the changes in G_{Na} and G_K are different. However, Kordaš (1969) reported that the voltage dependence of the endplate current did not agree with the changes of shape, decay, and amplitude of the endplate current predicted by the hypothesis of Gage and Armstrong. The time for half-decay of the current increased with membrane polarization. The relation between current amplitude and potential was slightly nonlinear at membrane potentials between 0 and -120 mV. This nonlinearity was later confirmed by Magleby and Stevens (1972b). It was proposed that the changes in the time course of the endplate current with hyperpolarization depended on the stability of the receptor-mediator complex rather than on different time courses of Δg_{Na} and Δg_K.

This hypothesis found support in recent results obtained independently by Magleby and Stevens (1972a, b) and Kordaš (1972a, b). It was confirmed that the rate of decay of the endplate current depends on the membrane potential, according to:

$$I(t)=I_0\exp[-\alpha(V)t] \tag{3.28}$$

where $I(t)$ is the epc, t is the time after an arbitrary reference time $t=0$, I_0 is the current at $t=0$, and $\alpha(V)=B\exp(AV)$, where B and A are constants. The voltage dependence of α persisted despite changes of the current amplitude induced by curare, increased $[Ca^{2+}]_o$, facilitation, depression, and inhibition of acetylcholinesterase with prostigmine. Therefore, a possible influence of V on the hydrolytic rate of acetylcholine was ruled out as the cause of the voltage-dependent decay of the endplate current. The instantaneous current–voltage relation was linear, suggesting that the nonlinearity of the current–voltage curve for peak current depended on the number of ionic channels open at a given potential, but each channel would have a linear current–voltage relation.

The results were analyzed with a model based on enzyme kinetics and indicated that the voltage sensitivity of the current decay depended on the stability of the complex acetylcholine-receptor, according to

$$\text{Release of acetylcholine (A)} \longrightarrow \overset{\substack{\text{hydrolysis} \\ \uparrow k_3}}{\underset{\substack{\downarrow k_2 \\ \text{diffusion}}}{\text{A}}} + \text{R} \underset{k_1}{\overset{k_{-1}}{\rightleftharpoons}} \text{AR}$$

Assuming that the increase in conductance is a linear function of [AR] and that it follows the variations of [AR] without appreciable time lag, Gabrovec et al. (appendix in Kordaš, 1972a) showed that the observed time course of current decay occurs only if the release of A is very fast, k_1/k_{-1} is relatively small, and $k_1 \ll k_2 + k_3$.

Magleby and Stevens (1972b) tried to explain the time course of the endplate current on a molecular basis. It was assumed that the endplate conductance g is proportional to a factor x that indicates receptors complexed to the transmitter in an open conformation and a factor γ that represents the conductance of one open channel, so that $g = x\gamma$. The change of the receptor from the close to the open conformation would be controlled by the rate constants α and β according to

$$\text{A} + \text{R} \underset{k_1}{\overset{k_{-1}}{\rightleftharpoons}} \text{AR} \underset{\alpha}{\overset{\beta}{\rightleftharpoons}} \text{AR}^*$$

where the open conformation is indicated by an asterisk. The experimental evidence suggested that the time course of the conductance change was determined by the rate of conformational change rather than the acetylcholine concentration in the synaptic cleft. Assuming that acetylcholine is rapidly removed from the cleft through binding, diffusion, and hydrolysis, [A] in the cleft falls soon to near zero, and the time course of the conductance change is given by $dg/dt = -\alpha g$. As the rate constant α determined the decay of the endplate current, the constant β characterizes the increase in conductance: $\beta(V) = b\exp(aV)$, where a and b depend on changes of the dipole moment associated with the conformational changes of the receptor molecule. The values of a and b can be experimentally determined. The time course of the current as a function of potential was computed with the empirically obtained values of A and B (Eq. 3.28). The resulting curves agreed well with measured values. These observations supported the view that only two voltage-dependent channels can account for the voltage dependence of the time course of the endplate current. The rate constants α and β are similar to the homologous rate constants of the Hodgkin-Huxley theory.

3.2 CARDIAC MUSCLE

Impedance measurements in cardiac muscle developed only after the microelectrode technique was available. This contributed to the understanding of myocardium properties such as automatic activity, propagation of the impulse, and long refractory period, properties that were described long ago (Marey, 1876; Eyster and Meek, 1921). Intracellular recordings of resting and action potentials from heart fibers were obtained soon after Ling and Gerard (1949) developed the glass microelectrode (Coraboeuf and Weidmann, 1949; Woodbury, Woodbury, and Hecht, 1950). A spontaneous diastolic depolarization was found in Purkinje fibers (Draper and Weidmann, 1951) and in pacemaker cells of rabbit hearts (West, 1955). Impedance measurements were more difficult to interpret than in nerve fibers or skeletal muscle, because of the two- or three-dimensional spread of current in most preparations.

3.2.1 AC Measurements in Cardiac Muscle

Relatively few ac measurements have been performed on cardiac muscle, and they lack the sophistication of ac measurements performed on cell suspensions or skeletal muscle. Rapport and Ray (1927) measured the conductivity of the whole tortoise heart at 200–300 Hz and reported that the increase of intraventricular pressure during the systole was accompanied by an increase in conductivity. Several decades later, Sperelakis and Hoshiko (1961) measured with extracellular electrodes the impedance of strips of cat ventricular muscle in a frequency range of 10–10^4 Hz. In isotonic sucrose, the impedance increased by a factor of 10. The relative impedance did not change with frequency for muscles incubated in Tyrode solution, but it decreased after incubation in isotonic sucrose. The results were interpreted to indicate the presence of high-resistance transcellular membranes. This was based on the assumption that incubation in sucrose reduced the ion concentration in the extracellular space tenfold, whereas the intracellular ion concentration remained constant. However, no experimental evidence was given to support this assumption.

Guilbault, Delahayes, and Paillard (1966) studied the influence of Ca^{2+} and Mg^{2+} on the action potential of the perfused guinea pig ventricle. Reduction of Ca^{2+} and Mg^{2+} by EDTA produced a delayed increase of the apparent resistance during the plateau. More recently, Antoni, Töppler and Krause (1970) investigated the influence of low-frequency alternating current on the isolated papillary muscle of rhesus monkeys. Subthreshold ac signals caused a delayed depolarization that, when threshold was reached, resulted in a series of action potentials. Further increase of the amplitude of the imposed signal caused disappearance of the action potential, then oscillations of the resting potential, and finally a stable steady-state level. The results were explained by an activation of the sodium system by alternating current.

Freygang and Trautwein (1969, 1970) proposed a three-time-constant model for Purkinje strands of sheep hearts, because the phase angle was larger than 45° at frequencies above 1 kHz, and 45° is the upper limiting value for the two-time-constant model proposed by Fozzard (1966, Sec. 3.2.2). Because of the poor development of the *T*-system in cardiac muscle and the presence of intercalated discs, two additional impedances were introduced and related to the disc system. Measurements of the longitudinal impedance resulted in impedance loci (25 Hz to 50 kHz) with two dispersions.

Paes de Carvalho et al. (1969) recorded phase-plane trajectories (Sec. 3.1.1) in rabbit atrial preparations. The different types of cardiac action potentials could be explained in terms of a fast and a slow phase of depolarization. The action potentials from the sinus and atrioventricular nodes lacked the initial sodium spike. However, Ruiz de Ceretti et al. (1971) reported that the initial phase of depolarization of action potentials from cells of the atrioventricular node was selectively depressed by low sodium concentrations.

3.2.2 Square-Pulse Measurements in Cardiac Muscle

In addition to the theoretical difficulties of interpreting input impedance in cardiac muscle, methodological problems have led to the wide use of methods employing a single microelectrode with bridge circuits. The quantity thus measured is an apparent resistance that has often been considered to be the cell input resistance; for example, values of 6 MΩ and 44–132 kΩ have been reported for cat ventricle. The first one is the apparent resistance of the cell.

According to the biological property investigated, the results are discussed here as follows: in Sec. 3.2.2.1, measurements of the electrical cell constants; in Secs. 3.2.2.2 and 3.2.2.3, measurements related to impulse propagation and pacemaker activity; and in Sec. 3.2.2.4, the changes in membrane resistance during activity, the rectifying properties of the membrane, and estimations of the ionic conductances.

3.2.2.1 Passive Electrical Properties of Cardiac Muscle. The cell constants of Purkinje fibers were first determined by Weidmann (1952). The application of the linear cable model was complicated because of branching of the fibers, so the appropriate equations for a short cable were derived. The values of the original quantities measured and the cell constants are listed in Table 3.11. The value of R_i corresponds to the resistivity of the core of a fiber strand, consisting of two to five cylindrical cells. The low value obtained indicated a cytoplasmic continuity between cells, but it is accepted today that low-resistance junctions exist between individual cells. The value of R_m for Purkinje fibers lies between the values found in nonmyelinated nerves and frog sartorius muscle (Tables 2.4 and 3.2). The membrane capacity is high compared to that of skeletal muscle and nerve. This was attributed to infoldings of the surface membrane. However, Fozzard (1966) showed that this was due to the contribution of the T-system to the total membrane capacity, because this system is rather well developed in some types of cardiac muscle (Porter, 1961; Simpson and Oertelis, 1962). Table 3.11 shows the capacitance values obtained by square-pulse analysis and from the foot of the action potential. The difference between them was attributed to the existence of two capacity components, which would make the two-time-constant model of Falk and Fatt (1964) also applicable to Purkinje fibers. Despite differences in the degree of development of the T-system in cardiac muscle (Muir, 1957 a, b), it seems that a large fraction of the combined membrane capacity is related to it. In frog ventricle, Van der Kloot and Dane (1964) showed that similar values of C_m were obtained with the square-pulse technique and the time constant of the foot of the action potential, as expected for this type of muscle, which lacks the transverse tubular system.

Recently, Mobley and Page (1972) analyzed quantitatively light and electron micrographs of sheep Purkinje cells, thus providing the morphological basis for the localization of the large capacity measured with square pulses. From these measurements, two surface areas were distinguished: (1) the total surface area of the fiber, which results from adding the cell

Table 3.11 Electrical Cell Constants of Cardiac Muscle Obtained with Square-Pulse Measurements

Method	Measured Quantities	Model	Derived Quantities	Reference
1 of Table 1.3	R_{inp}: 478 kΩ λ: 1.9 mm τ_{inp}: 19.5 ms	5 of Table 1.5	R_m: 1900 Ω cm^2 R_i: 105 Ω cm C_m: 12.4 μF/cm^2	Weidmann (1952)
1 of Table 1.3	R_{eff}: 130–760 kΩ λ: 1.3–2.6 mm τ_{inp}: 17.5–24 ms	5 of Table 1.5 Eq. 1.43	R_m: 1130–2300 Ω cm^2 R_i: 46–315 Ω cm C_m: 12.8 μF/cm^2 C_s: 2.4 μF/cm^2	Fozzard (1966)
2 of Table 1.3	R_{inp}: 187 kΩ R_{app}: 11 MΩ τ_{inp}: 3.2 ms λ: 375 μm	5 cf Table 1.5	R_m: 2600 Ω cm^2 R_i: 460 Ω cm C_m: 1.2 μF/cm^2	Van der Kloot and Dane (1964)
1 of Table1.3	R_{eff}: 160 kΩ λ: 0.83 mm τ_{inp}: 3.7 ms	5 cf Table 1.5	R_{inp}: 2 MΩ assuming $R_i = 150$ Ω cm and $R_m = 1900$ Ω cm^2	Matsuda (1960)
2 of Table1.3	$\Delta AP/I$		ΔR_{inp} (kΩ) Purkinje fibers: 84 (dog) 132 (cat) 256 (guinea pig) 104 (rat) Ventricular fibers: 44 (cat) 78 (guinea pig) 43 (rat)	Johnson and Wilson (1962)

Table 3.11 *Continued*

Method	Measured Quantities	Model	Derived Quantities	Reference
2 of Table1.3	Rat ventricle: ΔR_{inp}: 38.5 kΩ R_{inp}: 153 kΩ			Coraboeuf and Vassort, quoted by Coraboeuf (1969)
5 of Table1.3	R_{app}: 1.45 MΩ			
2 of Table 1.3	Guinea-pig ventricle: ΔR_{inp}: 14 kΩ R_{inp}: 50 kΩ			
1 of Table 1.1	τ_{inp}: 3.4–9.8 ms λ: 0.23–0.41 mm	4 of Table 1.5	R_m: 280 Ω cm^2 C_m: 3 μF/cm^2	Trautwein, Kuffler, and Edwards (1956)
2 of Table1.3	R_{inp}: 70–200 kΩ λ: 50–400 μm τ_{inp}: 1–8 ms			Woodbury and Gordon (1965)
Fig. 3.17a	λ: 1.23 mm τ_{inp}: 4.2 ms ($t^{1/2}$ vs x) τ_{inp}: 2.5 ms (τ of ΔV) τ_f: 1.13 ms (foot AP)	5 of Table 1.5	R_m: 3503 Ω cm^2 C_m: 1 μF/cm^2 C_m: 0.76 μF/cm^2	Sakamoto (1969)
1 of Table 1.3	R_{eff}: 160–250 kΩ λ: 880 μm τ_m: 4.4 ms τ_f: 380 μs d: 15 μm Extracellular space: 32%	4 of Table 1.5	R_m: 4000 Ω cm^2 R_i: 470 Ω cm R_e: 47 Ω cm R_m: 9100 Ω cm^2 C_m: 0.81 μF/cm^2 C_s: 0.59 μF/cm^2	Weidmann (1970)

5 of Table 1.3	R_{app}: 3.8 MΩ τ_{inp}: 0.32 ms R_{inp}: 24 kΩ	1 of Table 1.5 1 of Table 1.5	R_m: 1360 Ω cm^2 C_m: 1.6 μF/cm^2 R_m: 190 Ω cm^2	McCann (1966) Kriebel (1968)
1 of Table 1.3	R_{inp}: 134 kΩ λ: 1.17 mm τ_{inp}: 7.3 ms R_{inp}: 163 kΩ λ: 1.28 mm	5 of Table 1.5	R_m: 707 Ω cm^2 R_i: 202 Ω cm R_m: 1297 Ω cm^2 R_i: 300 Ω cm	Fozzard and Dominguez (1969)
5 of Table 1.3	R_{app}: 5.5 MΩ R_{app}: 6.4 MΩ	1 of Table 1.5 5 of Table 1.5 1 of Table 1.5	R_m: 315 Ω cm^2 (frog) λ: 385 μm R_m: 489 Ω cm^2 (cat)	Tarr and Sperelakis (1964)
1 of Table 1.3	$\Delta V = f(x)$ λ: 360 μm	10 of Table 1.5	R_m: 1316 Ω cm^2 R_d: 0.25–1.25 Ω cm^2 $(f=1)$ R_i: 502 Ω cm C_m: 1.3 μF/cm^2 R_m: 2632 Ω cm^2 $(f=2)$ R_d: 0.25–1.25 Ω cm^2 R_i: 502 Ω cm C_m: 0.65 μF/cm^2	Jongsma and van Rijn (1972)
5 of Table 1.3	R_{app}: 36 MΩ	1 of Table 1.5	R_m: 1440 Ω cm^2	Pappano and Sperelakis (1969 b)

area facing the connective tissue sheath, and (2) the internal surface, which is the total area formed by the portions of the surface of the cells that do not face the external sheath. The total surface area of a sheep Purkinje fiber was 10–12 times the area calculated for a smooth cylinder of similar dimensions. The ratio between the internal and external surfaces for a cylindrical fiber 100 μm in diameter was 4 or 5 to 1, which predicts a ratio for the two capacity components close to that obtained by Fozzard (5.3 to 1).

Matsuda (1960) investigated the passive electrical properties of terminal Purkinje fibers in the dog heart. The decay of the electrotonic potential was exponential for interelectrode distances larger than 100 μm. The length constant (Table 3.11) was measured in the whole bundle and calculated from the exponential part of the electrotonic decay. At smaller separation of the electrodes, the electrotonic decay was more abrupt. This was explained by the existence of low-resistance pathways between fibers. Using the Frank and Fuortes (1956) method, Johnson and Tille (1961) measured the electrical characteristics of cells from the rabbit ventricle. Weidmann (1955b) showed that shifts in membrane potential in Purkinje fibers influenced the size of the action potential and the overshoot. When the membrane is hyperpolarized, the amplitude of the action potential becomes constant at membrane potentials of -100 to -110 mV. The voltage–current curve obtained by Johnson and Tille in Purkinje fibers showed two parts: a linear relationship for membrane potentials higher than -100 mV and a nonlinear portion indicating that the input resistance at the crest of the action potential is not negligible. Two types of voltage–current curves were found in ventricle fibers. In one of them, current did not change the amplitude of the action potential but affected the activation mechanism because the depolarization rate changed. This indicated that the input resistance recovered its resting value promptly during the upstroke. A second type of ventricle fiber showed a curve with zero slope up to a certain value of hyperpolarizing current, beyond which the amplitude of the action potential increased linearly with current. The amplitude of this spike was the same as the increase in action potential amplitude.

Using the same principle, Johnson and Wilson (1962) performed a comparative study on different species. In rabbit and guinea pig hearts, the action potential amplitude reached a maximum as the hyperpolarization increased. Beyond this point, a further increase in current resulted in a negative slope of the voltage–current curve. It was interpreted that the input resistance at the crest of the action potential was higher than at rest. Table 3.11 lists these results and those of Coraboeuf and Vassort (quoted by Coraboeuf, 1969), who showed that in rat and guinea pig ventricle, the input resistance at crest is not negligible. Therefore, the value obtained with the Frank-Fuortes method is considerably lower than the resting input resistance.

Tanaka and Sasaki (1966) measured electrotonic potentials in strips of mouse ventricle and analyzed them with their lattice model (9 in Table 1.5). The two-dimensional electrotonic decay was not exponential, and apparent length constants of 70 and 700 μm were obtained for short and long interelectrode distances, respectively. The extrapolation to zero distance gave an input resistance between 200 and 600 kΩ. The sharp electrotonic decay found in the mouse ventricle agrees with similar findings in other species (Matsuda, 1960; Sakamoto, 1969). The resistance per unit area of intercalated disc was calculated as 2 Ω cm^2, assuming a cell length between 50 and 200 μm and considering that the cytoplasmic resistivity is lumped together with the resistance of the intercalated discs.

A similar electrotonic decay was reported by Woodbury and Crill (1961) in rat atria trabecula. The decay in the transverse direction was sharper than in the longitudinal direction. The resistance of the intercalated discs was estimated as 1 Ω cm^2. Woodbury and Crill (1970) later compared their model with the theoretical treatment developed by Heppner and Plonsey (1970). The values agreed fairly well with previous estimates. The cytoplasmic resistivity of rat atrial cells was directly measured by Schanne, Thomas, and Ceretti (1966), applying a combination of methods 2 and 3 of Table 1.3. The values of R_{inp} (177 kΩ) and R_i (120 Ω cm) are within the range reported by others (Table 3.11). The membrane resistance of atrial cells was estimated as 1000–2000 Ω cm^2.

The theoretical analysis of resistance measurements in cardiac muscle is simplified if the preparations are polarized with extracellular electrodes, because the current spreads as in a linear cable. However, Kamiyama and Matsuda (1966) showed that the input resistance of the cell cannot be calculated because the current flowing through the cell membrane is not known. Trautwein, Kuffler, and Edwards (1956) investigated the passive electrical properties of bundles of frog atrial muscle. The application of acetylcholine shortened the space constant by 30%, and it was concluded that this was due to a large decrease in R_m. Woodbury and Gordon (1965) used a current electrode of 50 μm diameter to polarize frog atrial trabecula of about 80 μm diameter. The length constant and the input resistance were obtained from plots like those shown in Fig. 3.6 and are listed in Table 3.11.

Kamiyama and Matsuda (1966) investigated the electrical properties of canine ventricular muscle with the experimental arrangement shown in Fig. 3.17a. The changes in membrane potential were recorded through a microelectrode at different distances from the partition, considered the point of current injection. The electrotonic decay was exponential, result-

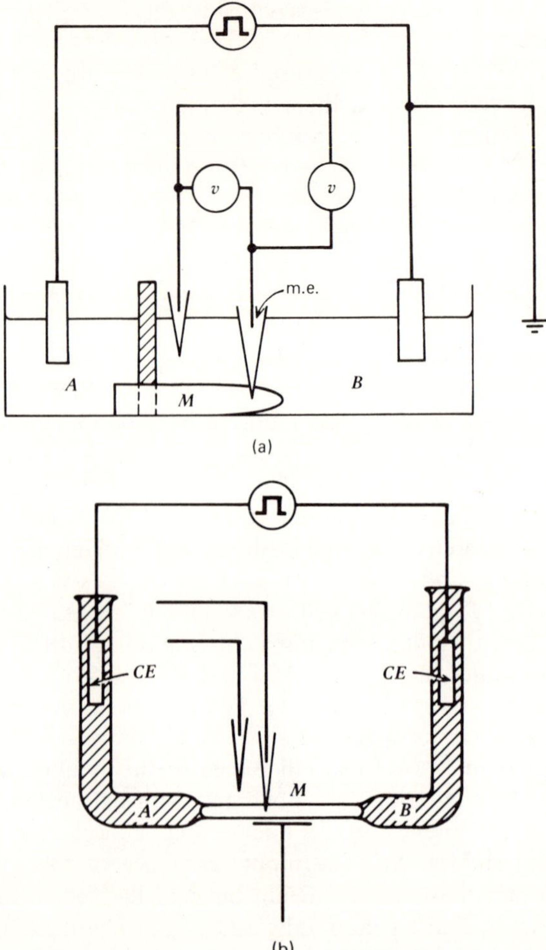

Figure 3.17. (a) Experimental arrangement to record electrotonic spread in papillary muscle. *A* and *B*, chambers of the tissue bath; m.e., recording microelectrode; *M*, muscle. (b) Experimental arrangement to determine the cell constants in trabecula of ventricular muscle. *M*, muscle; *A* and *B*, tissue holders filled with Tyrode solution and containing the current electrodes (*CE*).

ing in a λ of 1.3 mm, and the input time constant was 2 ms. The decrease of R_{inp} at the crest of the spike was very small, as in rabbit ventricle. During the repolarization phase, the input resistance had the same value as at rest. The short length constant and the difficulty of eliciting an action potential or inducing repolarization in ventricular muscle by intracellular injection of current were attributed to a small fiber diameter, a higher degree of ramification, and more numerous intercellular connections. Similar results were later obtained by Sakamoto (1969) in dog papillary muscle. The voltage–current curve was linear over the range ± 15 mV. The value of C_m (Table 3.11) is small compared to the 2.4 $\mu F/cm^2$ reported by Fozzard (1966) for Purkinje fibers, but this may be due to a different structural arrangement. Polarizing the cells by intracellular injection of current resulted in a very sharp electrotonic decay, suggesting a three-dimensional spread of current. Weidmann (1970) polarized ventricular trabecula with the method shown in Fig. 3.17b. The surface component of the membrane capacity was measured from the foot of the action potential, according to

$$C_s = (Ka/2R_i v^2)[r_i/(r_i + r_e)] \tag{3.29}$$

where the parallel longitudinal resistance is taken into account. The cell constants are listed in Table 3.11. The cytoplasmic resistivity is higher than for other cardiac cells. The value of λ is in good agreement with those published for ventricular muscle in other species. The value of R_m agrees well with the potassium conductance of the surface membrane as predicted from measurements of K^{42} efflux (Weidmann, 1966).

McCann (1966) studied the effects of current injection in single cells of the moth heart. These cells are permeable to large anions, which suggests a low-resistance membrane with high permeability coefficients (McCann, 1964). Voltage–current curves were linear within the range ± 20 mV, from which the apparent resistance was determined. The value of R_m is of the same order of magnitude as those found in other excitable structures, but no conclusion can be drawn on the permeability characteristics of the membrane, because the original quantity measured was not a true input resistance.

Kriebel (1968) determined the electrical constants of tunicate heart cells, *Ciona intestinales* and *Chelyosoma productum*. These hearts are tubular structures, and when cut along the longitudinal axis they yield a sheetlike preparation of a single layer of cells. The transversal resistance was measured as shown in Fig. 3.18a. An insulating ring sealed off a chamber of 1 mm^2 surface, so that the area for current flow was known. The longitudinal resistance was measured by the sucrose-gap technique shown in Fig. 3.18b. The cell constants (Table 3.11) were calculated using geometrical data obtained with histological methods. Assuming a cytoplasmic resistivity twice that of seawater, the resistance of the cytoplasm would account for only 3% of the total longitudinal resistance. The specific nexus resistance was 0.3 Ω cm^2 in *Chelyosoma* and 0.2 Ω cm^2 in *Ciona* hearts.

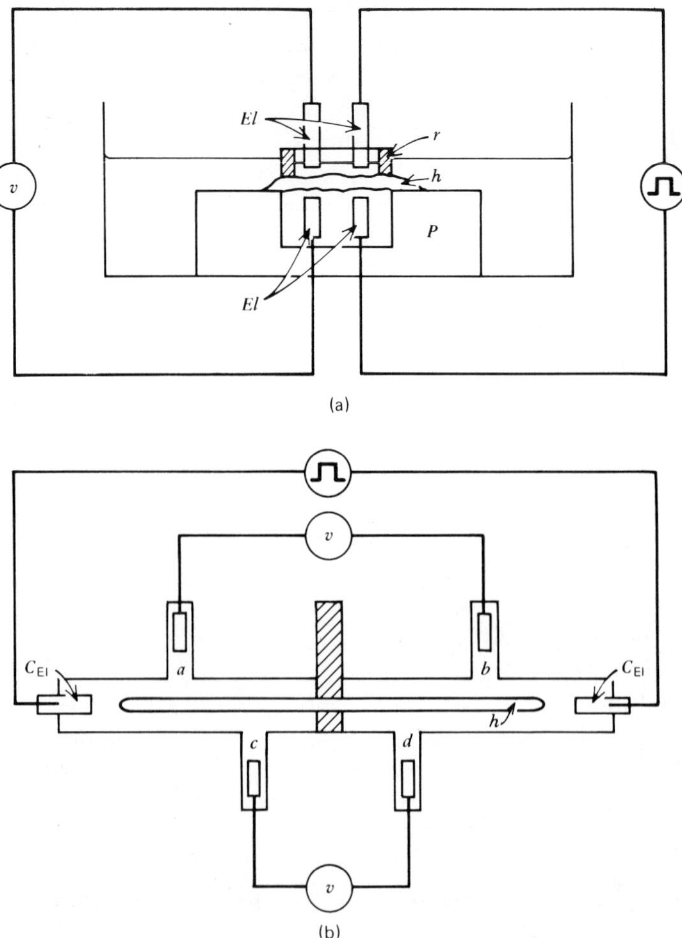

Figure 3.18. (a) Experimental arrangement to measure the transverse resistance in the sheet preparation of the tunicate heart: *r*, ring; El, electrodes; *h*, heart; *P*, Plexiglas block; *v*, voltmeter. (b) Sucrose-gap technique to measure the longitudinal resistance of the tunicate heart. Shaded area, sucrose; *h*, heart; C_{El}, current electrodes; a, b, c, d, branches containing sea water; *v*, voltmeters.

3.2.2.2 Resistance Measurements Related to the Mechanism of Impulse Propagation. Impulse propagation in the heart occurs as in a single cell (Eyster and Meek, 1921), and this view has to be reconciled with the fact that cardiac muscle consists of distinct cells (Sjöstrand and Andersson, 1954) surrounded by a membrane of rather high resistivity. Electrophysiological evidence for electrotonic spread between adjacent cells was supplied by Weidmann (1952), Woodbury and Crill (1961), and Trautwein et al. (1956), who showed that the value of λ exceeds one cell length. Electron-microscope studies of cardiac muscle revealed a region of close membrane apposition (the nexus), believed to be a site of low resistance, thus allowing the electrotonic spread between cells.* Woodbury and Crill (1961) estimated that the disc resistance would be between 1.2 and 12 Ω cm^2. This estimation agreed with the calculation of Weidmann (1965) made from measurements of the longitudinal flux of K^{42} in sheep ventricle strips. The distribution of K^{42} followed a profile corresponding to the electrotonic spread in a linear cable, with λ of 1.3–2.2 mm. From this, and the time constant of K^{42} efflux, a disc resistance of 3 Ω cm^2 was obtained. If this result is corrected for cytoplasmic resistivity, the disc resistance becomes 1 Ω cm^2 (Weidmann, 1966). The value for the nexus resistance reported by Kriebel (1968) for the tunicate heart agrees closely with these estimations.

Spira (1971) performed a morphological study of canine cardiac muscle to provide the necessary data to estimate the disc resistance from measurements of the cell constants. In a trabecula of cardiac muscle, the resistance of the core per unit length (r_c) is given by: $r_c = r_i + r_d$, where r_d is the resistance of the disc per unit length of fiber and is related to the resistance of the disc per square centimeter of fiber surface (R_d) by the relation $r_d - nR_d/\pi a^2$, where n is the number of discs per centimeter length of fiber. If the nexus is the only path for current flow, the nexus resistance (R_n) can replace R_d in this expression, and then λ is given by:

$$\lambda = R_m a / \left[2(R_i + nR_n/f) \right] \tag{3.30}$$

where f represents the fraction of the fiber cross-sectional area corresponding to the area of the nexus. The morphological factors a, n, and f were determined in electron micrographs. Using data from the literature for R_m, λ, and R_i, the nexus resistance was estimated as 1.4 Ω cm^2, in agreement with other reports (Table 3.11).

Barr and Berger (1964) showed that a sucrose gap between two pools of Ringer's solution prevented impulse propagation along an atrial strip placed across the gap. Conduction was restored when an external resistor bridged the gap. Similar experiments performed by Barr, Dewey, and Berger (1965) showed that when the sucrose solution in the gap was hypertonic, the longitudinal resistance of the tissue increased to the point

*The possible influence of frequency and extent of the nexus structure on the conduction properties of different regions of the heart has been reviewed by Challice (1971). DeFelice and Challice (1969) correlated structural features of the atrioventricular junction with its electrophysiological characteristics.

that propagation was blocked, and it was not restored by switching in the external shunt resistor. Electron micrographs showed disruption of the nexuses and lack of regions of close membrane apposition along the intercalated discs. These findings are consistent with the theory of propagation by electrotonic spread, but some other results apparently contradict that conclusion (Sperelakis, 1969a). Sperelakis, Hoshiko, and Berne (1960) used the circuit shown in Fig. 3.19 to investigate the intercellular resistance in frog ventricle. The value of the resistance between two cells (24.8 MΩ) was double the resistance measured in a single cell (12.4 MΩ), and this was

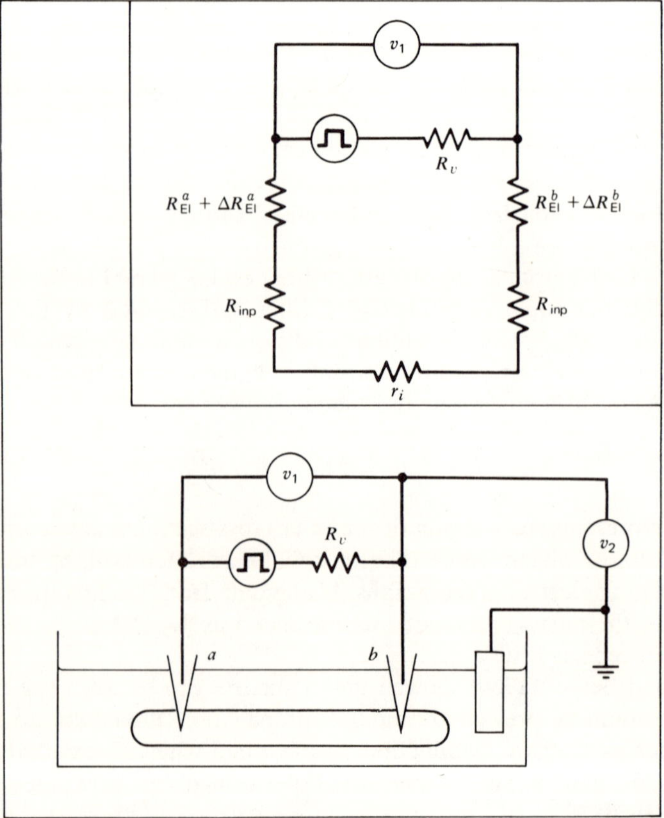

Figure 3.19. Determination of intercellular resistance. Current is injected simultaneously through electrodes a and b impaled in the tissue. v_1 records the potential difference between a and b, v_2 records the potential at b with respect to ground. Inset: equivalent circuit of the resistances measured when the two electrodes a and b are in the cells. r_i: longitudinal resistance of the core, R_v is a high resistance that, together with the pulse generator, forms a constant current source.

interpreted as an indication of electrical isolation of the cells. However, the electrical equivalent of Fig. 3.19 suggests that the reported result may be attributed to ΔR_{El} being much larger than R_{inp} and r_i so that the contribution of the latter to the measured impedance becomes negligible.

Tarr and Sperelakis (1964) examined the electrotonic spread in frog and cat ventricle at different interelectrode distances. A bridge circuit was used for current injection and potential recording at $x=0$, and a second microelectrode recorded the change in potential. The electrotonic spread was evaluated by the degree of interaction: the ratio of ΔV at the second electrode to ΔV recorded at the first electrode. At short interelectrode distances (7–45 μm), a large variation of the degree of interaction was found. At electrode distances of 60–390 μm, no interaction occurred. It was concluded that when strong interaction existed, the electrodes were in the same cell. At greater distances, strong interaction did not occur, and this was interpreted as no flow of current between cells. The resistance values agreed well with the apparent resistances reported by Kawata (1965) in frog ventricle. Tarr (1967) further investigated the cause of the weak electrotonic interaction observed even when the electrodes were in the same cell. He examined the validity of the bridge circuit to measure R_{inp} and found an average increase of 5 MΩ in R_{El} when the electrode was changed from a medium of 70 Ω cm resistivity to one of 150 Ω cm. This introduced a large error into the bridge measurements because the measured apparent resistances were of the order of 5–10 MΩ. This finding also explained that the R_{inp} calculated from the ΔV recorded with the second electrode was much lower than the one recorded with the first electrode, even when both electrodes were in the same cell. Additional measurements made by Tarr (1967) gave input resistances of 260–560 kΩ for cat heart and 250–1450 kΩ for frog heart. These input resistances were of the order of magnitude expected for a branching syncytium, if R_m is assumed to be 1000 Ω cm^2. It was concluded that a weak electrotonic interaction does not necessarily mean that the cells are electrically isolated. Tille (1966) reported some results that supported this interpretation. In rabbit hearts, the spatial electrotonic decay was much sharper in ventricle than in Purkinje fibers. Weak electrotonic interaction was sometimes found between ventricular fibers, but the electrotonic potential could be detected up to 800 μm in ventricular fibers and more than 1 mm in Purkinje fibers.

Jongsma and van Rijn (1972) showed that the disc model developed by Shiba (model 10 in Table 1.5) fit the potential decay measured in monolayer cultures of rat heart cells. Since the cells are attached to the culture dish, an uncertainty enters the application of the model because it is difficult to estimate the leakage resistance of the lower surface of the culture. The cell constants were calculated for a correction factor $f=1$ (no leakage) and $f=2$ (current path through the entire lower surface). The spatial distribution of potential was best fit by Eqs. 1.58 and 1.59 with the value of 360 μm for λ and 0.16 for $m=\rho_i/2\pi d$, where ρ_i is the lumped cytoplasmic resistivity and resistance of intercellular junctions, and d is the thickness of the monolayer. Assuming some geometrical parameters, the resistance of the disc was estimated. The values of R_m and C_m compare well with those reported by others.

Some of the experimental evidence reviewed here was used to support the hypothesis of the lack of electrical coupling between cardiac cells, but methodological considerations and geometrical factors furnish an alternative explanation: The branching and intercommunication of the fibers are responsible for the short length constant and the rapid falloff of the electrotonic potential when current is injected intracellularly (Matsuda, 1960; Sakamoto, 1969; Tanaka and Sasaki, 1966; Tarr, 1967).

3.2.2.3 Resistance Measurements Related to Pacemaker Activity. Before microelectrode methods were developed, the spontaneous rhythmic activity of the heart had already been ascribed to a slow spontaneous depolarization during diastole (Bozler, 1942). West (1955) postulated that the spontaneous diastolic depolarization was not an electrotonic spread but an inherent property of the cell membrane, implying that the membrane conductance undergoes cyclic changes. Weidmann (1951a) showed that the membrane resistance increased during the diastolic depolarization. Sperelakis and Lehmkuhl (1966) reported that the input resistance increased by 10–20% during the repolarization phase in cultured cardiac cells. A mean increase of 32% in R_m was estimated and attributed to a decrease of g_K, although the contribution of g_{Cl} and g_{Na} to the total membrane conductance could not be evaluated. Sperelakis and Lehmkuhl (1964) also studied the effect of current injection on pacemaker and nonpacemaker chick heart cells in culture. The frequency of discharge changed as a function of current, following a sigmoid curve; this change occurred mainly because of a change in the slope of the spontaneous depolarization. The frequency of discharge had a linear relationship with the logarithm of this slope. The exploration of the membrane properties in cells of the sinus node is difficult because they are confined to a very small region and packed in abundant connective tissue (James et al., 1966; James, 1967). That is why the pacemaker properties have been studied mainly in spontaneous beating cultured cells. Crill, Rumery, and Woodbury (1959) studied the effect of current injection on the spontaneous activity of clusters of cultured ventricle cells. Weak depolarizing currents increased the beating rate, whereas the opposite effect was obtained with hyperpolarizing currents. The results were explained in terms of changes in sodium conductance.

Trautwein and Kassebaum (1961) reported that in spontaneously beating atrial fibers, a reduction of $[Na^+]_o$ produced hyperpolarization. There was a negative correlation between the hyperpolarization and the value of the maximum diastolic potential. In nonbeating Purkinje fibers, an increase in depolarizing currents elicited repetitive activity. In spontaneously beating cells, depolarizing or hyperpolarizing currents had the same effects as in cells in culture. Lowering $[Ca^{2+}]_o$ decreased the current requirements to produce single or repetitive responses. Repetitive activity was abolished when $[Na^+]_o$ was reduced to 50% or less. The R_{inp} of Purkinje fibers in sodium-free medium increased with depolarization, and this was attributed to a decrease in potassium conductance. It was concluded that $[Ca^{2+}]_o$ affected the mechanism of sodium activation that would be responsible for repetitive discharges and that a time-dependent change in membrane resistance, as well as the voltage dependence observed in voltage–current curves, underlies automatic activity. Acetylcholine increased the resting potential and reduced the input resistance. Hyperpolarization of the membrane resulted in a depolarizing response to acetylcholine, with a reversal potential at -102 mV. The negative chronotropic effect of acetylcholine was then attributed to an increase in P_K (Trautwein and Dudel, 1958). Results supporting this view were obtained by Sperelakis and Lehmkuhl (1966), Sperelakis and Pappano (1969), and Sperelakis and Shigenobu (1972) in cultured cells with and without automatic activity. Addition of $BaCl_2$ (5–10 mM) depolarized both pacemaker and nonpacemaker cells, and electrical activity disappeared. Repolarization by current injection resulted in repetitive activity even in cells that had not formerly shown spontaneous firing. The apparent cell resistance increased by a factor of 5. $SrCl_2$ (5–10 mM) induced hyperpolarization and pacemaker activity in previously quiescent cells and antagonized the depolarizing action of Ba^{2+}. Increasing $[K^+]_o$ to 25 mM abolished pacemaker activity, and the apparent resistance fell slightly. Further elevation to 40 mM led to a rapid and large depolarization. If Ba^{2+} was added to the cells depolarized by high $[K^+]_o$ the cells remained depolarized, but pacemaker activity appeared when hyperpolarizing pulses were applied. The initiation of pacemaker activity by Ba^{2+} was attributed to its effect on g_K, and this was supported by results of Sperelakis and Shigenobu (1972).

Hermsmeyer and Sperelakis (1970) confirmed that Ba^{2+} decreased g_K in frog ventricle muscle. A fivefold increase of R_{app} occurred in 10 mM Ba^{2+}, and the membrane depolarized from -82 to -30 mV. During long exposure to Ba^{2+}, the apparent resistance returned to control values, despite a maintained depolarization, which suggested a voltage-dependent increase of g_K. Sperelakis and Pappano (1969) investigated the depolarizing effects of veratrine and veratridine on chick heart cells in culture. These alkaloids did not alter automatic activity of pacemaker cells, but they caused a membrane depolarization of -15 mV in nonpacemaker cells. The same results were obtained in the presence or absence of chloride. Apparent resistances increased only slightly when low values of membrane potential were reached, but when depolarization was prevented by prior addition of tetrodotoxin or in sodium-free medium, veratridine caused a 50% fall in apparent resistance. It was proposed that veratrine alkaloids act mainly through an increase in resting g_{Na}. The depolarizing effect of Ba^{2+} and veratridine would differ in that Ba^{2+} decreases g_K specifically, whereas veratridine mainly increases g_{Na}. The membrane properties of isolated single myocardial cells in culture were further investigated by Pappano and Sperelakis (1969b). In contrast to cell clusters or cells in sheets, many of those cells did not show spontaneous activity, and their resting potentials were lower (-14 mV) than that of cells in clusters (-40 to -70 mV). Ba^{2+} did not change the membrane potential or the apparent resistance, in contrast to the effect observed in cell clusters. Since g_K seemed to be the major component of G_m in resting conditions, it was concluded that the low resting potential of isolated cells is due to a low g_K, which results in a g_{Na}/g_K ratio higher than in other cells. However, it was reported that the commanding pacemaker cell spontaneously changed its firing characteristics and these changes were followed by the rest of the cells in a cluster. This indicated that current spread from one cell to another. Therefore, if cells in a cluster form a syncytial structure, the current injected in one cell will flow through a larger membrane surface. If uniform polarization is assumed, R_{inp} of a cluster of five or six cells may be as little as four or five times the R_{inp} of a single cell. Therefore, a difference in R_m between single and grouped cells cannot be concluded from this type of measurements.

Sperelakis and Shigenobu (1972) related the incidence of pacemaker activity in cells of embryonic chick hearts to changes in P_K with age. The increase of membrane potential with age was attributed to a progressive increase in P_K rather than to an increase of $[K^+]_i$, because cells from younger hearts showed: (1) a greater occurrence of hyperpolarizing afterpotentials and slow diastolic depolarization, (2) a larger difference between E_K and V, and (3) a lower sensitivity to increases in $[K^+]_o$. With age, the apparent resistance decreased, the resting potential approached the value of E_K, and pacemaker activity was reduced.

3.2.2.4 Resistance Measurements Related to the Action Potential and the Rectifying Properties of Cardiac Muscle. Because of its fast time course, the conductance changes underlying the upstroke could not be evaluated with square-pulse methods. By contrast, these methods could be used to investigate the slower time course of the plateau, the repolarization phase, and the voltage dependence of g_K associated with them.

Eyster and Gilson (1947) measured apparent resistances at rest and during the monophasic action potential in turtle hearts. The resistance dropped during the upstroke and rose steadily during the plateau, reaching values above the resting level. Cranefield, Eyster, and Gilson (1951) reported that during the plateau the injury current decreased steadily as a consequence of the rise in resistance. Tanaka (1959) found that the amplitude of the electrotonic potential at the crest of the action potential was the same as at resting level in toad atria. A progressive increase was observed during the plateau. Since apparent resistances were measured, the drop of R_{inp} at the crest of the action potential may have been masked by a large electrode component that remained constant during excitation. Coraboeuf, Zacouto, Gargouïl, and Laplaud (1958) measured a considerable drop of R_{inp} at the peak of the action potential in guinea pig ventricle. Weidmann (1951a) found that the half-time of the electrotonic potentials was 20 ms during diastole and only 1 ms at the crest of the action potential in kid Purkinje fibers. Assuming that the membrane capacity does not change upon activity, this finding indicates a considerable decrease in R_m. During the plateau, R_{inp} was comparable to that in the diastolic phase. The changes in R_m during the action potential were estimated from:

$$\ln(\Delta V_d/\Delta V_{ap}) = \ln\{\beta + \lfloor(\beta - 1)x/\lambda\rfloor\} \tag{3.31}$$

where β is the ratio between R_m at rest and at the peak of the action potential and x is the interelectrode distance. The value of β was 56:1. However, this value was considered to be an approximation, mainly because the value of λ at rest was used in Eq. 3.31, and during the action potential as R_m changes λ will change accordingly. Increasing the $[Ca^{2+}]_o$ did not alter R_m appreciably (Weidmann, 1955a). Using these data, Woodbury (1962) estimated the time course of the changes in relative slope conductance during the action potential.

Johnson, Robertson, and Tille (1958) observed that the time course of the changes in G_m underlying the action potential was different in Purkinje fibers than in ventricle cells. It was then proposed that the inactivation of the process producing the upstroke would occur earlier in ventricle fibers than in Purkinje fibers. This was supported by Woodbury (1961), and the same argument was used by Brady (1966) to explain the insensitivity of the action potential and the overshoot of guinea pig ventricular fibers to changes in $[Na^+]_o$ (Coraboeuf et al., 1959; Coraboeuf and Guilbault, 1960). It is known today that the overshoot in ventricular cells depends on a calcium inward current (see 3.2.3.1).

Johnson and Tille (1960) measured the changes in R_{inp} during the repolarization phase of the action potential in rabbit Purkinje fibers.

Within 10 ms after the upstroke, R_{inp} recovered its resting value. Thereafter, it increased and reached a maximum of 1.6 times the resting value at a level of about 50% of repolarization and returned to diastolic levels during the final phase of repolarization. The results were interpreted as an indication that g_K decreases during repolarization, to return to its resting value toward the end of the action potential. Similar results were obtained by Johnson and Wilson (1962) in several species: dog, cat, sheep, rabbit, and guinea pig. In the rat, however, R_{inp} did not reach higher values than at rest. Soon after the upstroke (within 10 ms), it returned to values very close to the resting level and did not change further until the end of the repolarization phase. Based on the linearity of the voltage–current relationship, it was assumed that the ionic conductances were independent of the membrane potential and that chord and slope conductances for each ion were equal. Johnson and Tille (1964) derived a set of equations for the repolarization phase of the cardiac action potential. They were based only on time-dependent changes of g_{Na} and g_K and were restricted to the cases where there was a linear relationship between the action potential and current. However, Noble (1962a) computed the current distribution in a two-dimensional model and remarked that in spite of nonlinearities of the current–voltage relation, the current–voltage curves for total polarizing current are linear at various distances from the polarizing electrode. It was then concluded that the linear voltage–current curves found by Johnson and Tille (1960, 1961) during the plateau of the action potential did not exclude the voltage dependence of G_m. A report of Fozzard and Sleator (1967) further suggested that g_K was voltage dependent. Potassium conductance decreased during the plateau, reaching a minimum of 25–30% of the resting value. An increase in g_{Na} of 250% of resting g_K occurred during the upstroke. The relationship between relative g_K and membrane potential agreed with the results of Hall, Hutter, and Noble (1963). Acetylcholine accelerated the time course of the conductance changes.

Weidmann (1955c) demonstrated the rectifying properties of Purkinje fibers by applying short square pulses to membranes polarized by a long pulse. Strong hyperpolarization (up to -150 mV) did not affect R_m, but moderate depolarizations up to -40 mV increased membrane resistance. Further depolarization decreased R_m but never below 50% of its resting value. Hutter and Noble (1960b) observed a fourfold decrease in conductance during depolarization in sodium-free medium. Since most of the resting G_m can be attributed to potassium (Hutter and Noble, 1959), these changes were interpreted as a decrease of g_K during depolarization. Over the range between the resting potential and the threshold, the conductance was always lower than at rest. The nonlinearity within the potential range at which spontaneous depolarization occurs is important in explaining automatic activity. The relative contribution of K^+ and Cl^- to the resting G_m was investigated by Carmeliet (1961a, b) and Hutter and Noble (1961). Carmeliet (1961a) reported that chloride permeability was low at rest but comparable to other ions during depolarization. An influence of chloride on G_m was found when depolarizing currents were applied at low potassium concentrations. It was concluded that Cl^- becomes quantitatively important as a charge carrier when g_K decreases. From flux measurements, Carmeliet (1961b) estimated g_K as 22.8×10^{-5} mho/cm^2. The decrease in R_m with increasing $[K^+]_o$ was attributed to an increase in g_K, assuming that g_{Na} and g_{Cl} remain constant. This value for g_K is only a fraction of the total G_m calculated from measurements of Weidmann ($50–83 \times 10^{-5}$ mho/cm^2). No explanation was given for this discrepancy. Hutter and Noble (1961) reported that the effective conductance decreased by only 11% upon removal of chloride, thus confirming the small contribution of G_{Cl} to G_m.

To explain the high membrane resistance during the plateau, a fall in g_K was suggested by Brady and Woodbury (1960). Furthermore, anomalous as well as normal rectification in Purkinje fibers had been reported by Weidmann (1955c) and Hutter and Noble (1960). The results of Hall et al. (1963) suggested that neither Na^+ nor Cl^- was responsible for nonlinearities. Voltage–current curves were measured in high $[K^+]_o$ to establish whether the change in input resistance depended on the membrane potential or on the current direction. Figure 3.20 shows that R_{inp} decreased in high $[K^+]_o$. Voltage–current curves were also calculated assuming a constant P_K, so that the theoretical G_m was equal to the experimental value at resting potential in $[K^+]_o = 4$ mM. For small currents, agreement was obtained between measured and calculated values, indicating that P_K was equal at the two K^+ concentrations, but there was a large discrepancy between experimental and calculated curves for large currents. This suggested that P_K remains constant as long as the membrane potential is close to the potassium equilibrium potential. For depolarizations greater than 30 mV, an increase in conductance was observed. These results were interpreted in the light of those of Adrian and Freygang (1962a), and two channels for potassium were then postulated. Furthermore, Noble (1962a) showed that anomalous rectification in Purkinje fibers accounts for the effects of $[K^+]_o$ on the action potential: P_K decreases when the outwardly directed electrochemical gradient increases, as in depolarization. When the membrane is depolarized by current, g_K decreases, but when the depolarization results from high $[K^+]_o$, g_K increases. Since P_K is constant when the membrane potential is close to E_K, it was assumed that P_K is a function of $(V - E_K)$. Current–voltage relations computed according to that hypothesis agreed with the experimental results.

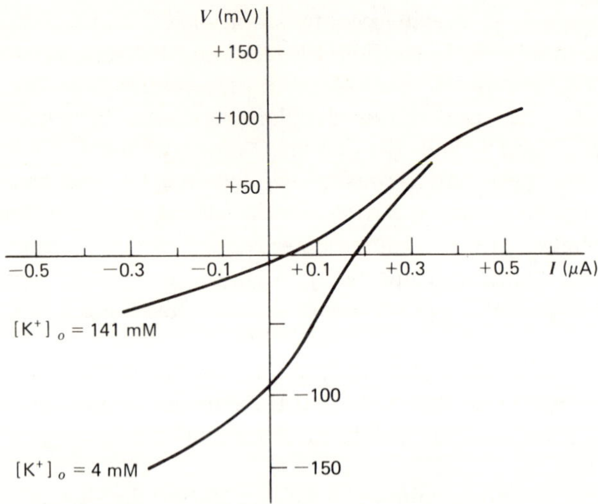

Figure 3.20. Voltage–current curves of Purkinje fibers obtained in low-Na^+ choline-chloride medium with $[K^+]_o = 4$ mM, and in a potassium–methylsulfate medium with $[K^+]_o = 141$ mM. Modified from Hall et al. (1963), by permission.

3.2.3 Voltage-Clamp Measurements in Cardiac Muscle

The development of voltage-clamp techniques for cardiac muscle is rather recent because of the difficulties involved in obtaining a spatially uniform clamp in a syncytial structure. This problem could be partially solved by reducing the surface of the preparation. This was accomplished in two ways: shortening a Purkinje fiber by placing two ligatures at distances of 1–2 mm ("short" Purkinje fibers), and creating an artificial node between two gaps of a nonconductive medium. These features characterize the voltage-clamp methods applied so far to cardiac muscle (Table 1.4, methods 5b, 6, and 7). A preliminary report (Trautwein et al., 1963) on the development of the short Purkinje fiber preparation and its electrical properties was followed by a detailed account of results showing that this preparation was suitable for study of the ionic currents under voltage-clamp conditions (Deck, Kern, and Trautwein, 1964). The electrical isolation of a small membrane area is based on the phenomenon of healing-over first observed in cardiac muscle by Engelmann (1877), which consists of the formation of a new high-resistance membrane over a cut surface within a few minutes after the injury. The same effects are found after ligating or cutting a bundle of Purkinje fibers (Hecht, Hutter, and Lywood, 1964). One-fourth of these preparations recovered their normal resting potentials within 30 min after ligature and had action potentials of normal configuration. No electrotonic decay was found along the short Purkinje fiber, suggesting that the preparation could be considered as a uniformly polarized cell. Circuit 7 of Table 1.4 was used to voltage clamp the preparation. The potential distribution was tested by impaling a third microelectrode at various distances from the current- and potential-controlling electrodes. The method was effective in clamping the membrane potential as long as R_{inp} did not fall below 50 kΩ. Therefore, it could be used to study currents flowing during the plateau and repolarization phase of the action potential.

Table 3.12 Electrical Cell Constants and Ionic Currents of Cardiac Muscle Obtained with Voltage-Clamp Measurements

Conditions	Derived Quantities	Remarks	Reference
Method 7 of Table 1.4	I_K: 7.4 $\mu A/cm^2$	At $V=-60$ mV	Deck and Trautwein (1964)
Measured: I_m vs t	I_{Na}: 2.6 $\mu A/cm^2$	At resting potential	
Model 7 of Table 1.5	I_K: 42 $\mu A/cm^2$	At $V=-10$ mV; $t=12.5$ ms	
	I_{Na}: 29 $\mu A/cm^2$	At $V=-10$ mV; $t=12.5$ ms	
	I_K: 17 $\mu A/cm^2$	At $V=-10$ mV; $t=300$ ms	
	I_{Na}: 8.4 $\mu A/cm^2$	At $V=-10$ mV; $t=300$ ms	
	g_K: 0.46 $mmho/cm^2$	At resting potential	
	g_{Na}: 0.024 $mmho/cm^2$	At resting potential	
Method 7 of Table 1.4	C_s: 4 $\mu F/cm^2$	Ramp signal: 30 V/s	Dudel, Peper, Rüdel, and Trautwein (1966)
Measured: I_m vs V	C_e: 8 $\mu F/cm^2$	$[Na^+]_o$: 5 mM	
Model: Fig. 3.2a	R_e: 150 Ω cm^2		
	C_m: 9.2 $\mu F/cm^2$		
	I_{Na}: 0.2 mA/cm^2		
	P_{Na}^{max}: 0.5×10^{-3} cm/s	Ramp speed: 30 V/s	
	E_{Na}: -47 mV	$[Na^+]_o$: 5 mM, short Purkinje fibers, sheep	
Method 7 of Table 1.4	C_s: 2.4 $\mu F/cm^2$	Short Purkinje fibers	Fozzard (1966)
Measured: R_{inp} 325–1200 kΩ	C_e: 7.0 $\mu F/cm^2$		

Table 3.12 *Continued*

Conditions	Derived Quantities	Remarks	Reference
Model: Fig. 3.2a	R_e: 298 Ω cm^2		
Method 7 of Table 1.4	R_m: 2490 Ω cm^2	Sodium-free, and [Ca^{2+}]: 1.8 mM	Reuter (1967)
Measured: R_{inp}			
275 kΩ	C_m: 11.5 μF/cm^2		
Model 1 of	R_m: 1920 Ω cm^2	Ca^{2+}-free	
Table 1.5	R_m: 2240 Ω cm^2	[Ca^{2+}]$_o$: 7.2 mM	
	R_m: 1810 Ω cm^2	Ca^{2+}-free Tyrode's	
	R_m: 1820 Ω cm^2	[Ca^{2+}]$_o$: 1.8 mM in Tyrode's	
	R_m: 1500 Ω cm^2	[Ca^{2+}]$_o$: 7.2 mM in Tyrode's	
Method 7 of Table 1.4	C: 1.9–2.3 μF	Ventricular trabecula (dog)	Beeler and Reuter (1970a)
Measured: I_c			
vs time	R_s: 565–608 Ω		
τ: 1.1–1.4 ms	G_s: 2 mmho	Computer model	
Model 2 of			
Table 1.5	C: 2 μF		
	E_{Na}: +50 mV		
	g_{Na}: 5 mmho		
Method 6 of	R_1: 7.7 kΩ	Frog atrial trabecula	Tarr and Trank (1971)
Table 1.4	R_2: 87.6 kΩ		
I_c vs time			
Model 2 of Table 1.5	C: 0.17 μF		
	R_m: 2600 Ω cm^2		
	C_m: 5.7 μF/cm^2		

3.2.3.1 Inward Currents and the Excitation Mechanism. Deck et al. (1964) and Deck and Trautwein (1964) reported the first results from voltage-clamped short Purkinje fibers. An early inward current flowed within the first 100 ms and in the voltage range between threshold and −10 to +10 mV. The outward current declined as a function of time, reaching a steady-state level at 200 ms at large depolarizations. E_{Na} was estimated between 25 and 50 mV.

The potassium and sodium chord conductances and their voltage dependence were calculated according to Eq. 3.13; the results are shown in Table

3.12. The mean value of the resting g_K corresponds to a membrane resistance of 2200 Ω cm^2, in agreement with that measured by the square-pulse technique (Table 3.11). Two findings marked the main difference between Purkinje fibers and the squid axon: a slow inactivation of I_{Na} and the appearance of a long-lasting outward current when the fiber was repolarized to the resting potential. Its reversal potential was −100 mV, and this current declined with a time constant of 250–300 ms.

Trautwein, Dudel, and Peper (1965) compared the current–voltage curves obtained with step and ramp signals. The curves obtained with ramp pulses at a depolarization rate of 100 mV/s were comparable to those measured after 100 ms of the onset of the step signal. At a ramp speed of 1 mV/s, the current–voltage curve corresponded to the steady-state outward current flowing at 800 ms of a step depolarization. Dudel, Peper, Rüdel, and Trautwein (1966) reported that, unless ramp signals were used, the surge of early ionic currents could not be separated from the capacitive artifact because of the high capacity of cardiac muscle. The capacitive current was analyzed according to the two-time-constant model (Fig. 3.2a), and the following values were obtained for the elements of the network: $r_s = 400$ kΩ, $r_e = 12$ kΩ, $c_s = 0.05$ μF, $c_e = 0.10$ μF. Figure 3.21 shows the current–voltage curves for the capacitive current as measured in Purkinje fibers and in a model network. These measurements of capacitive current confirmed previous results of Fozzard (1966) on the applicability of the two-time-constant model to cardiac muscle. The current flowing during the capacitive artifact showed two phases: an initial surge and a

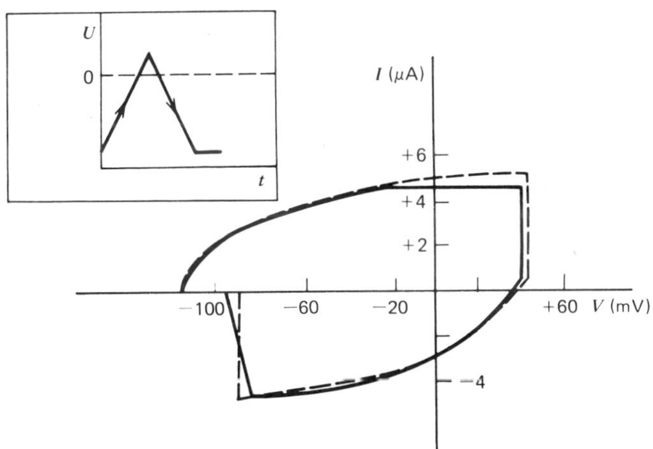

Figure 3.21. Current–voltage curves for the capacitive current in response to a ramp signal (see inset), measured in Purkinje fibers (full line) and in a model network (dashed line). Modified from Dudel, Peper, Rüdel, and Trautwein (1966), by permission.

much slower tail. The values for the elements of the network are listed in Table 3.12. The discrepancy between the two values reported for R_e is yet unexplained, especially since the measurements of Dudel et al. were performed in a medium of high resistivity and Fozzard found that R_e increased by a factor of 2 when the conductivity of the external solution was decreased by sucrose replacement.

Dudel, Peper, Rüdel, and Trautwein (1966) reported that a second component of inward current appeared between -40 and $+20$ mV in sodium-free medium. Previous results of Dudel, Peper, and Trautwein (1966) seemed to exclude the contribution of calcium ions to the current recorded in sodium-free media, but omission of Ca^{2+} had profound effects on current–voltage curves obtained with ramp signals. Their N-shape was lost, and hysteresis was less evident. Although the results did not prove the existence of a net Ca^{2+} influx during depolarization, they did not exclude it either. The possibility remained of a calcium current flowing during the upstroke or the plateau being masked by the capacitive current or the outward current. The second component of inward current remained under TTX (Dudel et al., 1967b). This suggested the existence of a second parallel ionic channel for inward current. These voltage-clamp data were used by Munk and George (1968) to develop a mathematical model for the action potential in Purkinje fibers. The Hodgkin-Huxley equations were adjusted to incorporate a slow sodium component. The changes in g_{Na} and g_K as a function of potential and time were plotted according to an equation similar to Eq. 2.12. The Hodgkin-Huxley expression for g_{Na} described well its changes during the upstroke, while a slow component of g_{Na} and a second inactivation constant had to be introduced to simulate the plateau. A slow component of g_K had to be incorporated as well, as in the model developed by Noble (1962b).

Reuter (1967) demonstrated that the slow inward current described by Dudel, Peper, Rüdel, and Trautwein (1966) was carried, at least partly, by calcium ions. C_m was calculated from the initial capacitive current during step voltages, and R_{inp} and R_m were obtained from the steady-state membrane currents flowing during small (5–10 mV) depolarizing steps. Changes in $[Ca^{2+}]_o$ did not affect R_m or C_m (Table 3.12). In sodium- and calcium-free Tyrode's, outward current flowed throughout the duration (500 ms) of a depolarizing step to -34 mV. In sodium-free medium at a $[Ca^{2+}]_o$ of 7.2 mM, an initial peak of outward current was followed by a steady inward current independent of $[Cl^-]_o$, thus excluding a chloride outflux as the cause of this current.

Subtracting the current recorded in calcium-free media from the total current obtained in the presence of Ca^{2+} gave the current–voltage curve for this component of inward current, as shown in Fig. 3.22, in comparison

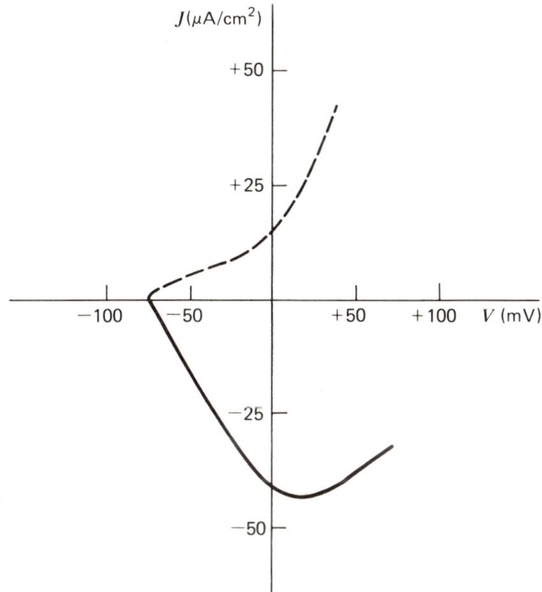

Figure 3.22. Current–voltage curves of potassium current (dashed line) and calcium current (full line) of Purkinje fibers, recorded in sodium-free solutions. Modified from Reuter (1976), by permission.

with the outward potassium current. The slow inward current declined at potentials of 20–40 mV. By extrapolation, it intersected the voltage axis at a potential between 90 and 180 mV; from this, an intracellular Ca^{2+} concentration of 10^{-5}–10^{-9} M was calculated. The threshold potential for the slow inward current was −20 mV in Tyrode's and −50 mV in sodium-free medium.

The presence of this slow inward current provided an alternative explanation for the plateau of the Purkinje fiber action potential, other than the slow inactivation of g_{Na} postulated by Deck and Trautwein (1964) or the time-dependent increase in g_K suggested by Noble (1962b). It would correspond to the slow increase in g_{Na} used by Munk and George (1968) to compute the action potential.

Reuter (1968) reported a slow component in the inactivation of the sodium conductance, which seemed to be an important feature of repolarization during the first 100 ms of the plateau. Complete repolarization could not be triggered by this mechanism, however, as it can be achieved later during the repolarization. The slow increase in g_K described by McAllister and Noble (1966b, 1967) should play an important role in the repolarization phase. The chloride outward current described by Dudel et al. (1967a) was also confirmed. Inactivation of g_{Cl} seemed to be similar to the h_∞ curve for inactivation of g_{Na} in the squid axon. As suggested by Dudel et al. (1967a), the chloride outward current would determine the fast repolarization phase preceding the plateau in Purkinje fibers. The voltage and time dependence of g_{Ca} together with the characteristics of the potassium rectifier would be the main features determining most of the plateau and the late repolarization.

McAllister (1968, 1969) analyzed the effects of applying the two-time-constant model on the computed action potential and evaluated the usefulness of ramp pulses in the study of the dynamic electrical properties of Purkinje fiber membranes. The electrical equivalent of Fig. 3.23 was used for computation. It includes the two components of g_{K} proposed by Noble and a series capacity resistance branch. The membrane current in this circuit is given by:

$$I_m = \{C_s + C_e[l - \exp(-t/R_eC_e)](dV/dt)\} + [(g_{\text{K1}} + g_{\text{K2}})(V - E_{\text{K}})]$$
$$+ g_{\text{Na}}(V - E_{\text{Na}}) + I_e\exp(-t/R_eC_e) \tag{3.32}$$

where I_e expresses the current through the R_eC_e branch. The effect of the

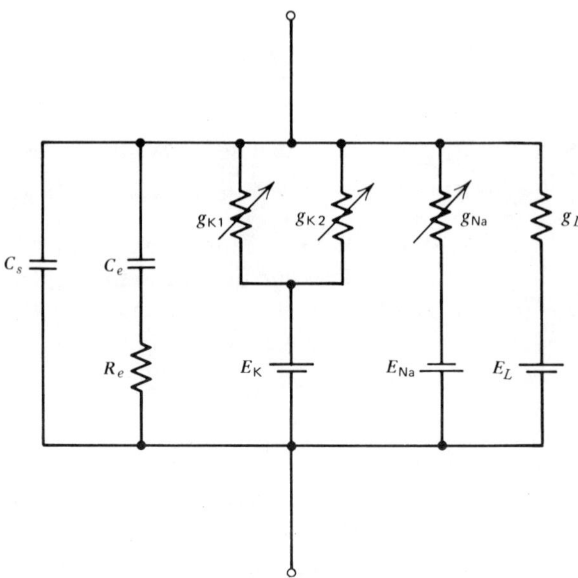

Figure 3.23. Equivalent electrical circuit for the Purkinje fiber membrane. From McAllister (1968), by copyright permission of The Rockefeller University Press.

series R_eC_e network was such that the repolarization rate would be controlled by the interrelation between I_e and the net ionic current. McAllister did not consider the slow inward current in this model, but this does not invalidate the conclusions concerning the influence of I_e on the shape of the AP.

Voltage–current curves were computed for ramp speeds between 0.01 and 10 V/s. Up to 0.1 V/s, the current–voltage curves were identical for both the one- and two-time-constant models. However, the curves were influenced by the ramp speed, as predicted by the Hodgkin-Huxley theory. I_{Na} increased with increasing rate of depolarization because the conductance parameters are furthest from their end values for a given potential when the potential is changing more rapidly. At very slow rates (0.01 V/s), all the conductance parameters maintained their steady-state values. Hysteresis appearing at higher ramp speeds was attributed to different degrees of activation of g_{K2} at the moment at which repolarization begins. It was concluded that it is difficult to obtain a general picture of the currents flowing during the action potential when they are inferred from results obtained with ramp signals, but that this technique can show a time integral of the total current flow.

The problem of an effective voltage clamp during large and fast currents arises because the amount of current that can be injected through a microelectrode is limited. This problem was overcome by adapting the double sucrose-gap technique to the voltage-clamp method. This technique was first used for potential measurements in cardiac muscle by Rougier, Ildefonse, and Gargouïl (1966), who showed that the input resistance as well as the effect of ionic changes on the action potential could be measured with this method. Voltage- and current-clamp measurements on frog atria applying the double sucrose-gap principle (method 5b in Table 2.5) were first performed by Rougier, Vassort, and Stämpfli (1967, 1968). Sodium-free solutions and TTX abolished all signs of electrical activity. Anomalous rectification was observed during long pulses under these conditions. Current–voltage curves for early and late currents (Fig. 3.24a, b) had the same general features as are found in the squid axon, but the late currents differed by their slow rise, smaller amplitude relative to the initial inward current, and spontaneous inactivation (Fig. 3.24a). Upon depolarizing steps larger than 50 mV, the inward current recorded in Ringer's solution inactivated in two phases with different time constants (Fig. 3.24c). The time constant for the rapid inactivation was 0.8–4 ms. The slow inactivation phase corresponded to the inward current remaining in TTX-Ringer's. Using TTX and Mn^{2+} as inhibitors of membrane conductances, Rougier, Vassort, Garnier, Gargouïl, and Coraboeuf (1968, 1969) separated the two components of membrane inward current and identified the ionic species involved (Fig. 3.24b). Manganese ions blocked selectively

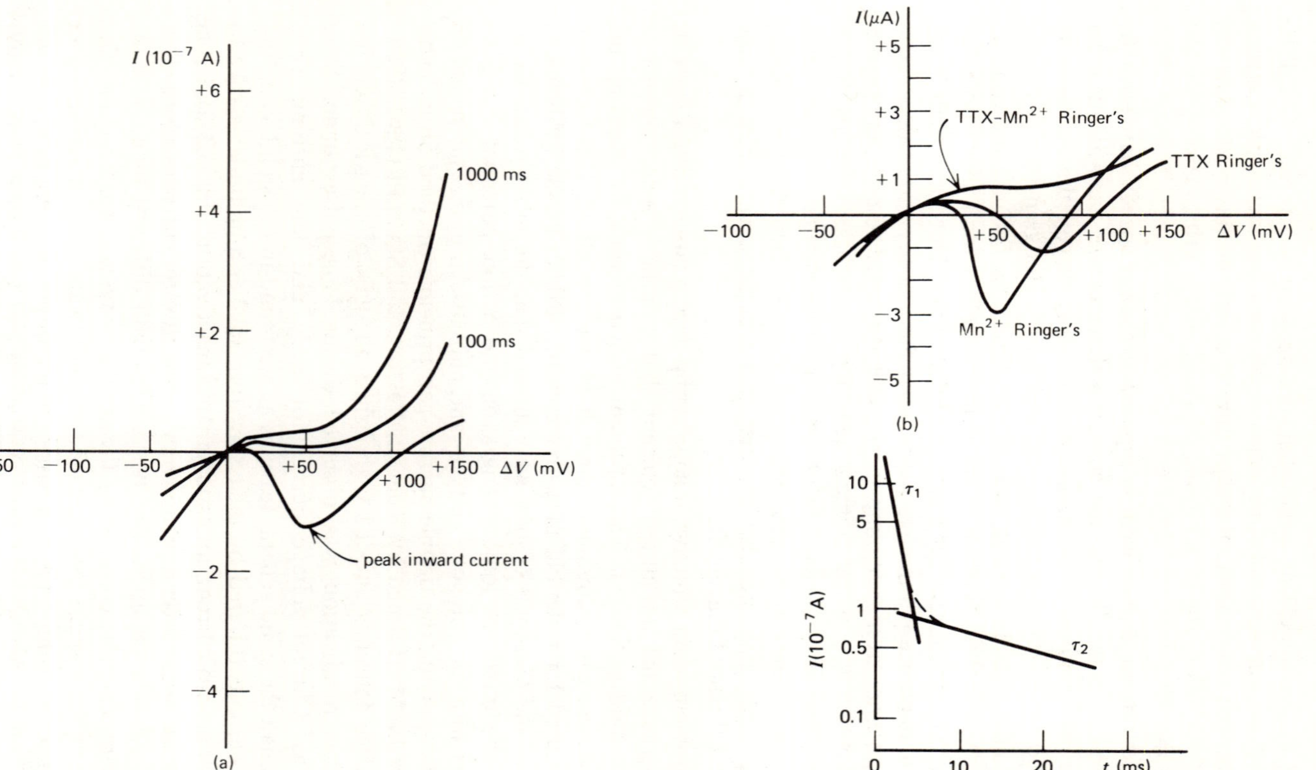

Figure 3.24. (a) Current–voltage curves from frog atrial trabecula measured with the double sucrose-gap technique in Ringer's solution (from Rougier, Vassort, and Stämpfli, 1968). (b) Current–voltage curves of frog atrial trabecula in TTX-Mn^{2+} Ringer's, TTX-Ringer's, and Mn^{2+} -Ringer's. (c) Time constant of inactivation of inward current in Ringer's. Modified from Rougier et al. (1969), by permission.

the slow phase of initial current. Addition of TTX to Mn^{2+}-Ringer's blocked the fast component, and the region of negative slope between depolarizations of 30–70 mV disappeared. In TTX-Ringer's the peak of slow current occurred at a ΔV of 70–80 mV and the current reversed at a ΔV of 110–120 mV. The slow current inactivated slowly, with a time constant 10–20 times higher than the time constant for inactivation of the fast current. In sodium-free medium, the slow current was carried by Ca^{2+}, because it was abolished by removal of Ca^{2+}. Its reversal potential agreed with the value found by Reuter (1967) for the calcium current in Purkinje fibers. Removal of Ca^{2+} did not suppress the slow current if Na^+ was present, and the reversal potential appeared at a ΔV of 80–100 mV. It was concluded that the depolarization phase of the frog atria action potential is due to the successive activation of two inward currents: a fast sodium component (blocked by TTX) and a slow sodium-calcium component (blocked by Mn^{2+}). These results also showed that these inhibitors are specific for a given channel but not for the ionic species involved. Tarr (1971) confirmed these findings and showed that the changes observed in the action potential configuration under TTX and after removal of calcium or sodium corresponded to the proposed role of the two ionic channels in the generation of the upstroke. Manganese at high concentrations (10 mM) abolished both inward currents, thus indicating that its blocking effect is not so selective for the slow channel as TTX is for the fast one. These results do not support the explanation suggested by Johnson and Lieberman (1971), who argued that the slow inward current was an artifact due to inadequacies in the spatial and temporal control of the membrane potential in voltage-clamp measurements.

Garnier et al. (1969) reported that perfused rat hearts and frog atrial trabecula submitted to a medium without bivalent cations in the presence of EDTA presented a progressive lengthening of the plateau and a final stabilization of the membrane potential at a level near zero, and proposed that this effect was due to the steady flow of a slow sodium inward current. This view was confirmed by results of voltage-clamp measurements on thin bundles of rat ventricular muscle, reported by Besseau and Gargouïl (1969) and Besseau, Léoty, and Gargouïl (1969). In addition, it was found that delayed rectification was absent or barely detectable, and that chloride ions participated in the genesis of the outward currents elicited by depolarization.

Mascher and Peper (1969) described the voltage-clamp method shown in Fig. 3.25. Beeler and Reuter (1970a) discussed its advantages and limitations. It was applied mainly to ventricular muscle from different species. Mascher and Peper (1969) used trabecula excised from the right ventricle of sheep hearts. The two components of the inward current were separated according to their dependence on: the previous level of potential, $[Na^+]_o$, $[Ca^{2+}]_o$, and their reversal potentials. The first inward transient observed during depolarization showed the inactivation characteristics of the sodium system, and the h_∞ curve was shifted by $[Ca^{2+}]_o$, as reported for Purkinje fibers (Weidmann, 1955a). The reversal potential was a function of $[Na^+]_o$ and agreed closely with the calculated value of E_{Na}. The second inward transient (slow) was independent of previous levels of potential, and the voltage range of its appearance extended beyond the value for E_{Na}. It developed only when the calcium concentration was higher than 1.8 mM and did not depend upon the presence of sodium ions. The voltage dependence of the slow inward current explained the positive plateau of the ventricular action potential in contrast to the plateau of the Purkinje fiber action potential, which lies normally in the region of −20 to −30 mV. These results confirmed and extended previous observations of Giebish and Weidmann (1967).

The ionic currents and their kinetics were measured in dog ventricular trabecula by Reuter and Beeler (1969a, b) and Beeler and Reuter (1970a, b). The time course of the capacitive current corresponded to a model network like number 2 in Table 1.5. The values of C_m and the series resistance are listed in Table 3.12. A second measuring microelectrode was used to test the potential distribution during the application of a command signal. During the occurrence of a fast, large inward current, the potential was not effectively clamped. This was interpreted as resulting from the existence of the finite extracellular resistance in series with the membrane. The validity of the measurements was assessed by computation of a set of

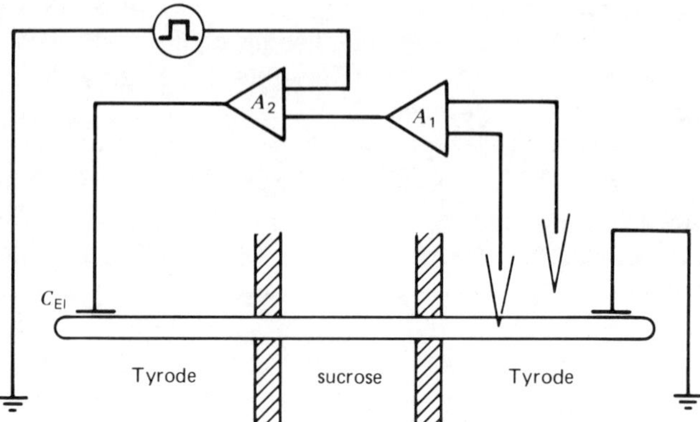

Figure 3.25. Voltage-clamp technique using a single sucrose gap (middle compartment). Step command signals are applied through the current electrode C_{EI}. A_1, differential electrometer amplifier; A_2, feedback amplifier. Partitions: two thin rubber membranes.

equations derived for the situation in which the recorded clamp potential is the sum of the voltage drops across the membrane and the series resistance.

A similar equivalent circuit for frog atrial trabecula was found by Tarr and Trank (1971). The capacitive current resulting from a 20 mV depolarizing step rose rapidly to a maximum (I_0) and then decayed exponentially with a time constant of 1 ms to a steady-state value (I_{ss}). Two equivalent circuits (model 2 in Table 1.5 and Fig. 3.2) can give such a pattern for I_c. The results were compatible with circuit 2 of Table 1.5. The following equations were used to determine the elements of the equivalent circuit

$$\left.\begin{aligned} R_1 &= \Delta V/I_0 \\ R_2 &= \Delta V/I_{ss} - R_1 \\ C_m &= \tau/(1 - I_{ss}/I_0)(R_1) \end{aligned}\right\} \tag{3.33}$$

The electrical constants of the membrane were calculated taking into account the structure of the preparation and the geometry of the artificial node in the experimental chamber (Table 3.12). They agree with values for R_m and C_m found in cardiac cells with other methods (Table 3.11). R_1 was considered to represent the resistance of the intercellular clefts.

Beeler and Reuter (1970a, b) determined the h_∞ characteristic of dog ventricular trabecula in Tyrode's. A mean value of −55.7 mV for half-inactivation of the sodium system was found for a $[Ca^{2+}]_o$ of 1.8 mM. At $[Ca^{2+}]_o$ of 0.2 and 7.2 mM, this value was −64 and −54 mV, respectively. Inactivation of the fast sodium current by a conditioning depolarizing step led to separation of the slow inward current, whose threshold occurred at potentials of −33 to −25 mV. This threshold shifted to more negative potentials when $[Ca^{2+}]_o$ was increased or in sodium-free solutions. The rate constants of decay of the slow inward current showed a steep dependence on potential with a rapid inactivation at negative potentials and a slow inactivation in the positive range. This characteristic confirmed the significance of the slow current in determining the plateau in the ventricular action potential. The activation of I_{Ca} did not depend on the potential level previous to the depolarizing step. The average E_{Ca} was estimated at +62 mV, but it was shown that it is not constant. Starting at an initial E_{Ca} of +132 mV, the equilibrium potential decreased as $[Ca^{2+}]_i$ increased during the inward current, and it approached a steady-state level after 30 ms, at a value of +60 mV. However, the decay of I_{Ca} was not determined by the change in E_{Ca}, but it was due to the decay in g_{Ca}. These observations confirmed the findings of Matsubara and Matsuda (1969) about the contribution of Ca^{2+} to the depolarizing current during the action potential of ventricular fibers of the dog heart.

Vitek and Trautwein (1971) reported that, as in frog atria, sodium and calcium carried the slow inward current in Purkinje fibers. The time constants for inactivation were 40 ms at -70 mV and 90 ms at -25 mV, showing a voltage dependence of inactivation similar to the one described in dog ventricular muscle. Ochi (1970) reported that in guinea pig myocardium the slow current was also related to the positive plateau of the action potential. Shigenobu and Sperelakis (1971) and Sperelakis and Shigenobu (1972) showed that the action potential of young embryonic chick hearts (2–4 days of development) seems to arise from the activation of the slow channel at a low membrane potential. In contrast, the mechanism of excitation in older hearts behaved like a fast sodium channel, TTX-sensitive, but this mechanism reversed to that of a slow channel when the cells were placed in culture.

The preceding discussion has shown that two independent current pathways are responsible for the upstroke and plateau of the cardiac action potential. Consequently, the kinetics of the excitation process are more complicated and cannot be fully described with the Hodgkin-Huxley equations in their original form. Weidmann (1955b) made the first attempt to study the inactivation characteristics of the sodium system in Purkinje fibers, by investigating the effects of a conditioning pulse on the maximum depolarization rate of an action potential evoked at the break of the pulse. These inactivation curves could be fitted by the equation:

$$h = 1/\left[1 + \exp(V_h - V)/k\right] \tag{3.34}$$

where V is the conditioning potential, V_h the potential at which h is half-maximal, and k had the value of 5. The resulting curve was 1.4 times as steep as the one found for the squid axon (Table 2.5). Increasing $[Ca^{2+}]_o$ to 4 times normal, the curve was shifted to more positive values of V (Weidmann, 1955a).

Dudel and Rüdel (1970) investigated the kinetics of I_{Na} in Purkinje fibers cooled to 5–10°C. The voltage dependence of the variables m and h was similar to that described for the squid axon. The rate of removal of inactivation was voltage-dependent: half the inactivation was removed in 4 ms at -180 mV, in 90 ms at -160 mV, and after 1 s at -140 mV. The plot h_∞ versus V at 8°C covered an unusual range of potentials, from -180 to -100 mV. Q_{10}'s of 3 and 3.8 were obtained for the rise and fall of the sodium current. In contrast, the temperature dependence of the maximum I_{Na} was relatively small: Q_{10} below 1.3. It was concluded that low temperature shifted the h_∞ characteristic toward more negative potentials. The shift was 15 mV per 10°C change of temperature.

In frog atria, Haas, Kern, Einwächter, and Tarr (1971) showed that a single inactivation constant h does not adequately describe the processes of inactivation and removal of inactivation. These findings were confirmed by DeHemptinne (1971a). In a depolarizing step, inactivation developed as an exponential process with a time constant of 1.5–2.7 ms. On repolarization or hyperpolarization, inactivation was removed slowly with time constants between 100 and 600 ms. The curve h_∞ versus V was fitted by Eq. 3.34 with V_h from -45 to -62 mV and k between 3.5 to 5.2. In the Hodgkin-Huxley equations, inactivation and its removal are described by a first-order equation with a single time constant for both (Eqs 2.13 and 2.15). The variable h in these equations was replaced by the product of two variables, p and q, having a similar voltage dependence (they both decrease on depolarization) but different rate constants. Accordingly, for $\tau_p \ll \tau_q$, Eq. 3.34 gives $pq = \exp(-t/\tau_p)$, indicating that inactivation is a fast process determined by the variable p. Upon repolarization, inactivation is removed according to the expression $pq \cong 1 - \exp(-t/\tau_q)$, for $t \gg \tau_p$.Therefore, this process is mainly determined by the variable q. This difference between the rate constants explains that the relative refractory period outlasts the duration of the action potential in frog atria. The same kinetics could probably be applied to cells of the atrioventricular junction of the rabbit heart where the same phenomenon has been described (Merideth et al., 1968).

3.2.3.2 The Outward Currents: Repolarization and the Pacemaker Potential. Voltage-clamp measurements in Purkinje fibers confirmed the rectifying properties of these membranes. Hecht et al. (1964) showed that the current–voltage curve obtained in sodium-free solutions presented a region of higher conductance on hyperpolarization than on depolarization. McAllister and Noble (1966a, b, c) measured the time and voltage dependence of the outward current. Depolarization to -60 mV produced a slow development of outward current after the inward surge. This current was much smaller and slower than in nerve fibers. Its reversal potential was roughly equal to E_K. The ratio of the conductances at the end of the pulse and on repolarization ranged between 0.3 and 0.5, indicating profound nonlinearities of the instantaneous current–voltage curve within that potential range. Depolarization beyond the plateau level of the action potential elicited another component of outward current that declined with time. This decrease was independent of E_K. The nonlinearities in the current–voltage relation showed that g_K before the end of a 50 mV depolarizing pulse is only one-third of that which occurs immediately after the pulse. This indicated that the two components of the potassium conductance (g_{K1} and g_{K2}) in Fig. 3.23 have rectifying properties. A reconstruction of the potassium currents during the repolarization phase of the action potential was proposed, but experimental evidence obtained later (see below) showed that a more complicated model was required to explain the repolarization process in Purkinje fibers. McAllister and Noble (1967) reported that depolarizations of 16 mV elicited the slow development of outward current. Up to two-thirds of the maximum current (I_{K2}) could be activated by these small depolarizations. In the region of the resting potential, less than 25% of the maximum g_{K2} was activated at the steady state. These results agreed with a previous report by Vassalle (1966), who clamped spontaneously active Prukinje fibers at the level of maximum repolarization and recorded an inward current increasing with time that resulted in an action potential upon releasing the clamp. Lowering $[K^+]_o$ from 5.4 to 2.7 mM resulted in an increase in the rate of diastolic depolarization. The results were consistent with the pacemaker potential being due to a delayed voltage-dependent increase followed by a time-dependent decrease of g_K.

A contribution of chloride to the outward current in Purkinje fibers was described by Dudel et al. (1967a). A time- and voltage-dependent component of outward current appeared at positive membrane potentials. It did not depend on $[Na^+]_o$ or $[K^+]_o$. Replacement of chloride by nonpermeant anions reduced markedly the positive current between -20 and $+40$ mV. This indicated that a major part of this current was carried by chloride ions. It was concluded that the rapid repolarization following the upstroke of the Purkinje fiber action potential was produced by this transient current. It disappeared in chloride-free solutions. This outward current would be a particular feature of Purkinje fibers because the fast repolarization phase does not exist in other types of cardiac fibers, where the plateau occurs at positive levels of membrane potential.

The kinetics of the component g_{K2} were further analyzed by Noble and Tsien (1968). Because most of the current attributable to a change in g_{K2} is activated at potential levels negative to the threshold for the sodium system, the g_{K2} system plays a leading role in the range of the pacemaker potential. The original model of Noble (1962b) has been modified, so that g_{K2} not only rectifies slowly in the outward-going direction, but also rectifies in the inward-going direction for instantaneous changes in potential. At any potential, g_{K2} is proportional to a first-order variable s according to $g_{K2} \propto s$. The curve s_∞ versus V was a sigmoid similar to the curves obtained for the variable h_∞ in squid axon. The reversal potential for I_{K2} was a function of E_K, and it followed the predicted shifts in E_K when the potassium concentration was varied between 6 and 2.6 mM, thus confirming the results of Peper and Trautwein (1969). In addition, the current–voltage relation had a region of negative slope, and the curves obtained at different $[K^+]_o$ crossed each other when depolarizing pulses were applied. The region of negative slope appeared beyond a potential 25 mV more positive than E_K. It was concluded that the time and voltage dependence of g_{K2} could be described by a model in which the conductance value results from the product of two factors: the variable s and the rectifier function $I_{K2}(V, E_K)$. This function describes the instantaneous I_{K2} versus V curve for any value of s. The shape of this current–voltage relation depends on the value of E_K, although it is independent of the previous potential level and the value of s. The mechanism of pacemaker activity was explained as shown in Fig. 3.26. The diastolic depolarization extends over the potential range in which s_∞ varies from 0 to 1. At the beginning of the pacemaker potential, at $t=0$, s_∞ is nearly 0. Therefore, s will decline slowly and so will I_{K2}. As the membrane depolarizes, s_∞ increases and equals s at some point in time. In the current–voltage curve, the negative slope appears at the potential in which $s=s_\infty$. Therefore, I_{K2} will continue to fall even though s starts rising. Finally, I_{K2} declines sharply at the end of the diastolic depolarization. The rapid increase in the depolarization rate at this point is attributed to the negative slope of the instantaneous current–voltage curve.

A complete reconstruction of the repolarization process in Purkinje fibers (Noble and Tsien, 1969b) was achieved by investigating the ionic currents within the potential range where the plateau occurs (Noble and Tsien, 1969a). Since s_∞ is nearly constant for potentials more positive than -65 mV, no time-dependent changes in current occurring on depolarization beyond this potential can be attributed to I_{K2}. Therefore, -30 mV

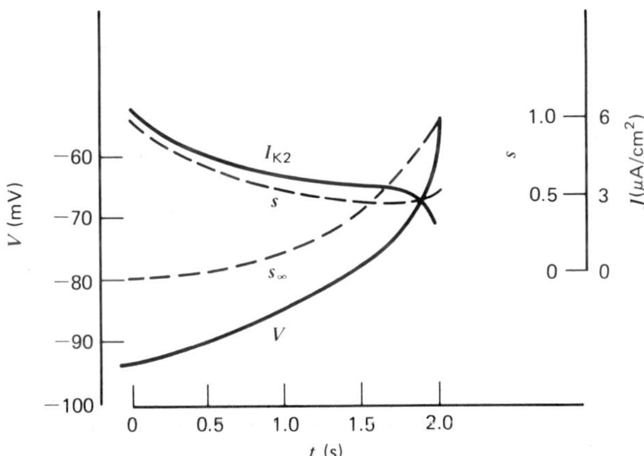

Figure 3.26. Reconstructed time course of membrane potential V, potassium current I_{K2}, and variables s and s_∞ during spontaneous depolarization in Purkinje fibers. Modified from Noble and Tsien (1968), by permission.

was used as holding potential between pulses, thus eliminating any possible interference from changes in g_{Na} and g_{K2}. The time course of the onset of current on depolarization and the return to the steady state upon repolarization could be fitted by a sum of two exponentials with time constants τ_1 and τ_2. At 0.2 s after repolarization to the holding potential, the tail currents were fitted by a single exponential, because at this time the activation of the slow component was negligible. After 17 s, two exponentials were required. From these results, the slow time-dependent current in the plateau range (I_x) was considered as the sum of two exponentially changing components (I_{x1} and I_{x2}) so that $\bar{I}_x = I_{x1} + I_{x2}$. If $\bar{I}_{x1}$ and $\bar{I}_{x2}$ are the maximum currents that may flow at a given potential,

$$\left.\begin{aligned} \bar{I}_{x1} &= I_{x1}(V, x_1 = 1) \\ \bar{I}_{x2} &= I_{x2}(V, x_2 = 1) \end{aligned}\right\} \tag{3.35}$$

where x_1 and x_2 are the fractional degrees of activation of each current. $\bar{I}_{x1}$ and $\bar{I}_{x2}$ depend only on voltage, and the slow changes are attributed to the degree of activation, which in turn depends on V and t. At the steady state, for each value of V:

$$\left.\begin{aligned} (x_1)_\infty &= \alpha_{x1}/(\alpha_{x1} + \beta_{x1}) \\ (x_2)_\infty &= \alpha_{x2}/(\alpha_{x2} + \beta_{x2}) \end{aligned}\right\} \tag{3.36}$$

where α and β are rate constants as in the Hodgkin-Huxley theory, having their correspondent time constants τ_1 and τ_2. Current–voltage relations of $\bar{I}_{x1}$ and $\bar{I}_{x2}$ showed anomalous rectification for $\bar{I}_{x1}$ but, unlike I_{K2}, no region of negative slope appeared. The reversal potential was -85 mV. $\bar{I}_{x2}$ was fitted by a straight line except for the most negative potentials, and its reversal potential was 20 mV more positive than for $\bar{I}_{x1}$. The value of the reversal potentials suggested that both current components may be carried largely, but not exclusively, by potassium ions.

Steady-state current–voltage relations for total current (I_t), total current after subtraction of I_{x2}, and total current minus ($I_{x1} + I_{x2}$) are shown in Fig. 3.27. The latter has a much more extensive region of negative slope conductance in the range -15 to -50 mV. I_{x2} is not necessary for repolarization because its subtraction does not produce a net inward current. In contrast, I_{x1} is required for repolarization because the current–voltage curve does not lose the region of inward current unless I_{x1} is at least partially activated. Moreover, Hauswirth, Noble, and Tsien (1972a) reported that I_{x1} is the main determinant of the known frequency dependence of action potential duration in Purkinje fibers. However, I_{x1} is

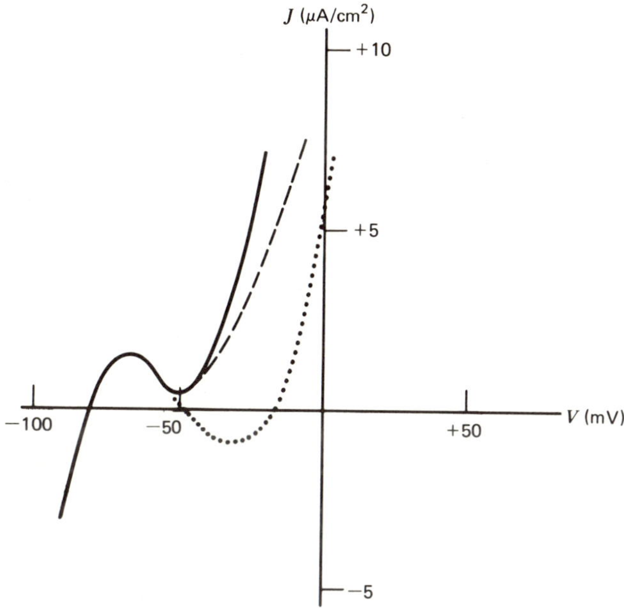

Figure 3.27. Steady-state current–voltage relations for total current I_t (full line), total current minus I_{x2} (dashed line), and $I_t-(I_{x1}+I_{x2})$ (dotted curve). Modified from Noble and Tsien (1969a), by permission.

not solely responsible for this effect, because of the influence of diastolic intervals on the magnitude of the transient outward chloride current described by Dudel et al. (1967a).

The repolarization process was reconstructed by Noble and Tsien (1969b). The dotted curve in Fig. 3.27 represents the current–voltage relation at the beginning of the plateau; it is the sum of the steady-state sodium current, the time-independent potassium current, and I_{K2} for $s=1$. The slow inward current was neglected in this analysis. Figure 3.28a shows the instantaneous current–voltage relations used to reconstruct the plateau and the fast repolarization phase of Purkinje fibers in terms of the time-dependent change of I_{x1}. They intersect the abscissa at two points during the first 260 ms. This period was called the stable point phase, having a positive slope at the farthest right intersection, called V_{sp}. The other intersection, V_{th}, is unstable and is considered to be the threshold for repolarization. If the membrane potential is suddenly displaced to a level just negative to V_{th}, a regenerative repolarization occurs (Weidmann, 1951a; Vassalle, 1966). As the current–voltage relation shifts in the outward direction, V_{th} and V_{sp} move toward each other until $V_{th}=V_{sp}=$

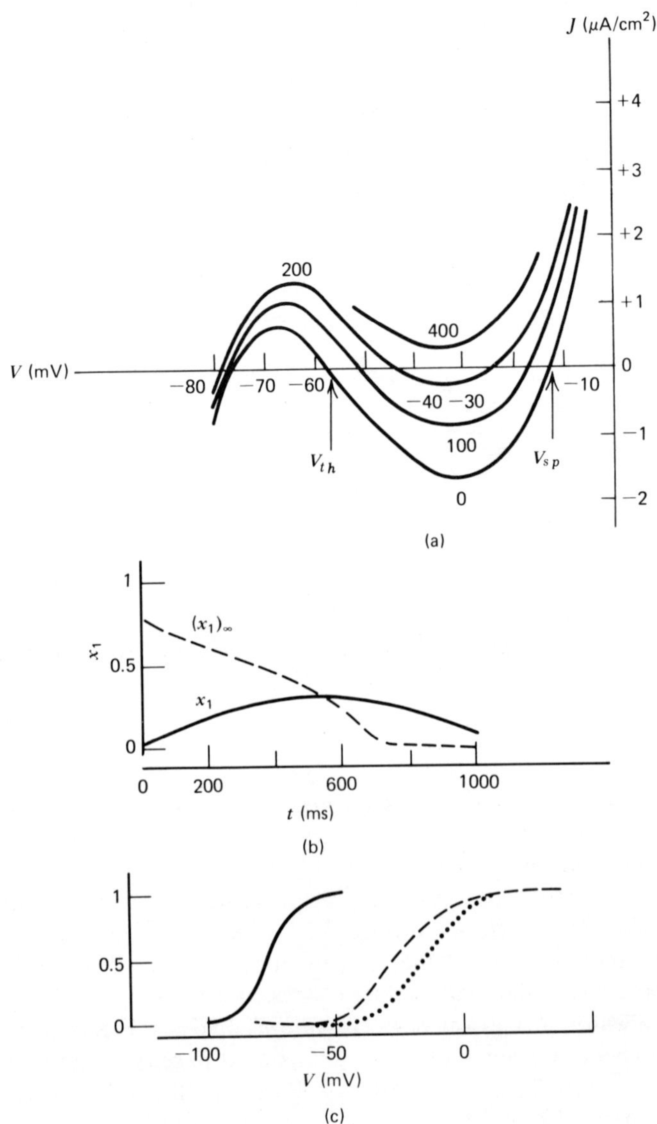

Figure 3.28. (a) Instantaneous current–voltage relations during the plateau phase of the action potential. Parameter: time after beginning of repolarization, in ms. The 0 ms curve is the same shown as the dotted curve in Fig. 3.27. The others are current–voltage curves during the activation of x_1. (b) Changes in x_1 and $(x_1)_\infty$ during repolarization. (c) Voltage dependence of the variables s_∞ (full line), $(x_1)_\infty$ (dashed line) and $(x_2)_\infty$ (dotted line) for the three components of outward current in Purkinje fibers. Composed from Noble and Tsien (1969b), by permission.

-34 mV. In the stable-point phase, the repolarization rate is determined mainly by the kinetics of x_1. However, since the net currents are very small, V lags behind V_{sp} by as much as 5 mV, and this difference generates the current required to trigger the fast repolarization. The disappearance of V_{sp} ensures that x_1 continues to activate, and then the membrane repolarizes at a higher rate, $(x_1)_\infty$ crosses x_1 (Fig. 3.28b), x_1 reaches its peak value and then declines. The repolarization occurs rapidly enough to produce a lag between $(x_1)_\infty$ and x_1, so that the resting potential is reached before x_1 is completely inactivated. The fast outward transient described by Dudel et al. (1967a) was included in this reconstruction.

Figure 3.28c shows the activation curves for s, $(x_1)_\infty$, and $(x_2)_\infty$, plotted as the fraction of the total current activated in the steady state. The s_∞ curve is quite steep and lies within the voltage range over which the pacemaker depolarization occurs, thus showing the leading role of the I_{K2} component in automatic activity. In contrast, x_1 and x_2 cover the potential range where the plateau occurs. There is also a potential range over which $s_\infty = 1$ while $(x_1)_\infty$ and $(x_2)_\infty$ are both zero. Since the rate constants for x_1 and x_2 are quite different, their contribution to the current responsible for repolarization will depend on the duration and potential level of the plateau phase. This in turn will depend on the duration and magnitude of the slow inward current.

In cardiac fibers showing a less developed plateau, the variable s will be quantitatively more important in initiating repolarization, and x_1 and x_2 will not be significantly activated, but, if the inward current is large, activation of x_1 will also be required for repolarization. Because of the voltage-dependence characteristics of x_2, this variable will be significantly activated in the action potentials where the plateau occurs at or beyond the 0 mV potential, as in ventricular muscle. On the other hand, the pacemaker potential will result from the deactivation of s rather than from deactivation of x_1, because I_{x1} is too small in the pacemaker range and its reversal potential is not sufficiently negative. Recently, Hauswirth et al. (1972b) reported that although I_{K2} and I_{x1} are kinetically separable, the results did not permit establishing whether they are controlled by the same membrane mechanism or not.

More than one component of outward current was also described by Brown and Noble (1969a) in frog atrial trabecula. Figure 3.29 shows that the current after a short depolarizing pulse obeyed a simple exponential function. Subtracting this current from the total current recorded after long pulses revealed a second exponential process. The time constants τ_1 and τ_2 were around 500 ms and 5.8 s, respectively. In Ringer's-TTX, the reversal potential for the first component of outward current was -40 to

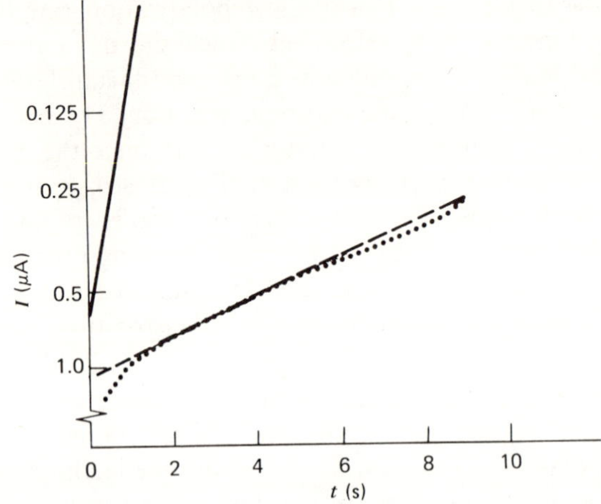

Figure 3.29. Semilogarithmic plot of the time course of the two components of the slow outward current in frog atrial trabecula. Full line: tail current after short depolarizations (1.6 s). The dots represent the current tail after a depolarization of 11.8 s. The full line was obtained by subtracting the dashed line from the dots. Modified from Brown and Noble (1969a), by permission.

−50 mV. In the range of potentials from −40 to −10 mV, the outward current was preceded by the slow inward current. The current–voltage curves of Fig. 3.30 show inward-going rectification. The intersection between them corresponds to the reversal potential for the first current component. The difference between the two current–voltage curves represents the time-dependent change of the current. These two components of slow current were compared to I_{x1} and I_{x2}. No current component similar to the s system was found. Since spontaneous firing in frog atrial fibers occurs only when the resting potential is lower than the value obtained in the trabecula used for these experiments, pacemaker activity was then ascribed to deactivation of the first current component. Brown and Noble (1969b) investigated the voltage dependence of the second component, called I_{x2}. Most of the current was activated at a potential of −25 mV. The curve of x_2 extended over the range −35 to −50 mV. The reversal potential was about −35 to −45 mV. $\bar{I}_{x2}$ did not show inward-going rectification, in contrast to the instantaneous I_{x2} versus V relation. The value of the time constant at the holding potential was 1.85 s. The role of I_{x2} in the action potential is not clear because the depolarization phase is too short to activate it.

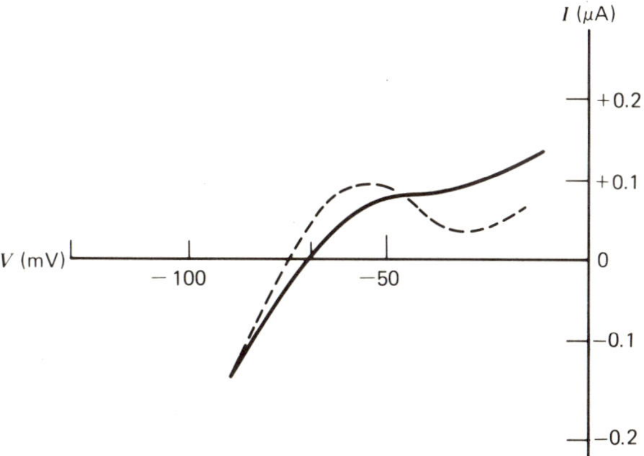

Figure 3.30. Current–voltage relations of the slow outward current in frog atria measured at the end of a short depolarizing pulse (full line) and right after the capacitive artifact has subsided (dashed line). Modified from Brown and Noble (1969a), by permission.

De Hemptinne (1971a, b) investigated the fast component of the delayed outward current and its frequency dependence in frog atria: g_K increased with delay when the membrane was depolarized and decreased exponentially upon repolarization, as shown in Fig. 3.31a, where the fit was obtained with an equation similar to Eq. 2.8. The activation constant increased only by a relatively small amount during the repolarization, but total repolarization was attained because a rather large fraction of outward current was carried by the leakage current. Figure 3.31c shows the frequency-dependent changes of I_K and its activation factor n^2. Repetitive stimulation after a rest period resulted in a faster activation of I_K and a reduction in action potential duration. The total I_K during both action potentials was the same, but during the second action potential the current started from an initial nonzero level, and the maximum I_K was reached earlier.

Brown, Clark, and Noble (1972) induced repetitive activity in quiescent atrial trabecula to study the ionic currents during the pacemaker potential. The pacemaker potentials resulted from the inactivation of an outward current that decayed as a single exponential. Its amplitude was only a fraction of the total delayed current that could be activated by depolarization. This pacemaker system differs from the s system in that its reversal potential does not coincide with E_K. The potential range of the pacemaker current in frog atria is also quite different from the one described for the s system.

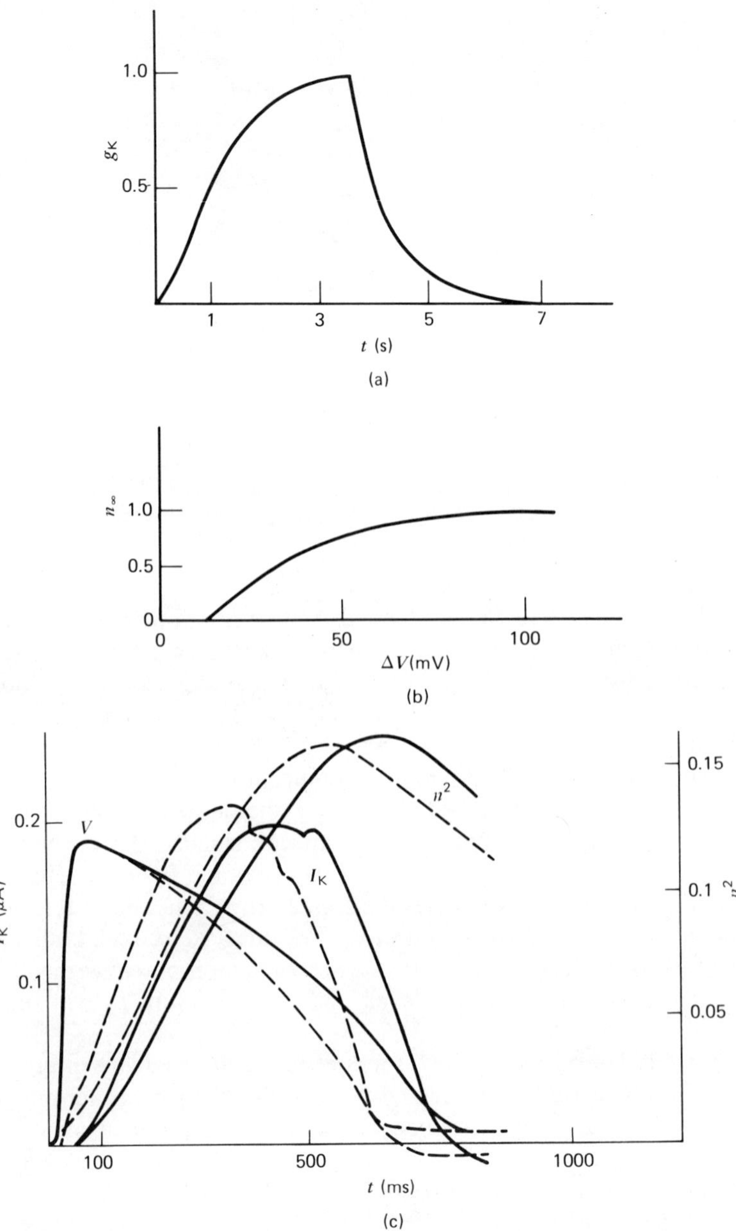

Figure 3.31. (a) Time course of g_K during a depolarizing step ($\Delta V = +120$ mV). (b) The voltage dependence of n_∞. (c) Variation of the activation factor n^2, and the resulting I_K during a normal action potential (full lines) and an action potential shortened by repetitive stimulation after a long rest period (dashed lines). Modified from De Hemptinne (1971a, b), by permission.

3.2.3.3 Voltage-Clamp Measurements Related to Electromechanical Coupling. Kavaler (1959) made one of the first attempts to correlate membrane depolarization with tension development in cardiac muscle, applying constant-current pulses to small bundles of calf and sheep ventricular muscle. Depolarizing current applied after the upstroke of the action potential held the membrane at positive potentials and produced a sustained contraction lasting as long as the current pulse. Wood, Heppner, and Weidmann (1969) observed that under steady-state conditions, the magnitude of the contractile response was related to the efficiency of the process of activation of the contractile elements rather than to their maximal capability to contract. It was postulated that the slow inward current contributes to the instantaneous $[Ca^{2+}]_i$ attained during a given action potential and this would give the correlation between the action potential configuration and its associated contraction. The dependence of a given contraction on the electrical and mechanical events preceding it is the consequence of the effect of repeated depolarizations on the processes of release and binding of calcium. A similar hypothesis had been proposed earlier by Furchgott and De Gubareff (1956).

Morad and Trautwein (1968) investigated the relationship between the duration of the action potential and contraction amplitude in cat, dog, and sheep papillary and trabecular muscle preparations. After the upstroke of a normal action potential the membrane was clamped to the resting potential or to a holding potential a few mV more negative. Within the initial 50 ms of the action potential, tension developed rapidly, and during the next 100–150 ms the increments of tension became smaller. Therefore, if the membrane was repolarized before 200 ms had elapsed the resulting contraction was smaller than the control.

A different result was obtained in sheep Purkinje fibers by Fozzard and Hellam (1968), who observed complete relaxation before the fiber was repolarized. Mechanical threshold was found at -55 mV. This work was extended by Gibbons and Fozzard (1971). The study was restricted to the first contraction following resting periods of 1.5 min. Two types of contractile responses appeared upon depolarizing clamps to or above -20 mV: a phasic response and a sustained tension of small amplitude relative to the twitch. The relationship between tension and potential was similar to the one obtained in frog atrial tissue (Fig. 3.32). The mechanical threshold was at -58 mV and the point of inflection was at -20 mV. Although these results are hardly comparable to those obtained under steady-state conditions, their interpretation suggests that electromechanical coupling in Purkinje fibers and in ventricular muscle have common features. The voltage dependence of the contractile response could be explained by the influence of membrane potential on release and uptake of activator Ca^{2+} by the sarcoplasmic reticulum and the contribution of the slow inward current. Experimental evidence has accumulated over the years to show the dependence of contractile activity on $[Ca^{2+}]_o$ and the role of Ca^{2+} in electromechanical coupling. In an analysis of the controversial aspects of the excitation-contraction mechanism in cardiac muscle, Brady (1964) predicted that "the plateau of the action potential would be the result rather than the prime cause of calcium entry into the cell," a hypothesis confirmed a few years later when the inward calcium current was found. The role of this current in activating contraction was investigated by Reuter and Beeler (1969b) and Beeler and Reuter (1970c) in strips of dog ventricle. A very small contraction appeared at the threshold potential for I_{Na}. Tension increased when the membrane potential was clamped at potential levels such that the slow inward current was activated. However, the time course of activation of the slow current and the degree of activation of contraction were different. Maximum contraction for a given membrane potential could be obtained only with voltage steps 10–20 times longer than required to fully activate the slow current. The results indicated that in Tyrode solution the calcium inward current and the tension developed are not directly coupled, but a direct coupling was found in the absence of Na^+. Under these conditions, maximum tension was observed upon the first depolarizing step, provided that the potential was such as to activate I_{Ca} and the duration of the pulse was longer than 40 ms, the time required for full activation of the calcium current. Moreover, the net gain in intracellular calcium calculated from the charge transferred during the initial slow inward current (10^{-6} to 5×10^{-5} M) would be sufficient to activate the contractile mechanism appreciably (Ebashi and Endo, 1968; Winegrad, 1971). The indirect effect of I_{Ca} on developed tension in sodium-containing solutions suggested that I_{Ca} contributes to fill intracellular stores from which Ca^{2+} is released during subsequent depolarizations. The mechanism of this release remains unknown, but it seems to be dependent on depolarization but independent of either I_{Na} or I_{Ca}. Within this hypothesis, the amount of calcium at the intracellular sites would be determined by the calcium current and the activity of the carrier mechanism that extrudes Ca^{2+} from the cell. The latter depends, in turn, on sodium concentration, since it has been shown that the net calcium outflow is inhibited in sodium-free solutions (Reiter, 1966; Reuter and Seitz, 1968; Langer, 1968). Supporting evidence for this hypothesis has been reported by Bravený and Súmbera (1970), who ascribed two different functions to the action potential: the filling of the pool by I_{Ca} and the release of activator Ca^{2+} from the same pool during depolarization.

The double role attributed to depolarization in electromechanical coupling was further investigated by Ochi and Trautwein (1971) in papillary muscles of guinea pig hearts. Simultaneous measurements of membrane current, membrane potential, and contraction permitted separation of the direct effect of depolarization from the effects mediated by the activation of the slow current. The rate of tension development depended on I_{Ca}. The potentiation phenomenon (positive staircase) occurred independently of the slow inward current, and its degree seemed to depend on the amount of depolarization. This phenomenon was attributed to the effect of membrane potential on the mechanism of calcium release from the intracellular pool. Experiments carried out in frog atria trabecula (Léoty et al., 1970a, 1971a, b; Léoty and Raymond, 1972) yielded results comparable to those obtained in mammalian ventricle. The threshold for contraction and the activation of I_{Ca} were identical. Peak tension and I_{Ca} showed similar voltage dependence, although the maximum value of current was obtained at potentials slightly more negative than the ones required to obtain maximum contractile response. Depolarizations longer than 100 ms elicited contractions not correlated to the voltage dependence of I_{Ca}. Beyond depolarizations of 80 mV, the developed tension became independent of the slow inward current. The contractile response of frog atria to depolarization seemed to consist of a phasic component directly coupled to I_{Ca} and a longer, slower tonic phase appearing at more positive values of membrane potential and showing voltage and time dependence. This component of contraction persisted in the presence of Mn^{2+}. Similar results in ventricular fibers were reported by McGuigan (1968). Vassort and Rougier (1972) obtained a slow increase in tension by imposing weak depolarizations of long duration but insufficient to activate any inward current. No threshold was found for this response, as shown by curve *a* of Fig. 3.32. Curve *b* shows the voltage dependence of the contraction component associated with I_{Ca}. For depolarizations beyond E_{Ca}, a second increase in tension appeared, showing the effect of large depolarizations on calcium release from intracellular binding sites. Curve *c* in Fig. 3.32 was obtained in the absence of Ca^{2+} or under Mn^{2+}. It was proposed that the contraction following a normal action potential is composed of an initial rise in tension triggered by the inflow of Ca^{2+}. While I_{Ca} subsides, the initial tension is added to the slow growth of the voltage-dependent contractile response.

Different results were obtained in frog ventricle by Goto (1971) and Goto et al. (1971). The threshold for the mechanical response and the slow inward current were very low and almost coincided with the threshold for I_{Na}. Maximum tension was obtained only with pulses 2 s long, almost

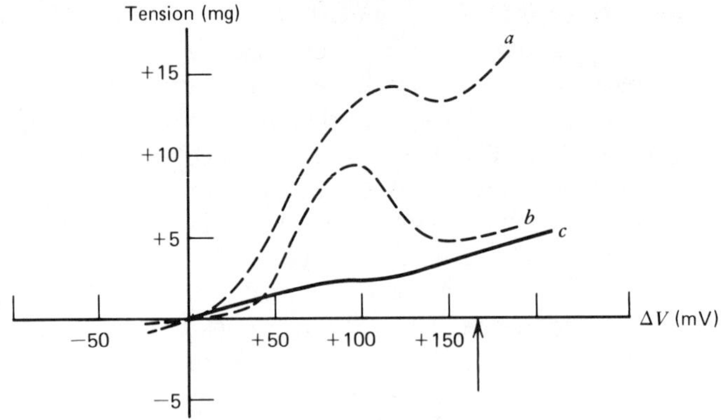

Figure 3.32. Voltage dependence of developed tension of frog atrial trabecula upon depolarizing pulses of (*a*) 620 ms and (*b*) 185 ms duration. Curve *c* obtained in Ca-free medium. The arrow shows the reversal potential for the slow inward current. Modified from Vassort and Rougier (1972), by permission.

twice as long as the normal action potential. In frog ventricle, the phasic and tonic responses seemed to be fused. This was attributed to the poor development of the sarcotubular system.

Hermsmeyer, Rulon, and Sperelakis (1972) reported that hypertonic solutions produced loss of the plateau of the action potential and depressed contractile activity in heart muscles with well-developed tubular systems. Species with poorly developed tubular systems did not show this effect. The apparent resistance decreased concomitantly with the loss of plateau. This was attributed to the large increase in surface area caused by dilatation of the T system, which offsets the decrease of surface due to shrinkage of the cells (Sperelakis and Rubio, 1971). The result could be explained if the tubular membrane participated in the increase of conductance of the slow channel that determines the plateau, similar to the active response suggested for the T system of skeletal muscle by Costantin (1970).

Goto, Kimoto, and Suetsugu (1972) reported evidence that sodium-calcium antagonism takes place at the ionic channels responsible for the slow inward current. Sodium depletion enhanced the slow inward current and the contraction associated with it, but both responses decreased later and contracture occurred. Two systems for excitation-contraction coupling were suggested: inward currents responsible for the phasic contraction and a sodium-calcium exchange mechanism responsible for tonic contraction and relaxation. The latter cannot be formally distinguished from the postulated intracellular calcium-pool. Moreover, Goto, Kimoto, Saito, and Wada (1972) reported that the relaxation occurring upon repolarization of frog atria consisted of three components: an initial delay in the decrease of tension, a rapid exponential decay, and a final slow decay of tension. It was proposed that repolarization would induce a transient increase in I_{Ca} by enhancing the driving force at a moment when g_{Ca} is not yet inactivated. This would explain the delay in initiating relaxation.

3.2.3.4 Effects of Various Agents on Cardiac Membranes. Peper and Trautwein (1967) investigated the mechanism by which aconitine applied to cardiac muscle produces repetitive activity (Scherf, 1947). In the presence of aconitine the membrane current was negative and showed a distinct inward peak over a wide range of potentials. TTX suppressed the effect of aconitine. Moreover, the drug seemed to increase the steady-state g_{Na}, because the inward current at slow ramp speeds persisted even during a repolarizing ramp. At a membrane potential of -60 mV, the outward current was strongly decreased, which explains the long plateau of the action potential at this level. Since a substantial I_{Na} was activated at this potential, the spontaneous repetitive activity arising from this second plateau (Schmidt, 1960) could be explained.

Several reports indicated that the ionic channels are not highly specific for a given ionic species (Tasaki et al., 1968; Pappano and Sperelakis, 1969a; Ochi, 1970). Carmeliet and van Bogaert (1969) reported that action potentials were generated in Purkinje fibers when Sr^{2+} replaced Ca^{2+} in a sodium-free solution. Vereecke and Carmeliet (1971a) showed that these action potentials were produced by a regenerative increase in g_{Sr}. The strontium action potentials resembled ventricular action potentials, with an overshoot of 20 mV and a very low depolarization rate (5–7 V/s). The amplitude of the overshoot showed a linear relationship with log $[Sr^{2+}]_o$, and $\dot{V}_{max}$ was also dependent upon $[Sr^{2+}]_o$. In voltage-clamp experiments, inward current was elicited upon depolarization, reaching a maximum at a membrane potential of -20 mV and reversing direction at $+20$ mV. For depolarizations between -50 and 0 mV, this current persisted for more than 1 s. This explained the high G_m measured during the plateau of the action potential.

Strontium action potentials were inhibited by other bivalent ions such as Ca^{2+} and Mn^{2+}. A theoretical analysis of these results (Vereecke and Carmeliet, 1971b) showed that this ionic channel does not obey simple first-order kinetics. The Hodgkin-Huxley theory and the cable model predict that the maximum depolarization rate ($\dot{V}_{max}$) should vary as a linear function of $[Sr^{2+}]_o$, according to $\dot{V}_{max} = -k\bar{g}_{Sr}(V - E_{Sr})$, provided k and $\bar{g}_{Sr}$ are independent of $[Sr^{2+}]_o$. k is a constant containing the variables m and h. The experimental plot was linear between 2 and 10 mM, but it leveled off above 20 mM. A relationship between $\bar{g}_{Sr}$ and $[Sr^{2+}]_o$ was derived that was similar to the Michaelis-Menten equation for enzyme kinetics:

$$\bar{g}_{Sr}=(\gamma_{Sr}\bar{g}_{max})/(1+K_{Sr}/[Sr]_o) \tag{3.37}$$

where $\bar{g}_{Sr}$ is the maximum g_{Sr}; γ_{Sr}, the intrinsic strontium-activity; $\bar{g}_{max}$, the maximum conductance for bivalent ions; and K_{Sr}, the dissociation constant for the complex strontium receptor. A plot of $1/\dot{V}_{max}$ versus $1/[Sr^{2+}]_o$ resulted in a straight line, as found in the Lineweaver-Burk plot. In different $[Ca^{2+}]_o$, the slope of the line relating $1/\dot{V}_{max}$ to $1/[Sr^{2+}]_o$ varied with $[Ca^{2+}]_o$, but the y intercept was not significantly altered. K_{Sr} and K_{Ca} were 7 and 0.2 mM, respectively, thus showing that Ca^{2+} has a 35 times larger affinity for the membrane site than Sr^{2+}.

Reuter (1967) showed that adrenaline shifted the plateau of the action potential to more positive levels and increased the amplitude of the slow inward current. These observations were confirmed by Vassort et al. (1968, 1969) in frog atria. In TTX-Ringer's, adrenaline produced a large increase in slow current, and the time constant of inactivation decreased from 30 to 23 ms. In every case, the threshold for the slow current appeared at 10 or 20 mV more negative. These effects did not depend on whether Na^+ or Ca^{2+} carried the charge across the membrane.

Kohlhardt et al. (1972a, b) reported that the pharmacological agents Verapamil and its methoxy derivative, compound D600, decreased the slow current to 30% of its control value. The effect was reversed by increasing $[Ca^{2+}]_o$. The current–voltage curve of the slow current indicated that these agents act on the conductance of the channel and also compete with Ca^{2+} for a common carrier mechanism.

Ruiz-Ceretti et al. (1976) reported that amphotericin B abolished the slow depolarizing phase of the upstroke of the action potential in the perfused rabbit heart. The effect was similar to the changes produced by Mn^{2+} (Ruiz-Ceretti and Ponce Zumino, 1976). Schanne et al. (1973) confirmed that this drug decreased the slow inward current in frog atria. The effect was not specific, because the fast inward current was also affected, but its decrease was completely reversed by hyperpolarization of the membrane. Schanne et al. (1972) reported that streptomycin depressed the fast inward current. Its threshold and the potential at which I_{Na} reached a maximum were shifted by about 20 mV toward more positive potentials, without change in the reversal potential for I_{Na}.

The relation between electrical activity and metabolism was investigated by Haas, Kern, and Einwächter (1970) using DNP as a metabolic inhibitor in frog atria. Within 15 min, DNP reduced I_{Na} to 10% of its control value, but it did not change the rate of removal of inactivation. In contrast to the depression of I_{Na}, the steady-state outward current increased by 200% at all levels of polarization. DNP did not change the current–voltage characteristic for I_{Na} (Fig. 3.33a), but E_{Na} was 20 mV more negative. Tracer

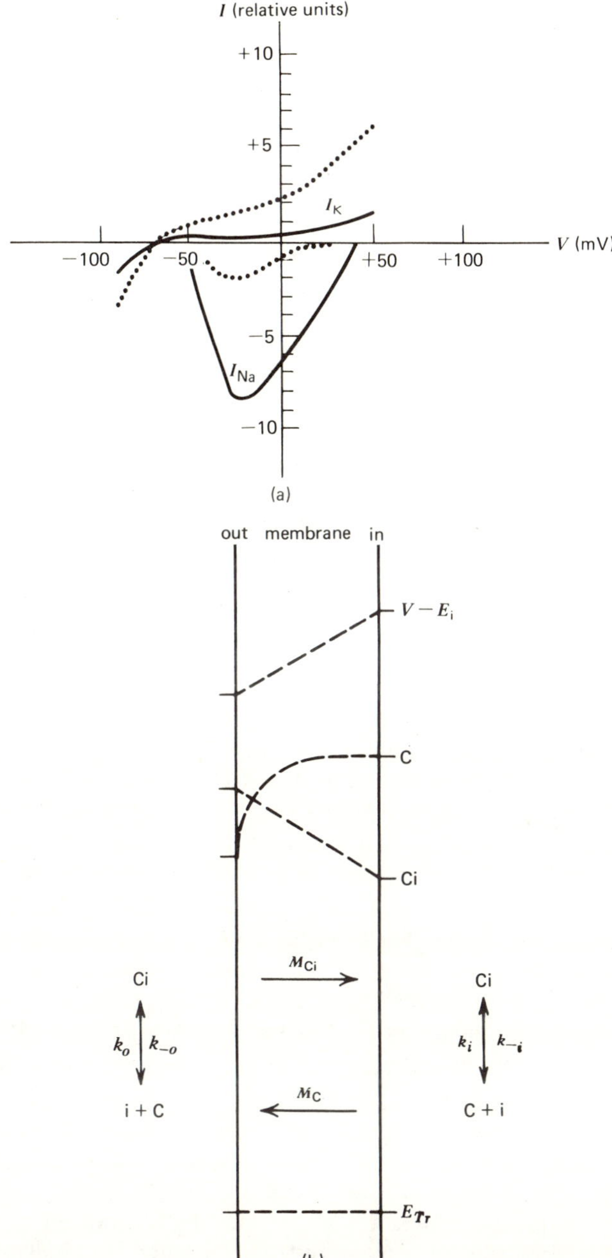

Figure 3.33. (a) Current–voltage curves obtained from frog atrial trabecula under voltage-clamp conditions in Ringer's (full lines) and after addition 0.5 mM DNP (dotted lines). (b) Schematic representation of the carrier system proposed by Haas. V and E_i represent the membrane potential and the equilibrium potential for the ion i; C, carrier in free form; Ci, carrier-ion complex; M, flux; E_{Tr}, reversal potential of the system. Modified from Haas et al. (1970), by permission.

experiments with K^{42} showed a reduction of potassium influx by 29%, while potassium efflux rose only 10% above control levels.

The reduction in $(V - E_{Na})$ partly accounted for the decrease of I_{Na}, but an incomplete block of sodium channels by DNP had to be assumed to fit the results quantitatively. However, a 300–400% increase in g_K had to be postulated to account for the measured current, and this increase could not be reconciled with the very slight augmentation of potassium efflux observed in tracer experiments. As an alternative, an electrogenic pump was proposed, so that the increase in outward current would result from its inhibition. The system was designed to have affinity for only one ionic species. The net movement of the ion resulted from a carrier-mediated influx, where the free and bound forms of the carrier are anion and electroneutral molecule, respectively. Since the transport system is potential dependent, it has a reversal potential E_{Tr} given by:

$$E_{Tr} = E_i + (RT/F)\ln(k_o k_{-i}/k_{-o}\, k_i) \tag{3.38}$$

where E_i is the equilibrium potential for the ion and k_o, k_{-o}, k_i, and k_{-i} represent the rate constants for the association and dissociation of the ion to the carrier at both sides of the membrane, as shown in Fig. 3.33b. Active transport (diffusion of Ci) will be opposite to the passive diffusion of the ion due to the driving force $(V - E_i)$. Within this range of potential, the system transports uphill (inwards). At other potential levels, the system transports in the downhill direction and can be regarded as facilitated diffusion. If the energy supply is impaired, the rate constants become equal (loss of asymmetry) and E_{Tr} shifts toward E_i. At total inhibition, E_{Tr} is equal to E_i, and the carrier system performs a downhill transport of the ion. The membrane current associated with the transport is M_C, the flow of the free form of the carrier. Since M_{Ci} and M_C are coupled, the change in M_{Ci} is reflected by extra current through the membrane. Therefore, the increased outward current measured under DNP reflects the decrease in inward movement of the anion C. E_{Tr} was calculated to be -105 mV, so that within the whole potential range of the action potential, potassium transport proceeds uphill and I_K is partly offset by the outward flow of the carrier. Assuming that the passive fluxes did not change during application of DNP, a passive potassium outflow of 0.4 was reduced by an inward transport of 0.29, resulting in a net current of 0.1 in normal atria. Under DNP, the same passive efflux was augmented by a downhill transport of 0.06 and gave a net outward current of 0.46. This gives an increase of 360% in outward current that fits the experimental results without assuming an enormous change in g_K.

Kern and Einwächter (1969), Haas et al. (1971), and Kern et al. (1971) reported that the neuroleptic agent Droperidol decreased the amplitude, upstroke velocity, and duration of the action potential by 50% after 30 min of exposure. The peak inward current was strongly reduced, but no appreciable effect on outward current was observed. The current–voltage curve showed that the depression of I_{Na} occurred at every potential level. The time constant of inactivation of I_{Na} increased by a factor of 2–3 during exposure to Droperidol. The time constant for recovery decreased from 125 ms to 110 ms during the first minutes under Droperidol, reflecting an early shortening of the refractory period, but the time course of reactivation was distinctly retarded. The normalized h_{∞} versus V curve under Droperidol was parallel to the control, but it was shifted 5 mV toward more negative potentials. Droperidol blocked about 40% of the available sodium channels, and the kinetics of the sodium system were also affected. The block of the sodium channels takes place probably at the outside surface of the cell, but the turnoff mechanism of the sodium system is apparently located at the inner surface (Narahashi and Haas, 1968), so this agent seems to act on more than one membrane site.

3.3 SMOOTH MUSCLE

Smooth muscles constitute a very heterogeneous group of cells that show great diversity in their properties. The visceral type shows spontaneous rhythmic contractions induced by action potentials originating in pacemaker regions within the muscle and conducted from cell to cell. As in cardiac muscle, it is difficult to determine the electrical cell constants because the pattern of current distribution is seldom known. Vascular smooth muscle does not present spontaneous action potentials or cell-to-cell conduction. Although single-probe techniques were often used for resistance measurements, the value of the apparent resistance should approximate the true input resistance of the cell, because the measured values are of the order of 50–100 MΩ, and usually the value of ΔR_{El} is less than 10 MΩ.

3.3.1 AC and Square-Pulse Measurements in Smooth Muscle

3.3.1.1 AC Measurements. Abe and Tomita (1968) reported ac measurements on strips of taenia coli of the guinea pig. The results were related to those of Jones and Tomita (1967) and Tomita (1969b). Using the ac formulation of the linear cable model as developed by Tasaki and Hagiwara (1957a), it was shown that at frequencies of 1–20 Hz, a strip of smooth muscle behaved electrically like a linear cable. The product $r_i c_m$ calculated for a linear cable was equal to $r_i c_m$ obtained from the foot of the action potential (Eq. 1.43). An electrical model of this preparation showed that the assumed equivalent circuit for the intercalated discs did not interfere appreciably with the exponential decay of the electrotonic potential as a function of electrode distance. To evaluate the junctional impedance, Jones and Tomita (1967) and Tomita (1969b) measured the longitudinal resistance of taenia coli and obtained a relation Z versus frequency that was interpreted according to model 2 of Table 1.5. The series resistance was identified with the longitudinal cytoplasmic resistance, and the impedance was considered to consist of a resistance and capacitance in parallel with the impedance of the junctional membranes. The values for the lumped junctional resistance and capacitance were 180 Ω cm and 1–3 μF/cm, respectively. The basic assumption of these experiments was that the longitudinal impedance depended on the cytoplasmic resistivity and the impedance of the intercalated discs. The theory of longitudinal impedance measurements shows, however, that the measured impedance also depends on the membrane impedance (Cole and Hodgkin, 1939; Cole and Curtis, 1936). Tomita (1969b) indicated that the frequencies applied were possibly not low enough to detect the impedance of the membrane.

3.3.1.2 Square-Pulse Measurements in Muscles from the Gastrointestinal Tract. Burnstock and Prosser (1960) showed that visceral and vascular smooth muscles had different values of R_{inp}. They related these values to the existence or absence of cell-to-cell conduction. Several smooth muscles were studied. It was found that the muscle strips that showed conducted action potentials had the lowest R_{inp}, and a correlation was obtained between propagation velocity and tissue resistance. Barr (1961) measured apparent cell resistances on cylinders of cat ileum and jejunum; ΔR_{El} was estimated as 3 MΩ, so the values in Table 3.13 are good approximations of R_{inp}. A very high input resistance was found, but the membrane resistance calculated by assuming the cells to be electrically isolated was of the same order of magnitude as that of other muscular tissues. The length constant was found to be one cell length, and this was considered compatible with the concept of electrically isolated cells.

These results do not agree with those of Nagai and Prosser (1963a), who reported a λ of 10 cell lengths in the longitudinal direction and more than 50 cell widths in the transverse direction. The time constant measured with extracellular electrodes (133 ms) was considered a composite time constant arising from the membrane resistance and capacity of several fibers. The hyperpolarizing branch of the voltage–current curve was linear, whereas the depolarizing side showed some degree of rectification. The cytoplasmic resistivity was estimated from measurements of the resistance between two microelectrodes cemented together and impaled in the same cell (Table 3.13). Measurements of interelectrode resistance suggested that adjacent cells were connected by a resistance one-fourth of that between one cell and the extracellular space. Morphological studies showed the existence of

Table 3.13 Electrical Cell Constants of Muscle from the Gastrointestinal Tract

Method	Measured Quantities	Model	Derived Quantities	Reference
1 of Table1.1 (modif.)	Ac measurements Z_{inp}: Z vs f	2 of Table 1.5	Whole tissue: Spec. res.: 370 Ω cm Intercalated discs: Spec. res.: 180 Ω cm Spec. cap.: 1 to 3 μF/cm	Tomita (1969b)
3 of Table1.3	R_{app}: 103.8 MΩ R_{intel}: 236 MΩ	1 of Table 1.5	R_m: 980 Ω cm^2 λ: 96 μm	Barr (1961)
5 of Table1.3	τ_{inp}: 31 ms R_{app}: 72 MΩ	1 of Table 1.5	R_m: ≈560 Ω cm^2 C_m: ≈56 μF/cm^2	Nagai and Prosser (1963a)
2 of Table1.3	R_{intel}: 0.3 MΩ (x=1 μm) R_{intel}: 5 to 7 MΩ (x=25 μm)			
5 of Table1.3	R_{app}: 30 MΩ τ_{inp}: 2 to 4 ms $R_{\text{app}}(V/I)$: 40 MΩ	1 of Table 1.5	R_m: 320 Ω cm^2 C_m: 10 μF/cm^2	Kuriyama and Tomita (1965)
1 of Table1.3 modif.	τ_{inp}: 70 to 100 ms λ : 1.6 mm τ_f : 6 ms v : 7 cm/s	5 of Table 1.5	C_m: 3 μF/cm^2 R_m: 20 to 30 kΩ cm^2	Tomita (1966a, b)
5 of Table1.3	R_{app}: 51.2 MΩ $R_{\text{app}}(V/I)$: 54 MΩ R_{app}: 66–71MΩ[a] R_{app}: 57–60 MΩ[b] R_{intel}: 18 MΩ R_{intel}: 53 MΩ τ_{inp}: 10 ms	1 of Table 1.5	R_m: 780 Ω cm^2 (longitudinal muscle) R_i: 222 Ω cm C_m: 12.7 μF/cm^2	Kobayashi et al. (1967)
		5 of Table 1.5	λ: 0.94 mm (longitudinal muscle)	
5 of Table1.3	R_{app}: 45 MΩ τ_{inp}: 3.5 ms R_{app}: 40 MΩ τ_{inp}: 35 ms	5 of Table 1.5	jejunum: R_m: 650Ω cm^2 C_m: 4.5 μF/cm^2 rectum: R_m: 600 Ω cm^2 C_m: 5 μF/cm^2	Kuriyama et al. (1967a)

[a] Circular muscle, quiescent; [b]circular muscle, active.

intercellular bridges that were considered to be shunts. The resistance of these regions was estimated to be as low as 2.38 MΩ, thus supporting the concept of conduction by local current flow. Sperelakis and Tarr (1965) opposed this view because electrotonic interaction was found only at very short interelectrode distances, but Tarr (1967) provided an alternative explanation for similar results obtained in cardiac muscle. Moreover, Barr's review (1963) supported the hypothesis of electrotonic spread as the basic mechanism for conduction.

Other mammalian smooth muscles have characteristics similar to cat intestinal muscle, as shown by Kuriyama and Tomita (1965) in guinea pig taenia coli. The apparent resistance decreased during the spike, in contrast to the results of Nagai and Prosser (1963a). Plots of the action potential amplitude versus current gave slopes of 10–30 MΩ (Table 3.13), which corresponds to the difference between apparent resistances at rest and at the peak of the action potential. The cell constants were calculated applying models 1 and 5 of Table 1.5. The latter gave values comparable to other excitable structures, and it was considered that representative values lie between 300 and 1500 Ω cm^2 for R_m and between 2 and 10 μF/cm^2 for C_m.

Tomita (1966a) succeeded in modifying spontaneous electrical activity through polarization of strips of guinea pig taenia coli by means of extracellular electrodes. The electrotonic potentials were recorded with a microelectrode. The values of λ and τ_{inp} in Table 3.13 agreed with the values reported by Nagai and Prosser (1963a) for circular intestinal muscle. Voltage–current curves showed anomalous rectification upon depolarization. The differences found between the responses to intracellular injection of current and to extracellular polarization were explained by the geometry of the tissue and by inhomogeneities of the membrane, which would have areas with different properties. The pattern of current flow within bundles formed by aggregates of interconnected cells was tested by Tomita (1966b) by means of model experiments. The experiments confirmed the nonapplicability of the linear cable model to calculations of cell constants from measurements made with intracellular polarization. The applicability of model 5 of Table 1.5 to measurements performed with extracellular polarization was later confirmed by Abe and Tomita (1968). The cell constants in Table 3.13 were estimated by applying Eq. 1.43 and assuming R_i = 125 Ω cm. The value of C_m is close to that found for the surface membrane of skeletal muscle, but it is smaller than that calculated from results with intracellular polarization (10 μF/cm^2, Kuriyama and Tomita, 1965). A relationship between a high R_m and automatic activity was postulated. This relationship does not hold for the intestinal muscle of the cat, where Kobayashi et al. (1967) found similar values of apparent resistance for longitudinal and circular intestinal muscle, whether they showed spontaneous activity or not.

The electrical characteristics of different sections of guinea pig intestine were investigated by Kuriyama et al. (1967a). As in the cat intestine, the longitudinal layer of muscle showed spontaneous activity and the circular layer was quiescent. The values of R_{app} and τ, as well as the cell constants, are listed in Table 3.13. The cell constants are comparable for longitudinal and circular muscle and nearly the same as for taenia coli. However, conduction velocities were different in the three types of tissue. This discrepancy may arise from the fact that the cell constants as estimated for single cells do not reflect the electrical properties of a bundle. This was further shown by Kuriyama et al. (1967a), who reported large differences in the value of τ for bundles of jejunum and rectum, whereas the apparent resistances of their cells were comparable.

The electrical properties of cells from the longitudinal layer of the guinea pig stomach were measured by Kuriyama, Osa, and Tasaki (1970). The time course of the foot of the action potential was exponential with a very large τ, which suggested a high C_m. The values of R_{app} and τ are listed in Table 3.13. The results indicated that stomach muscle also shows linear cable properties attributable to functional bundles. In addition, as in other types of smooth muscle, the action potential did not depend upon $[Na^+]_o$, but upon $[Ca^{2+}]_o$.

Brading, Bülbring, and Tomita (1969) determined the temperature dependence of the relative R_m in the guinea pig taenia coli. Cooling from 30 to 19°C increased the relative R_m. This effect was less marked in the presence of low $[Cl^-]_o$. Calcium deficiency depolarized the membrane and increased R_m at 37°C. These effects were reversed at 21°C. Measurements of fluxes of K^{42}, Cl^{36}, and Na^{24} showed a decrease in the efflux rate of K^+ and Cl^- at low temperatures. The increase of R_m at 21°C was interpreted as a reduction of g_K and g_{Cl}. Since the effects of Ca^{2+} on potential and conductance were different at 37°C than at 21°C (Bülbring and Tomita, 1968, 1969a), it was concluded that this cation might control independently the potassium and sodium conductances. However, Sakamoto (1971) reported profound effects of Ca^{2+} on membrane potential and resting conductances that were not mediated by Na^+. The experiments were performed at room temperature (20–22°C) in solutions where Na^+ had been entirely replaced by Ca^{2+} (Ca^{2+}-Locke's). After 30 min in this solution, the membrane hyperpolarized from -42.5 mV to -65 mV and the potential level at which rectification appeared was shifted to more negative potentials. The input resistance was consistently reduced. The ratio of R_m in Ca^{2+}-Locke's over control Locke's was 0.68. The results confirmed that calcium is the ion required for electrical activity in this muscle. In addition, it was found that Ca^{2+} contributes to the resting permeabilities and to the membrane potential.

Casteels, Droogmans, and Hendrickx (1971a, b) investigated the genesis of the resting potential and the changes in G_m with K^+ depletion in guinea pig taenia coli. The changes in ionic concentrations induced by incubation in K^+-free media were accompanied by changes in the membrane potential, P_K, and G_m (Fig. 3.34). In chloride medium (Fig. 3.34a), a slight depolarization was followed by a hyperpolarization, which in turn reversed to a final depolarization to about −10 mV. The time course of P_K was estimated with Eq. A3.1 of Appendix 3, assuming that after 10 min in K^+-free solutions, the extracellular potassium had been washed out of the extracellular space. The time course of the changes in P_K and G_m during potassium depletion in chloride solution (Fig. 3.34a) suggested that the hyperpolarization was caused mainly by an increase in P_K. In contrast, the slight hyperpolarization observed in propionate would result from the increase in the ratio $[K^+]_i/[K^+]_o$ because no appreciable changes were observed in neither P_K nor G_m (Fig. 3.34b). Ouabain did not affect the

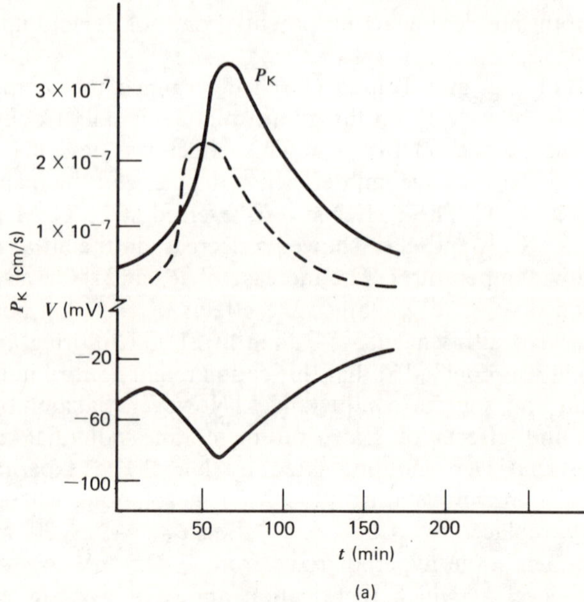

Figure 3.34. Changes of membrane potential P_K and relative membrane conductance (dashed lines) in arbitrary units during K^+-depletion in guinea pig taenia cole. (a) K^+-free chloride solution. (b) K^+-free propionate solution. (c) K^+-free chloride solution containing 10^{-5} M ouabain. Modified from Casteels et al. (1971a), by permission.

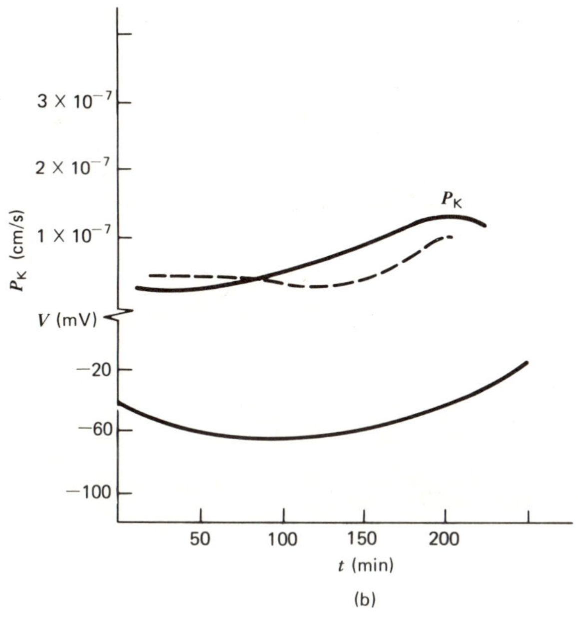

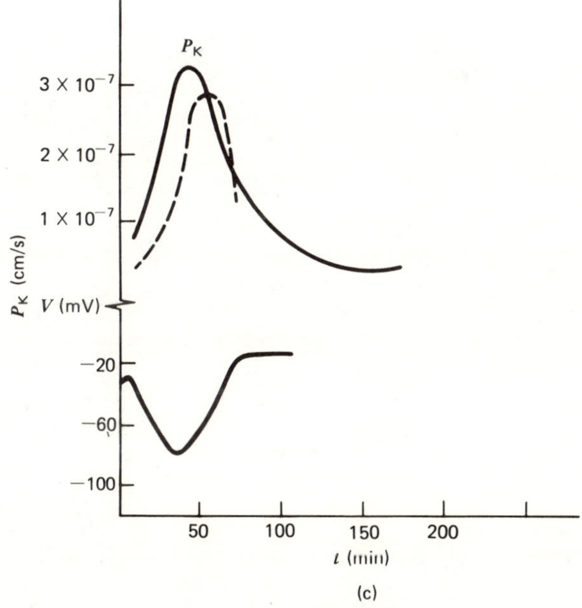

Figure 3.34. *Continued.*

membrane permeability. In order to estimate the contribution of the ionic pump to the repolarization induced by readmission of K^+ in K^+-depleted cells, Casteels et al. (1971b) measured the time course of the changes in E_K, membrane potential, and electrotonic potentials following addition of K^+ to the K^+-free solutions. Initially, the membrane potential followed closely the changes in E_K, but after 5–7 min V became consistently more negative than E_K (Fig. 3.35). The hyperpolarization with respect to E_K was absent

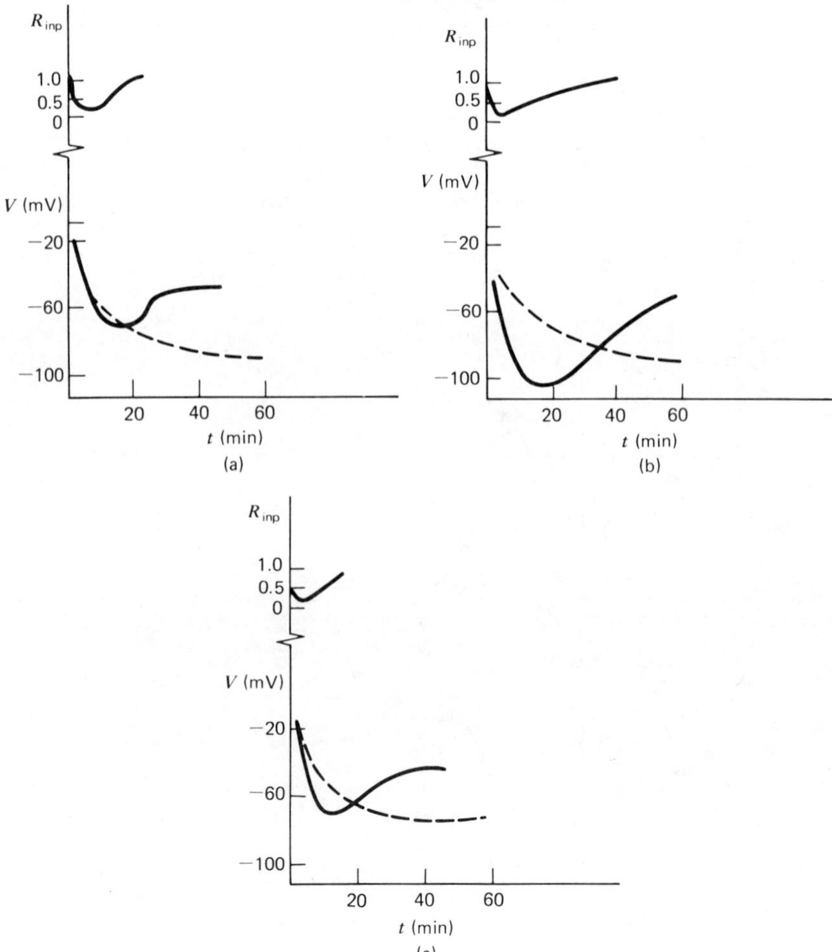

Figure 3.35. Time course of membrane potential (full lines), E_K (dashed lines), and relative input resistance upon readmission of K^+ to K^+-depleted cells of guinea pig taenia coli. $[K^+]_o = 5.9$ mM in (a) and (b), 11.8 mM in (c). Chloride solutions in (a) and (c); chloride replaced by propionate in (b). Modified from Casteels et al. (1971b), by permission.

in ouabain-containing medium. The results shown in Figs. 3.34 and 3.35 were analyzed with the equation derived by Finkelstein and Mauro (1963) and Kornacker (1969):

$$V = (\Sigma g_i)^{-1} I_p + (\Sigma g_i E_i / \Sigma g_i) \tag{3.39}$$

where g_i is the chord conductance for the ion i defined as in Table 2.5, and I_p is the current produced by the pump. The rather small initial difference between V and E_K was explained by assuming that the decrease of R_m reflects mainly an increase in g_K and that the first term of Eq. 3.39 remains very small because $(\Sigma g_i)^{-1}$ is very small. This explains the larger hyperpolarization found in nonpermeant anions. Therefore, the contribution of the electrogenic pump to the membrane potential will depend on the variations of G_m for the ion concerned. A pump current of 0.9 $\mu A/cm^2$ was estimated for steady-state conditions, and a value of -22 mV was found for the contribution of the electrogenic pump to the membrane potential.

Bülbring and Tomita (1969b) investigated the effects of adrenaline on the resting conductances of guinea pig taenia coli. Adrenaline produced hyperpolarization and a reduction of R_m. The hyperpolarization reversed to depolarization if the membrane had been previously hyperpolarized by 10–20 mV. Replacement of Cl^- by nonpermeant anions reduced G_m and the effect of adrenaline. Removal of external K^+ in the presence of large anions decreased G_m further and increased the hyperpolarization. The results suggested that adrenaline acts mainly on g_K and shifts the membrane potential toward E_K. The effect of adrenaline on g_K was confirmed by Magaribuchi and Kuriyama (1972).

3.3.1.3 Square-Pulse Measurements in Muscles from the Urogenital Tract. Kuriyama et al. (1967b) determined the cell constants in the guinea pig ureter. As in the taenia coli, extracellular stimulation elicited propagated spikes that were less often produced as a result of intracellular stimulation. Stretching induced spontaneous, repetitive discharges. The results are listed in Table 3.14. Assuming that 50% of the cell length overlaps the neighboring cells, the length constant covers more than 500 cells. Barium depolarized the cells and elicited spontaneous activity. These results supported the hypothesis of Tomita (1966b), who related spontaneous activity to a low g_K. The action potential was not affected by TTX but was abolished by Mn^{2+}. Bennett (1967a) reported that Ca^{2+} carried the charge during the action potential in the guinea pig vas deferens. Increasing $[Ca^{2+}]_o$ from 0.1 to 10 mM increased the amplitude and rate of rise of the action potential. The resting resistance increased and the crest resistance decreased in high calcium. This indicated that Ca^{2+} acts as charge carrier during activity.

Table 3.14 Electrical Cell Constants of Muscles from the Urogenital Tract

Method	Measured Quantities	Model	Derived Quantities	Reference
5 of Table 1.3	R_{app}: 15–23 MΩ τ_{inp}: 2–3 ms	1 of Table 1.5	R_m: 160 Ω cm^2 C_m: 15 $\mu F/cm^2$	Kuriyama, Osa and Toida (1967b)
1 of Table 1.3 modif.	λ: 2.5–3 mm τ_{inp}: 210–320 ms	5 of Table 1.5	R_m: 300 Ω cm^2 C_m: 5 $\mu F/cm^2$	
5 of Table1.3	R_{app}: 15 MΩ } quiescent τ_{inp}: 1.8 ms } quiescent R_{app}: 22 MΩ (resting) R_{app}: 18 MΩ (peak A.P.)			Bennett (1967b)
5 of Table 1.3	R_{inp}; 16.5 MΩ τ_{inp}: 1.5–6.7 ms	5 of Table 1.5	λ: 250 μm C_m: 1.3–5.5 $\mu F/cm^2$	Hashimoto, Holman, and Tille (1966)
5 of Table1.3	R_{app} (V/I): 18 MΩ τ_{inp}: 1.5–3 ms (63% of ΔV) τ_{inp}: 3.5–5 ms (84% of ΔV) λ: 2.1 mm			Tomita (1967)
5 of Table 1.3	$R_{app}(V/I)$: 7.5 MΩ τ_{inp}: 5–30 ms	1 of Table 1.5	R_m: 282 Ω cm^2 $(x=6\ \mu m)$ R_m: 377 Ω cm^2 $(x=8\ \mu m)$	Sperelakis and Tarr (1965)
		5 of Table 1.5	R_m: 400 Ω cm^2 λ: 176 μm	
1 of Table1.3 modif.	τ_f: 28 ms τ_{inp}: 160 ms			Kuriyama, Osa, and Tasaki (1970)
5 of Table1.3	R_{app}: 24 MΩ τ_{inp}: 2.3 ms λ: 1.1 mm			

Bennett and Merillees (1966) measured the apparent cell resistance of guinea pig vas deferens at rest and the changes produced by current injection on the excitatory junction potential (ejp). The effect of current injection on the ejp was variable: its amplitude sometimes followed a linear relationship with depolarization but in other cases it was not altered. Bennett (1967b) described two distinct populations of fibers in the guinea pig vas deferens. Seventy-five percent of the cells gave only passive responses to intracellular injection of depolarizing currents, and 20% gave an active response (spike). The apparent resistance of the latter was 30% greater than for the first group and decreased to less than 70% during the crest of the action potential (Table 3.14). The voltage–current curve was linear up to threshold potential, in contrast to the first group, where the voltage–current curve was linear up to a depolarization of 80 mV. Both cell types responded with action potentials to nervous and extracellular stimulation. This indicated that the different responses to intracellular injection of current may be due to geometrical factors and not to different membrane properties. This conclusion was supported by the findings of Hashimoto et al. (1966). Voltage–current curves were linear over the hyperpolarizing range and showed rectification upon depolarization. Only those cells with input resistances higher than 15 MΩ gave action potentials in response to intracellular stimulation. The length constant was estimated with the linear cable model. From those data, the transfer resistance would be between 29 and 41 MΩ, depending on the position of the electrode, whether it was impaled near the center or the end of the fiber. Comparable results were reported by Tomita (1967), who concluded that the different time courses of the ejp's and the electrotonic potentials obtained by extracellular polarization indicated that the fibers of the vas deferens are also aggregated in functional bundles.

Creed (1971a) investigated the electrical properties of the smooth muscle membrane in the guinea pig urinary bladder. The fibers are arranged in bundles, and the microelectrode was impaled into cells of the same bundle at several distances from a partition. The values of λ and τ were 1.23–1.95 mm and 127–132 ms, respectively. The time constant calculated from the foot of the action potential was 122 ms. Spontaneous activity was recorded, and the impulse was conducted as in other smooth muscles, but the resting potential (-30 to -45 mV) was 10–15 mV lower than in the ureter or the gastrointestinal tract. Creed (1971b) and Kurihara and Creed (1972) investigated the genesis of this resting potential. The curve relating the membrane potential to log $[K^+]_o$ was steeper at 36°C (31 mV per tenfold change of $[K^+]_o$) than at 12.5°C (24 mV per decade change), but much smaller than expected from the Nernst equation. Addition of ouabain or removal of potassium caused depolarization. During recovery, hyperpolarization was consistently observed. The results suggested that the membrane potential of the urinary bladder was mainly determined by ionic diffusion, but the contribution of an electrogenic pump could not be excluded.

3.3.1.4 Square-Pulse Measurements in Vascular Smooth Muscle. Ito and Kuriyama (1971) and Kuriyama, Ohshima, and Sakamoto (1971) investigated the electrical properties of cells from the guinea pig portal vein. The preparations showed cable-like properties, and the current–voltage curve was linear in the range -66 to -24 mV, where spikes were elicited. With hyperpolarization beyond -66 mV, anomalous rectification occurred. The value of τ_f was 220 ms, comparable to τ_{inp} measured from the electrotonic potentials (250 ms). Potential measurements in media of different composition and the effects of cold storage, rewarming, and ouabain strongly suggested the contribution of an electrogenic pump to the resting potential.

Mekata (1971) investigated the membrane properties of smooth muscle cells of the rabbit common carotid artery. Current–voltage curves showed an increase in R_m with hyperpolarization and a decrease with depolarization. The mean values of λ and τ were 1.13 mm and 212 ms, respectively, in isotonic Krebs. The current–voltage curve became linear on hyperpolarization in hypertonic medium. TEA abolished the nonlinearities of the current–voltage relation, but the effective resistance did not change significantly. The existence of low-resistance pathways between cells was suggested by the value of the length constant. The low conduction velocity was attributed to the large time constant.

3.3.2 Voltage-Clamp Measurements in Smooth Muscle

The introduction of the sucrose-gap technique permitted control of the current density through the preparation (space clamp) and investigation of the active properties of the membrane. The sucrose-gap method was first applied to smooth muscle to study the ionic basis of the resting potential. Bennett and Burnstock (1966) reported that increasing $[K^+]_o$ depolarized the membrane and abolished spontaneous electrical activity. The membrane potential decreased 53 mV for a tenfold change in $[K^+]_o$. Barr et al. (1968) used a single sucrose gap to prove that the action potential propagates electrically between cells in the guinea pig taenia coli. The ionic basis of the action potential of guinea pig taenia coli and ureter were studied by Kuriyama and Tomita (1970). Lowering $[Na^+]_o$ to 16 mM increased the threshold for excitation, but spikes could always be evoked. Spontaneous activity disappeared in Ca^{2+}-free solutions, and spikes did not reappear, even with external stimulation, unless Ca^{2+} was readmitted into the solution. It was concluded that the spike was produced mainly by an influx of Ca^{2+}, while the plateau of the ureter action potential was sodium dependent but was not abolished by TTX. These results confirmed previous findings of Nonomura, Hotta, and Ohashi (1966), who showed that the action potential of guinea pig taenia coli was abolished by Mn^{2+} but was not affected by TTX. These conclusions were supported by Kobayashi (1971), who reported that increasing $[Ca^{2+}]_o$ to five times normal shifted the $\dot{V}_{max}$ versus V curve to the left and increased its slope. Nagasawa and Suzuki (1970) reported that calcium influx was partly blocked in the presence of Mn^{2+}, a result that is compatible with the hypothesis of an inward I_{Ca} during the action potential. However, the results of Lammel and Golenhofen (1971a, b) were compatible with a Ca^{2+} spike only if the additional assumption was made that the increased Ca^{2+} influx during the spike was later compensated by a reduction of the resting influx.

Measurements of ionic currents in guinea pig taenia coli under voltage clamp were performed by Kumamoto and Horn (1970). The leakage current was estimated in the current–voltage curve by extrapolation of the steady-state current obtained during three steps of hyperpolarizing pulses under voltage-clamp conditions. On depolarization, the early inward current was less evident than in squid axon, unless the recordings were corrected for capacitive and leakage currents. The reversal potential for the inward current was found at +5 mV. Tetrodotoxin did not change the inward current or the steady-state outward current, but Mn^{2+} reduced the early current and shifted the reversal potential to more negative values. The activation of the early current was very slow compared to the activation of the delayed I_K, so that at early times the two currents overlapped and the peak transient of the total current represented only 70–80% of the maximum inward current. Correcting the current–voltage curve of total current for I_K gave a reversal potential of −25 mV.

Similar results were reported for rat myometrium by Anderson (1969), Anderson, Ramon, and Moore (1970), and Anderson, Ramon, and Snyder (1971). The rats were injected with estradiol benzoate for 5 days prior to the experiments. Under current clamp, hyperpolarizing pulses gave a linear current–voltage relation, and the electrotonic potentials had a time constant of 80 ms. Depolarization up to the threshold potential triggered local action potentials in the node. Under voltage clamp, depolarizing steps gave a family of current versus time curves composed of an initial capacitive and leakage current, a transient inward component, and a delayed outward current that reached a steady state. The curves were corrected for the leakage component; the resulting current–voltage relations are shown in Fig. 3.36. Correction for the leakage current resulted in a shift of the transient equilibrium potential from the negative to the positive side of the voltage axis, a larger peak inward current, and a well-defined region of negative conductance associated with the inward

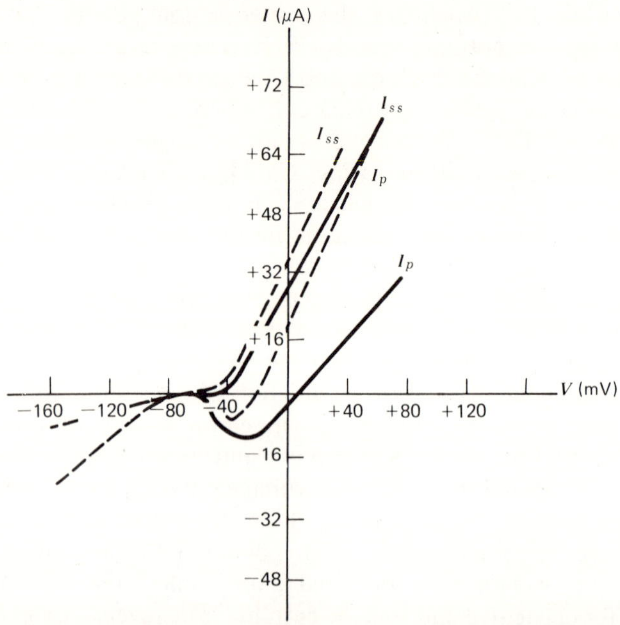

Figure 3.36. Current–voltage curves obtained in uterine smooth muscle under voltage-clamp conditions (dashed curves) without correction for leakage current and (full lines) after correction. I_p: inward peak current; I_{ss}: steady-state outward current. Modified from Anderson (1969), by copyright permission of The Rockefeller University Press.

current component. This shows that leakage current limits the effectiveness of the inward current to depolarize the membrane and suggests an important effect of leakage current on excitation of smooth muscle. In sodium-free solutions, the peak inward current decreased from 32 μA to 10 μA, and the equilibrium potential shifted from +30 mV to +20 mV after 1.5 min of exposure. Excitability was lost in sodium-free medium and also in Ca^{2+}-free solution with normal $[Na^+]_o$. Low $[Ca^{2+}]_o$ decreased the net inward current, and this effect was not reversed by depolarization or hyperpolarization of the membrane. This small residual current was not large enough to produce an action potential. Tetrodotoxin (10^{-5} M) had no effect on the magnitude of the peak inward current. Lanthanum and Mn^{2+} inhibited the net inward current and decreased the steady-state outward current. It was concluded that both Na^+ and Ca^{2+} are required for the transient inward current to develop, neither of them alone being capable of sustaining the excitation mechanism. Mironneau and Lenfant (1971) confirmed these findings in uterine strips from pregnant rats. The reversal potential for the early current was found at depolarizations of 100 to 130 mV, after correction for the leakage current. Anomalous rectification appeared at depolarizations larger than 90 mV. The inactivation curve for the slow current showed that inactivation is minimum at the level of the resting potential and maximum at a depolarization of 25 mV. The value of τ_h was 30 ms. Repetitive activity could be elicited by weak depolarizing pulses of several seconds duration (Mironneau and Lenfant, 1972a). This activity was not abolished by TTX but disappeared in the presence of Mn^{2+}.

The kinetics of the outward current were reported by Mironneau et al. (1971) and Mironneau and Lenfant (1972b). The outward current elicited by long depolarizations (600 ms) increased with potential and time. The current–voltage relation measured in Mn^{2+} medium showed Goldman rectification for the steady-state current and linear characteristics for the instantaneous current–voltage relation, which was considered to represent the leakage current. The equilibrium potential for the outward current was found at 10–15 mV more negative than the holding potential. Figure 3.37

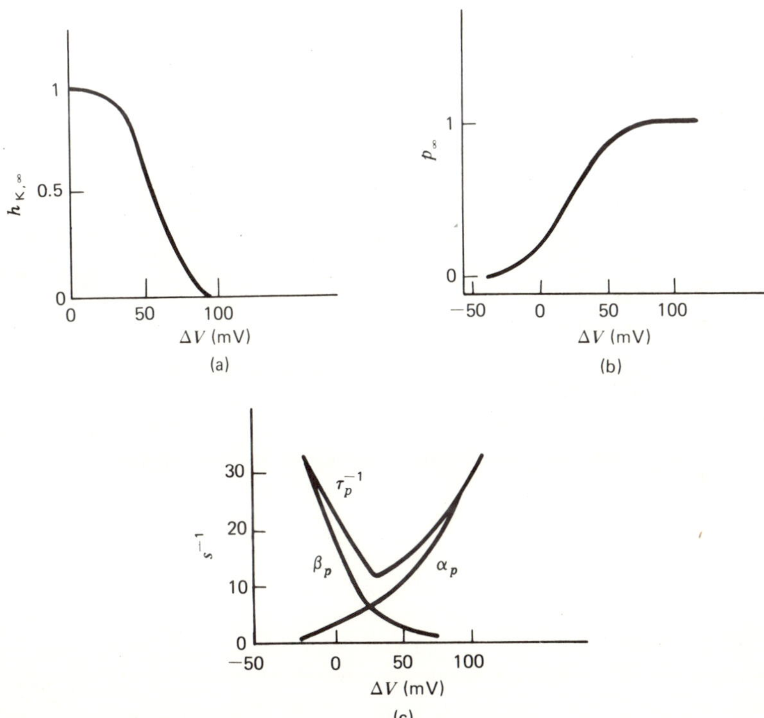

Figure 3.37. Kinetics of outward current in uterine muscle of pregnant rats. (a) Voltage dependence of the variable $h_{K,\infty}$. (b) Voltage dependence of the activation factor p_∞. (c) Voltage dependence of the rate constant (τ_p^{-1}) and the variables α_p and β_p. Modified from Mironneau and Lenfant (1972b), by permission.

shows the voltage dependence of the activation and inactivation variables, as well as the voltage dependence of the variables α_p, β_p, and the rate constant τ_p^{-1}. This system shows kinetics comparable to the I_{x1} current of cardiac muscle.

3.4 SUMMARY AND OUTLOOK

The results discussed throughout this chapter clearly show the importance of an adequate electrical model to the interpretation of impedance measurements. The lack of a suitable model for cardiac and smooth muscle explains the striking difference between the development of our knowledge of the electrical properties of skeletal muscle on the one hand and cardiac and smooth muscle on the other. For skeletal muscle, the inward current during excitation had been estimated from phase-plane trajectories and the rectifier properties were pretty well known before the ionic currents were directly measured under voltage-clamp conditions. The information obtained with the latter method did not modify substantially the concepts developed from more indirect estimates. Moreover, the resolution of two capacity components in the muscle cell and their further structural identification had not been possible without an appropriate model. From this, a fairly complete picture of the permeability characteristics of skeletal muscle was obtained. The information relative to the electrical properties of the tubular membranes that resulted from these investigations encouraged further research into the phenomenon of excitation-contraction coupling. One of the major points to be resolved in the future is whether depolarization of the tubular membranes following an action potential results from a passive electrotonic spread or from a dynamic response of these membranes.

In contrast, a comparative effort exerted in research on the properties of cardiac muscle did not yield as much information. When attempts were made to determine the mechanisms involved in the production of the plateau of the action potential from measurements of the input resistance, different conclusions were reached according to the electrical model applied. An analysis based on a simple, longitudinal spread of current led to the statement that the outward current depended only on time and not on voltage. This contrasted with the results obtained from a two-dimensional model that showed that this current was both voltage and time dependent. Confirmation of the latter was not obtained until the voltage-clamp method was applied to cardiac muscle. The existence of the slow inward current was postulated from analysis of the ionic dependence of the action potential but was confirmed only when the membrane currents could be directly measured. The structural arrangement and the complicated current spread occurring in cardiac muscle upon point injection of current constitute a great handicap to further investigation of the electrical properties or the excitation mechanism in specialized structures (sinoatrial and atrioventricular nodes) that cannot be electrically isolated to perform voltage-clamp measurements. A similar problem is encountered with respect to smooth muscle. The ionic dependence of the electrical activity of smooth muscles was known for a long time, but only recently, has direct exploration of membrane currents contributed to elucidate particular features of this heterogeneous tissue. The situation is further complicated by the several types of smooth muscle, which show striking differences, such as the presence of automatic activity, the existence of action potentials or graded responses, and cell-to-cell conduction. Derivation of adequate and applicable theoretical models and development of other voltage-clamp methods are the possibilities now open to allow pursuit of more advanced research on the electrical properties of cardiac and smooth muscle.

4

Impedance Measurements in Epithelia

Epithelial tissues were extensively studied mainly because they are the site of highly developed transport functions. The simplest cases deal with multicellular structures, and the measurements represent the overall result of at least two membranes in series. For this reason, such structures are referred to, in modern literature, as composite membrane systems (Dowben, 1969, p. 405). In this chapter, the discussion is restricted to those studies in which electrical variables were measured. Because of the large variety of tissues studied, the chapter is subdivided as follows: 4.1, frog and toad skin; 4.2, urinary and gall bladder tissue; and 4.3, gastrointestinal mucosa. Section 4.4 deals with the electrical properties of renal tubules in which electrical measurements were related to the permeability characteristics of the tubular wall, and 4.5 includes a homogeneous group of results mainly reported by the group of Loewenstein, dealing with electrical coupling between glandular and other epithelial cells. Each section may include results obtained with alternating current, square pulses, or voltage clamp. Most of the work related to transport phenomena was performed with a voltage–current method (Fig. 4.1), whereas impedances have been measured with alternating current, square pulses, or voltage clamp using extracellular electrodes or microelectrodes. In Table 4.1 and following, R_m expresses the resistance of the tissue in Ω cm^2, usually called total resistance. Although this value does not represent a true cell constant, it is used here in the context of model 1 of Table 1.5, the model that has been the most widely applied. The interpretation of this resistance in terms of a single membrane is difficult because of the structural complexities of epithelia. The sign conventions used throughout this chapter for ionic fluxes correspond to those used by the authors of the articles discussed; they do not necessarily correspond to the conventions used in Appendix 3 of this book.

4.1 FROG AND TOAD SKIN

Du Bois Reymond (1849) first described the existence of an electrical potential across frog skin, and Galeotti (1907) associated it with the fact that sodium ions did not permeate the skin from the inside to the outside. Huf (1935) reported that frog skin, bathed on both sides by identical saline solutions, performed a net transfer of sodium and chloride from the outside to the inside. Ussing (1948) advanced the hypothesis that the skin potential resulted from the active transport of sodium causing a passive net inward movement of chloride. This was confirmed by Ussing and Zerahn (1951), who performed electrical and ion flux measurements in frog skin. They developed the widely used short-circuit technique illustrated in Fig. 4.1, which permits the measurement of the actively transported current in epithelia. The tissue is placed in a bath, dividing it into two chambers, *A* and *B*, which contain identical saline solutions at equal hydrostatic pressure. The electrodes connected to meter v measure the potential across the structure. This potential may be reduced to zero by applying an opposite dc voltage through a second pair of electrodes. The current needed in the external circuit to reduce the skin potential to zero is measured by the galvanometer *G* and is equal in magnitude to the current generated by the skin. The existence of such a short-circuit current (I_{sc}) suggests an active

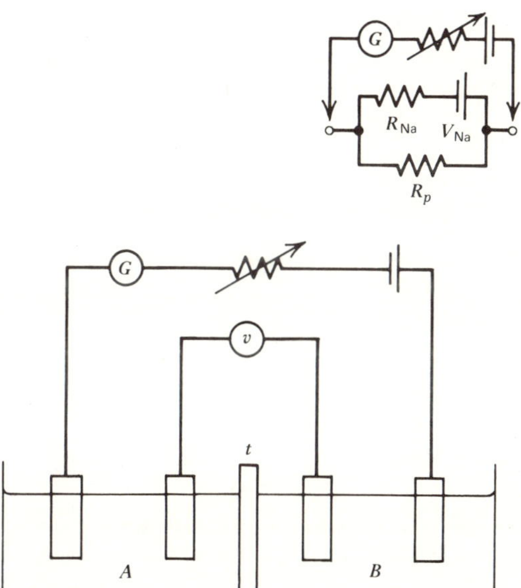

Figure 4.1. Method to measure the short-circuit current through epithelia. *G*, galvanometer; *t*, tissue; *v*, voltmeter. Inset: equivalent circuit.

transport mechanism across the skin. Measurements of ionic fluxes under short-circuit conditions will then determine the ionic species actively transported, and the net ionic flux is equal to I.

4.1.1 Electrical Measurements Related to Ionic Transport Across Amphibian Skin

Ussing and Zerahn (1951) reported that the transepithelial Na^+ influx (M_{in}^+) exceeded the short-circuit current of the frog skin by 10%. This difference corresponded to the transepithelial outflux (M_{out}^+), and so the measured current could be identified with the net sodium flux (M^+). In the equivalent circuit shown in the inset of Fig. 4.1, V_{Na} represents the electromotive force generated by the sodium net flux,* R_{Na} is the resistance to sodium flux, and R_p is the resistance to the flux of ions diffusing passively. When the potential across the skin is reduced to zero, the driving force for passive ion movements disappears. Under these conditions, the current flow depends solely on R_{Na} and V_{Na}. Ussing and Zerahn proposed a three-compartment model of the frog skin (outside solution–cell–inside solution) and assumed that (1) the membrane facing the outside solution presents a high resistance to the flow of the actively transported ion but is permeable to other ions and (2) the active transport occurs across the inner-facing membrane. Under these conditions the electromotive force V_{Na} can be calculated from Eq. A3.3 (see Appendix 3), provided V_{Na} is the net force acting on the ion and the resistance R_{Na} can be defined as in Eq. 1.79. Then, V_{Na} and R_{Na} can be calculated from the measurements of bidirectional ionic fluxes and short-circuit current from the equation:

$$R_{Na} = V_{Na}/I_{sc} = \frac{(RT/zF)\ln(M_{in}^+/M_{out}^+)}{M^+F} \tag{4.1}$$

Moreover, the mechanism of action of agents that are able to alter the ionic flux can be studied in terms of changes of the skin resistance or changes of I_{sc}, as shown in Table 4.1. Both adrenaline and neurohypophyseal extract increased the sodium influx (Na_{in}), but they acted through different mechanisms. While adrenaline decreased the electromotive force and the resistance across the skin, the hormonal extract only increased the conductivity without affecting the electromotive force (Ussing and Zerahn, 1951).

Frog and toad skins are very stable preparations in which the potential remains constant for many hours. Brown (1962) carried out comparative

*Here, V_{Na} represents the potential energy built up by the metabolic process during the ionic flux, since under actual experimental conditions, the potential across the skin is zero but the metabolic driving force exists (Dowben, 1969, p. 456).

measurements of potential and short-circuit current of frog skin in vivo and in vitro. The skin surface was controlled in the in vivo experiments by placing a section of Lucite tubing on the abdomen and filling it with the test solution. In both conditions the potential and the current depended on the sodium concentration; the values of current were higher in vivo than in vitro, while the potential difference across the skin was lower in vivo. The values of the skin conductance for the different test solutions indicated that the permeability for passive ion fluxes was reduced in the in vitro conditions. Removal of chloride reduced but did not eliminate the difference in conductance. Another in vivo preparation of frog skin was described by Bianchi et al. (1969). An incision in the skin at the level of the shoulders permitted the introduction of a composed concentric voltage–current electrode. The method had a serious drawback, however: It was not possible to calculate the total skin resistance because the surface was not accurately known.

Microelectrode measurements were performed in frog skin to study the electrical profile across the skin and to determine the site of the genesis of the potential. These measurements served to verify a model proposed by Koefoed-Johnsen and Ussing (1958). It consisted of three compartments. The cellular compartment was tentatively located at the stratum germinativum, the deepest layer of cells of the epithelium. The model postulated that the membranes facing the external solutions are selectively permeable for sodium and impermeable for potassium, whereas the membranes facing the internal solution (internal barrier) are almost impermeable for sodium and selectively permeable for potassium. In addition, a low intracellular sodium concentration as well as a high $[K^+]_i$ must be assumed for this model. According to the model, the net transport of Na^+ across the skin results from the passive movement of Na^+ from the external solution to the cellular compartment, whereas sodium is actively extruded in exchange for potassium at the internal barrier. The potential difference across the skin results from the diffusion of Na^+ through the external barrier plus the potential originating from the diffusion of K^+ from the cellular compartment to the internal solution. The low level of $[Na^+]_i$ and the high level of $[K^+]_i$ are assured by the operation of the pump.

The potential profile across the skin is still a matter of controversy mainly because of difficulties in microelectrode impalements through the whole skin. Hoshiko (1961) and Whittembury (1964) described two potential steps relative to the inside surface, where the indifferent electrode was placed. The input resistance values listed in Table 4.1 were attributed to the cell membrane across which the voltage steps appeared. The first potential step was attributed to the outer side of the cells in the stratum

Table 4.1 Electrical Constants of Frog and Toad Skin

Method	Measured Quantities	Model	Derived Quantities	Reference
Fig. 4.1	I_{sc}: 49 mC/cm²h	1 of Table 1.5	V_{Na}: 56 mV (Ringer's)	Ussing and Zerahn (1951)
	Na_{in}: 57 mC/cm²h		R_{Na}: 1800 Ω cm²	
	I_{sc}: 76 mC/cm²h		V_{Na}: 35 mV (epinephrine)	
	Na_{in}: 87 mC/cm²h		R_{Na}: 1010 Ω cm²	
	I_{sc}: 158 mC/cm²h		V_{Na}: 76 mV (neurohypophyseal extract)	
	Na_{in}: 168 mC/cm²h		R_{Na}: 1600 Ω cm²	
1 of Table 1.3	R_{inp}: 1.7 kΩ	1 of Table 1.5	R_m: 1.53×10^5 Ω cm² (skin)	Whittembury (1964)
	R_{inp}: 0.4 kΩ		R_m: 0.44×10^5 Ω cm² (str. germinativum)	
	R_{inp}: 0.2 kΩ		R_m: 0.14×10^5 Ω cm² (sub-epithelial)	
1 of Table 1.3	Skin potential	1 of Table 1.5	SO^{2-}-Ringer's:	Cereijido and Curran (1965)
			G_m: 0.77 mmho/cm² (outer)	
	I_{sc}		G_m: 1.88 mmho/cm² (inner)	
			Cl^--Ringer's:	
			G_m: 1.13 mmho/cm² (outer)	
			G_m: 4.76 mmho/cm² (inner)	
Fig. 4.1		Fig. 4.1	G_p: 24 to 54 μmho/cm² (from Na outflux)	Ussing and Windhager (1964)
			G_p: 30 to 54 μmho/cm² (from SO_4^{2-} influx)	
1 of Table 1.3 (modified)		1 of Table 1.5	G_m: 126 μmho/cm² (K outside)	
			G_m: 276 μmho/cm² (Na outside)	

Voltage–current	Z_{inp}: 343 Ω	2 of Table 1.5	R_m: 707–3240 Ω cm^2 (Ringer's) C_m: 1.8 μF/cm^2	Cuthbert and Painter (1969a)
	Z_{inp}: 288 Ω		R_m: 626–2278 Ω cm^2 (vasopressin) C_m: 2.06 μF/cm^2	
	R_{inp}: 70 Ω		C_m: 16 μF/cm^2 (inner barrier)	
	R_{inp}: 560 Ω		C_m: 2 μF/cm^2 (outer barrier)	
	R_{inp}: 630 Ω		C_m: <2 μF/cm^3 (whole skin)	
Voltage–current	Z_{inp}: 374 Ω Impedance locus	2 of Table 1.5	C_m: 2.14 μF/cm^2 (Ringer's)	Cuthbert and Painter (1969b)
	Z_{inp}: 120 Ω		C_m: 2.3 μF/cm^2 (theophylline)	
Voltage–current	ΔV I_{sc} (μA/cm^2)	1 of Table 1.5	R_m: 0.2–1 kΩ cm^2 C_m: 2.5 μF/cm^2	Finkelstein (1964)
Voltage–current	I: μA/cm^2 V: mV	1 of Table 1.5	R_m: 4551 Ω cm^2; $[Na]_o = 2.4$ mM R_m: 1029 Ω cm^2; $[Na]_o = 115$ mM	Bueno and Corchs (1968)
Voltage–current	V: 22.6 mV I_{sc}: 48.6 μA/cm^2	1 of Table 1.5	R_m: 150–600 Ω cm^2 R_m: 14.5 Ω cm^2; Ca^{2+}-free, EDTA	Fishman and Macey(1968)
1 of Table 1.4	I vs t τ: 50–120 μs	1 of Table 1.5	R_m: 150–1250 Ω cm^2 C_m: 1.3–5 μF/cm^2 subthreshold voltage steps	Fishman and Macey (1969a)

germinativum (-40 to -70 mV in chloride-free medium) and the second (-40 to -20 mV) occurred across the inner side of the same cells. The total skin potential ranged from 80 to 150 mV, inside positive.

Based on Whittembury's (1964) results, Cereijido and Curran (1965) measured the potential and conductance of the inner and outer barriers of the frog skin. Microelectrodes located in the stratum germinativum recorded about 50% of the total skin potential. In agreement with Whittembury (1964), they also found an intracellular potential of -17 to -18 mV when the skin was short-circuited. This potential remained constant despite large changes in the outside sodium concentration (3–115 mM). Table 4.1 shows that the conductance of both barriers decreased in sulfate Ringer's and that this effect was considerably greater at the inner barrier. Additional results suggested that the potential across the outer barrier was determined mainly by the sodium concentration. On the other hand, but in contrast to the model of Koefoed-Johnsen and Ussing (1958), the inner membrane did not respond as a potassium electrode. Ussing and Windhager (1964) also reported two voltage steps across the skin, but the morphological localization of the barriers was different from that proposed by Whittembury. The external sodium barrier was identified with the outer layers of the stratus spinosum. In contrast to these observations, Chowdhury and Snell (1965) reported a continuously changing potential across the skin. At any point along the potential profile, the potential changed with external sodium or internal potassium.

More recently, Rawlins et al. (1970) reexamined this matter in the isolated epithelium of the toad skin. The electrical profile was not uniform, and a variable number of voltage steps were found in different impalements. The results located the barrier with the highest resistance in the outer layers of the skin. In addition, the resistance values measured in different media and after addition of ADH confirmed the results of Cereijido and Curran. Indications that the external barrier is located near the external anatomical boundary of the skins were also reported by Kidder et al. (1963). In these studies, the time course of the shift in total skin potential was followed when sodium was changed in the external solution. The location of the external barrier was estimated from analysis of the kinetics of the potential change, and the sodium-sensitive barrier was located at 20–25 μm from the external surface of the skin.

None of the existing transport models explains satisfactorily the experimental evidence available, particularly the fact that the net active transport of sodium persists at sodium concentrations as low as 10^{-5} M in vivo and 10^{-3} M in vitro (Biber, Chez, and Curran, 1966). Under open-circuit conditions, the electrochemical potential for sodium cannot account for passive diffusion of sodium from the outer solution to the cell compartment, and therefore a compartmentalization of sodium was proposed (Rotunno et al., 1966; Biber et al., 1966; Cereijido and Rotunno, 1967; Cereijido, Reisin, and Rotunno, 1968; Nagel and Dörge, 1971). This model can explain the electrical profile described above (Biber et al., 1966). An alternate explanation for the existing experimental evidence has been put forward by Cereijido and Rotunno (1968).

The conductance of the toad skin for the passive effluxes of sodium and chloride was studied by Larsen (1971). The skin conductance undergoes cyclic changes related to the process of regeneration, keratinization, and shedding of the stratum corneum. After desquamation, an increase in active sodium influx occurred, and Larsen attributed this component to an increased permeability for sodium influx corresponding to the diffusion component of the sodium uptake described by Biber and Curran (1970). Quantitatively, the changes in total skin resistance differed whether the toads were kept in terrestrial or aquatic environment before isolation of the skin. Histological differences were found between the skins of these two groups of animals (Bani, 1966). It seems likely that the cement layer between the keratinized and the granulosa cells together with the occlusion belts between keratinized cells would constitute the structural basis for the diffusion barrier within the stratum corneum, thus determining the conductance of the outer region of the skin. This interpretation is based on the assumption that the passive fraction of the total transport across the skin is exerted through extracellular pathways that open during desquamation.

Candia and Zadunaisky (1964) studied the potassium influx in skins from the frog *Leptodactylus ocellatus*, which amounted to one-ninth of the net sodium transport. The difference was attributed to the active chloride transport present in these skins (Zadunaisky, Candia, and Chiarandini, 1963). However, Candia and Zadunaisky (1972) later reported that the short-circuit current was significantly larger than the K^+ influx in skins showing an active Cl^- transport as well as in skins where the latter is not present. The K^+ flux did not vary when the short-circuit condition was changed to open circuit or upon removal of Cl^-. Garcia Romeu et al. (1969) studied the in vivo sodium and chloride uptake in certain species of Chilean frogs. They found that both fluxes were independent. Two ionic exchange mechanisms were postulated, and it was demonstrated that endogenous hydrogen exchanges for sodium and endogenous bicarbonate for chloride.

4.1.2 Passive Electrical Properties of Amphibian Skin

Ussing and Windhager (1964) determined the passive electrical properties of the frog skin in order to estimate the magnitude of the ionic shunt (R_p, inset of Fig. 4.1) and the value of the driving force V_{Na}. The shunt was estimated by two methods. The first one is based on the assumption that in sodium-free outside solution the conductance $1/R_{Na}$ is negligible compared to $1/R_p$ and the total resistance becomes equal to the shunt resistance. The voltage drop across the skin was measured while the current was varied in steps from short- to open-circuit values. Most skins behaved as a linear resistance. Table 4.1 shows that the resistance of the skin is much higher with potassium than with sodium outside.

The second method depended on the assumption that the active pathway cannot conduct sodium in the reverse direction, and that sodium and the anions are the major contributors to the shunt. Therefore, the conductance to sodium and the anions can be estimated from the measured passive fluxes under conditions where the backflux is negligible (skin potential clamped to 100 mV, positive inside). Table 4.1 lists the fractional sodium and sulfate conductances obtained from measurements of sodium efflux and sulfate influx under those conditions, where the backfluxes were estimated to be as low as 0.04%. These results indicated that sodium contributes 46% and sulfate 34% to the total shunt resistance. Addition of urea to the outside solution increased both sodium and sulfate conductances. The resistance and the potential across the skin fell, whereas the short-circuit current remained constant. However, the drop in total skin resistance (from 3.3 kΩ cm^2 to 2.08 kΩ cm^2) was much smaller than the fall of the shunt resistance (from 24 kΩ cm^2 to 3 kΩ cm^2). These results suggested that the conductance increase was entirely attributable to the shunt pathway and an extracellular localization for the shunt. The potential profile indicated that the external barrier was located closer to the skin outer surface than postulated before, a finding later confirmed by Cereijido and Curran (1965).

Schultz (1972) analyzed the importance of shunts in determining the potential profile across epithelia, trying to explain the opposite profiles found in high-resistance epithelium like toad skin (Fig. 4.2a) and low-resistance epithelia like rabbit ileum (Fig. 4.2b). In the first one, the total transepithelial potential results from two voltage steps in the same direction; in the second, two steps in opposite directions are found. The analysis of the equivalent circuit in the inset of Fig. 4.2 shows that the potentials across the mucosal barrier (V_{mc}) and across the whole epithelium (V_{ms}) are:

$$V_{mc} = \left[(R_3 B + R_5) V_m A - R_1 A V_s B\right] / R_t \tag{4.2}$$

and

$$V_{ms} = R_5 (V_s B + V_m A) / R_t \tag{4.3}$$

where the quantities A, B, and R_t are defined as

$$\left.\begin{aligned} A &= R_2/(R_1+R_2) \\ B &= R_4/(R_3+R_4) \end{aligned}\right\} \tag{4.4}$$

and

$$R_t = R_1B + R_3A + R_5 \tag{4.5}$$

where the subscripts c, m and s mean the cell and the two barriers, mucosal and serosal respectively. V is the electromotive force operating across the mucosal or serosal membranes, R_1 and R_3 are the internal resistances of these batteries, R_2 and R_4 are shunt resistances across the respective membranes, and R_5 represents the extracellular shunts. From these equations, it follows that only when $R_5 \gg (R_1A + R_3B)$ or $R_5/R_t \approx 1$ is the transmembrane potential $V_{mc} \approx V_mA$. On the other hand, the existence of a high-conductance extracellular shunt ($R_5/R_t = 0.05$) will give the potential profile of Fig. 4.2b despite the batteries being oriented in the same direction. In rabbit ileum, a maximum value of 0.15 for R_5/R_t has been obtained. The analysis is consistent with the findings of Cereijido and Curran (1965). Moreover, Richardson (1972) derived the current–voltage equations for membrane systems composed of two, three, and four membranes from the flow equations of irreversible thermodynamics. These equations showed that bimembranes rectify in one direction, whereas triple and quadruple membranes can be double rectifiers. The theoretical analysis that led to the voltage–current relations was based on the assumption of highly charged membranes. It was found that nonpermeant ions play a significant role in the deviation of the resting potential from the logarithmic relation given by the Nernst equation, and that the direction of rectification would be determined by the polarity of the fixed charges in the membrane systems, not by the ionic concentrations in the bathing solutions.

Cuthbert and Painter (1968) investigated the mechanism and site of action of different agents in frog skin. Sodium transport was inhibited by the addition of ouabain, but the skin was voltage clamped at the initial potential to avoid changes in ionic concentrations arising from the asymmetric permeabilities. To determine whether a given change in total skin potential was due to the outward- or inward-facing membranes, a flap of skin was allowed to hang between the chambers and dip into saturated KCl solution. A reference electrode connected to the KCl was considered to be inserted between the outer and inner facing membranes and allowed to measure the voltage drops across them separately. The input resistance

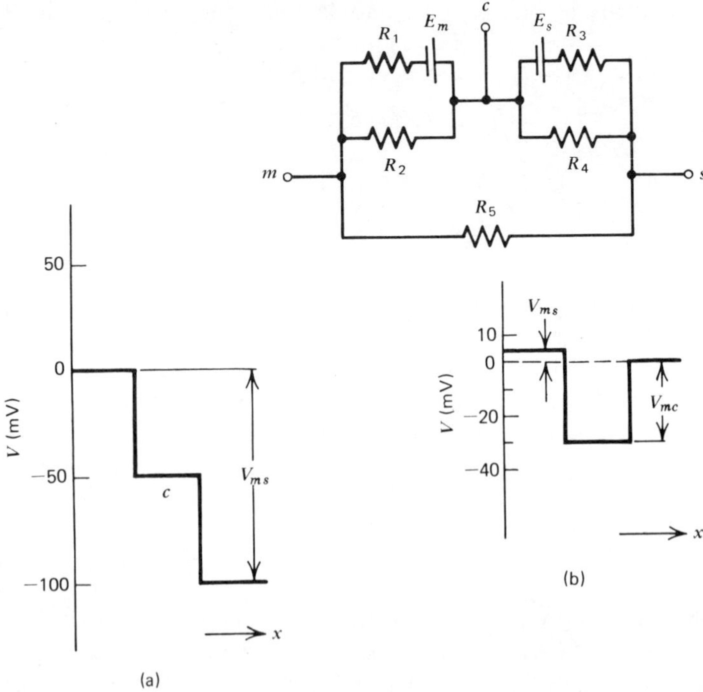

Figure 4.2. Schematic representation of the electrical profile found (a) across high-resistance epithelia like frog skin and (b) across low-resistance epithelia like rabbit ileum. x represents the inside-outside direction in (a) and the serosal-mucosal direction in (b). Reference electrodes are positioned at the inner surface of the skin and at the serosal side of the ileum. Inset: electrical equivalent circuit for epithelia. The symbols m, s represent the mucosal and serosal barriers, and c, the intracellular compartment. R_1 and R_3 are the internal resistances of the batteries E_m and E_s. For explanation, see text. After Schultz (1972), by copyright permission of The Rockefeller University Press.

was measured by applying square pulses under nonclamped conditions. Vasopressin decreased the skin input resistance and increased the potential, acting at the outer facing membranes. Theophylline (10^{-2}–10^{-3} M) caused a 50% fall in skin input resistance and a decrease in potential at the outermost membranes. This response depended on the presence of chloride ions. Cyclic AMP caused an increase in skin resistance. These results suggested that (1) vasopressin selectively increases the sodium permeability at the outer barrier and (2) the reduction in skin potential by theophylline can be due to an increased permeability to potassium or chloride.

The effect of the antidiuretic hormone was further studied by the same authors (Cuthbert and Painter, 1969a) using the square-pulse analysis described by Teorell (1946). From model 2 in Table 1.5, resistances R_2 and R_1 could be measured, representing, respectively, the input resistance of the skin and the combined resistance of electrodes, extracellular solutions, and cytoplasm. The graphical analysis of the transient of the imposed step voltage permitted the construction of impedance loci. The dc capacitance of the skin was calculated from the time constant of the voltage transient, and it decreased when vasopressin or pure antidiuretic hormone was added to the solution bathing the inside of the skin. Under control conditions, the impedance loci were perfect semicircles, and in every case the ac capacitance and the skin conductance were increased under the hormone (Table 4.1). Immersion of part of the skin in saturated KCl showed that the structures responsible for the observed voltage transient were mainly located in the outer facing membranes, Table 4.1. The participation of the outer barrier as a site of action for ADH was also reported by Cereijido and Rotunno (1971). Since the increase in conductance appeared only in the presence of sodium ions, the effect of vasopressin was postulated to be linked to the opening of pores or channels selectively permeable to sodium. If water-filled polar pores appeared in regions of the membrane where they had not existed previously, the increase in capacitance could also be explained. It was concluded that the effects on sodium transport and water flow were not linked processes, because the changes in capacitance did not follow the same time course as the increase in conductance and the hormone also increased water flow in the absence of sodium ions. A similar analysis was applied to the action of theophylline on frog skin (Cuthbert and Painter, 1969b). Theophylline also increased the skin conductance, but there was only a very slight increase in capacitance (Table 4.1). As with vasopressin, the conductance increase was almost complete before the capacitance changed.

In addition to the electrical properties related to ionic transport, it has been found that under certain conditions, amphibian skins generate a response that shows some features of an action potential. Finkelstein (1964) proposed that this type of electrical activity depended upon a time-variant resistance. For small values of current, there was a linear voltage–current relationship. The values of R_m and C_m are listed in Table 4.1. Increasing the current strength, the curve departed from linearity, with larger responses for outward currents. The threshold was reached at a level of about 300 mV (negative inside), and with a small increase in current the skin responded with an action potential. This was followed by a long refractory period of 5–20 s. The behavior of the toad bladder was qualitatively the same, the resistance being higher and the threshold about 200 mV (serosal side negative).

A characteristic of the response in both structures is that it fails to complete its course if the current is interrupted. The refractory period does not develop if the current is cut off during the spike. In this case, the potential falls rapidly to the resting level, and therefore it was attributed to resistance variations during constant current flow. Impedance measurements with an ac bridge showed an increase in impedance during the early and middle part of the rising phase of the spike, followed by a fall in impedance below its value at rest. The response was not significantly affected by drastic changes in the ionic composition of the inside solution, but it was greatly influenced by changes in the outside solution, except for chloride. Decreasing the sodium concentration to less than 20 mM resulted in the action potential changing to a graded response. If calcium was omitted from the solution the threshold rose and the response also became graded. All the effects were reversible. The analogy was made with the electrogenic mechanism in nerve membrane, where the action potential results from a change in membrane conductance. However, in frog skin the activation mechanism is not voltage dependent.

Similar results were obtained from toad skins by Bueno and Corchs (1968), who investigated relative changes in skin resistance. An increase in sodium concentration in the outside solution reduced the skin resistance, and, accordingly, stronger currents had to be used to polarize the tissue and induce single responses or the repetitive firing usually obtained with long-lasting pulses. Measurements of relative changes in membrane resistance were made by superimposing short pulses on the action potentials. The results were strikingly similar to those taken by Weidmann (1951a) in Purkinje fibers. Minimum values of membrane resistance were found immediately after the spike and also at the foot of the ascending limb of the potential. A time-variant resistance seemed to play a role in this phenomenon, and the results agreed with those of Finkelstein (1964). However, this interpretation does not agree completely with that of Lindermann and Thorns (1967), who attributed the spike potential observed in frog skin to delayed potassium rectification. Although the spikes seemed to generate at the outermost layer of cells, the authors did not reject the possibility that cell membranes below this layer contributed to this phenomenon. This possibility seems unlikely, in the light of the evidence supplied by Fishman and Macey (1968). They used a preparation called split frog skin, which consisted of only the epithelial layers of the skin after the dermis had been peeled off. It was confirmed that the source of the short-circuit current occurred in the epithelial structure (Aceves and Erlij, 1971). The values of potential, current, and resistance measured in the split frog skin are listed in Table 4.1. Also, it was found that the presence of calcium ions in the outside solution was essential for the excitation phenomenon. Total removal of calcium from the solution (calcium-free Ringer's plus EDTA) resulted in disappearance of the all-or-none response and a decrease of 5–30% in transepithelial resistance. The effect was much slower if EDTA was applied to the inside surface of the preparation, indicating that the excitation process occurs near or at the outward side of the preparation.

A detailed study of the electrical properties underlying excitation in frog skin was carried out by Fishman and Macey (1969a), applying current-clamp and voltage-clamp techniques. Essentially, the method used was a modification of method 1 of Table 1.4. Taking into account the resistances of solution, electrodes, and skin, the authors used an equation developed by Cole (1968, pp. 348) to estimate the potential distribution for a circular flat membrane backed by one planar-disc current electrode and with another disc electrode on its opposite side. The passive electrical properties of the skin were measured by imposing subthreshold voltage steps. The values listed in Table 4.1 were calculated according to model 2 of Table 1.5, assuming that the elements of the circuit are passive, linear, and time invariant. For the frog skin, R_1 represents conducting pathways across the skin (probably intercellular) other than the cell membranes, represented by the parallel resistance-capacitance network R_2C in the model. The time course of the current response for subthreshold pulses and the calculation of R_1 for the frog skin (Fig. 4.3a) showed that its value is not negligible (50 Ω cm^2). Consequently, the potential across R_2 and C was not clamped until 300 ms after the application of the voltage step. Current–voltage curves plotted at constant times after the voltage clamp (Fig. 4.3b) showed the characteristic N shape. The values of maximum negative resistance ranged between -150 and -430 Ω cm^2. This negative resistance during a voltage clamp occurred for only 20 ms after applying the voltage step. After this period, the skin lost this property, and the time during which the property was lost corresponded to the refractory period found by Finkelstein (1964). At high potentials (larger than 300 mV, outside positive), the skin passed into a low-resistance state where the N-shaped current–voltage relation disappeared and was replaced by a linear function. This finding was further confirmed by Fishman and Macey (1969b) with current–voltage curves obtained with triangular pulses. No hysteresis was found with pulse amplitudes below that at which the negative slope resistance developed. The ionic dependence of the characteristic current–voltage curve was also investigated by Fishman and Macey (1969c). Following replacement of calcium by EDTA in the outside solution, the negative slope was transformed into a linear positive resistance. This coincided with disappearance of the spike.

However, Candia (1970) reported a correlation between the voltage–cur-

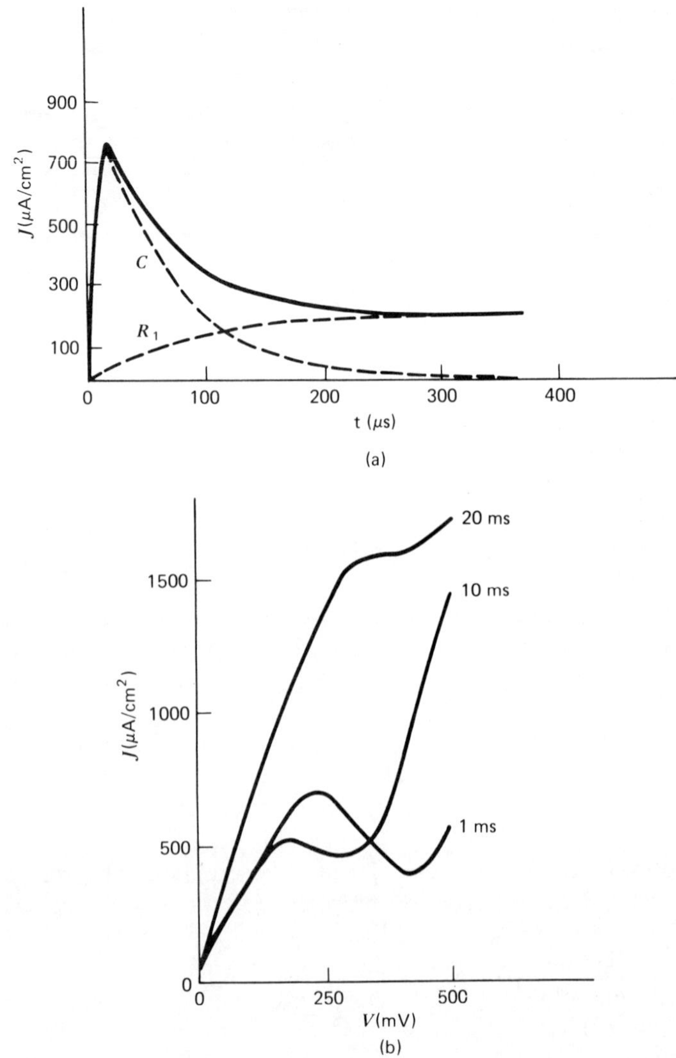

Figure 4.3. (a) Time course of the skin current upon clamping with a subthreshold voltage step. Full line, actual current recording; dashed line, its resistive (R_1) and capacitive (C) components as given by an equivalent circuit like model 2 of Table 1.5. (b) Current–voltage curves of frog skin 1, 10, and 20 ms after onset of voltage pulse. Abscissa, skin potential, outside with respect to inside. Modified from Fishman and Macey (1969a), by copyright permission of The Rockefeller University Press.

rent curve on the hyperpolarizing side and the sodium transport. The voltage–current curves were recorded under current clamp with slow (6 μA/s cm^2) and fast (0.015–7.5 μA/ms cm^2) triangular current pulses, using an automatic voltage-clamp recorder (Candia, 1966). Unidirectional sodium outflux was measured for different transepithelial potentials. Figure 4.4a shows a typical voltage–current curve obtained with slow ramps in skins bathed in Ringer's solutions. The relation was linear between the point of I_{sc} and 200 mV hyperpolarization (inside positive). From there on, a gradual breakdown occurred. Some of the skins showed a sharp increase in resistance with hyperpolarization (rectification) before the breakdown. This effect appeared consistently when Na_2SO_4-Ringer's replaced normal Ringer's. In Fig. 4.4b, the slope resistance is 2700 Ω cm^2 at $V=V_s$ and 12,000 Ω cm^2 at the breaking point. Moreover, the slope resistance at $V=V_s$ was 1.5 times larger than the resistance at $V=0$. Lowering the $[Na^+]_o$ to 31 mM in sulfate-Ringer's resulted in curves qualitatively similar to that of Fig. 4.4b, but the slope resistance was higher at every point of the curve. If the direction of the current is reversed before the breakdown, the resulting voltage–current curve is identical to the one obtained in the hyperpolarizing direction. However, the rectifier characteristic is lost if the current is reversed after the breakdown (dashed line, Fig. 4.4b). Inhibitors of the active Na^+ transport significantly modified the voltage–current curves. The equivalent circuit shown in Fig. 4.5 was proposed to explain the electrical characteristics of frog skin. The circuit is similar (Fig. 4.6a) to the one proposed by Civan (1970) for the toad bladder, but more complicated, and takes into account the two barriers described for the skin. The resistances R_1^- and R_2^- represent the passive paths for anions, mainly Cl^-; R_1^+ and R_2^+ are the passive paths for cations, including the passive component of Na^+ movement. The third path includes the rectifier and an oscillator that generates the spontaneous skin potential and represents also the energy source for the active transport of Na^+. Without an external current, net transport occurs toward the inside. The series resistance R_s represents the connective tissue. This circuit is consistent with the experimental findings that locate the highest potential difference, the action potential production, and most of the skin resistance at the outer barrier, here represented by R_1^-, R_1^+, and the rectifier. Moreover, the pump is here composed of two functionally different elements: the rectifier and the energy source. The effect of amiloride (which suppressed active Na^+ transport and the nonlinearities of the voltage–current curve, causing an increase in slope resistance) was interpreted as an increase in forward resistance of the rectifier.

Mandel and Curran (1972) investigated the permeability properties of the shunt pathways of frog skin by measuring unidirectional fluxes of urea,

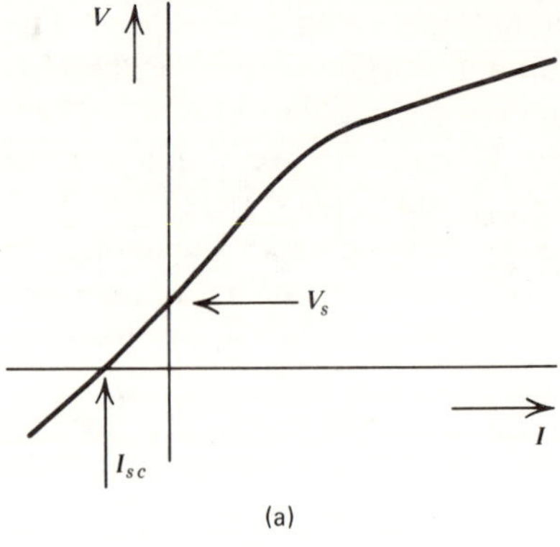

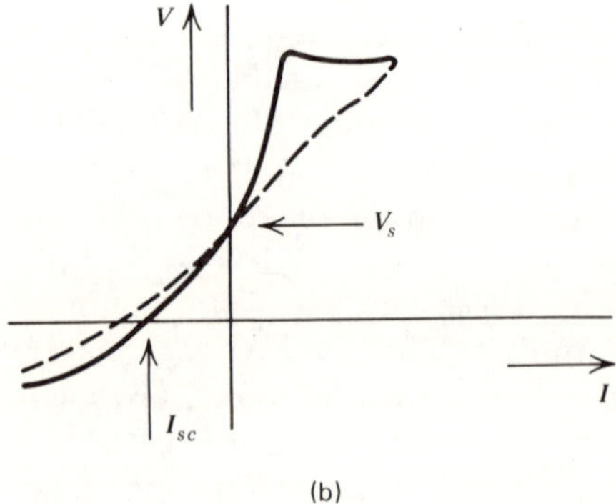

Figure 4.4. Voltage–current curves of frog skins bathed in (a) Ringer's and (b) Na_2SO_4-Ringer's. The positive side of the abscissa corresponds to hyperpolarizing current. I_{sc}, point of short-circuit current; V_s, spontaneous skin potential. Dashed curve in (b), depolarizing voltage–current curve initiated after the breakdown occurred. Modified from Candia (1970), by copyright permission of The Rockefeller University Press.

Na^+, K^+, and Cl^- under conditions of voltage clamp and suppression of the active Na^+ flux by ouabain. Steady-state fluxes were measured at different skin potentials. The permeability to urea and the electrolytes increased with depolarization, and the sodium influx increased linearly with urea influx. Relative permeabilities for the electrolytes with respect to urea were obtained with an equation derived from the Goldman theory:

$$\frac{M_{\mathrm{i}}^{in}}{M_{\mathrm{U}}^{in}} = \frac{\alpha_{\mathrm{i}} C_{\mathrm{i}}^{o}}{C_{\mathrm{U}}^{o}} \left[\frac{-FV/RT}{1-\exp(FV/RT)} \right] \tag{4.6}$$

where the indexes in and o stand for influx and outside, respectively, and the subscripts i and U apply to the ion concerned and urea, respectively. The symbol α_{i} expresses the relative permeability ($P_{\mathrm{i}}/P_{\mathrm{U}}$). The assumption that α_{i} does not vary with potential was confirmed by the experimental results. The values of α were 0.54 for Na^+, 0.72 for K^+, 2.41 for Cl^-, and 0.42 for mannitol. Therefore, the shunt pathway seemed to have a selectiv-

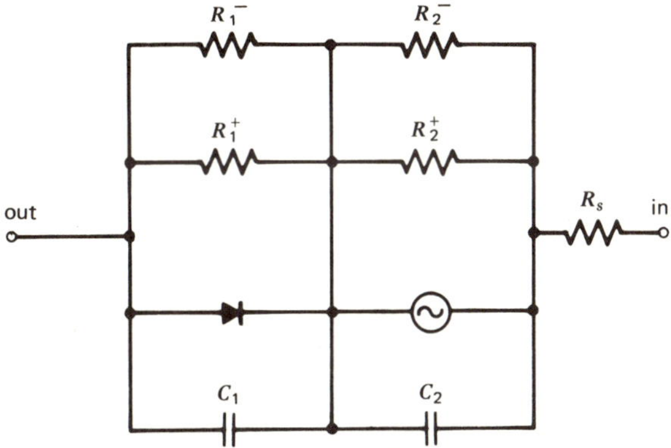

Figure 4.5. Equivalent circuit proposed for the electrical properties of frog skin. Out and in stand for outside and inside solutions, respectively. Modified from Candia (1970) by copyright permission of The Rockefeller University Press.

ity in the order $Cl^- >$ urea $> K^+ > Na^+ >$ mannitol. The permeability characteristics were symmetrical, with similar values of α for potentials of $+50$ and -50 mV. This finding supports the equivalent circuit proposed by Candia for the passive pathways, although the results indicate the existence of a common, single path for passive fluxes. The fact that the movement of solutes is fairly well described by the Goldman equation was taken as evidence for an extracellular localization of this shunt, as described for kidney tubules (see Sec. 4.4). This is supported by the measured fluxes of mannitol, which are known to penetrate the epithelial cells very slowly (Biber et al., 1972).

4.2 URINARY BLADDER AND GALL BLADDER TISSUE

The toad bladder is a structure similar to frog and toad skins: Active sodium transport occurs from the mucosal to the serosal side. However, it is a simpler structure, because the epithelium is composed of a single layer of cells. In both frog skin and toad bladder, lack of oxygen impairs sodium transport, and the potential across the structure falls to zero (Leaf, Anderson, and Page, 1958; Leaf, Page, and Anderson, 1959; Zerahn, 1956; McDougal and Sullivan, 1971). As in frog skin, the sodium in the tissue is supposed to be separated in two compartments (Zerahn, 1969) and only one pool intervenes in the transport process (Finn and Rockoff, 1971). The mucosal-facing side of the transport compartment was more than 10 times as permeable to sodium as the serosal, or pump, side. Using these results, Finn (1971) showed that vasopressin stimulated sodium transport by increasing both the pool size and the pump rate coefficient. These results agreed with those obtained by Wright and Snart (1971), who showed that small doses of vasopressin increased the mucosal permeability to sodium and water.

4.2.1 Electrical Measurements Related to Ionic Transport Across the Bladder

The bladder of fresh water turtles, in contrast to the toad bladder, maintains a certain level of potential under anaerobic conditions. Klahr and Bricker (1964) measured the short-circuit current and the sodium flux in isolated turtle bladders in aerobic and anaerobic conditions. When anoxia was produced by bubbling of N_2, the potential and the short-circuit current fell progressively, but positive values persisted for as long as 3 hr and a twofold increase in resistance was found. These changes were rapidly reversible upon readmission of O_2. Simultaneous measurements of ionic fluxes in aerobic conditions showed that the net sodium flux averaged only 82% of the short-circuit current. The difference between the two was statistically significant and increased in anaerobic conditions where the sodium influx decreased, but the current fell much more rapidly, so that the net sodium flux exceeded the short-circuit current. Finally, the total current amounted to only about 50% of the net flux. Cyanide applied to the serosal side produced a similar result, but the effect occurred through a great increase in sodium influx while the current remained unchanged. The data in Table 4.2 were calculated from the figures in Klahr and Bricker (1964). These results suggested that in the turtle bladder the active transport of another cation in the same direction as Na^+, or of one of more anions in the opposite direction, contributes to the potential across it. This problem was further investigated by Gonzalez, Shamoo, and Brodsky (1967) and by Gonzalez et al. (1967). Short-circuit current, bladder resistance, and chloride fluxes were measured in paired halves of turtle bladder. In sodium-free medium the net chloride flux and the short-circuit current were stable for 2 hr and then decreased gradually. In choline-Ringer's solutions, the transbladder potential was opposite in polarity to that observed in normal Ringer's (serosal fluid positive, relative to mucosal side, in presence of sodium). When the bladder was short-circuited, the net chloride flux accounted for 50% of the current, indicating an active transport of chloride, which was corroborated by the inversion of the potential polarity in sodium-free medium. Removal of bicarbonate from the bathing solutions restored the short-circuit current to the values given by the net chloride flux, indicating that bicarbonate contributes up to 50% of the short-circuit current. If the sodium-free system is replaced by a sodium-rich system, the potential reverses its polarity and reaches stable

Table 4.2 Electrical Properties of Bladder

Method	Measured Quantities	Model	Derived Quantities	Reference
Fig. 4.1	V: 30–60 mV I_{sc}: 5.7 μA/cm^2 R_{inp}: 251 Ω R_{inp}: 500 Ω	1 of Table 1.5	R_m: 1900 Ω cm^2 (aerobiosis) R_m: $\cong$3500 Ω cm^2 (anaerobiosis)	Klahr and Bricker (1964)
Fig. 4.1	I_{sc}: 152 μA R_{inp}: 390 Ω I_{sc}: 184 μA R_{inp}: 440 Ω	1 of Table 1.5	I_{Na}: 197 μA (Ringer's) I_{Na}: 189 μA I_{Cl}: 12 μA I_{Na}: 185 μA (Cl^-- and HCO_3^--free)	Gonzalez, Shamoo, Wyssbrod, Solinger, and Brodsky (1967)
Fig. 4.1	I_{sc}: 202 μA R_{inp}: 377 Ω I_{sc}: 213 μA R_{inp}: 384 Ω I_{sc}: 96 μA R_{inp}: 583 Ω I_{sc}: 73 μA R_{inp}: 443 Ω I_{sc}: 103 μA R_{inp}: 600 Ω	1 of Table 1.5	I_{Na}: 204 μA (pH 6.0–8.0) I_{Na}: 198 μA (pH 4.0–5.6) I_{Na}: 85 μA (pH 3.6–3.9) I_{Na}: 50 μA (pH 3.0–3.5) I_{Na}: 91.5 μA (pH 3.0–3.5) I_{H}: 11.6 μA	Gentile and Brodsky (1969)
Fig. 4.1	$\Delta V/\Delta I$: 1957 Ω $\Delta V/\Delta I$: 2190 Ω $\Delta V/\Delta I$: 680 Ω	1 of Table 1.5	R_m: 2500 Ω cm^2 (0–143 mV) beyond 143 mV R_m: 2300 Ω cm^2 (-10 to 142 mV)	Civan (1970)
1 of Table 1.3 modified		Fig. 4.1	R_t: 307 Ω cm^2 R_m: 4470 Ω cm^2 R_b: 2880 Ω cm^2 R_s: 324 Ω cm^2 $(R_m+R_b)/R_s$: 25.7 Ω cm^2	Frömter (1972)

values after 1–2 hr. At the same time, the transbladder resistance reaches an average of 470 Ω, about 40% of the value in sodium-free media (1090 Ω). Chloride fluxes were not significantly diminished by the presence of sodium.

Based on these results, an equivalent electrical network was proposed to represent the ionic movements in turtle bladders. The three ionic channels would be arranged in parallel so that under conditions of no potential difference across the tissue, the net ionic movements would depend only on the resistance of the channel and the energy supplied by the transport mechanism. This model was justified by the findings that chloride and bicarbonate movements were independent of each other. Also, chloride flux was independent of the presence of sodium. To validate the model further, Gonzalez, Shamoo, Wyssbrod, Solinger, and Brodsky (1967) showed the independence of the sodium and chloride fluxes. Simultaneous determinations of I_{sc}, sodium fluxes, and resistance (Table 4.2) indicated that with Ringer's as bathing solution the net sodium flux exceeded the short-circuit current by 44 μA. A mean transport potential of 75 mV for the active movement of sodium was calculated from Eq. 4.1. Comparison of the values of sodium and chloride fluxes with the measured current showed that chloride current accounted for over 30% of the difference between I_{sc} and I_{Na}. Finally, the short-circuit current became equal to the sodium current when both anions, chloride and bicarbonate, were removed from the solution bathing the mucosal side.

The turtle bladder spontaneously acidifies the mucosal fluid under both open-circuit and short-circuit conditions. Gentile and Brodsky (1969) investigated the effect of pH on the sodium transport by measuring sodium fluxes, short-circuit current, bladder input resistance, and hydrogen fluxes. The experiments were carried out under two different conditions: (1) an asymmetrical system, serosal side bathed by bicarbonate Ringer's at pH = 7.6 and mucosal side bathed by non-buffered sodium methylsulfate Ringer's; (2) a symmetrical system, both sides of the preparation bathed by the same solution, usually HCO_3^-- and Cl^--free, so that the only permeant ions were sodium and hydrogen. Table 4.2 lists the results. Lowering the pH on the mucosal side resulted in a decrease of I_{sc} and I_{Na}, but whereas at normal pH, I_{sc} and I_{Na} were equal, at low pH I_{Na} was significantly lower than I_{sc}, the residual current being carried by H^+. This hydrogen net flux could not be detected at pH > 5.5. The results were qualitatively similar in the symmetrical system, but I_{H} was always lower than in the asymmetrical system. This indicated that bicarbonate transfer contributed significantly to the rate of removal of hydrogen ions from the mucosal fluid. Changing the serosal pH did not have any effect. From these results, the physiological mechanism underlying the minimum spontaneous mucosal pH of 4.0 observed in the isolated bladder could be explained: The acidification mechanism lowers the pH from 8.0 to 4.0, an accumulation of hydrogen ions in the mucosal fluid occurs, but there is no migration of H^+ to serosal fluid yet. When the mechanism brings the pH to values lower than 4.0, the hydrogen ions are transferred from mucosal to serosal side, thus counteracting the acidification effect.

To determine whether or not hydrogen ions move down an electrochemical gradient during the process of acidification, Hirschhorn and Frazier (1971) measured the potential profile and the input resistance across the turtle bladder under open- and short-circuit conditions. In open circuit, the electrical profile consisted of two positive steps when a microelectrode was advanced from the mucosal to the serosal side. The mucosal cell border accounted for 66% of the total bladder potential and 82% of the total resistance. Under short-circuit conditions the cell interior was negative with respect to outside solutions. Again, about 90% of the total bladder resistance was measured across the mucosal cell border. These fractional resistances were estimated from the ratio of voltage drops obtained when square pulses were applied across the whole bladder or across the mucosal-facing membrane. The results indicated that hydrogen ions move against an electrical gradient under short-circuit conditions, but the interpretation of the results was complicated by the fact that the serosal chamber was kept at a pressure 10–15 cm H_2O below atmospheric pressure during the measurements. Water and ionic fluxes due to this difference in hydrostatic pressure may have influenced the results.

The experimental evidence that the epithelium is the site where the active sodium transport occurs was supplied by Nakagawa et al. (1967), who separated the layer of epithelial cells from smooth muscle and connective tissue of the turtle bladder. This procedure resulted in a preparation consisting of one or two cell layers. When this tissue was placed in a measuring chamber, a potential difference existed across it, and a net transfer of sodium could be measured with radioactive tracers. However, upon anaerobiosis, the short-circuit current and the potential decayed more rapidly in the isolated epithelium than in the intact bladder.

4.2.2 Passive Electrical Properties of the Bladder

Civan, Kedem, and Leaf (1966) investigated the mechanism of action of vasopressin. The relation between sodium fluxes and bladder resistance remained unchanged. The hormone stimulated the active transport of sodium, and this effect was accompanied by a decrease in the resistance of the toad bladder. The site of vasopressin action was studied by Civan and Frazier (1968) with a voltage–current method, using a microelectrode as potential electrode. Extracellular electrodes were employed to measure the transbladder potential and to polarize the preparation. The bladder was impaled from the mucosal side, and the microelectrode was left in a position that, according to the potential profile, would be across the apical barrier (borders of epithelial cells facing the mucosal side). Fractional resistances were determined as in Hirschhorn and Frazier (1971); see Sec. 4.2.1. When the microelectrode was advanced far enough to record a potential equal to the total transbladder potential, current injection produced equal voltage drops at both the microelectrode and the external potential electrodes. Vasopressin increased the transbladder potential with little or no change in the potential across the apical barrier, while the input resistance of the bladder and that of the apical barrier decreased. The contribution of the apical barrier to the change in input resistance of the whole bladder was 98%. Therefore, it was concluded that the hormonal effect occurred at this site.

A preparation developed by Janácěk and Rybová (1970) permitted the selective study of the properties of the serosal-facing membranes of the toad bladder. One lobe of the bladder was tied off to form a vesicle and was filled with paraffin. After incubation in Ringer's solution, cells shrunk and the sodium and chloride concentrations decreased. This proved that sodium was actively pumped out, followed by chloride ions and osmotically driven water. This process was stimulated by vasopressin. Measurements of serosal transmembrane potential (32 mV, negative inside the cell) supported the view that sodium was being transported against its electrochemical gradient. Since vasopressin does not change the total transbladder potential (Civan et al., 1966), it was postulated that it reduces the internal resistance of the pump (R_{Na} in inset of Fig. 4.1). These results do not contradict those of Civan and Frazier if one considers that the resistances of the serosal- and mucosal-facing membranes are in parallel and that the mucosal membrane is 10 times as permeable to sodium as the serosal membrane (Finn and Rockoff, 1971). Therefore, a small change in the higher resistance might have not been detected by Civan and Frazier (1968). Using the same method, Janácěk, Rybová, and Slavíková (1971) showed that aldosterone stimulated independently sodium entry and sodium extrusion in frog bladders.

Civan (1970) measured current–voltage curves of toad bladders with long pulses. Provided the channel containing the sodium pump was operative, a breaking point in the current–voltage characteristic was consistently found. At potentials more positive than this breaking point, the slope resistance increased. The mean bladder potential at which the transition point occurred was 177 mV, which corresponds to the driving force of the pump (V_{Na} in inset of Fig. 4.1). The following experimental results corroborated this hypothesis: (1) when the sodium pump was inactivated by either sodium substitution, ouabain, or anoxia, the transition point disappeared; (2) addition of vasopressin did not change the potential of this transition point but reduced the resistance at potentials positive to the transition point. An equivalent circuit (Fig. 4.6a) was proposed to account for this data. The resistance branches in the vicinity of the transition point could be approximated by linear functions, and the following relation was developed to describe the current–voltage curve of the toad bladder:

$$\left(\frac{dV}{dJ}\right)_{V>V_{Na}} \Big/ \left(\frac{dV}{dJ}\right)_{V<V_{Na}} = \frac{1}{1-(V_0/V_{Na})} \tag{4.7}$$

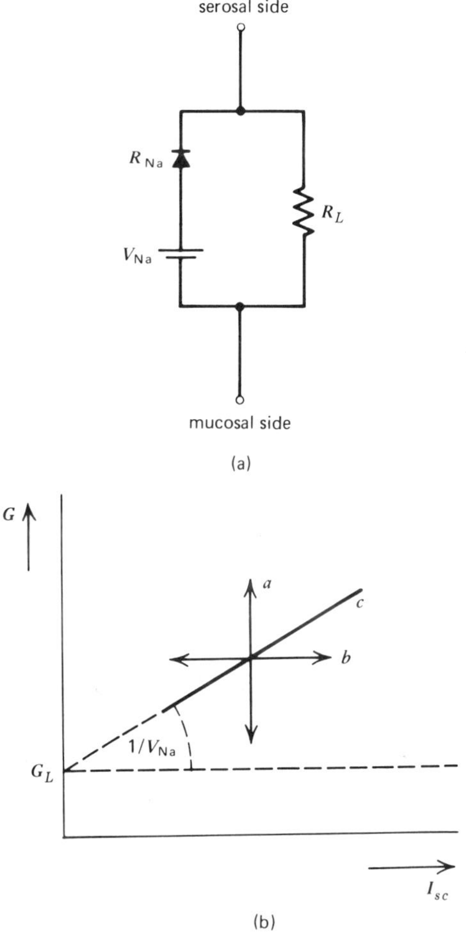

Figure 4.6. (a) Equivalent circuit for the toad bladder. V_{Na}, driving force of the sodium pump; R_{Na}, resistance of the active sodium channel; R_L, parallel-leak resistance. Modified from Civan (1970), by permission. (b) Theoretical curves of G versus I_{sc} used to experimentally determine V_{Na}. Modified from Yonath and Civan (1971), by permission.

where the left side of the equation is the ratio of the slope resistances in the vicinity of the transition point, and V_0 is the potential under open-circuit conditions. This cquation described well the experimental results.

Yonath and Civan (1971) used vasopressin as a tool to determine V_{Na} because this drug stimulates active sodium transport by decreasing R_{Na} without affecting V_{Na}. Moreover, with vasopressin as a control, the mechanism of action of other substances can be tested. The voltage–current relation of the circuit of Fig. 4.6a is given by:

$$V = \frac{I + V_{\mathrm{Na}} G_{\mathrm{Na}}}{G_{\mathrm{Na}} + G_L} \tag{4.8}$$

for the conditions $V > 0$ (serosal positive) and $I > 0$ (positive current flowing from serosa to mucosa). Considering that the short-circuit current (I_{sc}) is given by the product $V_{\mathrm{Na}} \times G_{\mathrm{Na}}$, it was postulated that a plot of total transbladder conductance (G) versus I_{sc} describes the electrical properties of the tissue. Figure 4.6b illustrates the trajectories along which G as a function of I_{sc} varies with changes of the leak pathway alone (a), the driving force (b), and the conductance G_{Na} without changes in V_{Na} (c). Under short-circuit conditions, the total conductance is given by:

$$G = G_L + (I_{sc}/V_{\mathrm{Na}}) \tag{4.9}$$

Therefore, the inverse of the slope of curve c in Fig. 4.6b gives the value of V_{Na}. For a transbladder potential V_x across the bladder, the total conductance is:

$$G = V_{\mathrm{Na}} G_L / (V_{\mathrm{Na}} - V_x) + I_x / (V_{\mathrm{Na}} - V_x) \tag{4.10}$$

Equation 4.10 reduces to Eq. 4.9 when $V_x = 0$. The method was evaluated by measuring the current flowing when V_x was alternately clamped at zero ($I = I_{sc}$) and 10 mV ($I = I_x$) by pulses of 10 ms duration, before and after adding vasopressin to the serosal medium. The conductance–current curves obtained at the two values of V_x were identical. Addition of vasopressin increased both the conductance and the current, but the slope of the curve G/I remained constant. The absolute value of V_{Na} was 74–186 mV (serosal positive) and the total tissue conductance was 0.5–1.5 mmho. The value of G_L (intercept to the ordinate) ranged from 50 to 90% of the total conductance at the onset of the experiments. Addition of cyclic-AMP produced G/I curves falling on the same line observed with vasopressin, an expected result if AMP is the sole mediator of vasopressin response. Moreover, ouabain reduced proportionally I_{sc} and V_{Na}, a result consistent with the known effect of the glycoside, which selectively depresses the active transport. In addition, the slope of the G versus I_x trajectory depended on the clamping voltage and became negative for clamping voltages higher than V_{Na}.

The epithelium as the site of the major contribution to tissue resistance was also observed in other structures. Wright and Diamond (1968) found that the resistance of the isolated gall bladder fell to less than 6% of its control value when the epithelium was scraped off. With identical solutions on both sides of the gall bladder, there is no potential difference because the ionic pump is not electrogenic and the serosal and mucosal membranes have symmetrical ionic permeabilities. A potential difference observed across such a structure is either a diffusion or a streaming potential.* When a KCl or NaCl gradient existed across the gall bladder, pH-dependent relative permeabilities were found: $P_{Cl}/P_K = 0.12$ at pH 7.4; 1.95 at pH 2.4; $P_{Na} > P_{Cl}$ at pH 7.0; $P_{Na} < P_{Cl}$ at low pH.

An interesting approach was used by Wright and Diamond (1969a, b), Diamond and Wright (1969), and Hingson and Diamond (1972) to determine the reflection coefficient for nonelectrolytes from measurements of streaming potentials. Moreover, Barry, Diamond, and Wright (1971) reported that the permeability and conductance properties of rabbit gall bladder were different from those currently attributed to cell membranes. It was concluded that the passive ion permeabilities described by Wright and Diamond (1968) represented predominantly high-conductance shunts, probably located at the intercellular spaces. This high-conductance shunt was investigated by Frömter and Diamond (1972) and Frömter (1972) in the gall bladder of *Necturus*, chosen because of its rather simple structure. It consists of a single layer of cells arranged like a flat sheet without villi, folds, or crypts. The cells are uniform and are large enough to be visible through a stereomicroscope, thus facilitating microelectrode impalements. When the electrodes are positioned in the lumen and the interstitium, the following current paths are possible (Fig. 4.7a): (*a*) from the lumen to interstitium via the luminal and basal membranes or via luminal and lateral membranes; (*b*) through the terminal bars (connections between cells) to the intercellular space; (*c*) along other paracellular routes. These can be represented by the lumped resistance of Fig. 4.7b: R_m, the resistance of the luminal membrane; R_b, the lumped resistance of basal and lateral membranes; and R_s, in parallel to $(R_m + R_b)$, the resistance for the current bypassing the cells. To determine these resistances, constant current pulses were injected between lumen and interstitium and voltage drops were measured across R_m and R_b with an intracellular microelectrode and across the epithelium (R_t) with extracellular electrodes. In addition, current was injected intracellularly, and the spread of the electrotonic potential in neighboring cells was measured (see Fig. 4.9). Moreover, the resistance at the point of current injection (R_z) was calculated with model 10 of Table 1.5. For the other measurements, model 1 of Table 1.5

*Streaming potentials are created when liquid is forced by pressure through capillaries. Therefore, they can be found as a consequence of the movement of liquid caused by an osmotic gradient between two compartments (MacInnes, 1961, p. 437).

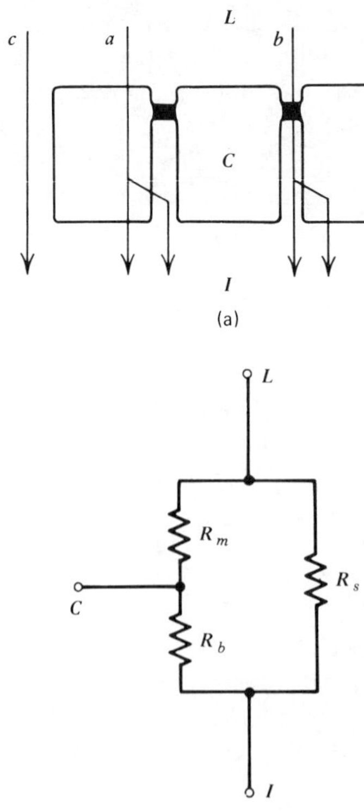

Figure 4.7. (a) Schematic represention of *Necturus* gall bladder epithelium and the possible transepithelial current pathways. *L*, *I*, and *C* stand for lumen, interstitium, and cell, respectively. The arrows *a*, *b*, and *c* are the current paths described in the text; dark zones are terminal bars. (b) Lumped equivalent circuit. Modified from Frömter (1972), by permission.

was used. R_m, R_b, and R_s are related to measurable quantities by:

$$\left.\begin{aligned} 1/R_t &= (1/R_s) + [1/(R_m + R_b)] \\ \Delta V_m/\Delta V_b &= R_m/R_b = a \\ 1/R_z &= (1/R_m) + (1/R_b) \end{aligned}\right\} \qquad (4.11)$$

Voltage–current curves for the transepithelial resistance were linear only for instantaneous values and low current densities. The mean value of R_t was 307 Ω cm^2. The electrotonic spread measured with method 1 of Table 1.3 was consistent with the theoretical pattern predicted by the model; λ was 290 μm and R_z was calculated as 1020 Ω cm^2. Average values from six bladders are listed in Table 4.2. The sum of the resistances of the luminal and basal cell membranes amounted to 7400 Ω cm^2. Since the total transepithelial resistance is much lower, these results indicate that 96% of the total current flows through the extracellular path identified with the terminal bars and lateral spaces. The total transepithelial resistance decreased when hydrostatic pressure was applied or an osmotic gradient across the epithelium existed, thus opening the intercellular spaces (microscopic observation).

4.3 EPITHELIUM OF THE GASTROINTESTINAL TRACT

Hogben (1955) demonstrated that the potential across the gastric mucosa arose from the active movement of chloride that is transported from the serosal to the mucosal side and secreted into the stomach lumen as HCl. The experiments were performed on gastric mucosa from the frog *Rana temporaria*. The mucosal epithelium was isolated from the serosal muscular layers and mounted in an experimental arrangement similar to that of Fig. 4.1. The preparation spontaneously secreted acid for several hours. Input conductances were measured from current–voltage curves that were linear over a range from +90 mV to −30 mV. Application of the flux-ratio equation and the correlation between measured fluxes, short-circuit current, and conductance permitted the establishment of the mechanism by which different ions move across the gastric epithelium. There is practically no net sodium flux in the short-circuited preparation. The results obtained with potassium are more difficult to interpret because of errors arising from the high intracellular potassium concentration: a net flux of 0.05 $\mu eq/cm^2$ hr was found under conditions of zero potential, indicating that potassium could contribute slightly to the current. For chloride, under conditions of zero potential, there was a net flux of 2.2 $\mu eq/cm^2$ hr, which compares to a current of 2.1 $\mu eq/cm^2$ hr. When the mucosal fluid was a mixture of bicarbonate and chloride, the net chloride flux doubled. At a spontaneous potential of 30 mV, serosal positive, the net chloride flux was directed toward the mucosal side, although the electrical field favored the movement in the opposite direction. Since the preparations secreted acid spontaneously, at least two ions of opposite charge were assumed to be actively transported in the same direction: chloride and hydrogen ions. The total conductance was 2 $mmho/cm^2$, and sodium and potassium contributed 25–30% to it, but the results did not permit the identification of the remaining conductance components.

The electromotive force of the hydrogen-secreting mechanism of the frog stomach mucosa was estimated by Sanders, Kurfees, and Rehm (1968), who clamped the mucosa at progressively higher voltages (mucosal side positive) until the hydrogen secretion was reduced to zero (44 mV in the presence of chloride and 102 mV in chloride-free solutions). Their results indicated that there is an electrical and chemical coupling between the transport of both ions. However, previous results from Rehm and

Lefevre (1965) showed that in chloride-free media the resistance of the mucosa increased from 571 to 1350 Ω cm^2. Since passive and active pathways can contribute to this resistance in complex structures (Civan, 1970), it is difficult to interpret the results.

Flemström (1971) found that variations of oxygen tension changed the electrical and transporting properties of isolated frog gastric mucosa. Impedance loci were obtained with frequencies between 20 Hz and 50 kHz. At low oxygen tensions, two impedance loci were found, the separation point being in the range 50–60 Hz. The same impedance loci were found at all the oxygen tensions below 700 mm Hg, so they were considered characteristic of the hypoxic state. An active transport of sodium occurred at 300 mm Hg. This effect was attributed to a stimulation of the sodium-potassium-activated ATPase produced by the decrease in intracellular pH that resulted from the inhibition of the H^+ secretion. The low intracellular pH would be within the optimum range for the activity of the enzyme.

The effect of solutions with different ionic compositions on the potential, short-circuit current, and resistance of mucosa from the sheep stomach was investigated by Ferreira et al. (1966). The mucosal layer of the rumen was separated from the muscular and serosal and mounted as shown in Fig. 4.1. The resistance of the preparation was measured with a voltage–current method. The first data in Table 4.3 were obtained using a Ringer's solution with only glucose and acetate as substrates, but the potential was unstable. The stability of the preparation improved when propionate and butyrate were added to the bathing solutions. However, the values of the resistances are comparable to the first ones. A definite higher epithelium resistance was obtained when buffers were added and sulfate replaced chloride in the bathing solution. Increasing the potassium concentration in either the mucosal or the serosal sides increased the potential difference. If sodium was omitted from the bathing solutions, no potential difference was observed. Ouabain applied on the serosal side of the tissue abolished the potential, but no effect was observed when it was applied on the mucosal side. Therefore, the spontaneous potential across the sheep rumen epithelium can be attributed to the active transport of sodium ions, but diffusion of potassium probably contributes to the measured potential.

Edmonds and Marriot (1968) investigated the contribution of a diffusion potential to the open-circuit potential existing across the rat colon mucosa. The epithelial layer, stripped of the muscular and serosal layers, was mounted in a chamber like that used by Ussing and Zerahn (1951). The input resistance was determined by measuring the shift in potential produced by a known current. At the onset of the pulse, there was a brief transient (2–3 s), and afterwards the potential reached a stable level. A

Table 4.3 Electrical Properties of Gastrointestinal Mucosa

Method	Measured Quantities	Model	Derived Quantities	Reference
Fig. 4.1	I_{sc}: 16–21 μA/cm^2 V : 10.3 mV	1 of Table 1.5	R_m: 837–909 Ω cm^2(Ringer's, glucose, and acetate)	Ferreira et al. (1966)
	I_{sc}: 11.6 μA/cm^2 V: 9.3 mV		R_m: 1100 Ω cm^2 (as above, + propionate, butyrate)	
	I_{sc}: 17.3 μA/cm^2 V: 17.6 mV		R_m: 1501 Ω cm^2 (SO^{2-}-Ringer's, substrates, buffers)	
Fig. 4.1	V: 4.9 mV	1 of Table 1.5	R_m: 158 Ω cm^2(12°C)	Edmonds and Marriot (1968)
	I_{sc}: 38 μA/cm^2 V: 11 mV		R_m: 136 Ω cm^2 (22°C)	
	I_{sc}: 80 μA/cm^2 V: 15 mV		R_m: 108 Ω cm^2 (32°C)	
	I_{sc}: 143 μA/cm^2			
Fig. 4.1	I_{sc}: 2.94 μeq/cm^2 hr	1 of Table 1.5	I_{Na}: 1.94 μeq/cm^2 h	Lew (1970)
Fig. 4.1	I_{sc}: 28 μA/cm^2	1 of Table 1.5	R_m: 89 Ω cm^2(unstripped ileum)	Field, Fromm, and McColl (1971)
	I_{sc}: 61 μA/cm^2		R_m: 45 Ω cm^2 (mucosa)	
	I_{Na}: 0.9 μeq/cm^2 hr		25 mM HCO_3^-, no glucose	
	I_{Na}: 3.7 μeq/cm^2 hr I_{Cl}: 1.4 μeq/cm^2 hr		25 mM HCO_3^-, 1–5 mM glucose	
	I_{Na}: 4.7 μeq/cm^2 hr I_{Cl}: 1.3 μeq/cm^2 hr		HCO_3^--free 1–5 mM glucose	
1 of Table 1.3	ΔV_{muc}: 11.1 mV	1 of Table 1.5	R_m: 21 Ω cm^2 (transmucosal) R_m: 28 Ω cm^2 (intercellular) R_m: 56 Ω cm^2 (serosal) R_m: 28 Ω cm^2 (mucosal) KCN and iodoacetate: R_m: 34 Ω cm^2 (transmucosal) R_m: 35 Ω cm^2 (intercellular) R_m: 600 Ω cm^2 (serosal) R_m: 300 Ω cm^2 (mucosal)	Rose and Schultz (1971)

linear current–voltage relation was found at 32, 22, and 12°C. The potential and the short-circuit current rose with increasing temperature while the tissue resistance decreased (Table 4.3). The activation energy (calculated from the Q_{10}) was consistent with a tissue resistance based on a diffusion process of ions in water ($Q_{10}=1.2$) and with an enzymatic process for the short-circuit current ($Q_{10}=1.8$–2.1), thereby linking the latter to active transport mechanisms. After several hours, the potential and current dropped to zero, and changes in temperature did not affect them while the tissue resistance still decreased with increasing temperature. Anoxia resulted in a rapid fall of the potential and, consequently, of the current. In contrast to its effect on toad bladder, vasopressin did not increase the potential. The rat colon mucosa was less sensitive to ouabain than other tissues, and it required fairly large concentrations (10^{-3} M) on the serosal side to reduce the potential appreciably. Other metabolic inhibitors such as cyanide and DNP reduced the potential to zero in 15 min. Removal of sodium on the mucosal side resulted in a drastic fall of potential and current. These results suggested that the potential across the colon mucosa is determined by an active sodium transport and does not depend on a diffusion process. The potential value depends on intraluminal sodium concentration. Potassium transport is not coupled to sodium transfer. A more detailed analysis of the origin of the transmucosal potential of rat colon showed that the situation is more complicated. The results of Edmonds and Nielsen (1968) indicated some discrepancies with the theoretical predictions for passive, independent fluxes (Appendix 3), which could be explained by (1) active transport of potassium and sodium at the serosal side of the cells and (2) particular pathways for sodium and chloride transport through special cells or between cells.

As in the rat, the colon mucosa of other species shows a potential that depends on active sodium transport. Lew (1970) investigated the correlation between short-circuit current and ionic fluxes in the isolated colonic mucosa of the toad *Bufo arenarum*. The mucosa was mounted in symmetrical chambers, so that one half of the preparation could serve as control. Current–voltage curves were linear over a wide range of potentials (-50 to -10 mV), if transients were avoided by changing the current only by small steps. Anoxia reduced the short-circuit current to zero with a half-time of 8 min. Removal of bicarbonate from the solution in the serosal side reversibly eliminated the potential and the current, independent of the presence of CO_2 or changes in pH. Current and potential were also reduced when ouabain was applied to the serosal side. This effect was absent in sodium-free medium. Removal of sodium from the solution bathing the mucosal side resulted in a sharp drop in potential and current. If sodium-free media bathed both sides of the preparation, the input resistance increased by 150 Ω (equivalent to a decrease in total conductance of 0.5 mmho/cm^2). Potential and short-circuit current were not reduced to zero unless sodium and bicarbonate were replaced by choline and chloride. A slight decrease in mucose conductance (0.3–0.7 mmho/cm^2) occurred in chloride-free solutions.

Measurements of ionic fluxes (Table 4.3) showed no significant net flux of chloride or bicarbonate, and only 65% of the short-circuit current was due to sodium net flux. The proportion of current due to the non-sodium component declined with time. The fraction of current remaining after sodium removal also decreased with time. Both current components were sensitive to the presence of bicarbonate in the serosal side. However, it was difficult to attribute the nonidentified fraction of the current to active transport of bicarbonate, because the net flux of this ion was not significantly different from zero. Lew advanced the hypothesis that the transported ions might be formed within the epithelium in a compartment that does not have isotopic exhange with the bathing solutions. In this case, a net flux could occur between this compartment and the lumen and still be undetected by the classical method of flux measurements. Results reported by Carlisky and Lew (1970) supported this hypothesis. Transport of bicarbonate ions from mucosa to lumen may also account for part of the short-circuit current in intestinal mucosa of mammals. Field et al. (1971) analyzed the ionic fluxes, resistance, and short-circuit current of rabbit ileum stripped off muscular layers. Mucosa resistance was measured by imposing square pulses on the non-short-circuited mucosa using a voltage–current method. The values were corrected for electrode and fluid resistance. Stripped preparations had lower mucosal resistance and higher net ionic fluxes (sodium and chloride) and short-circuit current than nonstripped preparations. This finding indicated that the passive pathways of the tissue are also determinants of net fluxes. Flux and current measurements were performed under different experimental conditions: with or without bicarbonate added to the basal side, and in the absence or presence of glucose on the mucosal side (Table 4.3). Addition of glucose increased the short-circuit current and net ionic fluxes substantially. It was found that the short-circuit current was higher than expected from sodium and chloride net fluxes. This residual flux was greater than the chloride net flux. It could be reduced but not completely suppressed in bicarbonate-free solution. The chloride flux seemed to be independent of the presence of bicarbonate. It was concluded that the bicarbonate excretion known to exist in the ileal mucosa was not due to a chloride-bicarbonate exchange. On the other hand, the residual current could be attributed to bicarbonate ions composed of two different fractions: (1) bicarbonate transported from serosal to mucosal side (fraction of I_{sc} disappearing in bicarbonate-free Ringer's), and (2) bicarbonate formed in the cell from CO_2 and water, as metabolic product. Rose and Schulz (1971) showed that the sodium influx coupled to the transport of amino acids and sugars in rabbit ileum is electrogenic and determines part of the transmucosal potential. The changes of epithelial resistance in the presence of metabolic inhibitors (Table 4.3) indicated that a low-resistance transepithelial shunt contributed to the decrease in transepithelial potential observed under these conditions.

From the material presented in this chapter, it is evident that the interpretation of determinations of short-circuit current or resistance in epithelia is difficult, unless the measurements are complemented by ion-flux measurements and analysis of the passive electrical properties. Ginzburg and Hogg (1967) derived a set of equations for the model shown in Fig. 4.8 that demonstrates that the measured short-circuit current may or may not represent the current of an actively transported ion, according to whether or not the membranes involved in the process are symmetrical. The model is applicable to epithelial structures where the basal and apical membranes are in series and may have different properties. The equations describe the parameters that determine the active current and the short-circuit current. In Fig. 4.8 and in the following equations, g, I and E have the usual meaning. The subscripts 1, 2, and 3 indicate the ion species, and a indicates the current generated by active mechanisms. The passive current is the product of the conductance and the driving force (Table 2.5). It is assumed that in the steady state, the net flux across each membrane for each ion is the same. Across membrane A (inset), all the fluxes are passive, while across C the fluxes are active and passive. Compartment B separates membranes A and C. Ion 1 contributes to the short-circuit current I_{sc1} with its net flux, and the total short-circuit current is the sum of the contributions of each ion. Assuming that I_3 and I_3' are zero because ion 3 is in equilibrium, the short-circuit current and the total active currents for ions 1 and 2 are given by the following expressions

$$I_{sc1}+I_{sc2}=I_{a1}+I_{a2}-g_1'(V-E_1)-g_2'(V-E_2) \tag{4.12}$$

$$I_{sc1}+I_{sc2}=g_1(V-E_1)+g_2(V-E_2) \tag{4.13}$$

$$I_{a1}+I_{a2}=(g_1+g_1')(V-E_1)+(g_2+g_2')(V-E_2) \tag{4.14}$$

where $(I_{sc1}+I_{sc2})$ is the total short-circuit current, V is the potential across the structure, and $(I_{a1}+I_{a2})$ is the total active current. Equation 4.12 applies to the short-circuit current through membrane C, and Eq. 4.13 to the short-circuit current through membrane A. The equations were applied to three cases: (1) Membrane A is impermeable to cation 1 and C is

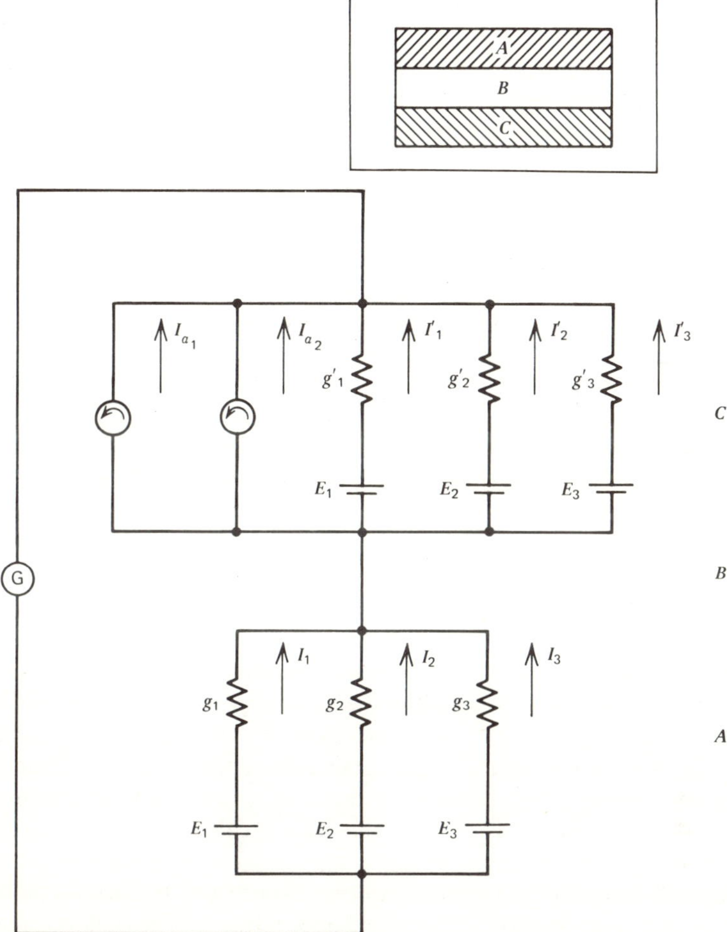

Figure 4.8. Model representing two membranes in series (*A* and *C* in the inset). Circular arrows, active transport mechanisms. G: galvanometer. Modified from Ginzburg and Hogg (1967), by permission.

impermeable to cation 2. Here, the conductances g_1 and g'_2 become zero, and the measured short-circuit current is equal to the active current for ion 2. This applies to frog skin if ion 1 is identified with *K* and ion 2 with Na. (2) The system consists of a single homogeneous membrane, where no diffusion potential is present. No current arises in the system unless an electrogenic pump exists, and the total short-circuit current equals the total active current, (3) Two identical membranes are in series, with the conductances equal for corresponding ionic channels. A short-circuit current would not be obtained in this case if the ionic pumps are neutral. In conclusion, the relation between active current and short-circuit current depends on the degree of asymmetry of the system.

4.4 ELECTRICAL PROPERTIES OF RENAL TUBULES

Hegel et al. (1967) measured the transfer resistance between lumen and interstitium in proximal tubules in the rat kidney in situ. The standard square-pulse analysis with two microelectrodes was used, together with double-barreled microelectrodes and a modification of the latter, in which one barrel was a standard microelectrode and the second was a sharpened 11-μm diameter electrode. Treating the tubule as a linear cable, the length constant was determined by three different methods: (1) the standard two-microelectrode technique, using a drop of oil in the lumen to control the current distribution; (2) using an internal electrode along the lumen, which was shifted at various distances from the current electrode, and (3) using two potential-recording microelectrodes and one current microelectrode. The value of 86 μm for λ was considered representative. A tubular radius of 13–15 μm was calculated, while the measured value was 10 μm. This difference was explained by inhomogeneities of the tubular wall. The value of R_m (Table 4.4), lower than the value reported for *Necturus* (640 Ω cm^2) by Windhager et al. (1966), was attributed to uncertainties in the estimation of the membrane surface because of the presence of microvilli on the cell border. The resistivity of the tubular wall, estimated from flux measurements, agreed well with the value calculated from electrical measurements.

The leaky nature of the tubular wall was confirmed by Hoshi and Sakai (1967), who used the two-microelectrode method. The microelectrodes were inserted either in the tubular cells or in the lumen of the newt kidney. The electrotonic potential decayed exponentially along the peritubular cells over many intercellular boundaries. Therefore, the value listed as R_{inp} for tubular cells in Table 4.4 does not represent the input resistance of a single cell, but rather the input resistance of interconnected cells forming a tubular structure. The high values of R_{inp} and λ suggest that the surface membranes conduct poorly while the intercellular membranes have low resistances. The values of R_m and R_i were calculated using the linear cable model: R_i represents here the resistivity within the interconnected cells, including cytoplasmic resistivity and the resistance of intercellular membranes. Values of 250 kΩ for R_{inp} and 2.5 mm for λ of the tubular wall were calculated on the assumption that the tubular lumen was completely

Table 4.4 Electrical Properties of Renal Tubules

Method	Measured Quantities	Model	Derived Quantities	Reference
1 of Table 1.3	R_i: 33 kΩ	5 of Table 1.5	R_m: 4.9–5.7 Ω cm^2 (proximal tubule)	Hegel et al. (1967)
2 of Table 1.3 Ionic fluxes	R_{inp}: 100 kΩ		R_i: 620 Ω cm (tubular wall)	
1 of Table 1.3	λ: 86 μm		R_i: 645 Ω cm (tubular wall)	
1 of Table 1.3	R_{inp}: 310 kΩ λ: 400 μm	5 of Table 1.5	R_m: 836 Ω cm^2 R_i: 625 Ω cm	Hoshi and Sakai (1967)
1 of Table 1.3	λ 0.14 cm R_{inp}: 1.2 MΩ	5 of Table 1.5	R_m: 2700 Ω cm^2 (collecting tubules)	Burg et al. (1968)
1 of Table 1.3	R_{inp}	5 of Table 1.5	R_m: 867 Ω cm^2 d: 23.8 μm (core diameter) R_m: 867 Ω cm^2 (luminal area)	Helman et al. (1971)
1 of Table 1.3 2 of Table 1.3	R_{inp}: 27.5 kΩ λ: 9.4 mm	5 of Table 1.5	R_m: 5.58 Ω cm^2 (proximal) d: 1.73 mm P_K/P_{Na}: 1.1 P_K/P_{Cl}: 1.58 G_{Cl}: 70 mmho/cm^2	Boulpaep and Seely (1971)
	R_{inp}: 528 kΩ		R_m: 600 Ω cm^2 (distal)	
3 of Table 1.4	R_{inp}	1 of Table 1.5	(instantaneous value) R_m: 43.2 Ω cm^2 (Steady-state value) R_m: 155.7 Ω cm^2	Spring and Paganelli (1972)

surrounded with a homogeneous membrane whose characteristics were described above, assuming a core resistivity (intratubular fluid) of Ringer's solution and using a measured tubular diameter of 80 μm. However, measurements with the electrodes in the lumen showed such a sharp decay of the electrotonic potential along the tubule that λ could not be determined. It was concluded that actually the tubular wall was leaky, due to the presence of extracellular shunt pathways within the wall.

Burg et al. (1968) used microelectrodes to measure the transtubular potential of isolated collecting and proximal tubules of the rabbit kidney. No potential difference was measured in the proximal tubules, while the collecting tubules had a transtubular potential of 25 mV, lumen negative. Ouabain abolished this potential; vasopressin decreased it by 50%. The decay of the electrotonic potential was linear over the tubular length, and this was interpreted as resulting from the cable finite length, ending at a distance smaller than λ, and looking into a low terminating resistance. The equations for a linear cable (modified for the condition where $\lambda \geqslant$ total length) were used to calculate the values listed in Table 4.4.

Helman et al. (1971) modified the technique developed by Burg et al. (1968) to eliminate the short-circuiting effect of the tubular end. The lumen was insulated by means of a liquid insulator. Improvement of the experimental conditions resulted in a value of R_m approximately one-third the value previously reported by Burg et al. (1968) (see Table 4.4). The fact that vasopressin seems to stimulate sodium transfer in this structure but fails to change the measured resistance was interpreted as an indication of the existence of shunt pathways between the tubular cells.

The electrical properties of proximal and distal tubules of the dog kidney were investigated by Boulpaep and Seely (1971). The tubular lumen and the peritubular capillaries were perfused, thus permitting fast changes of the composition of the interstitial fluid. In addition, pulsatile movements, which complicated microelectrode impalements in the kidney in situ (Seely and Boulpaep, 1971), were eliminated. Transtubular potentials were measured with a single microelectrode, and the cable properties of the tubules were investigated by simultaneous impalement of a double-barreled microelectrode and two single microelectrodes. The mean transepithelial potential of proximal tubules was -2.0 mV (lumen negative). The ionic permeabilities of the tubular wall were estimated from potential measurements with different concentrations of sodium and chloride, from which transport numbers were calculated: $T_{\mathrm{Na}}=0.58$ and $T_{\mathrm{Cl}}=0.42$, which correspond to a $P_{\mathrm{Na}}/P_{\mathrm{Cl}}=1.38$ (see Sec. 1.6.1.4). The value of the relative permeability $P_{\mathrm{K}}/P_{\mathrm{Na}}$ was 1.10. The contribution of streaming potentials arising from bulk flow due to active sodium transport was estimated from measurements of potential during perfusion with hypertonic solutions. The curve relating potential and osmolarity had a slope of 5 mV/osmole for water efflux and 3 mV/osmole for water influx. The point at which net flow across the tubules was zero corresponded to an osmotic gradient of 180 mosmoles sucrose. The maximum contribution of streaming potentials to the transtubular potential was estimated to be -0.5 mV. The electrical constants of proximal and distal tubules are shown in Table 4.4. The input resistance increased by 80% during hyperosmotic perfusion when water flowed into the lumen and fell by 37% during outward flow. The mean potential of distal tubules was 42.7 mV. The electrical constants were also significantly higher than in proximal tubules. The data were insufficient to explain the genesis of the potential measured across the tubular walls. However, the low membrane resistance found in the proximal tubule indicated important extracellular shunt pathways. The chloride conductance (Table 4.4) is such as to account for all the net chloride flux.

Spring and Paganelli (1972) investigated the relation between transtubular potential, rate of transport, and sodium permeability in proximal tubules of *Necturus* under voltage-clamp conditions. Assuming an iso-osmotic reabsorption of solution, the sodium flux was estimated from measurements of water absorption. The rate of fluid transport increased under short-circuit conditions, and the calculated flux increased from 107 to 227 pmol/cm^2. The passive plasma-to-lumen flux could be estimated by subtraction of the measured values in open-circuit conditions (part of the active flux offset by passive flux) and short-circuit conditions (only active lumen-to-plasma flux remains). The transtubular potential was clamped at values ranging from -25 to $+10$ mV (lumen relative to interstitium). The relationship net sodium flux versus voltage consisted of two linear components whose slopes were 15 pmoles/cm^2 s mV between -15 and $+10$ mV, and -49 pmoles/cm^2 s mV for the section between -15 and -25 mV. The corresponding transepithelial permeabilities (3.73 and 7.05×10^{-6} cm/s) were calculated with the Goldman equation. Current–voltage curves measured with square pulses of short duration did not show either voltage or time-dependent changes. However, current–voltage relations measured with triangular pulses under voltage-clamp conditions showed marked hysteresis. Table 4.4 summarizes the resistance values obtained from voltage-clamp data and those calculated from potential measurements and Na^+ flux. The asymmetry of sodium fluxes suggests different pathways according to the direction of the flow or a dependence of P_{Na} on the transepithelial potential. The lower P_{Na} at potentials more negative than the potential at flux equilibrium was attributed to solvent-drag effects on fluid movement from lumen to plasma (Appendix 3), and the hysteresis was considered to be consistent with the development of an osmotic gradient within the epithelium. The influence of the transepithelial potential on fluid movement across the tubular wall agreed with the findings of Boulpaep and Seely (1971). Moreover, Maruyama and Hoshi (1972) reported that active transport of sugars alters considerably the potential profile across the proximal tubule of the Japanese newt. It was suggested that the primary event induced by sugar transport might be a depolarization of the luminal membrane. The existence of an extracellular electrical shunt would be responsible for a secondary depolarization of the basal membrane.

4.5 ELECTRICAL PROPERTIES OF JUNCTIONAL MEMBRANES

Extensive work on electrical coupling between epithelial cells was carried out recently mainly by Loewenstein's group. Close attachments between epithelial cells have long been known (Loewenstein and Kanno, 1964). The permeability characteristics of these regions of close membrane apposition were first studied in salivary glands of *Drosophila flavorepleta* larvae. The electrotonic spread of current between cells was measured on isolated glands mounted in a tissue bath. The experimental arrangement used throughout the work of this group, together with a typical set of results, are shown in Fig. 4.9. Current is injected in cell I, and the voltage drop is recorded simultaneously in cell I and an adjacent one (II). Normally, the ratio $\Delta V_{II}/\Delta V_I$ is higher than 0.5 and lower than 1. If the

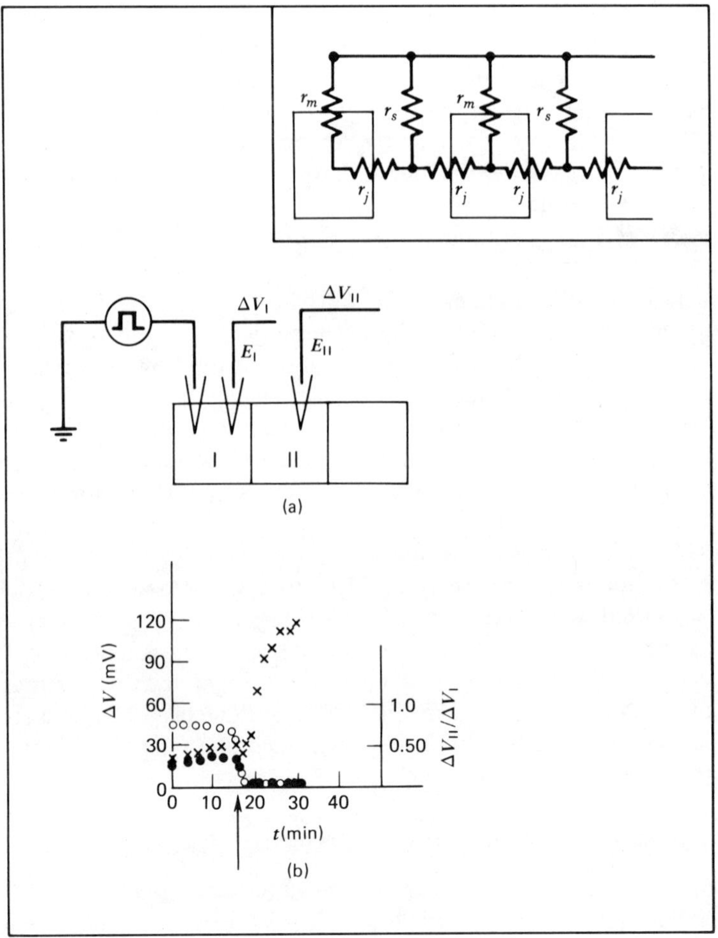

Figure 4.9. (a) Experimental arrangement and (b) typical result of electrotonic spread between epithelial cells. (X) ΔV recorded in cell I; (●) ΔV recorded in cell II; (O) ratio of $\Delta V_{II}/\Delta V_I$. Arrow: uncoupling occurs. Inset: electrical equivalent circuit for electrically coupled cells. Modified from Loewenstein (1966), by permission.

cells are electrically uncoupled, the ratio $\Delta V_{II}/\Delta V_I$ drops to zero. Here, the microelectrodes E_I and E_{II} record the voltage drops across the surface membrane. If electrode E_{II} is connected to ground, E_I records the voltage drop across the junctional membrane so that its resistance can be estimated. The value of the input resistance in Table 4.5 was obtained from the slope of voltage–current curves that were linear for currents in the region $\pm 1.5 \times 10^{-6}$A. The equivalent circuit (inset of Fig. 4.9) was used to interpret the measurements: r_j represents the resistance of the junctional membrane; r_m, the resistance of the surface membrane; and r_s, the resistance between intercellular space and the exterior. The experimental results can be fitted to this model, provided that $r_m, r_s \gg r_j$. Electron micrographs showed the presence of extensive infoldings of the membrane. Correcting for this larger surface, R_m became as large as 10,000 Ω cm^2. This value corresponds to the combined resistance of the surface membrane in parallel with the resistance of the intercellular space. The relative magnitude of these two resistance components could not be evaluated, but from additional evidence Lowenstein and Kanno estimated that they are of the same order of magnitude, 10^4 Ω cm^2. Immersing the gland in oil and bypassing the surface membrane by connecting E_{II} to ground, a linear voltage–current curve was obtained (equivalent to r_i in the linear cable model) whose slope was 500 kΩ/cm. From this value, R_i in Table 4.5 was calculated. Its value was only slightly higher than the resistivity of extruded cytoplasm. The permeability of the junctional membranes to molecules other than ions was demonstrated by Kanno and Loewenstein (1966). Similar cell communications were described for several other epithelia by Loewenstein, Socolar, et al. (1965). The values of the coupling ratios ($\Delta V_{II}/\Delta V_1$), length constant, and derived membrane resistances are listed in Table 4.5. The estimate of the resistance of the junctional membranes (R_m^j) takes the entire contact area as the electrical junction and attributes the core resistance entirely to junctional barriers. Therefore, the value must be considered as an upper limit. In the case of the toad urinary bladder (a two-dimensional sheet of cells), current spread followed only particular paths or channels. Each channel consisted of a chain of tightly connected cells, about one cell wide and several cells long.

The capability of forming communicating junctions seems to be a very extended and general characteristic of cell membranes. Michalke and Loewenstein (1971) reported the existence of electrical coupling between several different types of mammalian cells in culture. The highest coupling ratios were between cells of the same type, but some combinations of cell strains such as rabbit lens and rat liver showed ratios as high as 0.4.

Dissociated sponge cells also form aggregates and establish electrical communication within minutes of their first mechanical contact (Loewenstein, 1967a). Ca^{2+}, Mg^{2+}, and an organic factor known to play a role in

Table 4.5 Electrical Properties of Gland and Other Epithelia

Method	Measured Quantities	Model	Derived Quantities	Reference
1 of Table 1.3	R_{inp}: 4–5 kΩ λ: 1.2 mm	5 of Table 1.5	R_m: 380 Ω cm² R_i: 110–190 Ω cm	Loewenstein and Kanno (1964)
1 of Table 1.3 Fig. 4.7	$\Delta V_{II}/\Delta V_I$: 0.93		R_i: 150 Ω cm	Loewenstein, Socolar, et al (1965)
			R_j: 0.3–12 Ω cm²(junctional membrane)	
	λ: 1.2 mm		R_s: 10^4 Ω cm² (surface membrane)	
	$\Delta V_{II}/\Delta V_I$: 0.8		R_i: 140 Ω cm (salivary gland)	
	λ: 0.75 mm		R_j: 0.6–1 Ω cm²	
			R_s: 9×10^3 Ω cm²	
	$\Delta V_{II}/\Delta V_I$: 0.5		R_s: 420–5900 Ω cm² (Malpighian tubule)	
	λ: 0.26 mm		R_j: 30 Ω cm²	
			R_s: 3000 Ω cm²	
	$\Delta V_{II}/\Delta V_I$: 0.4–0.8		R_i: 180 Ω cm (urinary bladder)	
			R_j: 0.02–0.04 Ω cm²	
			R_s: 100 Ω cm²	
1 of Table 1.3	R_{inp}: 2.7 MΩ R_{inp}: 3.0 MΩ		Normal thyroid Cancerous thyroid	Jamakosmanović and Loewenstein (1968)
1 of Table 1.3	R_{inp}: 5.5 MΩ R_{inp}: 26 MΩ		Normal stomach Cancerous stomach	Kanno and Matsui (1968)
1 of Table 1.3	R_t	10 of Table 1.5	R_m: 250–2000 Ω cm² R_j: 5–50 Ω cm² C_m: 5–50 μF/cm²	Siegenbeck van Heukelom (1971)
1 of Table 1.3	R_t	10 of Table 1.5	R_i: 1–1.5×10³ Ω cm R_m: 700–900 Ω cm² λ: 300–400 μm	Shiba (1971)

cell adhesion intervene in this process. The role of calcium and magnesium ions on the permeability of the junctional membranes was further studied by Loewenstein, Nakas, and Socolar (1967) in *Chironomus* salivary glands. Chelating agents (EDTA and EGTA) caused complete uncoupling of the gland cells, with ΔV_{II} falling to zero and ΔV_{I} rising severalfold. This was interpreted as indicating a very large increase of r_j (inset of Fig. 4.9), equivalent to the sealing of the junctional membrane. Alkalinity (pH 7–10) also caused uncoupling, the input resistance of the junctional membrane rising from values lower than 0.1 MΩ to 3–6 MΩ. Similar results were obtained with trypsin and hypertonic media. Magnesium ions can substitute for Ca^{2+} in maintaining junctional coupling. If chelating agents were used in Ca^{2+}-free media, uncoupling occurred (shown by a fall of $\Delta V_{\text{II}}/\Delta V_{\text{I}}$), but ΔV_{I} did not rise unless calcium was added. Addition of calcium did not reverse the uncoupling process. A complete discussion of the role of calcium and magnesium ions and the model for the intercellular junction can be found in two review papers by Loewenstein (1966, 1967b).

Cellular uncoupling was reported to occur spontaneously in cancerous thyroid epithelium (Jamakosmanović and Loewenstein, 1968) and cancerous stomach epithelium (Kanno and Matsui, 1968). The coupling ratio for these anomalous tissues was lower than 0.002. Input resistances were also measured, and Table 4.5 shows that stomach cancerous tissue had a higher input resistance than normal tissue. The interpretation of these data is subject to the comments mentioned above, since in the case of normal tissue, the measured resistance is the input resistance of the whole preparation, while in the uncoupled cells it is the input resistance of a single cell.

Electrical coupling between cells of epithelial origin was also reported by other researchers. Penn (1966) studied the electrotonic spread in mouse liver. Measurements were made on in situ and in vitro preparations with interelectrode distances between 8.3 and 33 μm, but electrotonic potentials could be detected up to 332 μm from the current electrode, a distance of approximately 20 cells. Mapping of the current distribution showed a nearly uniform flow in all directions along the liver surface and into its depth. Voltage–current curves were linear in a current range of $\pm 6 \times 10^{-7}$ A. The slope (calculated from the published data) gives a value of 100 kΩ for the tissue input resistance. Similar experiments were carried out by Loewenstein and Penn (1967) in regenerating liver. No differences were found compared to normal liver.

Electrical coupling was also described between cells in the early chick embryo incubated for 18–22 hr (Sheridan, 1966), and between fibroblasts in tissue culture (O'Lague et al., 1970). The latter showed the existence of low-resistance pathways between normal interphase and mitotic fibroblasts from chick and mouse embryos. Input resistances of single cells not in contact with other fibroblasts ranged from 2 to 20 MΩ. Phase contrast micrographs and scanning electron micrographs showed the existence of cytoplasmic prolongations from the mitotic cell in contact with the interphase fibroblasts. Cells fixed immediately after the electrical measurements showed evidence of low-resistance pathways.

Siegenbeek van Heukelom et al. (1970) used monolayer cultures of 10-day-old chick embryo intestine to measure the electrotonic spread of an injected low-frequency square-wave current to evaluate the resistance of junctional and nonjunctional membranes. The results were analyzed by applying a discrete model based on the following assumptions: (1) the cells are arranged in a honeycomb structure; (2) all cells have equal heights h and equal sides d; (3) the contribution of the resistance of extracellular and intracellular fluids is negligible with respect to the resistances of the junctional (R_j) and nonjunctional (R_m) membranes; (4) the cells are arranged in concentric rings. The authors estimated the ratio R_m/R_j to be about 100. An equivalent continuous model gave a length constant of 70 μm. In this model, the monolayer is thought of as a continuum bounded by the nonjunctional membrane, and the resistance of the junctional membranes is included as part of the intracellular fluid resistance. Measuring the injected current I_0, and the voltage drops ΔV_0 and ΔV_1 obtained from the innermost cell and from one cell of the first ring, the following equation was derived:

$$I = \frac{Gdh}{R_j}(\Delta V_0 - \Delta V_1) + \tfrac{3}{2}\sqrt{3}\,\frac{2^2}{R_m}V_0 \tag{4.15}$$

From Eq. 4.15 and the evaluation of the voltage transient following the onset of the current pulse, absolute values for R_m, R_j, and C_m were estimated. A detailed analysis of these results led to the values listed in Table 4.5 (Siegenbeek van Heukelom, 1971). The values of the membrane constants were not corrected for membrane folding and microvilli.

Shiba (1971) studied the pattern of current distribution in the newt gastric epithelium. The electrotonic decay with distance was steeper than predicted by an exponential function because of the two-dimensional spread of current. Electrotonic potentials were detected up to 20 or 30 cells distant from the point of current injection. The cell constants listed in Table 4.5 were calculated according to model 10 of Table 1.5. They are within the range found for nonjunctional biological membranes, junctional membranes, and cytoplasmic resistivity.

4.6 SUMMARY AND OUTLOOK

The discussion in this chapter is centered on two topics where impedance measurements have made considerable contribution to present knowledge: the study of electrogenic transport mechanisms in epithelia and the investigation of intercellular contacts. By studying electrogenic pumps, it became clear that single-sided electrical or tracer approaches could not produce the insight in a biological mechanism that a combination of both methods could give. Comparing values of membrane resistance obtained by electrical measurements with those calculated from determinations of net flux permits analysis of the nature of a transport process. Moreover, the identification of the effect of vasopressin with a decrease in the internal resistance of the pump mechanism enables the analysis of pump energetics in electrical terms. This method of analysis, developed for epithelia, proved later to be profitable for the analysis of electrogenic processes in single cells as evidenced by the work of Haas's group on cardiac muscle.

In the study of intercellular propagation, a combination of electrical, mathematical, and morphological methods produced results that had, in a rather short time, a profound impact on the concepts of modern biology. In epithelial tissues we are no longer dealing with an isolated cell but with interconnected cell conglomerates with points of contact that are at least permeable to small molecules and ions. These points of contact are dynamic in nature, and their impact on cell growth and regulatory processes has still to be fully explained.

Further exploitation of the concept of an electrogenic pump, elaborated in epithelia, will have considerable impact on the analysis of transport processes in isolated cells for some time to come, especially since electrogenic components of membrane potentials are found in an increasing number of cells. The same can be stated for the analysis of the nature of intercellular connections. Intracellular potassium concentration is related to protein synthesis, and intracellular calcium is important for the contractile process in muscle and for the slow current during excitation. Both ions pass easily through the dynamic low-impedance pathways between epithelial and some muscular cells. The future potential of these studies is evident, provided a meaningful collaboration between the pertinent specialties can be assured.

5

Impedance Measurements on Cell Suspensions

In this chapter, measurements on an inhomogeneous group of cells will be described. Most of the measurements were performed with alternating current, and the suspension equation (Eq. 1.12) was used to relate the input resistance to the electrical cell constants. To test the applicability of the suspension equation, some model studies were carried out with nonbiological materials such as sand. Since these investigations were performed to verify the applicability of the suspension equation to biological problems, they are included here.

The task of describing impedance measurements of cell suspensions is complicated by the volume of published data, much of which is only of historical interest. Therefore, a selection of papers was made. Relevant model studies on suspensions will be presented first, followed by a discussion of measurements according to the category of cells. Reviews of measurements on cell suspensions can also be found in Rajewski (1938), Dänzer (1938), Cole and Curtis (1950), Schwan (1957, 1963), and Cole (1968, part 1).

Most of the measurements discussed in this chapter were performed in measuring cells using large platinized electrodes (see Sec. 1.2.1 and Appendix 1). The main technical problem with such measurements is to separate the input impedance from the polarization impedance, which represents the interaction impedance in this type of measurement (see Sec. 1.1). Therefore, information about the type of measuring cell employed and the frequency range used are included in the discussion to provide the elements necessary to the assessment of the technical soundness of the different experimental approaches.

5.1 THEORETICAL AND MODEL STUDIES OF SUSPENSIONS

Fricke (1924) investigated the applicability of the suspension equation (Eq. 1.12) to suspensions of biological material and proposed its generalized form:

$$(\sigma/\sigma_1 - 1)/(\sigma/\sigma_1 + \gamma) = \rho'(\sigma_2/\sigma_1 - 1)/(\sigma_2/\sigma_1 + \gamma) \quad (5.1)$$

which was derived originally by Maxwell (1873) for a suspension of spheres with conductivity σ_2. The spheres are suspended in a medium of conductivity σ_1, and σ is the conductivity of the resulting suspension, with ρ' the volume concentration of the spheres. Treating the case of suspended ellipsoids, because their geometry approximates that of blood corpuscles, Fricke was able to identify γ as a form factor whose numerical value changes according to the geometry of the suspended phase. Fricke calculated graphs (Fricke, 1924) and tables (Fricke, 1925b) to determine γ, considering the ellipsoid geometry of erythrocytes and assuming that two of the three main axes are equal. This theory was expanded later to ellipsoids with three unequal axes by Velick and Gorin (1940), who also investigated the effect of the orientation of the ellipsoids in the electric field.

A further problem for the application of Eq. 5.1 to suspensions of biological cells was the uncertainty over which range of volume concentration the suspension equation was applicable. It was already known that it was not valid for high values of ρ'. Fricke (1924) verified his mathematical formalism with results obtained by Stewart (1899) from low-frequency measurements of dog erythrocytes. Fricke considered the erythrocytes to be nonconducting ellipsoids with $\sigma_2 = 0$. Since the ratio σ/σ_1 can be measured, ρ' could be calculated assuming an appropriate form factor, and this result was compared with the volume concentration measured with an independent method. For dog erythrocytes, the error in the determination of ρ' was less than 2% for a volume concentration between 15 and 80%. Results of sand suspensions obtained by Oker-Blom (1900) were treated similarly. Again, Eq. 5.1 with an adequate form factor described the results accurately at volume concentrations between 20 and 50%.

A second paper (Fricke, 1925b) contains the derivation of a formula that permits determination of the static membrane capacity from cells in a suspension. If particles of any shape (n particles per cm^3) are surrounded by a membrane of thickness d having a dielectric constant ε_r, the membrane capacity C_m in $\mu F/cm^2$ of such a particle is:

$$C_m = (\varepsilon_r / 4\pi d)(10^{-5}/9) \quad (5.2)$$

If one considers (1) that at low frequencies the resistivity inside the particles is small compared to the impedance of the cell membrane; (2) that the energy dissipated in a suspension is proportional to the average value of the square of the electric force $\overline{V}'^2$, and (3) that the energy dissipated in the suspension is equal to the energy dissipated by the same driving voltage V in a homogeneous conductor with conductivity σ, one obtains:

$$(1-\rho')\sigma_1 \overline{V}'^2 = \sigma V^2 \tag{5.3}$$

where σ_1 is the conductivity of the suspending phase. The average electrostatic energy present at the surface of a particle in a homogeneous field with a force $\overline{V}'$ is then equal to $NC_m\overline{V}'^2$, where N is a constant of proportionality dependent upon the geometry of the particle. Consequently, the equation for the energy content of the suspension is:

$$\frac{1}{2}C^* V^2 = nNC_m\overline{V}'^2 \tag{5.4}$$

where C^* is the specific capacity of the suspension measured in F/cm. Combining Eqs. 5.3 and 5.4, we have:

$$C^* = 2NC_m(\sigma/\sigma_1)[n/(1-\rho')] \tag{5.5}$$

Further treatment of this equation to evaluate N leads to:

$$C_m = C^*/\alpha q(1-\sigma/\sigma_1) \tag{5.6}$$

where q is half the length of the major axis of the spheroid and α is a tabulated dimensionless value (Fricke, 1925b) that depends on the ratio of the spheroid's half axes. It is of historical importance that Fricke used Eq. 5.6 to estimate the thickness of the erythrocyte membrane (Fricke, 1925a).

Later, Fricke (1933) reported a method that extended the applicability of the suspension equation to volume concentrations up to 84% for determinations of the membrane capacity: The apparent capacity $C^*_{100\%}$ for a volume concentration of 100% is calculated from the measured capacity C^* as follows:

$$C^* = C^*_{100\%}(1-\sigma/\sigma_1) \tag{5.7}$$

Equation 5.7 becomes identical with Eq. 5.6 when the $C^*_{100\%}$ is replaced by αqC_m. It is remarkable that Eq. 5.7 does not depend upon the geometry of the spheroid. In the same article, Fricke developed the concept of a polarization impedance located at the cell surface, and this idea later influenced the work of Cole.

The electrical models used to interpret the polarization impedance are modifications of model 3 of Table 1.5, where the meaning of the elements changes according to whether high or low frequency is applied. According to Fricke (1933), at low frequency R_0 represents the bulk resistance of the intercellular liquid and C'_{pol} in series with R'_{pol} stand for the bulk polarization impedance of the cell surface (Fig. 5.1a). C'_{pol} can be related to a polarization capacity of 1 cm^2 by an equation similar to Eq. 5.6. The quantities R'_{pol} and R_{pol} (the resistance of 1 cm^2 of surface at different frequencies) can be evaluated with the equation:

$$\omega R_{pol} C_{pol} = \omega R'_{pol} C'_{pol} = m \tag{5.8}$$

where m was found to be a dimensionless constant. It is remarkable that the relation between C_{pol} and C'_{pol} and that between R_{pol} and R'_{pol} involve only the form, size, and volume concentration and not frequency. Moreover, the measured values R_{inp} and C_{inp} can be related to C'_{pol} and R'_{pol} by:

$$\left.\begin{aligned} C_{inp} &= C'_{pol}/(1+m^2) \\ 1/R_{inp} &= 1/R_0 + \omega R_{inp} C_{inp} \end{aligned}\right\} \tag{5.9}$$

Combining Eqs. 5.8 and 5.9 results in a relation for C'_{pol}:

$$C'_{pol} = C_{inp}\left[1 + (1/R_{inp} - 1/R_0)^2/(\omega C_{inp})^2\right] \tag{5.10}$$

and for m we have:

$$m = (1/R_{inp} - 1/R_0)/\omega C_{inp} \tag{5.11}$$

The right side of Eq. 5.11 contains only measurable quantities, but when the term $m\omega C_{inp}$ is small, it simplifies to:

$$R_0 = R_{inp}(1 - m\omega\tau_{inp}) \tag{5.12}$$

Equations 5.11 and 5.12 permit the calculation of C'_{pol} and m.

For high frequencies, however, the diagram in Fig. 5.1b describes the situation. Here the additional resistance R'_i represents the sum of the resistances of intra- and intercellular liquid traversed by the current that

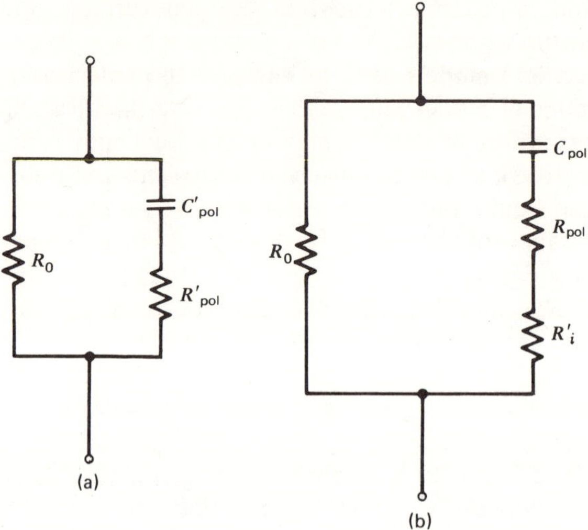

Figure 5.1. Electrical model for cell suspensions, (a) at low frequencies and (b) at high frequencies. For details, see text.

passes also through the interior of the cells. In this situation, one obtains:

$$\left.\begin{aligned} C'_{\mathrm{pol}} &= C_{\mathrm{inp}}\left[1+(1/R_{\mathrm{inp}}-1/R_0)^2/(\omega C_{\mathrm{inp}})^2\right] \\ R'_{\mathrm{pol}}+R'_i &= (1/R_{\mathrm{inp}}-1/R_0)/\omega^2 C'_{\mathrm{inp}} C'_{\mathrm{pol}} \\ m &= \omega C'_{\mathrm{pol}} R'_{\mathrm{pol}} \end{aligned}\right\} \tag{5.13}$$

Most of the results reported in the older literature attach much importance to the frequency dependence of C'_{pol}. In fact, this is the criterion for identifying C_{pol} with a polarization capacity. It has been found that this frequency dependence can be described as a power function:

$$C'_{\mathrm{pol}} = C'_{\mathrm{pol}}(1)\omega^{-\alpha} \tag{5.14}$$

where α is constant and empirically found to be unity. $C'_{\mathrm{pol}}(1)$ represents the polarization capacity for $\omega = 1$. It has been shown (Fricke, 1932) that in this case m is also constant and one obtains:

$$m = \tan(\pi/2)\alpha = \tan\Phi \tag{5.15}$$

Equations 5.8–5.15 represent the basic tools available for evaluating ac measurements on cell suspensions. It is interesting to note that the quantities R_0, R_i', R_{pol}, and C_{pol} have purely electrical dimensions, meaning that the information thus obtained cannot be directly related to the cell constants. The significance of the models in Fig. 5.1 is that they allow an interpretation of electrical measurements in terms of the relative contributions of the solvent, the bulk of the nonconducting suspended particles (low frequency), and the bulk of the combined extra- and intracellular liquid (high frequency) to an experimentally determined input impedance. The studies of the bulk properties of suspensions are significant for their attempts to clarify the nature of biological impedances, and the most important result was the interpretation of the membrane impedance as a polarization impedance. Although it is generally accepted today that the membrane impedance itself can be approximated by frequency-independent circuit elements, an interesting question remains unanswered: What is the molecular basis of the impedance loci with a depressed center?

Cole (1928a) considered an inhomogeneous suspended phase and refined the suspension equation further. He treated the suspended cells as spheres containing a resistive material and surrounded by a concentric shell of material with impedance properties. The resulting equation containing the cell constants R_i and C_m is given in Table 1.6, and here only some additional data will be mentioned. Figure 5.2 and Eqs. 5.16 and 5.17 show some important geometrical properties of the impedance locus with a depressed center. In Fig. 5.2, the chords u and v are related to the absolute value of the membrane impedance $\mathbf{Z}_m$ by:

$$\left.\begin{aligned} u/v &= |\mathbf{Z}_m|/\xi \\ \xi &= aR_e\left[1+\frac{R_i(1+\gamma\rho')}{R_e\gamma(1-\rho')}\right] \end{aligned}\right\} \tag{5.16}$$

Moreover, the intervals b and c can be obtained as

$$\left.\begin{aligned} b &= (R_0-R_\infty)m/2 \\ c &= (R_0-R_\infty)(1-m)^{1/2}/2 \end{aligned}\right\} \tag{5.17}$$

where m is defined by Eq. 5.15. The symbols used in Eqs. 5.16 and 5.17 correspond to those defined in Table 1.6. It is also possible to check whether the phase angle is constant with a plot of $\log|Z_m|$ versus $\log\omega$, which should result in a straight line (Bozler and Cole, 1935).

Later, Cole and Curtis (1936) recognized that it was unsatisfactory to interpret the complex impedance locus as a polarization resistance. Looking for an alternative explanation for a power factor $\alpha=1$, they found that a statistical distribution of the fiber diameters and the membrane capaci-

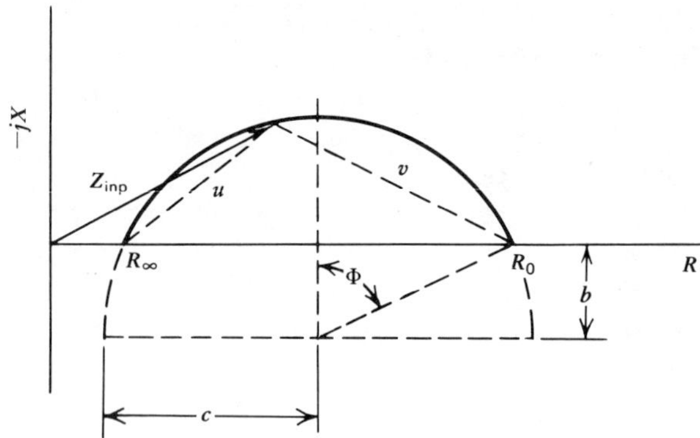

Figure 5.2. Impedance locus for the circuits of Fig. 5.1. For details, see text.

tance in skeletal muscle could semiquantitatively account for the observed results. This explanation is based on the following argument. Equation 1.12 can be transformed into:

$$\frac{Z^*_{\text{inp}} - R_e}{Z^*_{\text{inp}} + R_e} = \frac{\rho'(R_i - R_e + X^*_m)}{R_i + R_e + X^*_m} \tag{5.18}$$

with $X^*_m = 1/aj\omega C^*_m$ as the specific membrane reactance. Now Eq. 5.18 can be rewritten for a heterogeneous suspension where X_i is the membrane reactance of the ith fiber, whose volume concentration is ρ'_i:

$$\frac{Z^*_{\text{inp}} - R_e}{Z^*_{\text{inp}} + R_e} = \frac{\Sigma\rho'_i(R_i - R_e - j\Sigma X_i)}{R_i + R_e - j\Sigma X_i} \tag{5.19}$$

Equating the right sides of Eqs. 5.18 and 5.19 gives the following expression for the specific membrane reactance:

$$X^*_m = (R_e + R_i)(1 - f)/f \tag{5.20}$$

with

$$f = \int_0^\infty \left[F(x)/(1 + jx)\right] dx \tag{5.21}$$

where x equals $X^*_m/(R_e + R_i)$ and $F(x)$ represents the distribution function for the fractional volume concentration of the fibers falling in the group between x and $(x + dx)$.

Assuming that the fiber diameters and the fiber capacities have a logarithmic Gaussian distribution, it can be shown with Wagner's (1913) method that the impedance of the whole muscle closely approximates the impedance of a system containing a polarization impedance element. Here, as the distribution curve widens, the parameter equivalent to the power factor α decreases. In the case of a Wagner distribution, Cole and Curtis found that the characteristic frequency of the muscle was related to the diameter and the capacity of the fibers by:

$$\bar{C}_m = (1 - R_\infty^*/R_0^*)/\left[aR_0^*\bar{\omega}(1 - R_e/R_0^*)(1 + R_e/R_0^*)\right] \tag{5.22}$$

where $\bar{\omega}$ is the characteristic frequency and $\bar{C}_m$ is the most probable membrane capacitance of the population of muscle fibers. Using Eq. 5.22, the authors calculated the $\bar{C}_m$ of sartorius muscle from the results obtained by Bozler and Cole (1935) and obtained a value of 1 μF/cm^2 assuming a fiber diameter of 100 μm. This is remarkable, since it took 30 years more until this value of the capacitance of the surface membrane of skeletal muscle was confirmed by square-pulse measurements.

Another approach was chosen by Cole and Cole (1941) when they suggested an analogy of biological membranes to dielectrics. According to Debye's (1929) theory for the dielectric behavior of a dilute suspension of free dipoles, such a dielectric can be described in terms of a complex dielectric constant (see Sec. 1.5.2.3). Cole and Cole proposed an empirical equation

$$\varepsilon^* = \varepsilon_\infty + \left[(\varepsilon_0 - \varepsilon_\infty)/(1 + j\omega\tau)^\alpha\right] \tag{5.23}$$

which describes the experimental results better than Eq. 1.83. Here α is defined as in Eq. 5.15. Equation 5.23 represents a semicircle with a depressed center, and it has become known as the Cole-Cole equation. Recent measurements of physicochemical systems (Fig. 5.3) have shown that the loci of the complex dielectric constant can be even more complicated (Cole, 1965). Several relaxation models have been developed (Cole, 1965) to account for these difficulties, and it is interesting to note that they all are hinged on finding a physically meaningful explanation for the complex impedance locus with depressed center. The choice for one or the other representation of the dissipative and conservative properties of a biological structure depends upon the desired information: When the researcher is concerned with the state of ions in a solution, the representation of the complex dielectric constant is suitable. This will generally be the case for high-frequency measurements (γ dispersion, Table 1.9). How-

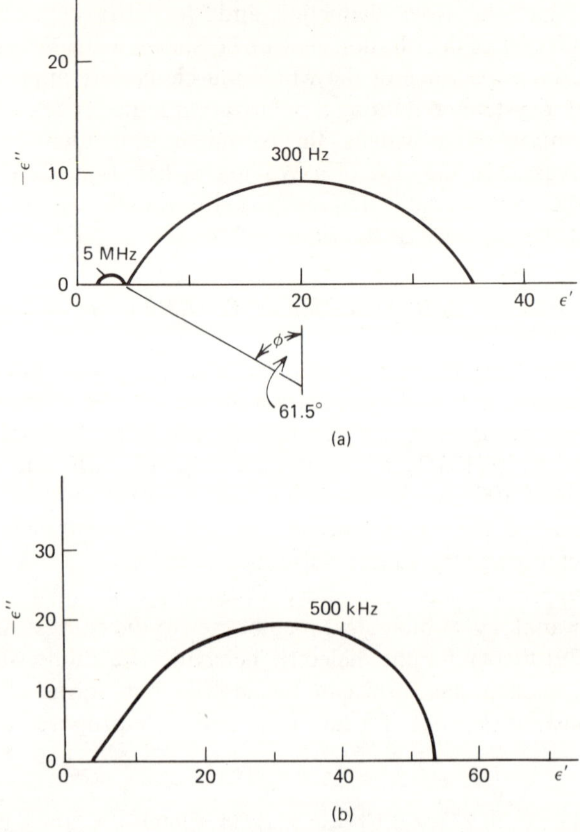

Figure 5.3. Complex dielectric constants for physicochemical systems. (a) HBr at −200°C. This is an example of the complex dielectric constant being represented by an arc with a depressed center; the small arc is neglected in the discussion. (b) Glycerol at −20°C, an example for a skewed arc function of the complex dielectric constant. Modified after Cole (1965), by permission.

ever, if low-frequency measurements are involved, the interest is centered on the relation between impedance and cellular structure (β dispersion). In this case, notation of the complex impedance is preferable.

Schwan (1957) developed a unifying hypothesis for the three dispersions (Table 1.9) found in biological cells and cell suspensions. He proposed an alternative to the Cole-Cole equation to describe a relaxation system with multiple time constants. Assuming a distribution function $f(\tau)$, one has:

$$\varepsilon^* = \varepsilon_\infty + (\varepsilon_0 - \varepsilon_\infty)\int\left[f(\tau)/(1+j\omega\tau)\right]dt \qquad (5.24)$$

The distribution function $f(\tau)$ can be approximated by a histogram of the time constants τ with the class interval $d\tau$. Schwan demonstrated also that the Cole-Cole equation corresponds to Eq. 5.24 when the distribution function $f(\tau)$ is written as follows:

$$f=\sin\beta\pi/\{2\pi\cos h[(1-\beta)\log(\tau/\tau_0)]-\cos\beta\pi\} \tag{5.25}$$

where β equals $\pi/2-\Phi$ and τ_0 represents the inverse of the characteristic frequency (see Eq. A2.13 in Appendix 2). Moreover, Schwan demonstrated that the plot of the complex dielectric constants as obtained for the Wagner distribution and for the Cole-Cole distribution were almost identical. This shows that the same set of measurements can be described by assuming different distribution functions, which, however, implies different physical interpretations of the same measurement.

5.2 IMPEDANCE MEASUREMENTS ON EGGS

The eggs of echinoderms, mainly from *Arbacia* (purple sea urchin) and the sea urchin *Hipponoë* (Tripneustes), attracted the attention of investigators early because they are nearly spherical, which greatly facilitates the theoretical analysis of impedance measurements.

Cole (1928a) began to measure impedances in *Arbacia* eggs suspended in seawater using a voltage–current method. Measurements were taken with frequencies ranging from 1 kHz to 15 MHz. The suspension equation was used, and it was found that R_i had 3.6 times the resistivity of seawater (Table 5.1). In the following years, these measurements were extended to marine eggs from *Hipponoë* (Cole, 1935) and from the starfish *Asterias* (Cole and Cole, 1936). The results of these experiments were summarized by Cole (1937) in a review paper. A Wheatstone bridge was used for the impedance measurements, and the results were expressed as impedance loci that had two remarkable features: (1) Their low-frequency part ($f<$ 250 kHz) was a semicircle indicating a frequency-independent membrane capacitance, and (2) there was an indication for an additional high-frequency dispersion. While the low-frequency dispersion confirmed Fricke's concept of a cell membrane behaving as a simple dielectric capacitance, the nature of the high-frequency dispersion was never completely explained (Cole, 1968). Since such a dispersion is absent in human red cells, it could have been caused by the nuclear membranes, but the dispersion of proteins also occurs in this frequency range (Oncley, 1943).

Measurements of fertilized eggs resulted in an even more complicated impedance locus, indicating yet another dispersion. Its origin became clear only after Cole and Spencer (1938) reexamined the impedance of fertilized

Table 5.1 Electrical Constants of Eggs from Several Species

Method	Measured Quantities	Model	Derived Quantities	Reference
Voltage–current	$Z=f(\omega)$ 1 kHz to 15 MHz	3 of Table 1.5	*Arbacia* R_i: 90 Ω cm	Cole (1928a)
5 of Table 1.1 modified	Impedance locus 1 kHz to 16 MHz	3 of Table 1.5	*Hippopnoë*, unfertilized: C_m: 0.87 μF/cm² *Asterias*, unfertilized: C_m: 1.1 μF/cm² *Arbacia*, unfertilized: C_m: 0.73 μF/cm² C_m obtained at 250 kHz	Cole (1937)
5 of Table 1.1, modified	Impedance locus	3 of Table 1.5	*Arbacia*, unfertilized: R_i: 142 Ω cm C_m: 0.86 μF/cm² Φ: 90° *Arbacia*, fertilized: R_i: 180 Ω cm C_m: 3.3 μF/cm² Φ: 90°	Cole and Spencer (1938)
5 of Table 1.1, modified	Impedance locus 1 kHz to 2.5 MHz	3 of Table 1.5	*Arbacia*, single eggs, unfertilized: R_i: 185 Ω cm C_m: 1.13 μF/cm² Φ: 90° Fertilized: R_i: 215 Ω cm C_m: 2.8 μF/cm² Φ: 88.5° C_m obtained at $\bar{\omega}$	Cole and Curtis (1938c)
5 of Table 1.1	Impedance locus 30 Hz to 2.5 MHz	1 of Table 1.5	Grasshopper, precuticular membrane R_m: 84 Ω/cm² C_m: 0.027 μF/cm² Φ: 79–83° Postcuticular membrane R_m: 45 kΩ/cm² C_m: 0.08–0.1 μF/cm² Φ: 82–88° $[K]_o$: 1 M	Cole and Jahn (1937); Jahn (1936)

Table 5.1 *Continued*

Method	Measured Quantities	Model	Derived Quantities	Reference
5 of Table 1.1, modified	Impedance locus 50 Hz to 10 kHz	3 of Table 1.5	Frog, single egg Unfertilized, in 10% Ringer's R_i: 480 Ω cm R_m: 70 Ω cm^2 C_m: 2.7 μF/cm^2 Fertilized, in 10% Ringer's: R_i: 625 Ω cm R_m: 140 Ω cm^2 C_m: 1.8 μF/cm^2 Unfertilized, in spring water: R_i = 510 Ω cm R_m: 430 Ω cm^2 C_m: 1.7 μF/cm^2 Fertilized, in spring water: R_i: 580 Ω cm R_m: 390 Ω cm^2 C_m: 1.4 μF/cm^2	Cole and Guttman (1942)
5 of Table 1.1, modified	Impedance locus 100 Hz to 10 kHz	3 of Table 1.5	Trout, single egg Unfertilized R_i: 202 Ω cm C_m: 0.56 μF/cm^2 Φ: 83° Fertilized R_i: 159 Ω cm C_m: 0.57 μF/cm^2 Φ: 83° C_m obtained at $\overline{\omega}$	Rothschild (1946)
1 of Table 1.3	R_{inp}: 350 kΩ m.p. −20 to −30 mV R_{inp}: <50 kΩ m.p. becomes positive	2 of Table 1.5	Frog, single egg Untreated: R_m: 14 kΩ cm^2 Ribonuclease treated: No data	Gingell and Palmer (1968)

and unfertilized *Arbacia* eggs with a refined technique. It was found that the impedance loci for fertilized and unfertilized eggs were almost perfect semicircles, which resulted in frequency-independent values for their membrane capacity. These results were confirmed by measurements on single eggs (Cole and Curtis, 1938c; Cole, 1938). In light of this evidence, Cole attributed the more complex impedance loci found in earlier measurements to a contamination of the suspension of fertilized eggs with unfertilized ones.

At the same time, Cole's group investigated the electrical properties of the egg membrane of the grasshopper *Melanoplus differentialis* (Cole and Jahn, 1937). These eggs are surrounded first by a particular protein membrane and subsequently develop three additional layers, of which a chitinlike cuticular layer is the most important. The measurements were interpreted with model 1 of Table 1.5, and the results are expressed in terms of parallel resistance and capacitance. This model could be used because the eggs were split open and the membrane was sandwiched between two glass plates, each with a central hole 1.6 mm in diameter. A complex impedance locus was found with a depressed center for the pre- and postcuticular membrane. In all membranes, independent of their stage of development, the resistance increased with the resistivity of the medium in contact with both sides of the membrane (3 M KCl, 1 M KCl, diluted seawater), but the membrane capacity and phase angle were not influenced. At 1 kHz the postcuticular membrane had a much higher resistance and a lower reactance than the precuticular one (see Table 5.1). This study confirmed results reported earlier by Jahn (1936).

Cole and Guttman (1942) studied the impedances of single unfertilized and fertilized frog eggs, mainly because they are large and have a regular shape. The principle of the measuring cell is shown in Fig. 5.4. Since the diameter of the hole in the central diaphragm is small compared to the size of the electrodes, there is an apparent increase of the resistance of the hole due to the convergence resistance or end-effect (see Sec. 1.2.2.1). This resistance increase depends on the diameter of the hole in the central diaphragm and the resistivity of the solution in the measuring cell. Cole and Guttman corrected their measurements by determining experimentally the resistance increase due to the end-effect. They measured the interelectrode resistance of the measuring cell as a function of channel length in the central diaphragm after filling the cell with the experimental solution but with no egg in place. The results of these measurements were extrapolated to a channel length of zero, which yielded the resistance component attributable to the end-effect, thereby permitting application of a correction for it.

Impedance loci were determined in fertilized and in nonfertilized eggs,

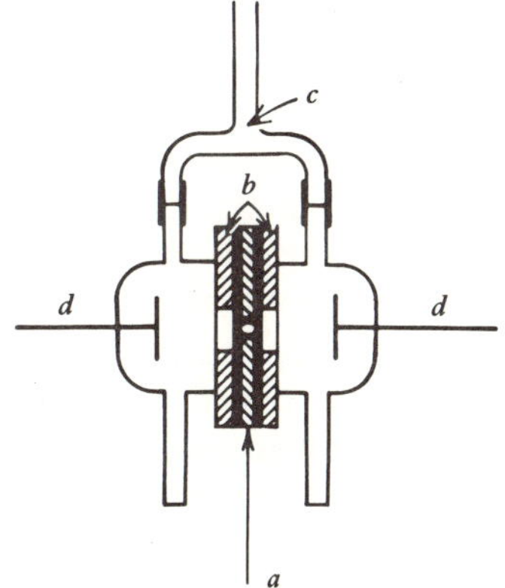

Figure 5.4. Measuring cell for single frog eggs. (a) Central diaphragm holding the egg in a central hole. (b) Supporting diaphragms. The egg is held in place by pieces of cheesecloth sandwiched between the central and the supporting diaphragm. (c) Inlet for medium. (d) Electrodes. Modified after Cole and Guttman (1942), by copyright permission of The Rockefeller University Press.

and no difference was found. Cole and Guttman also succeeded in evaluating the cell constants: R_i was obtained by Eq. 1.16 and the effective volume concentration ρ' was evaluated by equating the left side of Eq. 5.1 with $\rho'/2$. Similarly, R_m was obtained from Eq. 1.14, replacing the term $1/ja\omega C$ by R_m/a and measuring Z^*_{inp} at low frequency. However, calculation of C_m was more difficult because a finite R_m was found. Therefore, Eq. 1.14 cannot be used in its original form to calculate C_m, and the term Z^*_c from Eq. 1.13 must be written as:

$$\frac{1}{Z^*_c} = \frac{(1/R_i)(a/R_m + ja\omega C_m)}{(1/R_i) + (a/R_m + ja\omega C_m)} \tag{5.26}$$

since R_m is in parallel with C_m. One now obtains for the input impedance of the egg by combining Eq. 1.12 and Eq. 5.26 with $\gamma = 2$:

$$Z^*_{\text{inp}} = R_e \frac{\begin{array}{l}(1-\rho')R_e + (2+\rho')R_i \\ \quad + \left[(1-\rho')R_e + (2+\rho')(R_i + R_m/a)\right]/j\omega C_m R_m\end{array}}{\begin{array}{l}(1+2\rho')R_e + 2(1-\rho')R_i \\ \quad + \left[(1+2\rho')R_e + 2(1-\rho')(R_i + R_m/a)\right]/j\omega C_m R_m\end{array}} \tag{5.27}$$

This equation is equivalent to Eq. 1.14 when a finite R_m is considered in parallel with the dielectric capacity of the membrane. Assuming the phase angle to be 90° and introducing R^*_0, R^*_∞ and the characteristic frequency $\bar{\omega}$ in Eq. 5.27, one obtains for the membrane capacitance:

$$C_m = \left\{ a\bar{\omega}\left[R_i + \frac{R_e(1+2\rho')}{2(1-\rho')} \right] \right\}^{-1} + \frac{1}{\bar{\omega} R_m} \tag{5.28}$$

However, this is only an approximation, because actually phase angles between 80° and 88.5° were found. The results are summarized in Table 5.1. No significant difference was found between the cell constants of fertilized and nonfertilized eggs, but there was an increase in R_m and a possible decrease in R_i in cells exposed to 10% Ringer's solution and spring water, respectively, but C_m remained constant. Moreover, no changes of the cellular impedance with time was observed.

In an earlier report, Hubbard and Rothschild (1939), working on the egg of the rainbow trout, found periodic changes in its impedance. The frequency of the changes was about 1.5/min, and the impedance was constant in dead eggs. Moreover, the changes, which were most evident at 4–5 kHz, disappeared after several days in fertilized eggs. Later, Rothschild (1946) investigated the electrobiological properties of the trout egg. The major part of his paper consists of a comprehensive summary of the theory of the complex impedance locus used for representing biological impedance data. The quantitative information he obtained from the trout egg is listed in Table 5.1. The technique was similar to that described in the previous paper; parallel resistance and capacity were measured and then converted into series resistance and reactance[†] under the assumption of an infinite R_m and taking into account the phase angle of the polarization element:

$$C_m = 2(R_0^* - R_\infty^*)/\left[a\bar{\omega}R_0^{*2}(2 + R_e/R_0^*)(1 - R_e/R_0^*)\sin\Phi\right] \quad (5.29)$$

The experimental results indicated that there was no difference in R_i and C_m between fertilized and nonfertilized eggs.

Rothschild (1957) offered an explanation for an apparently anomalous behavior of the membrane capacity in marine eggs exposed to hypotonic solutions. The eggs swell, and it was reported by Cole (1935) for unfertilized eggs (*Hipponoë*) and by Iida (1943) for fertilized sea urchin eggs that C_m decreased. Since the membrane capacitance was identified with a dielectric capacitance, an increase in the capacitance was expected during swelling due to a thinning out of the membrane. However, according to Cole and Iida, C_m is inversely related to the cell radius (Eq. 5.29). Rothschild produced electron microscope evidence that the membranes of fertilized and unfertilized eggs are folded; moreover, these folds disappear when the membrane is stretched. Such a finding could explain that the value of the membrane capacitance apparently decreases when the cells swell. The same arguments have been used to explain the apparently high C_m measured in crab muscle (see Table 3.3).

[†]The conversion equations from parallel values of resistance and capacitance into series values are often mentioned in the literature. Some of the more accessible references are Schwan (1963) and Geddes and Baker (1968, p. 156).

After these studies, it seems that interest in the electrical parameters of marine eggs was lost for almost a decade until microelectrodes could be employed to explore the quantity that is the least accessible when models based on the suspension equation are used: the membrane resistance. Gingell and Palmer (1968) investigated the electrical changes accompanying the cortical contractions in the egg of *Xenopus laevis*. They used the square-pulse method with microelectrodes (circuit 1 of Table 1.3). To avoid the effects of cell division on the measured parameters, contraction was induced by ribonuclease, and both the membrane resistance and the membrane potential decreased. Under this treatment the latter was even reversed to positive values. The effects depended on the presence of divalent cations in the extracellular medium, and the results are discussed in the context of a transducer function of the membrane to mediate external stimuli to the cytoplasm.

5.3 SUSPENSIONS OF BLOOD CELLS

The ready availability of blood cells, especially erythrocytes, made them a preferred object for impedance measurements even though the aims of the measurements changed with time. Höber (1910, 1912, 1913) had already performed impedance measurements up to frequencies of 10 MHz and had found that the impedance of red cell suspensions decreased with frequency. This led to the assumption that the cells were surrounded by a poorly conducting membrane and that they contained a cytoplasm of relatively low resistivity. Höber estimated that this resistivity was equal to that of a 0.1–0.4% NaCl solution. Later, Philippson (1921) confirmed this result, measuring the impedance of packed erythrocytes between 500 Hz and 3 MHz using a voltage–current method.

Fricke (1925a) reported measurements of the membrane capacitance of dog red cells (Table 5.2). For these, he developed a bridge capable of measuring from 800 Hz to 4.5 MHz. His description of calibrating it is still worth reading today. The actual measurements were performed at volume concentrations between 10.6 and 43.9%, and the quantity $C_{100\%}$ was calculated with Eq. 5.7 using a ratio of 1/4 for the half-axes of the erythrocytes whose shape was approximated by an ellipsoid (see also Sec. 5.1). From the equation:

$$C_m = C^*_{100\%}/\alpha q \tag{5.30}$$

he obtained a value of 0.81 μF/cm^2 for the membrane capacitance. Assuming a dielectric constant of 3, he calculated a membrane thickness of 33 Å.

In the same year, Fricke and Morse (1926), working with calf blood, confirmed the interpretation of C_m as a static capacity because it did not

Table 5.2 Electrical Constants of Blood Cells

Method	Measured Quantities	Model	Derived Quantities	Reference
5 of Table 1.1, modified	C_{inp}: 73–232 pF $10.6\% < \rho' < 43.9\%$ 87 kHz	Eqs. 5.7, 5.30	Dog erythrocytes: C_m: 0.81 μF/cm^2	Fricke (1925a)
5 of Table 1.1, modified	R_{inp}: $R(\omega)$ 87 kHz to 4.5 MHz C_{inp}: $C(\omega)$	3 of Table 1.5, modified	Calf erythrocytes: R_i: 313 Ω cm	Fricke and Morse (1926)
3 of Table 1.1	C_{inp}: $C(\omega)$ 500 Hz to 16.4 MHz	3 of Table 1.5	Rabbit erythrocytes, lecithinated: R_i: 150 Ω cm	Fricke (1933)
5 of Table 1.1, modified	R_{inp}: $R(\omega)$ C_{inp}: $C(\omega)$ 250 Hz to 16.0 MHz	3 of Table 1.5	Rabbit leukocytes: R_i: 140 Ω cm C_m: 1.0 μF/cm^2	Fricke and Curtis (1935)
5 of Table 1.1, modified	R_{inp}: $R(\omega)$ below 16 MHz	3 of Table 1.5	Various species, erythrocytes R_i: 140 Ω cm	Fricke and Curtis (1934a)
5 of Table 1.1, modified	R_{inp}: $R(\omega)$ R_i at 90 MHz ε_c at 250 MHz	3 of Table 1.5	Erythrocytes Man: R_i: 193 Ω cm ε_c: 50 Beef: R_i: 230 Ω cm ε_c: 51 Sheep: R_i: 228 Ω cm ε_c: 50 Dog: R_i: 216 Ω cm ε_c: 51 Cat: R_i: 190 Ω cm ε_c: 53 Rabbit: R_i: 170 Ω cm ε_c: 55 Chicken: R_i: 204 Ω cm ε_c: 52	Pauly (1959); Pauly and Schwan (1966)
5 of Table 1.1, modified	$C_{inp} = C(\rho')$ 16 kHz	Eq. 5.35	Dog erythrocytes: C_m: 0.82 μF/cm^2 Rabbit erythrocytes: C_m: 0.82 μF/cm^2	Fricke (1953c)

Table 5.2 *Continued*

Method	Measured Quantities	Model	Derived Quantities	Reference
			Sheep erythrocytes: C_m: 0.82 μF/cm^2 Rabbit erythrocytes, lecithinated: C_m: 1.07 μF/cm^2 Rabbit leukocytes: C_m: 0.97 μF/cm^2	
3 of Table 1.3, modified	R_{app}: 6.5 MΩ	1 of Table 1.5	Human erythrocytes: R_m: 7 Ω cm^2	Lassen and Sten-Knudsen (1968)
Voltage–Current	R_{app}: 240 MΩ	1 of Table 1.5, modified	Human erythrocytes: R_m: 10.6 Ω cm^2	Johnson and Woodbury (1964)
5 of Table 1.1, modified	$R_{inp}=R(\omega)$ $C_{inp}=C(\omega)$ 10 Hz to 100 kHz	Eq. 1.83	Beef erythrocytes, lysed: C_m: 0.9 μF/cm^2	Schwan and Carstensen (1957)

change with frequency. Moreover, they determined R_i with the relation:

$$(1-R_{\infty}^{*}/R_e)(R_e/R_i-1)/(1-R_{\infty}^{*}/R_i)=\beta\rho'/(1-\rho') \qquad (5.31)$$

where β is a tabulated constant (Fricke, 1924) that depends on the ratio of the two major axes of an ellipsoid, and R_∞ was obtained by extrapolation of the high-frequency values of R_{inp}.

One year later, McClendon (1926) reported impedance measurements on beef erythrocytes at frequencies between 300 Hz and 1 MHz. Although considerable attention was given to the technical aspects of the measurements, their interpretation in terms of cell constants is only of historical interest. Fricke (1933) extended the upper frequency limit of his measurements to 16.4 MHz using a resonance method to cover the range from 2 MHz up, see circuit 3 of Table 1.1. Lecithine was used to make the rabbit erythrocytes spherical. This eliminated geometrical complications, and a form factor of 2 could be employed in the suspension equation. In addition, it provided a more precise extrapolation to the infinite frequency resistance. Therefore, Eq. 5.1 could be employed with $\sigma=1/R_{inp}^{*}$ to calculate R_i. The power factor α, calculated with Eq. 5.14, was found to be nearly constant, indicating a complex impedance locus with a suppressed center. Moreover, it was found that rabbit red cells hemolyzed with water retained the impedance properties of the membrane, whereas these were not distinguishable from the incubating medium when the cells were lysed with saponin. Therefore, the membrane of the red cell ghosts retained its insulating properties, whereas hemolyzing with saponin led to destruction of the cell membranes. Using the same approach, Fricke and Curtis (1934a) published a value of 140 Ω cm for the cytoplasmic resistivity of erythrocytes of sheep, rabbit, and chicken. This value is lower than those reported previously by Fricke and Morse (1926), and the difference was attributed to a more precise value of R_∞, since the frequency range was extended to 16 MHz. Along the same lines, Fricke and Curtis (1935) investigated the impedance of rabbit leukocytes. Volume concentrations between 10 and 42% were used in a buffered 0.95% NaCl solution. Cytoplasmic resistivity and membrane capacitance were determined, and their values were comparable to those found for the erythrocytes (Table 5.2). Assuming a dielectric constant of 3 for the membrane, its thickness was calculated to be 27 Å.

As already stated in Sec. 5.1, Velick and Gorin (1940) extended Fricke's analysis of the geometry of erythrocytes to ellipsoids with three different axes. They verified their theory with duck erythrocytes. This experimental model was chosen because it approximates an ellipsoid more closely than mammalian erythrocytes. Since these authors were interested in the theory of conductance measurements in colloidal solutions to determine the shape of the suspended particles, the measurements were restricted to frequencies where the erythrocytes could be considered as nonconducting ellipsoids. Therefore, they simplified Eq. 5.1 ($\sigma_2 = 0$) to obtain:

$$\rho' = \left[(R^*_{\text{inp}}/R_e) - 1\right] / \left[(R^*_{\text{inp}}/R_e) - 1 + f\right] \tag{5.32}$$

where the form factor f is related to the quantity γ in Eq. 5.21 by:

$$\gamma = 1/(f-1) \tag{5.33}$$

The new form factor f was tabulated for spheres, ellipsoids, oblate ellipsoids, and rods. For each geometry, different axial ratios and orientations in the electric field were considered. Since R^*_{inp} and ρ' can be obtained experimentally, f can be calculated from Eq. 5.32. Alternatively, if a suspension is first measured at a high volume concentration and then diluted consecutively, one obtains:

$$v/v_D = \rho'_D \left\{1 + f/\left[(R^*_{\text{inp}}/R_e) - 1\right]\right\} \tag{5.34}$$

where v_D is the total volume to which 1 ml of the original suspension has been diluted at maximum dilution and v is the total volume containing 1 ml of the original suspension at any dilution. Since v/v_D and $R^*{}_{\text{inp}}/R_e$

can be determined experimentally, it follows from Eq. 5.34 that ρ'_D and f can be determined from a plot of v/v_D versus $1/[(R^*_{\text{inp}}/R_e)-1]$, which gives a straight line with a slope of $\rho'_D f$ and an intercept ρ'_D with the ordinate. The values for f so determined corresponded to those determined from the measured ratio of the axes and the tabulated values of f. It is now possible to use Eq. 5.32 to determine ρ'.

Gougerot and Foucher (1972) reinvestigated the problem of why in erythrocyte suspensions the phase angle is usually found to be less than 90°. They used a bridge technique together with a conductivity cell where the electrode distance could be varied (see Appendix 1, Fig. A1.1c). Measurements were performed between 50 kHz and 3 MHz. The authors investigated the behavior of the phase angle as a function of different experimental conditions: (1) the influence of sedimentation time on the complex admittance locus, (2) the influence of an increased volume concentration of cells suspended in plasma, and (3) the influence of a change in the suspending medium from plasma to isotonic NaCl when ρ' was kept constant. Interesting results, represented in the complex admittance plane, show: (1) an initial decrease of the phase angle during sedimentation and then a return to control values (Fig. 5.5a), (2) an increase in phase angle when the erythrocytes, suspended in plasma, are diluted (Fig. 5.5b), and (3) an increase in phase angle by changing the suspending medium from plasma to isotonic NaCl (Fig. 5.5c). The results are interpreted as being due to an aggregation of erythrocytes during sedimentation and a disaggregation of the rouleau when the erythrocytes are suspended in NaCl instead of plasma. Such an interpretation attributes the change in the phase angle to a change in the form factor. On the other hand, a possible influence of the plasma proteins on the results of Fig. 5.5b, c was not considered. However, in their theoretical analysis, the authors did not consider the results of Velick and Gorin (1940) and Schwan (1957), which would probably have permitted a deeper analysis of the results.

Fricke (1953b) used his earlier approach (Fricke, 1925a) to calculate the static membrane capacitance to demonstrate the proportionality between the dielectric constant and the volume fraction of a suspension of ellipsoids:

$$\varepsilon_{\text{inp}} = \alpha' g(\varepsilon_m/d)\rho' + \varepsilon_s \sigma_{\text{inp}}/\sigma_s \tag{5.35}$$

where ε_{inp} is the dielectric constant of the suspension, α' is half the length of the largest axis, g is a tabulated value related to the geometry of the ellipsoids, d is the thickness of the membrane, and ε_s is the dielectric constant of the suspending medium. Since α' and g are constant, the ratio of the dielectric constant of the membrane (ε_m) and its thickness can be

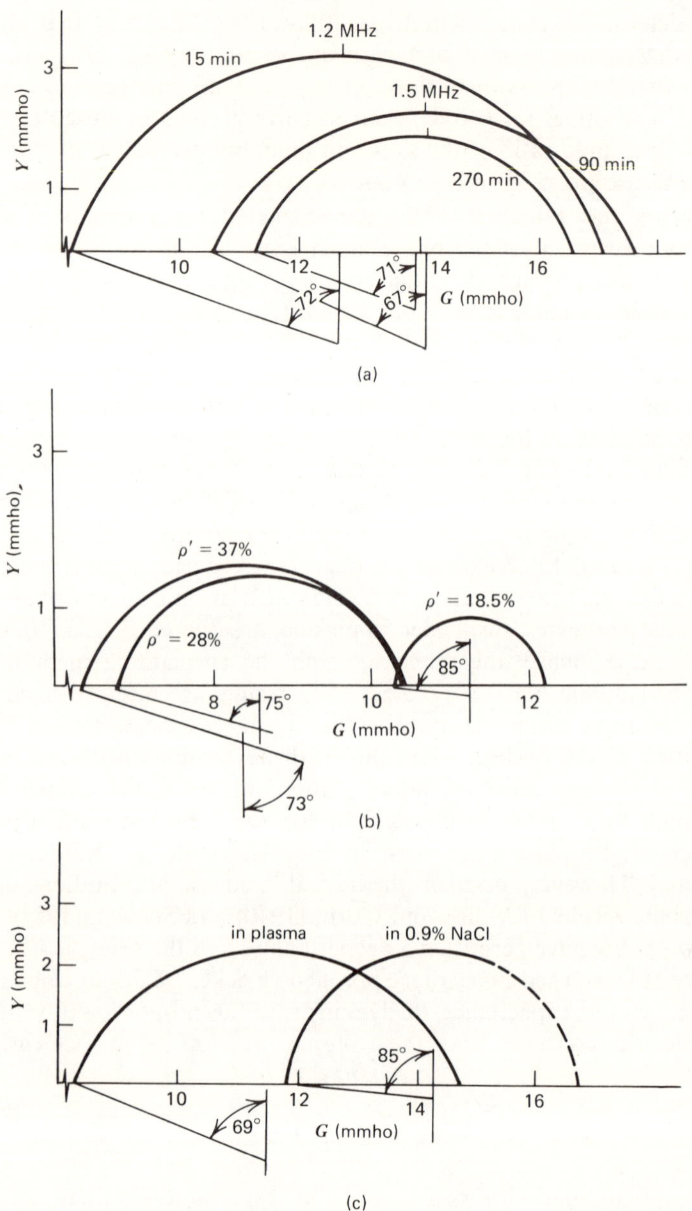

Figure 5.5. Loci in the complex admittance plane of erythrocyte suspensions under different experimental conditions. (a) Changes of phase angle as a function of time (sedimentation). (b) Changes in phase angle as a function of volume concentration. (c) Effect of a change in suspending medium (plasma to 0.9% NaCl) on phase angle. Modified after Gougerot and Foucher (1972). By permission.

obtained from the slope of a diagram $\varepsilon_{inp}=f(\rho')$. In the same year, Fricke (1953c) reported graphs of ε_{inp}, the dielectric constant of the suspension, versus ρ', obtained for erythrocytes and leukocytes of different species (see Table 5.2). He obtained the C_m and the ratio d/ε_m from the slope of the ε_{inp} versus ρ' relation extrapolated to $\rho'=0$ (Fig. 5.6). The measurements were made at 16 kHz in a region where ε_{inp} is constant, just between the α and β dispersions. The graphs depend on the shape of the cells, but they are independent of the absolute cell size. As can be seen from the figure, rabbit and dog erythrocytes (which have a similar shape) gave identical curves, whereas the graph obtained for spherical rabbit erythrocytes is different. The conclusion of these experiments is that the membrane capacitance of blood cells can be related to the membrane thickness d by the relation:

$$d=10\varepsilon_m \tag{5.36}$$

Bothwell and Schwan (1956) measured the impedance of tightly packed beef erythrocytes ($\rho'=80$–90%) between 10 Hz and 200 kHz. Extending their measurements to such low frequencies required special precautions to control errors due to the electrode polarization artifacts (Schwan, 1951). Therefore, a measuring cell with variable electrode distance was used (see Appendix 1, Fig. A1.1c). A substitution method permitted the elimination of distributed capacitances and leakage conductances. These precautions, together with the use of an elaborate bridge (Schwan and Sittel, 1953),

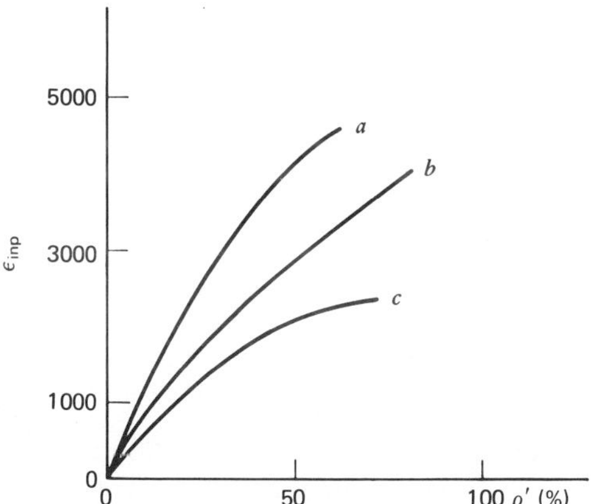

Figure 5.6. Relation between dielectric constant and volume concentration in different suspensions of blood cells. (a) Leucocytes (rabbit). (b) Erythrocytes (rabbit and dog). (c) Lecithinated rabbit erythrocytes (spherical). Modified after Fricke (1953c).

resulted in a precision of $\pm 0.0005\%$ for the resistance measurements (above 10 Hz) and a precision of $0.5/f \pm 0.5\%$ for the capacity measurements, the frequency being expressed in kHz. The authors found no frequency dependence of the capacitance between 500 and 5000 Hz, as had been reported earlier for rabbit blood (Fricke and Curtis, 1935). The observed increase of the conductivity above 1 kHz and the decrease of the dielectric constant above 10 kHz was attributed to a dispersion originating in the cytoplasm. Similar results were obtained by Bothwell, Schwan, and Wiercenski (1954) for packed beef erythrocytes.

The theoretical implications of a frequency-independent membrane capacitance were later summarized by Schwan (1957): (1) The capacitance of the plasma membrane is frequency independent. If a nondetectable frequency dependence is assumed, the phase angle will be at most 0.4° less than 90°. (2) Considering the uncertainty of the measured values of C_m, the membrane resistance at 1 kHz should be higher than 20 kΩ cm². (3) Based on the maximum permissible error in the resistance measurements, one can calculate a change in membrane resistance with frequency of less than 10^4 Ω/cm². In addition, if one assumes an infinite R_m at zero frequency, the membrane resistance at 1 kHz will be larger than 10 kΩ cm². Conclusions 2 and 3 are based on a set of formulas derived from relaxation theory:

$$\left.\begin{aligned} \varepsilon_r &= q\rho' a C_m/(4\varepsilon_v) \\ \sigma &= (1-3\rho'/2)/R_e + q\rho' a/4R_m \end{aligned}\right\} \tag{5.37}$$

A condition for the applicability of Eqs. 5.37 is that R_m/a is much larger than R_e and R_i, a condition generally fulfilled for biological cells.

It is interesting to compare these estimates for R_m with those obtained later from microelectrode measurements (Lassen and Sten-Knudsen, 1968). They used a piezoelectric device to drive a microelectrode into human erythrocytes. Injecting a square pulse through a single microelectrode, they measured an apparent resistance, which they then corrected for the change in microelectrode resistance due to the difference in resistivities of the intracellular and extracellular medium. This resulted in a membrane resistance of 7 Ω cm².

Schwan and Carstensen (1957) and Schwan (1957) investigated the dielectric properties of beef erythrocytes lysed with distilled water. They obtained an unchanged membrane capacitance and an estimate for R_m of 100 Ω cm² above 100 kHz. The estimate was based on the assumption of an infinite membrane resistance at zero frequency. The study also confirmed an earlier report that the membrane capacity remains constant after lysis by distilled water (Fricke, 1933). This indicated that the membrane structure of the ghost is not considerably altered by lysis in distilled water, in contrast to lysis with saponin, which leads to a breakdown of the membrane and disappearance of the β dispersion (Fricke and Curtis, 1935).

There is an apparent discrepancy between Schwan's estimation of R_m and that obtained by Lassen and Sten-Knudsen. The different experimental procedure alone cannot explain the results. Inspection of the records of Lassen and Sten-Knudsen shows that the pulses used were of short duration (< 300 μs) and the electrotonic potential did not reach a steady state. Consequently, the values for the apparent resistance were probably obtained at a time when the transient had not leveled off, and the apparent resistance was so underestimated. On the other hand, values published for the membrane resistance of four human erythrocytes with extracellular electrodes resulted in a mean value for R_m of 10.6 Ω cm^2 (Johnson and Woodbury, 1964). The pulses used were of longer duration (60 ms), and again the electrotonic potential did not reach a steady state, but it is not very probable that this error can fully explain the discrepancy between the results. The most probable explanation for the discrepancy may be Schwan's assumption of an infinite membrane resistance at zero frequency.

In the late 1930s, measurements of the cytoplasmic resistivity of erythrocytes of several species were available (see Table 5.2 in this book and Table 4 in Rajewski, 1938). The values were quite different, and therefore Pauly (1959) undertook a comparative study of R_i in several species. Working with an improved bridge method, measurements could be performed between 500 kHz and 250 MHz. It was found that at 90 MHz the membrane was completely short-circuited and the conductivity of the internal medium was independent of the frequency. This was checked by low-frequency measurements on cell suspensions where the membranes had been destroyed by toluene or saponin. The internal resistivity of erythrocytes of seven species was found to be twice the value expected for an electrolyte solution of the same ionic strength.

Later, Pauly and Schwan (1966) continued this study; they were particularly interested in finding an explanation for the apparent low ionic mobility in cytoplasm. The measuring method was similar to that of Pauly (1959), but now the authors introduced a dilution method to determine experimentally the difference between actual cytoplasmic resistivity and the theoretical conductivity (ions of the cytoplasm considered as being in aqueous solution). The cytoplasmic resistivity was determined at 90 MHz, where all membrane components are short-circuited. The theoretical conductivity was determined by diluting packed erythrocytes consecutively in distilled water and by measuring the limiting conductivity at high dilutions (Fig. 5.7). With this method, they obtained 13.5–14.0 mmho/cm for the theoretical and 5.18 mmho/cm for the actual cytoplasmic conductivity of human erythrocytes. These values differ by a factor of 2.7. Taking into

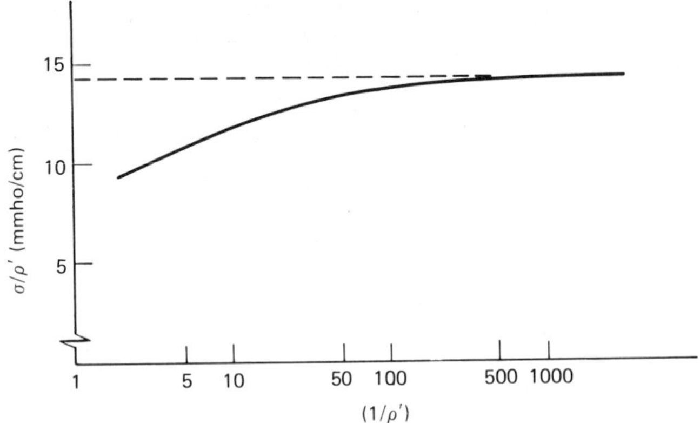

Figure 5.7. Theoretical conductivity of erythrocytes as determined by a dilution method. After leveling off, the drawn-out curve gives the value of the theoretical conductivity. The curve was corrected for volume fractions of ghosts and hemoglobin. Modified after Pauly and Schwan (1966), by copyright permission of The Rockefeller University Press.

account that hemoglobin binds about 30% nonsolvent hydration water brings the ratio down to 2.4, and correcting for the dependence of the equivalent conductance on concentration lowers it further to 2.0, but the fact remains that the actual conductivity of the cytoplasm is only one-half of the theoretical conductivity, and this difference has to be explained.

Binding of ions could be ruled out, but there remain two possible causes of the reduced mobility: (1) an electrostatic interaction between the charged cations and the charged groups of the hemoglobin molecule and (2) a hydrodynamic interaction resulting from a viscous drag on the small ions caused by the proximity of the surface of the macromolecules. To elucidate further the relative contribution of electrostatic and hydrodynamic interaction to the reduced ionic mobility of the cytoplasm, Lachance (1972) and Lachance and Schanne (1972), using the same methods as Pauly and Schwan (1966), worked with erythrocytes whose volume was increased and reduced by an osmotic challenge. Simultaneously, the extracellular pH was varied between 6 and 8 to alter the number of charges on the hemoglobin molecule. It was found that the change in pH did not influence the ratio between measured and theoretical conductivity, and therefore the role of electrostatic interferences was considered negligible. Taking into account recent results about the availability of bound water as solvent, the contribution of hydrodynamic interaction to the reduction of ionic mobility has been considered equal to that of ions immobilized in the water shell of the hemoglobin molecules.

Krupa et al. (1972) developed a method for determining the cytoplasmic resistivity in suspensions of erythrocytes that is independent of the erythrocyte's volume concentration. It is based on an equation relating the conductivity of a suspension to the cell constants of the suspended cells (Terlecki and Fiutak, quoted as personal communication in the above-mentioned report):

$$\frac{\sigma_{\text{inp}}}{\sigma_e} = 1 - \rho' \left[\frac{3(\sigma_e - \sigma_i)}{2\sigma_e + \sigma_i} + \frac{d}{a}(\sigma_m - \sigma_i) \frac{9(2\sigma_m - \sigma_i)\sigma_e}{(2\sigma_e + \sigma_i)^2 \sigma_m} \right] \tag{5.38}$$

where d is the thickness of the cell membrane and the subscripts have the usual meaning. In contrast to the suspension equation, Eq. 5.38 predicts a linear relationship between $\sigma_{\text{inp}}/\sigma_e$ and ρ'. The authors derived an equation

$$\sigma_i = \sigma_e \frac{\sigma_\infty - \sigma_0}{\sigma_e - \sigma_0} \tag{5.39}$$

which relates the conductivity of the cytoplasm with the conductivity of the suspending medium and the limiting conductivities of the suspension. The method was validated with suspensions of human erythrocytes ($\rho' =$ 40–67%), and values from 185 to 190 Ω cm were reported for the cytoplasmic resistivity.

5.4 SUSPENSIONS OF CELLS OTHER THAN BLOOD CELLS

For practical reasons, impedance measurements were performed on readily available cells or on cells with a simple geometry. Besides the extensive work performed with blood cells and marine eggs, there are some interesting contributions to the field obtained with other cell types.

Fricke and Curtis (1934b) reported impedance measurements on suspensions of yeast cells in the range of 250 Hz to 16 MHz (see Table 5.3). The cells were suspended in media of NaCl concentrations between 0.01 and 1.0%. There was a strong increase in capacity at low frequencies, but the overall picture of the dielectric behavior of these cells resembled that of red blood cells. At 16 Hz, a value for C_m of 0.6 μF/cm^2 was obtained. Several years later, Ruhenstroth-Bauer and Zeininger (1956) performed measurements on liver cells, their nuclei, mitochondria, and microsomes. The cells of rat liver were mechanically disintegrated, and the subcellular particles were obtained by fractionated centrifugation and subsequent resuspension. Measurements of $Z(\omega)$ made between 12 kHz and 11 MHz showed a β-dispersion with a characteristic frequency between 3 and 4 MHz. This is interpreted as evidence that all the investigated structures are surrounded by a closed membrane.

Pauly (1963) reported dielectric measurements at 1 kHz and from 500 kHz to 200 MHz in suspensions of ascites tumor cells suspended in

Table 5.3 Electrical Constants of Cells and Subcellular Particles

Method	Measured Quantities	Model	Derived Quantities	Reference
5 of Table 1.1, modified	$R_{\text{inp}} = R_p(\omega)$ 250 Hz to 16 MHz	3 of Table 1.5	Yeast C_m: 0.6 μF/cm^2 ρ': 63%	Fricke and Curtis (1934b)
5 of Table 1.1, modified	$R_{\text{inp}} = R_p(\omega)$ 1 kHz, 500 kHz to 200 MHz	3 of Table 1.5	Ascites cells C_m: 1.7 μF/cm^2 R_i: 73.6 Ω cm	Pauly (1963)
5 of Table 1.1	$C_{\text{inp}} = C(\rho')$ 500 Hz to 16.4 MHz	Eq. 5.35	*Saccharomyces cerevesiae* C_m: 0.78 μF/cm^2	Fricke (1953c)
5 of Table 1.1 modified	$C_{\text{inp}} = C_p(\omega)$ 50 Hz to 150 MHz	Eq. 5.35	*E. coli* C_m: 0.7 μF/cm^2	Fricke, Schwan, Li, and Bryson (1956)
5 of Table 1.1, modified	$R_{\text{inp}} = R_p(\omega)$ $C_{\text{inp}} = C_p(\omega)$ 500 kHz to 250 MHz	3 of Table 1.5	Liver mitochondria C_m: 0.5 μF/cm^2	Pauly et al. (1960)
5 of Table 1.1, modified	$R_{\text{inp}} = R_p(\omega)$ $C_{\text{inp}} = C_p(\omega)$ 500 kHz to 250 MHz	3 of Table 1.5, modified	PPLO R_i: 91 Ω cm (σ_e: 66 Ω cm) R_i: 89 Ω cm (σ_e: 73 Ω cm) R_i: 125 Ω cm (σ_e: 187 Ω cm) C_m: 1.3 μF/cm^2	Schwan and Morowitz (1962)
5 of Table 1.1, modified	$R_{\text{inp}} = R_p(\omega)$ $C_{\text{inp}} = C_p(\omega)$ 500 kHz to 200 MHz	3 of Table 1.5	*M. lysodeikticus* R_i: 67 Ω cm C_m: 1 μF/cm^2	Pauly (1962)
5 of Table 1.1, modified	$R_{\text{inp}} = R_p(\omega)$ $C_{\text{inp}} = C_p(\omega)$	3 of Table 1.5, modified	*E. coli* R_i: 333 Ω cm ρ_w: 333 Ω cm Micrococcus, nonspecified Fresh R_i: 125 Ω cm ρ_w: 91 Ω cm Aged R_i: 125 Ω cm ρ_w: 62.5 Ω cm	Carstensen (1967)
5 of Table 1.1, modified	$R_{\text{inp}} = R_p(\omega)$ $C_{\text{inp}} = C_p(\omega)$	3 of Table 1.5, modified	Rod outer segments, frog C_m: 1.54 μF/cm^2 C_m: 0.91 μF/cm^2 High-frequency dispersion C_m: 1.54 μF/cm^2 Distributed parameters C_m: 0.91 μF/cm^2	Falk and Fatt (1968a)

Krebs solution with 5% of gelatine added. The measurements were originally expressed as parallel resistance and capacitance, and they were converted into conductance and dielectric constant as a function of frequency (Fig. 5.8). While the conversion of R_p into σ_p is straightforward, converting C_p into values of ε_r is less obvious. The conversion formula can be obtained by combining Eqs. A1.4 and A1.6 and solving for ε_r:

$$\varepsilon_r = kC_p/\varepsilon_v \tag{5.40}$$

Figure 5.8 represents a relaxation typical for heterogenic dielectrics. This was interpreted as follows: The suspended ascites cells were considered as a suspension of spheres each having a dielectric constant ε_i and conductivity σ_i. These spheres are surrounded with a shell characterized by ε_m and σ_m, and the suspending medium is characterized by ε_e and σ_e. Such a model has been rigorously treated by Pauly and Schwan (1959). However, in the case of the ascites cells, the membrane is very thin and is considered to be an insulator. Therefore, the time constant resulting from the membrane can be neglected, and the system reduces to one with a single relaxation time constant and can therefore be described by Eq. 1.83. Using the suspension equation with $\gamma=2$, one obtains the following relations:

$$\tau_r = 1/\bar{\omega} \approx aC_m(\sigma_i + 2\sigma_e)/2\sigma_i\sigma_e \tag{5.41}$$

where σ_i and σ_e are the conductivities of the cytoplasm and the medium, respectively. Moreover, one can define the input values for the dielectric constants and conductivities at very low and very high frequencies as

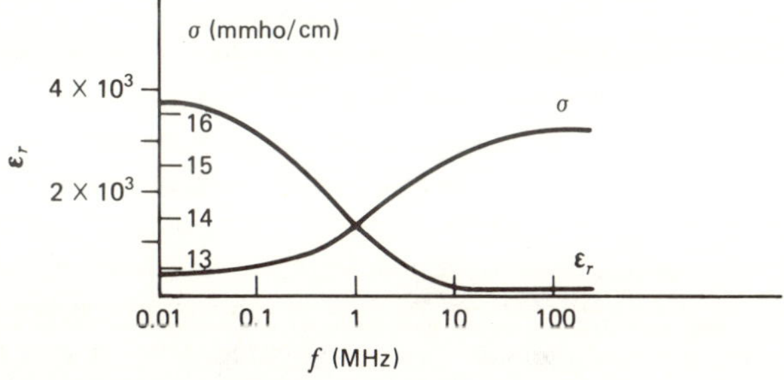

Figure 5.8. Change of ε_r and σ with frequency in a suspension of ascites cells: $\rho' = 13\%$, $T = 28°C$. Modified after Pauly (1963), by permission.

follows:

$$\varepsilon_0 \approx \frac{1-\rho'/2}{1+\rho'}\varepsilon_e + \frac{9}{4\varepsilon_v}\left(\frac{\rho'}{1+\rho'}\right)aC_m \tag{5.42}$$

where

$$C_m = \varepsilon_m \varepsilon_v / a \tag{5.43}$$

and a represents the thickness of the membrane. In addition, one obtains:

$$\varepsilon_\infty \approx \varepsilon_e \left[\frac{(1+2\rho')\varepsilon_i + 2(1-\rho')\varepsilon_e}{(1-\rho')\varepsilon_i + (2+\rho')\varepsilon_e}\right] \tag{5.44}$$

$$\sigma_0 \approx \sigma_e \frac{1-\rho'}{1+\rho'/2} \tag{5.45}$$

$$\sigma_\infty \approx \sigma_e \left[\frac{1+2\rho'(\sigma_i-\sigma_e)/(\sigma_i+2\sigma_e)}{1-\rho'(\sigma_i-\sigma_e)/(\sigma_i+2\sigma_e)}\right] \tag{5.46}$$

Since ρ' and σ_e can be determined independent of Eqs. 5.41–5.46, σ_i can be obtained from Eq. 5.46, C_m results from Eq. 5.41 or, independently, from Eq. 5.42. The quantities $\bar{\omega}$, ε_0, ε_∞, σ_0, and σ_∞ can be obtained from Cole-Cole plots of the dielectric constants and the conductances. These plots, which are shown in Fig. 5.9, can be obtained from data measured in the frequency domain. It is interesting to note that although the values on the ordinate of the conductivity plot (Fig. 5.9a) have the dimension of a conductivity, the numerical values were obtained from capacity measurements; the dimensionless values on the ordinate of the plot of the dielectric constant (Fig. 5.9b) were obtained from conductivity determinations. It is also evident from the figure that the characteristic frequencies for the conductivity $\bar{\omega}_\sigma$ is larger than that for the dielectric constant $\bar{\omega}_\varepsilon$, which indicates an anomalous dispersion (see Sec. 1.5.2.4). Pauly based the interpretation of these results on the work of Schwan (1957), who showed that the Cole-Cole plot was quite insensitive to the type of distribution function responsible for the suppression of the center of the admittance locus. Assuming a constant distribution function, Pauly tested whether the experimental distribution of time constants could be explained by the variation of the cell diameters in the investigated cells. It was found that the experimentally determined spectrum of time constants was larger than that expected for the distribution of cell diameters.

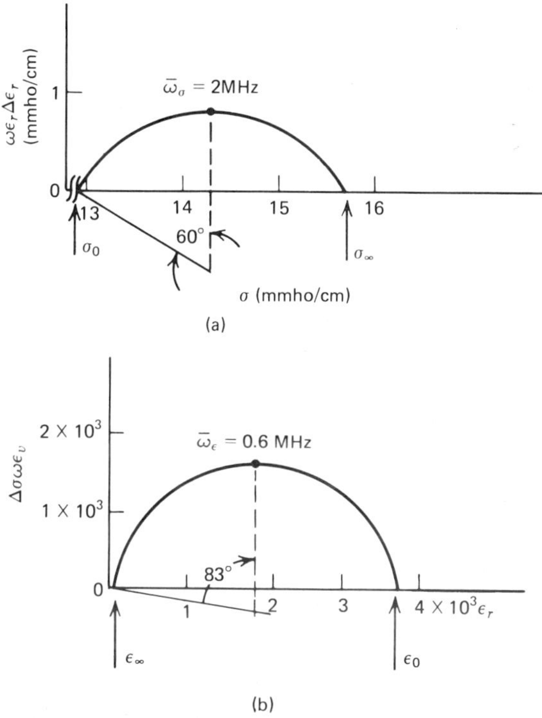

Figure 5.9. Cole-Cole plots of (a) the conductivity and (b) the dielectric constant of a suspension of ascites cells. Modified after Pauly (1963), by permission.

Another interesting result of this study is the value for the cytoplasmic resistivity (Table 5.3), which is very low compared to values obtained for other cell types. The resistivity of the cytoplasm of ascites cells is very close to the theoretical resistivity expected for an electrolytic solution with an ion content similar to that of the ascites cells. This can be explained by the low protein content of tumor cells, which imposes a minimum constraint on ionic mobility in the cytoplasm of ascites cells.

5.5 SUSPENSIONS OF MICROORGANISMS AND SUBCELLULAR PARTICLES

There were some early reports that the conductivity of bacteria was well below that of the suspending medium (Shearer, 1919; Brooks, 1923; Green and Larson, 1922; Zoond, 1927), but systematic investigations of the impedance of subcellular particles and microorganisms began only later.

Fricke (1953c) used an approach discussed in Sec. 5.3 to evaluate the membrane capacitance of fungi (*Saccharomyces cerevesiae*) from the initial slope of the relation $\varepsilon_{inp} = f(\rho')$. The fungi were suspended in 0.25% NaCl solution, and their shape was spherical with a diameter of 5.5 μm. Several years later, Fricke et al. (1956) studied the dielectric properties of suspensions of *E. coli* and found α and β dispersions. The frequency range of the β dispersion was larger than that expected for an uncomplicated dispersion of the Debye-Hückel type. The oblong form and the heterogeneity in size of the bacteria were used to explain the discrepancy. In the range below 2 MHz, the dielectric constant was found to be constant and a value of 0.7 $\mu F/cm^2$ was obtained for the membrane capacitance. Using the same approach as in an earlier investigation of blood cells (Fricke, 1953b), the ratio of membrane thickness to its dielectric constant was determined to be 13 Å, compared to a value of 10 Å (Eq. 5.36) for blood cells. This could indicate either an increased membrane thickness or an increased dielectric constant in the wall of the bacterial membrane.

Pauly et al. (1960) reinvestigated the electrical properties of mitochondrial membranes. Rat liver mitochondria were swollen by incubation in hypotonic KCl to render them spherical. The distribution of their diameters was determined by light microscopy, and the dielectric properties were determined using a bridge circuit operated between 500 kHz and 250 MHz in a conductivity cell shown in Fig. A1.1c (Appendix 1). When the functions $\varepsilon(\omega)$ and $\sigma(\omega)$ were plotted, they showed an anomalous β-dispersion (see Sec. 1.5.2.4). Representing the results as Cole-Cole plots in the admittance plane (see Appendix 2) resulted in semicircles with a depressed center. Considering that the suppression of the center in the Cole-Cole plot can be caused by a variety of distribution functions (Schwan, 1957), Pauly et al. tested whether the following factors were responsible for the observed anomalous dispersions: (1) a distribution of mitochondrial diameters; (2) a distribution of time constants due to a variability of internal conductivity in the mitochondria (Eq. 5.41), which could be caused either by an electrically isotropic or an electrically anisotropic interior of the mitochondria; (3) a constant phase angle of the membrane; and (4) a combination of factors 1–3. The analysis was based on Eqs. 5.41–5.46, which were simplified by assuming $\rho' \ll 1$ and $\varepsilon_i = \varepsilon_e$. The theoretically derived equation for a dispersion caused by a single time constant (Eq. 1.83) can be generalized for an experimentally determined distribution of time constants:

$$\varepsilon_{inp} = \varepsilon_{\infty} + \sum \Delta\varepsilon_{inp} / \left[1 + (\omega\tau_{r(i)})^2\right] \tag{5.47}$$

where $\Delta\varepsilon_{\text{inp}}$ represents the dielectric increment ($\varepsilon_{0i} - \varepsilon_{\infty i}$) associated with the relaxation time constant $\tau_{r(i)}$ of a given individual membrane.

The possibility that the depressed center in the admittance plane depends on a distribution of the diameters of the mitochondria was tested by the following approach. Assuming the membrane capacity of the mitochondria to be static, one can use the superposition principle to evaluate the contribution of the mitochondria in a given size group to the total experimentally obtained dielectric increments. The problem is to evaluate the dielectric increment $\Delta\varepsilon_{\text{inp}}$ caused by each size group of mitochondria using Eq. 5.47. Considering Eqs. 5.41–5.46, this requires the evaluation of the frequency-independent parameters ε_0, ε_∞, σ_0, σ_∞, σ_e. In the calculations a_i represents the mean diameter of a size group (Fig. 5.10a) and ρ_i' is the volume concentration of the mitochondria within size group i. This parameter can be obtained from the experimentally determined distribution of mitochondrial radii (Fig. 5.10a). Moreover, a mean value for C_m can be obtained from Eq. 5.41. Once these parameters are known, the evaluation of the characteristic frequency $\bar{\omega}$ and the relaxation time constant τ_r is possible, and consequently the theoretical input dielectric constant ε_{inp} for the given size group can also be obtained from Eq. 5.47. Finally, with these data, the contribution of each size group to the total dielectric dispersion is known and can be represented graphically as shown in Fig. 5.10b.

Further analysis of Fig. 5.10b reveals that the theoretically expected distribution of time constants due to the distribution of diameters is smaller than that found experimentally. Similarly, the observed spectrum of time constants cannot be explained by a variation of σ_i in an electrically isotropic interior of the mitochondria. However, good agreement between experimental and theoretical values was obtained, considering a two-time-constant system resulting from an electrically anisotropic interior of the mitochondria caused by the presence of the cristae.

Schwan and Morowitz (1962) measured the dielectric behavior of suspensions of the pleuropneumonia-like organism A 5969 (PPLO) at frequencies between 500 kHz and 250 MHz in a conductivity cell similar to that shown in Fig. A1.1c (see Appendix A1). The particles are spherical, and the Cole-Cole plot of these measurements showed a phase angle of 90°. The measurements permitted the calculation of C_m and R_i. The value of the membrane capacitance (Table 5.3) was comparable to that found in other cells, and R_i changed little with a change in ionic strength of the suspending medium.

In the same year, Pauly (1962) reported the dielectric behavior of *Micrococcus lysodeikticus*. The measuring procedures were similar to those in previous papers from the same group. This bacterium is known to have

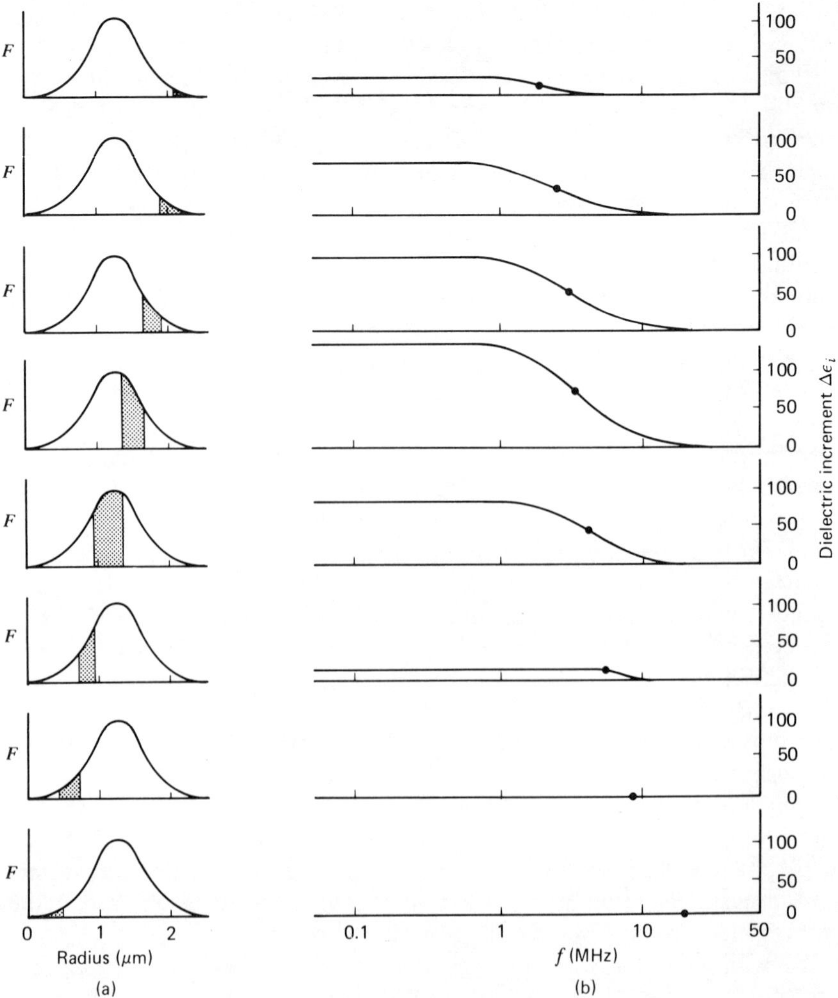

Figure 5.10. (a) Experimentally determined distribution of radii of swollen mitochondria. The shaded regions correspond to the eight size groups analyzed. (b) Dielectric increment contributed by the different size groups. The points represent the characteristic frequency. It can be seen that the characteristic frequency increases with smaller mitochondria, and at the same time the contribution of the smaller size groups to the dielectric increment becomes negligible. Modified after Pauly et al. (1960), by copyright permission of The Rockefeller University Press.

a rather thick cell wall (500 Å) that can be enzymatically digested by lysozyme. After digestion, a spherical body remains surrounded by an osmotic barrier, the protoplast. Measurement of σ and ε as a function of frequency yielded the same membrane capacitance before and after digestion. This led to the conclusion that the structure responsible for the membrane capacitance was the osmotic barrier surrounding the protoplast. Also, the conductivity of the interior of the protoplast was found to be higher (15 mmho/cm) than in most other cells, due to the high ionic concentration of the cell interior. Dilution experiments have shown that the ratio of theoretical and measured conductivity is about 2 (Pauly and Schwan, 1959).

Some years later, the group of Carstensen published a series of papers investigating the dielectric properties of several types of bacteria (Carstensen et al., 1965; Carstensen, 1967; Carstensen and Marquis, 1968; Einolf and Carstensen, 1969). The aim of the study was to interpret the electrical measurement by a geometrical model. Carstensen et al. (1965) used low-frequency measurements (1.6 kHz, well below the β dispersion), to develop a geometrical model of the bacterial cell and to estimate the surface charge of the cell wall. For their measurements, they used a strain of *E. coli* and *M. lysodeikticus*. Using Eq. 5.1, they determined σ_2 as a function of the conductivity of the suspending medium σ_1. In this approach, σ_2 represents the effective conductivity of a bacterium, but this does not imply that the bacterium has a homogeneous electric conductivity. The volume concentration ρ' was determined independently using radioactive dextran. The measurements showed an effective homogeneous conductivity of *M. lysodeikticus* larger than that of *E. coli*. Moreover, there was a proportionality between σ_2 and σ_1 at higher values of σ_1. Such a result could be attributed to a surface charge at the exterior of a nonconducting sphere (Fricke and Curtis, 1936; Schwan et al., 1962). However, a control study of the electrophoretic mobility of the two bacterial strains ruled out this possibility. An attempt to attribute the effective conductivity to the resistance of a thin membrane surrounding a cell with a high cytoplasmic conductivity (Pauly, 1962) led to unrealistically low values for the membrane resistance (30–200 mΩ cm^2). Therefore, the effective conductivity was tentatively attributed to a spherical model consisting of a nonconducting core surrounded by an outer conducting shell, the cell wall. At low frequencies, the cell interior enclosed by a nonconducting membrane appears, as a whole, like a nonconducting core. It was assumed that the experimentally observed effective conductivity was related to mobile counterions situated in the cell, the presence of which is due to fixed charges in the cell wall. Assuming that the cell wall is in equilibrium with the extracellular medium, one obtains for the ion concentrations in the cell wall:

$$C_w^+ + C_w^- + C_w^f = 0 \tag{5.48}$$

where C_w^+ and C_w^- are co- and counterion concentrations and C_w^f is the concentration of fixed charges, with the subscript w standing for cell wall. If the chemical potential is the same in the extracellular medium and within the cell wall, we may write:

$$C_w^+ C_w^- = C_o^2 \tag{5.49}$$

where C_o is the total ion concentration in the extracellular medium. Now we can combine Eqs. 5.48 and 5.49 and write for the conductivity of the cell wall σ_w:

$$\sigma_w \propto u_w^+ C_w^+ + u_w^- C_w^- = \frac{1}{2} C_w^f \left[(u_w^- - u_w^+) + (u_w^- + u_w^+) \sqrt{1 + (2C_o/C_w^f)^2} \right] \tag{5.50}$$

This relation can be simplified if the cation mobility within the cell wall u_w^+ is comparable to the anion mobility u_w^-:

$$\sigma_w \propto u_w C_w^f \sqrt{1 + (2C_o/C_w^f)^2} \tag{5.51}$$

where u_w is the ionic mobility of the counterions in the membrane. It can be seen from the equation that for a small conductivity in the extracellular medium ($C_o \to 0$), σ_w is constant, and for a high conductivity ($C_o \to \infty$), σ_w depends linearly on C_o. This has been experimentally confirmed.

Equation 5.51 can now be used to estimate C_w^f, using an equation formally equal to Eq. 5.1 that reduces to:

$$\sigma_2 = \sigma_w (1 - \rho')/(1 - \rho'/2) \tag{5.52}$$

provided the conductivity of the cell interior is large compared to the conductivity of the membrane. Combining Eqs. 5.51 and 5.52, we obtain an expression of the form:

$$\sigma_2 = A \sqrt{1 + (2C_o/C_w^f)^2} \tag{5.53}$$

where the constant A depends on the effective mobility of the ions, the magnitude of the fixed charge concentration, and the geometry of the cell. Equation 5.53 can now be fitted to the measured values of σ_2, which allows us to evaluate C_o, using the condition $(2C_o/C_w^f)=1$. We can now estimate the values of the fixed charge concentration: 0.07 eq/liter cell wall H_2O for *E. coli* and 0.2 eq/liter cell wall H_2O for *M. lysodeikticus*.

In a subsequent paper, Carstensen (1967) tested further the validity of his model for a bacterial cell. Although it was previously found that attributing the effective conductivity of the bacterial cells to a membrane resistance led to very low values for this quantity, there was no experimental proof that the membrane resistance was actually higher. Measurements were performed on *E. coli* and on a nonspecified type of micrococcus between 1 and 200 MHz. It was found that, assuming the appropriate electrical cell constants, the results could be fitted either to a sphere with a conducting interior or a sphere with a nonconducting core surrounded by a conducting shell. Only with the latter model, however, was it possible to describe meaningfully the finding of an increased conductivity with age in the micrococcus (Table 5.3) and therefore this model was retained.

Carstensen and Marquis (1968) determined the conductivity of the cell wall by performing dielectric measurements on isolated bacterial walls (*M. lysodeikticus*). Their technique was the same as the one used in the paper discussed above. Using an equation reported by Fricke (1953c), they derived a limiting expression for the conductivity of the suspended phase (provided the dielectric constant is the same in the solvent and the dispersed phase):

$$\sigma_w^\infty = \sigma_w \rho' + \sigma_e(1-\rho') \tag{5.54}$$

which permitted the calculation of the effective conductivity of the cell walls. σ_w^∞ was determined at frequencies between 100 and 200 MHz. Using an approach similar to that of Carstensen et al. (1965) and taking into account that with increasing concentration of the suspending medium the cell walls shrank from a volume ρ_0' to ρ', they rewrote Eq. 5.53 as:

$$\sigma_w = U_w C_w^{fo}\,(\rho_0'/\rho')\left[1+(2C_o\rho'/C_w^{fo}\rho_0')^2\right]^{1/2} \tag{5.55}$$

where u_w is the ionic mobility of the fixed charges in the wall and the index o indicates parameters determined at low ion concentration in the outside medium. Equation 5.55 could be fitted to the experimentally obtained function σ_w (σ_e), and from this a fixed charge concentration of the cell wall C_w^{fo} of 90 meq/liter was obtained, assuming the u_w of sodium was equal to that in aqueous solution. In another approach, the charges of the cell walls were directly titrated and it was shown that there was qualitative agreement between net charge density obtained with titration and C_w^f determined from conductivity measurements.

Along the same line of research, Einolf and Carstensen (1969) analyzed the dielectric properties of *M. lysodeikticus*. The cell wall was digested with lysozyme as described by Pauly (1962). Pauly had postulated that the structure responsible for the dielectric phenomena was mainly the osmotic barrier surrounding the protoplast. Einolf and Carstensen carried out measurements in a higher frequency range (20 Hz to 200 MHz) than Pauly using three different bridges. With this approach, the following results were obtained: (1) The results of Pauly (1962) were confirmed, (2) the model for a bacterial cell consisting of a conducting shell surrounding a nonconducting core was confirmed, and (3) a low-frequency (α) dispersion was discovered that was tentatively explained by the theory of Schwarz (1962) for the low-frequency dielectric dispersion of colloidal particles. Moreover, the appendix to Einolf and Carstensen's paper contains a useful demonstration of the interrelation of electrophoretic mobility and the homogeneous effective conductivity of a suspended particle.

Zivy, Lévy, and Hènon (1968) reported measurements of the input resistivity of suspensions of several bacterial strains as a function of temperature. However, no attempt was made to relate changes in the input resistivity to the electrical cell constants of the microorganisms.

Schwan (1965) applied microwave techniques to hemoglobin to obtain information about the binding of water by macromolecules. Measurements were made between 0.5 and 1000 MHz. At the lower frequencies bridge methods were used, and above 150 MHz transmission-line techniques were applied (Schwan and Li, 1955). Solutions of either hemoglobin or erythrocytes lysed with toluene were used at concentrations of 5–20%. It was found that above 20 MHz a dispersion persisted; this was attributed to an additional relaxation mechanism caused by water bound to the macromolecules. Modifying Eq. 1.12, the following expression was used:

$$\frac{\varepsilon_{\text{inp}}-\varepsilon_1}{\varepsilon_{\text{inp}}-2\varepsilon_1}=\rho'\frac{\varepsilon_2-\varepsilon_1}{\varepsilon_2-2\varepsilon_1} \tag{5.56}$$

where ε_1 is the dielectric constant of the suspending electrolyte and ε_2 the effective dielectric constant of the macromolecule; the subscript 2 represents the form factor for spherical particles. Equation 5.56 was fitted to the experimentally obtained dispersion, and hydration values of the hemoglobin molecule between 0.3 to 0.4 were obtained.

Using another approach, Falk and Fatt (1968a, b) investigated the electrical properties of a completely different experimental model: the rod outer segments from frog retinae. These cell particles are circular cylinders 6 μm in diameter and 50 μm long. The interior of the rods consists of a stack of discs closely enveloped by a surface membrane. The rods contain the pigment rhodopsin, and the investigation was carried out with dark-adapted rods. They were spun down in a specially constructed conductivity cell that could be inserted directly into the centrifuge to avoid damage to the rods during the transfer from the electrolytic cell. The measurements were performed with a bridge similar to that described by Cole and Gross (1949) using frequencies between 15 Hz and 500 kHz. The results were expressed as a locus in the complex impedance plane, which showed a high-frequency dispersion between 3.4 and 500 kHz with a deviation from the semicircle at frequencies below 3.4 kHz. The characteristic frequency was obtained by extrapolating the circular arc and using the relation:

$$\bar{\omega}_z = 2/\left[(R_0^* - R_\infty^*)C^*\right] \tag{5.57}$$

where C^* is the specific capacitance of the high-frequency dispersion. When the same data were plotted in the admittance plane, the high-frequency dispersion was much less obvious; this is explained by the reciprocal relation between resistance values and conductances. The characteristic frequencies found for the same preparation in the impedance and admittance plane are related to each other (Bode, 1945):

$$\bar{\omega}_y / \bar{\omega}_z = (R_0^* / R_\infty^*)^{-1/\alpha} \tag{5.58}$$

where $\bar{\omega}_y$ and $\bar{\omega}_z$ are the characteristic frequencies in the admittance and impedance plane, respectively, and the power factor α is related to the phase angle by Eq. 5.15. It could be shown that the experimentally obtained data were consistent with Eq. 5.58, and this was taken as evidence that the high-frequency measurements were free from error. Moreover, when the rods were incubated in a solution of high conductivity, the observed changes in the impedance locus could be described by an expression, equivalent to the suspension equation, that was derived by Bruggeman (1935). It holds for suspensions of spheres and cylinders with an axial ratio (length/diameter) of about 100, and its important feature is that it applies to high values of ρ'. The equation has been used to calculate the internal resistivity of the rods R_i according to:

$$R_i = R_e \frac{1-(1-\rho')(R_0^*/R_e^*)^{1/2}}{1-(1-\rho')(R_e^*/R_0^*)^{1/2}} \tag{5.59}$$

Close agreement with the experimental results was found for the relation $1/R_i = 0.5/R_0^* + 280$ μmho/cm. Assuming that the high-frequency dispersion can be represented by a single capacitance network, the capacitance of the rod surface calculated from the characteristic frequency was

1.54 μF/cm^2, while the assumption of a distribution in properties of the suspension according to Bruggeman's theory gave a value for the membrane capacitance of 0.91 μF/cm^2. Falk and Fatt considered the latter value more valid. Basically, they assumed just another distribution function for the distribution of parameters.

In the same year, Falk and Fatt (1968b) extended their investigations to an analysis of the passive electrical properties of isolated rod outer segments under the influence of light pulses. In the first part of their report, the methods and techniques used were basically the same as described in the previous paper. The electrical changes were measured as follows: The admittance of the packed rods was determined using a bridge circuit, the output of which was displayed on an CRT oscilloscope screen as Lissajou's figures that were reduced to a straight horizontal line by means of a phase-shifting network. In response to a light pulse, a new ellipse was obtained that reflected the change in passive electrical parameters of the suspension. The results were represented in the complex admittance plane or plotted as increments of B and G as a function of frequency. To test the internal consistency between the $B(f)$ and $G(f)$ plots, the Kramers-Kronig relation was used:

$$\Delta G(f) = \Delta G(0) + 2f^2/\pi \int_0^\infty \frac{\Delta B(\xi)}{(f^2-\xi^2)\xi} d\xi \tag{5.60}$$

Here ξ is a variable of integration that replaces f in $\Delta B(f)$. An additional advantage of the use of Eq. 5.60 is that the behavior of the Kramers-Kronig relation is known and therefore measurements at fewer frequencies are needed to describe the system.

Figure 5.11 shows the $G(f)$ and $B(f)$ relations of a rod suspension in media of different conductivities. These experimentally obtained functions could be approximated by a model consisting of a variable conductance G_a in series with a capacitance C_a and both shunted by a variable conductance G_b (inset of Fig. 5.11). Based on this model, the results in the figure were interpreted as representing two components: Component 1 is frequency independent over the whole range of frequencies, and component 2 represents the frequency-dependent (above 1 kHz) part of the rods. It is obvious that the conductance of component 1 (G_1) varies with the conductance of the suspending medium, and it could be shown that this dependence was linear. Component 1 was ascribed either to a nonspecific change in the permeability of the rod surface (ΔG_1 was not influenced by ionic changes of the medium) or to a volume change of the rods in response to a light flash. The dependence of component 2 on frequency was much less pronounced than that of component 1 and it was identified with a conduction change along the surface membrane.

To further elucidate the physiological implications of the observed

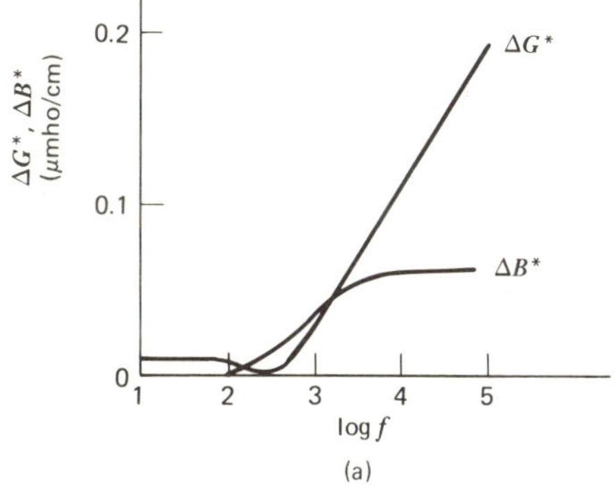

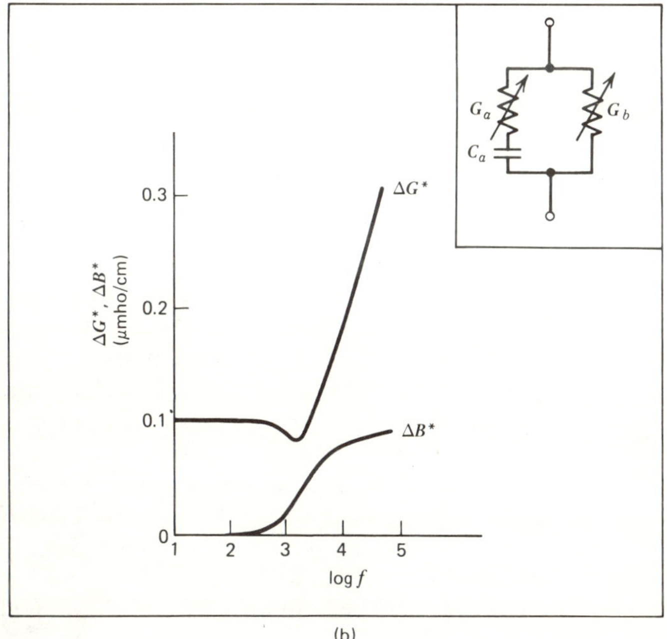

Figure 5.11. Frequency dependence of ΔG^* and ΔB^* of an outer rod suspension in a suspending medium of (a) $\sigma = 729$ μmho/cm and (b) $\sigma = 4400$ μmho/cm. Inset: equivalent circuit for rod suspensions. Modified after Falk and Fatt (1968b), by permission.

admittance changes, their time course was observed in reaction to light flashes. The change in Z was measured with a voltage–current method using constant-current conditions between 100 kHz and 1 MHz. These experiments showed a transient increase of conductance after a light flash due to the rapid degradation of the light energy into heat, and a conductance component prominent at high frequencies, which reflected the conductivity of the rod interior and caused by the uptake of H^+ by the visual pigment. Consequently, this process influenced the ionization of the buffer system. It also yielded Q_{10} values of the amplitude of components 1 and 2. This permitted correlation of component 2 with the formation and continued presence of metarhodopsin II, whereas component 1 depended on an earlier stage in the thermal conversion of the rhodopsin photoproducts.

5.6 SUMMARY AND OUTLOOK

It has been shown in this chapter how the ingenious use of a single formalism, the suspension equation, made significant contributions to biophysical knowledge during a half-century of research. Beginning with Fricke's calculation of the thickness of the red cell membrane, work on cell suspensions established that the value for the membrane capacitance was constant for biological cells. Moreover, there was evidence that the resistivity of the cell interior was about twice the value it should theoretically have been if the cytoplasm were an aqueous electrolyte solution. When very high values for the membrane capacitance were published in the 1950s and 1960s they were checked against the old claim that C_m was about 1 $\mu F/cm^2$. This resulted in the attribution of the apparently high C_m in crustacean muscle to infoldings of the membrane, and in the case of frog skeletal muscle it led to discovery of the importance of the *T*-system for the interpretation of electrophysiological phenomena. Moreover, recent attempts to treat the cytoplasm as a suspension of macromolecules in an electrolyte solution have contributed to the development of theories of interaction between macromolecules that have contributed significantly to the discussion of the state of water in the cytoplasm. It is amazing to see how the use of a basically simple equation contributes to our knowledge when it is combined with a traditional measuring method. The results thus obtained can presently be equalled, but not surpassed, by those obtained with such modern and sophisticated methods as nuclear magnetic resonance or tracer measurements.

Speculating on the potential of this approach for the near future, one can presume that the application of the suspension equation to the analysis of the physical state of the cytoplasm may prove extremely profitable. This implies that the cytoplasm is considered as a suspension of macromolecules. Attempts could be made to determine the surface charge of macromolecules as a function of biological variables, and the problem of the possible contributions of surface charges of the muscular filaments to the contractile process could be studied. Further studies of the cytoplasmic resistivity of erythrocytes, together with other independent biological approaches, may give further insight into the state of water bound to hemoglobin molecules. Finally, modern biochemical separation techniques hold promise for the study of the electrical properties of subcellular elements.

APPENDIX 1

Measuring Cells

Measuring cells for impedance determinations of biological material are derived from conductivity cells used in physical chemistry. In such a conductivity cell the conductance of the cell filled with a given solution is related to the conductivity of the test solution by the equation:

$$R = k\rho \tag{A1.1}$$

or its equivalent:

$$G = (1/k)\sigma \tag{A1.2}$$

Here R and G are the measured resistance and conductance, respectively; ρ and σ are the resistivity and conductivity of the solution; and k is the cell constant with the dimension cm^{-1}. The cell constant is determined by the geometry of the measuring cell. In practice, the unknown constant k' of a cell can be determined by measuring the resistance of a given electrolyte in it and in a cell whose constant k is known. If R is the resistance found in the cell with the known constant and R' that found in the other cell, k' can be calculated as:

$$k' = k(R'/R) \tag{A1.3}$$

The basic types of measuring cells are shown in Fig. A1.1. The cells shown as (a), (b), and (c) can be used with tissue or with fluid; cell (f) is usable with tissue only, and cells (d) and (e) are suitable only for solutions.

The most interesting feature of the cells shown in the figure is their electrode assemblies, which are usually made of platinum covered with platinum black. Cells (a), (c), (e), and (f) have large-surface main electrodes (A, B). These are suitable for low-frequency measurements, generally in the range from 10 Hz to 10 kHz. Since the resistance of platinum

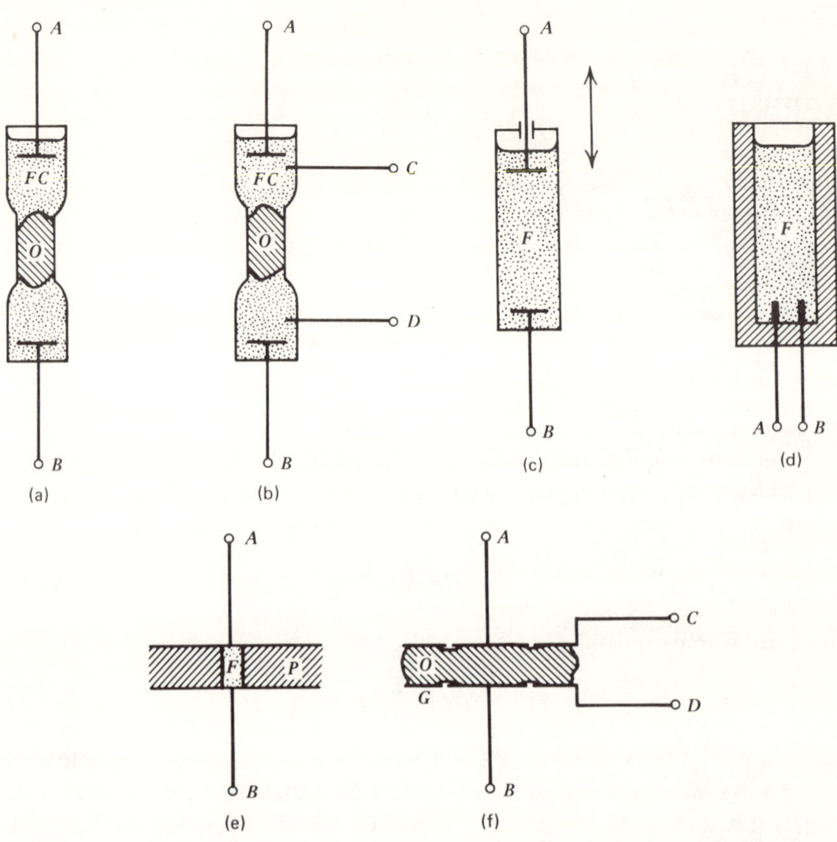

Figure A1.1. Different types of measuring cells for large electrodes. *A*, *B*, main electrodes; *C*, *D*, auxiliary electrodes; *FC*, contact fluid; *F*, biological fluid to be measured; *G*, guard ring; *O*, tissue to be measured; *P*, Plexiglas block with central port. (a), (b), (c), Cells for low-frequency measurements. (d), (e), Cells used in the range of radiofrequency. (f) Special cell for measurements where the electric field can be controlled (guard-ring technique using a three-terminal bridge).

black–covered electrodes at low frequencies is high, one has to work with large electrode surfaces and large electrode distances to obtain a sample impedance comparable to the electrode impedance. Cell (e) can also be used for higher frequencies, provided the port in the Plexiglas block *P* is kept small. Since cell (d) has a small interelectrode distance and small-surface electrodes, it can be used in a frequency range up to about 250 MHz.

Cell (a) represents the simplest type of cell with a device to take up the measured object in its narrow part. Contact with the electrodes is obtained by means of a fluid column *FC*, usually a physiological saline. Practically, a measurement is performed by first determining a blank reading when the cell is filled with saline, then taking a second measurement after introduction of the tissue. The tissue impedance is then obtained by subtracting the results of the two measurements (substitution method). A prerequisite for the applicability of the substitution principle is that the tissue be far enough away from the electrodes that no shadow effect can occur (Schwan, 1951). Cell (b) is a modification of cell (a). The auxiliary electrodes *C* and *D* are connected to a voltmeter with a high input impedance. This permits the recording of the voltage drop across the tissue. When the current flowing from *A* to *B* is measured, the measurement becomes independent of the electrode polarization impedance (four-terminal method). In cell (c) the electrode *A* can be moved along the vertical axis. Provided that the conductivity cell has a cylindrical cross section and the apparent impedance of a suspension of biological material is measured at different electrode distances, it has been shown that the difference between two impedance values obtained at different electrode distances is independent of the polarization impedance. This method is considered to be the most powerful one to eliminate an influence of the polarization impedance on the determination of biological impedances (Schwan, 1951). The same is possible for cell (e), where the thickness of the exchangeable Plexiglas block *P* and the geometry of its central port permit the cell to be adapted to various volumes of fluid or tissue as well as to different frequency ranges. Cell (d) uses small electrodes, which also permit the use of a relatively small volume of test fluid. Since this cell is used in the range of radiofrequencies, the small electrode distance becomes necessary for the control of the stray field. Cell (f) is designed to produce an electric field perpendicular to the surface of the main electrodes. This is achieved by keeping the concentric guard-ring electrode *G* at the same potential as the central electrodes *A*, and *B*. With this electrode arrangement, the following equations hold:

$$C = \varepsilon_r \varepsilon_v A/d \tag{A1.4}$$

and

$$G = \sigma A/d \tag{A1.5}$$

where C is the capacitive component of the tissue impedance and G its conductance, A is the surface of the main electrodes, d is the interelectrode distance, and ε_v and ε_r represent the absolute and relative dielectric constants. Equation A1.5 corresponds to Eq. A1.2, since:

$$k = d/A \tag{A1.6}$$

Moreover, Eq. A1.4 is interesting because it permits the determination of the dielectric constant ε_r with a conductivity cell, using the guard-ring principle.

Generally, measuring cells with large electrodes have not been constructed with many accessories for keeping the tissue under study in its physiological state. The tissue is usually incubated in a saline solution, and sometimes a water jacket around the cell is used to control the temperature. Devices for gassing a solution in order to maintain the pO_2 or to control the pH are rarely employed.

APPENDIX 2

Impedance and Admittance Loci

The basic theory of a special case of impedance locus and its conversion into an admittance locus is developed here because these diagrams are widely used for the interpretation of biological measurements. The theory of impedance and admittance loci can be found in textbooks of electrical engineering.* One of the most commonly used circuits to interpret an impedance measurement on a biological structure is the circuit shown in Fig. A2.1a (see also model 2 of Table 1.5). The impedance Z of this circuit, measured as a function of the frequency, is shown in Fig. A2.1b. Inspection of the circuit in Fig. A2.1a shows that there are two limiting resistances. At low frequencies, when the reactance of C is very large, the impedance is:

$$R_0 = R_1 + R_2 \tag{A2.1}$$

and at high frequencies, when the reactance of C becomes negligible, one obtains:

$$R_\infty = R_1 \tag{A2.2}$$

A graphical representation of the limiting resistances R_0 and R_∞ is shown in Fig. A2.1b. A plot of this type can be obtained directly from measurements of the tissue impedance, and therefore R_0 and R_∞ are known. Assuming that the circuit in Fig. A2.1a represents adequately the measured object, its impedance can be written directly in complex notation:

$$Z = R_1 + \frac{R_2/j\omega C}{R_2 + 1/j\omega C} \tag{A2.3}$$

*See, for example, Electrical Engineering Staff, Massachusetts Institute of Technology, 1963. Electric circuits, Cambridge: MIT Press, pp. 478–513.

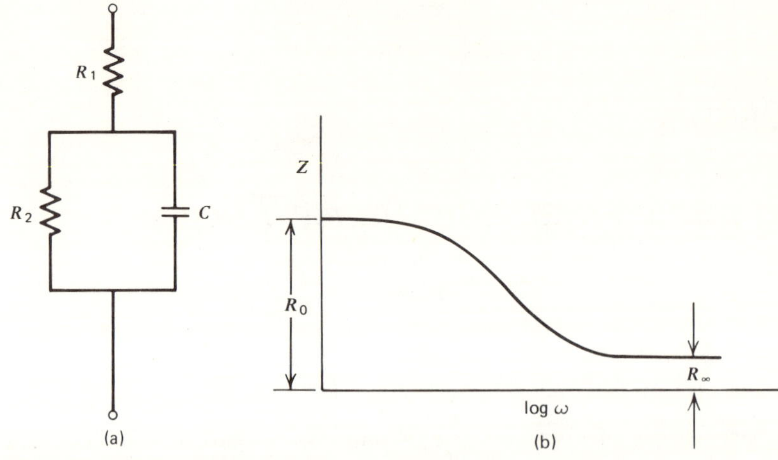

Figure A2.1. (a) Basic circuit used for demonstrating the theory of impedance locus, and (b) its impedance Z as a function of angular frequency ω.

with j the complex operator $\sqrt{-1}$. Separating real and imaginary parts, one has:

$$Z = R_1 + \frac{R_2}{1+\omega^2C^2R_2^2} - j\frac{\omega CR_2^2}{1+\omega^2C^2R_2^2} \tag{A2.4}$$

Using the symbol τ for the time constant R_2C as well as R_0, and R_∞ as defined in Eqs. A2.1 and A2.2, Eq. A2.4 can be simplified with the additional advantage of relating the impedance Z to the directly measurable values R_0 and R_∞; hence:

$$A = \left(R_\infty + \frac{R_0 - R_\infty}{1+\omega^2\tau^2}\right) - j\left(\frac{\omega\tau(R_0 - R_\infty)}{1+\omega^2\tau^2}\right) \tag{A2.5}$$

Here, the first term in parentheses is the real component R of the tissue impedance, and the second term in large parentheses represents the reactance X in the RX plane. Real and reactive terms are functions of ω, and thus Z is also. When the frequency is changed between 0 and ∞, Z will change continually along a curve in the RX plane. This curve is called an impedance locus (Fig. A2.2), such an impedance locus can also be obtained analytically. From Eq. A2.5, it follows that:

$$R = R_\infty + \frac{R_0 - R_\infty}{1+\omega^2\tau^2} \tag{A2.6}$$

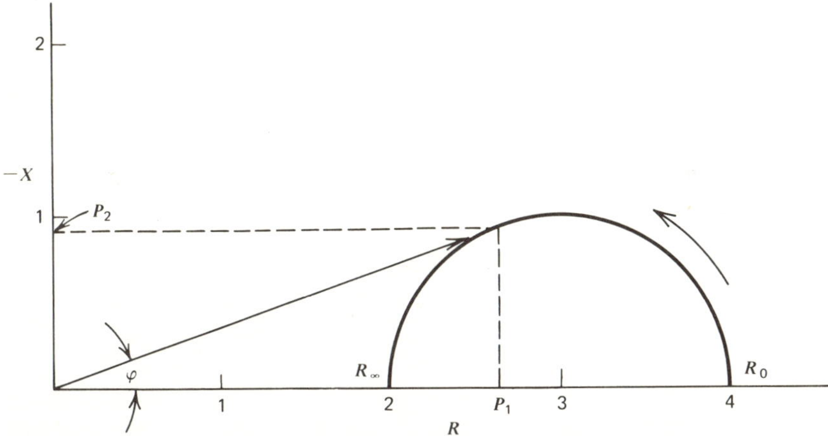

Figure A2.2. Impedance locus for the circuit in Fig. A2.1a for $R_0=4$ and $R_\infty=2$ in arbitrary units. The angular frequency ω increases counterclockwise as indicated by arrow.

and

$$X=-\frac{\omega\tau(R_0-R_\infty)}{1+\omega^2\tau^2} \tag{A2.7}$$

R and X can be considered functions of the parameter $\omega\tau$. They are related to Z by:

$$Z=(R^2+X^2)^{1/2} \tag{A2.8}$$

and to the phase angle φ (Fig. A2.2) by the relation:

$$\tan\varphi=X/R \tag{A2.9}$$

Elimination of $\omega\tau$ from Eqs. A2.6 and A2.7 yields the expression for the impedance locus in the RX plane:

$$X=\frac{(R_0-R_\infty)[(R_0-R)/(R-R_\infty)]^{1/2}}{(R_0-R_\infty)/(R-R_\infty)} \tag{A2.10}$$

Rewriting Eq. A2.10 yields the expression:

$$X^2+R^2-R(R_0+R_\infty)=-R_0R_\infty \tag{A2.11}$$

Inspection of Eq. A2.11 suggests completion of a quadratic expression in R

at the left side by adding the term $(R_0+R_\infty)^2/4$ to both sides, which results finally in:

$$X^2+\left[R-(R_0+R_\infty)/2\right]^2=\left[(R_0-R_\infty)/2\right]^2 \qquad \text{(A2.12)}$$

This is the analytical expression for the impedance locus that represents a circle with the center on the R axis at $(R_0+R_\infty)/2$ and a radius of $(R_0-R_\infty)/2$. However, inspecting Eq. A2.4 shows that its right side cannot change its sign in a physical system, and therefore the impedance locus of the circuit in Fig. A2.1a exists only as a semicircle in the $-X$ region as shown in Fig. A2.2. Here, the impedance locus is drawn in the RX plane for $R_0=4$ and $R_\infty=2$ in arbitrary units. The impedance vector **Z** turns around the origin 0, and its point moves along the impedance locus. At a given angular frequency, P_1 and P_2 are the respective values of R and $-X$. With increasing ω, **Z** moves counterclockwise along the locus.

The characteristic frequency $\bar{\omega}$ is defined as:

$$\bar{\omega}=1/\tau \qquad \text{(A2.13)}$$

and at this frequency, the reactance has the value

$$X_{\bar{\omega}}=(R_0-R_\infty)/2 \qquad \text{(A2.14)}$$

From this example of a simple impedance locus, rules can be derived for its derivation in other cases:

1. Choose a suitable equivalent circuit that fits the experimentally obtained curve $Z=f(\omega)$.
2. Derive the equation for Z in complex notation and separate real and imaginary parts.
3. Eliminate the parameter ω from the equation for reactive and ohmic components. The resulting equation in R and X represents the impedance locus. Sometimes it is more convenient to eliminate $\omega\tau$ instead of ω as was the case in the given example.
4. Rearrange the equation obtained in step 3 so that the resulting equation is readily interpretable in terms of analytical geometry. It is remarkable that most practically important impedance loci represent either circles or straight lines.

In the analysis of biological impedance measurements, a representation of the results in the admittance plane is sometimes used. Therefore, the

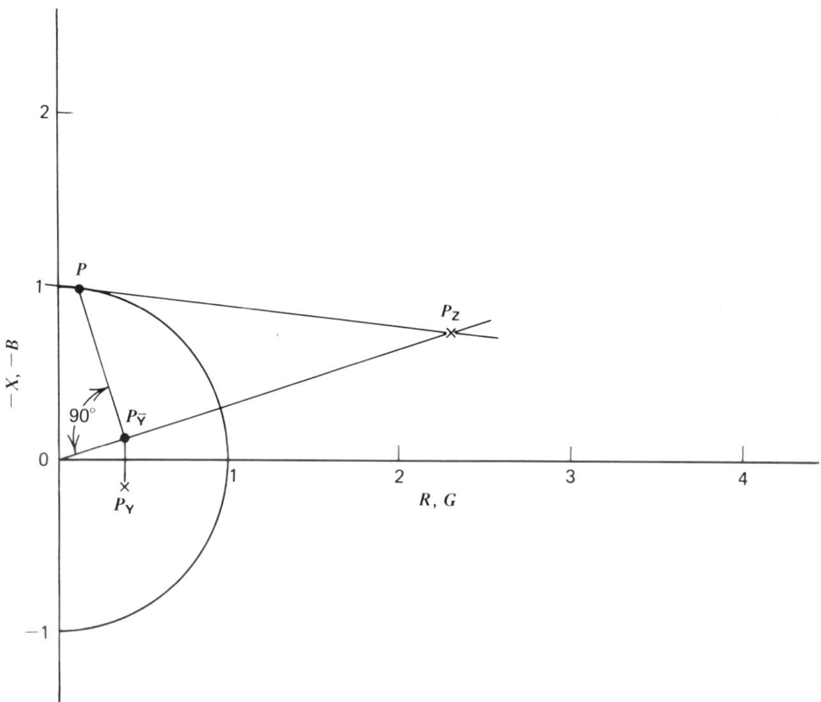

Figure A2.3. Method of geometrical conversion of a point P_Z of the impedance function **Z** into a point P_Y of the corresponding admittance function **Y**. For explanation, see text.

relation between the representation of a set of measurements in the impedance plane and the admittance plane will be shown by converting the impedance locus shown in Fig. A2.2 into a locus in the admittance plane. The relation between the impedance and the admittance Y is:

$$Z = 1/Y \tag{A2.15}$$

and the admittance Y can be written as:

$$Y = G + jB \tag{A2.16}$$

Here, G is a conductance and B the susceptance. The problem is now to find the reciprocal of the locus in Fig. A2.2 taking into account that although the locus is represented in the real GB plane, it is actually the locus of a complex function. The problem is therefore to find the inverse of a complex function, Eq. A2.5. This conversion can be achieved by geomet-

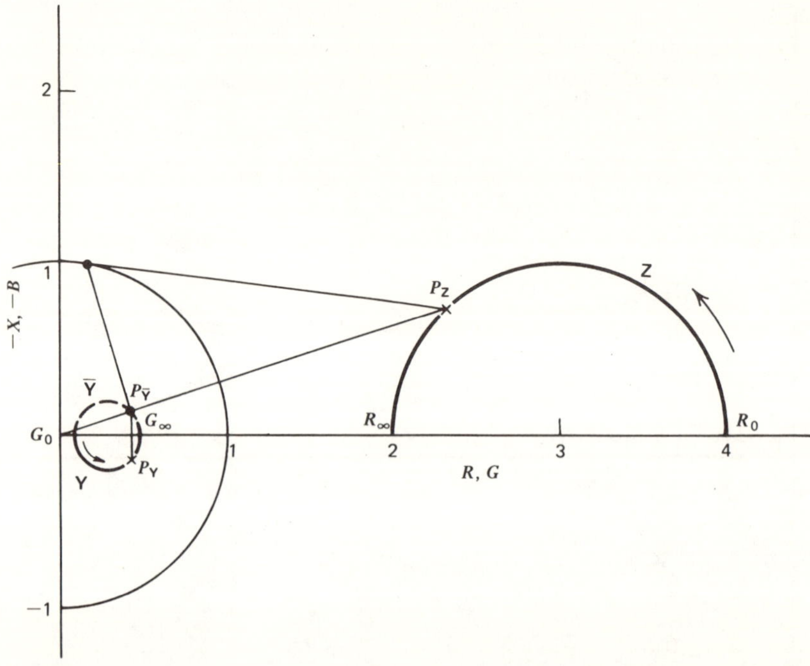

Figure A2.4. Conversion of the impedance function **Z** of Fig. A2.2 into its corresponding admittance function. Arrows: direction of change of impedance (admittance) with increasing frequency. For explanation, see text.

rical methods in two steps*: (1) Geometrical inversion of **Z** into an auxiliary locus $\bar{\mathbf{Y}}$, which is the conjugate of the sought locus **Y**; (2) consequently, **Y** is the image of $\bar{\mathbf{Y}}$ about the real axis according to a theorem from the theory of complex functions.

The function $\bar{\mathbf{Y}}$ can be obtained from **Z** considering (a) that the distance from any point of the $\bar{\mathbf{Y}}$ function to the origin represents the reciprocal of the distance of the corresponding point of the **Z** function from the origin, and (b) that the angles of the radius vectors to corresponding points of the **Z** and $\bar{\mathbf{Y}}$ function are equal. Figure A2.3 makes use of (a) and (b) to demonstrate how a point of the $\bar{\mathbf{Y}}$ function ($P_{\bar{\mathbf{Y}}}$) can be obtained from the corresponding point of the impedance function **Z**, $P_{\mathbf{Z}}$. The *GB* plane is superimposed on the *RX* plane. A circle of unit radius is then drawn

*Shown in detail in Electric circuits, loc. cit.

around the origin, and the tangent drawn from P_Z touches the circle at P. The perpendicular from P on $0P_Z$ gives $P_{\bar{Y}}$. The image of $P_{\bar{Y}}$ about the real axis is P_Y, which represents P_Z in the admittance plane. Using this method for each point of the impedance locus, one obtains the corresponding admittance locus.

In Fig. A2.4, the method demonstrated in the previous figure is used to convert the impedance function of Fig. A2.2 into its corresponding admittance function **Y**. $\bar{\mathbf{Y}}$ is the auxiliary locus resulting from the geometrical inversion of **Z**. To demonstrate the application of the method, the conversion of a point P_Z into P_Y is also shown. It is obvious from the figure that the points on the admittance locus move counterclockwise with increasing frequency as in the impedance locus.

APPENDIX 3

The Role of the Flux Ratio Equation in Interpreting Electrical Resistance Measurements

The proportionality between an ionic flux and an ionic current (Eq. 1.80) suggests ionic flux measurements as an alternative for determinations of the membrane resistance. A comparison of results from flux determinations with those from electrical measurements allows a deep insight into the nature of transport mechanisms. Most of this research has been done by using a formalism based on the Goldman equation (Eq. 1.87) that can be derived from the following expressions for the net passive ionic flux for an univalent ion species:

$$\left.\begin{aligned} M^{+} &= \frac{VF}{RT} P^{+} \left(\frac{C_i^{+} - C_o^{+}(-VF/RT)}{1-\exp(-VF/RT)} \right) \\ M^{-} &= \frac{VF}{RT} P^{-} \left(\frac{C_o^{-} - C_i^{-}(-VF/RT)}{1-\exp(-VF/RT)} \right) \end{aligned}\right\} \tag{A3.1}$$

where + and − denote monovalent cations and anions, respectively, and the subscripts i and o mean intra- and extracellular concentrations. Equation A3.1 can be separated into the components for the unidirectional fluxes to give the net flux for a cation:

$$\begin{aligned} M^{+} &= M_{\text{out}}^{+} - M_{\text{in}}^{+} \\ &= \frac{VF}{RT} P^{+} \left(\frac{C_i^{+}}{1-\exp(-VF/RT)} \right) - \frac{VF}{RT} P^{+} \left(\frac{C_o^{+}\exp(-VF/RT)}{1-\exp(-VF/RT)} \right) \end{aligned} \tag{A3.2}$$

where the subscript in denotes the flux from the outside toward the inside, and out means a flux in the opposite direction. For an anion a similar separation is possible. Dividing the two flux components we obtain:

$$\frac{M_{\text{in}}}{M_{\text{out}}} = \exp -(zV_j'F/RT) \tag{A3.3}$$

where z represents the number of charges and their sign for a given ion, and the electrochemical potential (driving potential) V_j' is defined as:

$$V_j' = V - E_j \tag{A3.4}$$

with the equilibrium potential (Nernst potential) E_j given by

$$\left.\begin{aligned} E^+ &= (RT/F)\ln(C_o^+/C_i^+) \\ E^- &= (RT/F)\ln(C_i^-/C_o^-) \end{aligned}\right\} \tag{A3.5}$$

Equation A3.3, known as the flux ratio equation, was independently derived by Ussing (1949a) and Teorell (1949). Its considerable importance for the interpretation of transport processes becomes clear when the conditions that have to be fulfilled before Eq. A3.3 is satisfied are listed (Schultz, 1969):

1. The passage through the membrane must be the rate-limiting step in the transfer process.
2. The chemical state of the transferred ion in the membrane has to be the same as that in the solution.
3. The flow of the transferred ion is independent from other flows through the membrane.
4. The membrane is in a steady state with respect to the transferred ion.
5. Since unidirectional fluxes can only be determined with isotopes the flows of the abundant species and the tracer must be independent. Coupling between them results in isotope interaction, thus complicating the interpretation of the flux ratio equation.
6. Both unidirectional fluxes permeate the membrane through pathways having identical properties. If the flux ratio equation is fulfilled, it can be concluded that the mechanism of transfer is not distinguishable from simple diffusion (Ussing, 1960; Curran and Schwartz, 1960).

Failure to satisfy Eq. A3.3 indicates that one or more of conditions 1–6 are violated. Several interpretations are possible in such a case.

(*a*) *Flow directly coupled to a metabolic reaction*. This is the definition of Kedem (1961) for active transport and is much narrower than that of Rosenberg (1948), who considered any transport against the electrochemical potential as active. The direct coupling of an ion flow with a metabolic reaction will favor one of the unidirectional fluxes. It is important to state that this is not necessarily the flow against the electrochemical potential. Consequently, any deviation from Eq. A3.3 can indicate the presence of an active transport mechanism. Such a statement is in contrast to some reports that specify the direction in which the flux ratio must deviate to be suggestive of an active transport mechanism (Ussing, 1960; Kotyk and Janáček, 1970, Chapter 5). The discrepancy reflects a lack of generality in the definition of an active transport process, because the analysis was either restricted to a given experimental model or the Rosenberg definition of an active transport process was used.

(*b*) *Solvent drag*. When there is a net flow of water across the membrane through narrow pores, the rate of flow of the solvent is larger than that of the solute. Koefoed-Johnson and Ussing (1953) expanded the flux ratio equation to account for this solvent drag:

$$\frac{M_{\text{in}}}{M_{\text{out}}} = \exp - \left[zV'F/RT + (\Delta_w/D) \int_0^{x_o} dx/A \right] \tag{A3.6}$$

where x is the distance across the membrane bounded by $x=0$ and $x=x_o$, D is the diffusion constant of the ion, A is the fraction of unit area available for the flow, and Δ_w is the rate of volume flow of water per unit area of the membrane. The volume flow is defined by:

$$\Delta_w = (A/g'_w)(dP/dx) \tag{A3.7}$$

where g'_w represents the frictional force exerted upon 1 mole of water moving at unit velocity in the membrane, and dP/dx stands for the pressure gradient, which can be either hydrostatic or osmotic. Inspection of Eq. A3.6 shows that the solvent drag is independent of the charge of the ion but inversely related to its diffusion coefficient. Therefore, the importance of the solvent drag increases with the size of the hydrated ions. This property is used to measure the solvent drag by determining the flux of uncharged molecules of a size similar to the ions under study. It has been found that in cases where the flux ratio equation is used to find an indication for active transport, the term for solvent drag can often be neglected (Ussing, 1960), with the exception of amphibian skins treated with antidiuretic hormone (Ussing, 1961).

(*c*) *Single file diffusion*. Hodgkin and Keynes (1955) found that in the giant axon of the squid the downhill transport of potassium ions was largely favored over the uphill transport, a result inconsistent with Eq. A3.3. Since coupling to a metabolic reaction could be excluded, they constructed a mechanical model to explain their results (Fig. A3.1). When the model with the short gap connecting the two compartments was shaken (Fig. A3.1a), the result approximated the theoretically calculated downhill flow, which was double the uphill flow because of the concentration gradient between both compartments. When the gap was made much longer, the downhill movement was found to be 18 times that in the uphill direction. Based on this model, Hodgkin and Keynes (1955) concluded that the deviation of their results from the flux ratio equation was due to a violation of the independence principle underlying it. This phenomenon

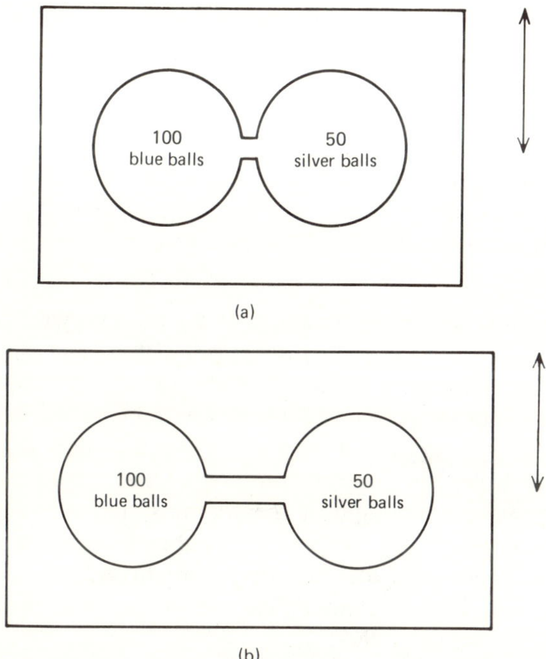

Figure A3.1. Mechanical model to demonstrate single-file diffusion. Two flat compartments of equal dimensions are connected by (a) a short gap and (b) a long gap. The width of the gap is just large enough to let a single ball pass. The balls in both compartments have the same diameter and their number represents ion concentration in the compartments. Arrows: direction of shaking. Modified after Hodgkin and Keynes (1955), by permission.

has been called single-file diffusion or long-pore effect, and it is possible to quantify the phenomenon by modifying Eq. A3.3:

$$\frac{M_{\text{in}}}{M_{\text{out}}} = \exp\left[-(znV'F/RT)\right] \tag{A3.8}$$

where $n > 1$. Experimentally, n was found to be 2.5 in the squid axon. This means that 2 to 3 potassium ions are within a long pore at any given time. The model does not imply that the actual transport mechanism in the membrane corresponds physically to the long-pore effect, but any alternative explanation has to satisfy Eq. A3.8.

(*d*) *Exchange diffusion.* Levi and Ussing (1948) described a mechanism that is not coupled directly to metabolic reactions and results in a deviation from the flux ratio equation. This mechanism has been termed exchange diffusion. The term was later applied to different transport mechanisms, but it is used here strictly for the following: An ion is bound to a saturated carrier and transported through the membrane; inside the membrane, the ion dissociates from the carrier and another ion of the same species is bound and transported toward the outside. Such a mechanism adds equal and additive components to both unidirectional fluxes driven by the electrochemical potential, with the effect that the values of the flux ratio will be driven closer to unity than predicted by Eq. A3.3. When exchange diffusion becomes predominant, the flux ratio will approach unity.

(*e*) *Composite membrane systems.* It has been shown that if a complex and nonhomogeneous system consisting of several membranes is lumped together, the resulting equivalent membrane may exhibit properties in disagreement with the predictions of the flux ratio equation. In such a case one cannot apply criteria 1–4 to infer a given transport mechanism in the lumped membrane. It is impossible to predict the complications introduced by lumping together a composite membrane system without having specified the properties of the component membranes. The phenomenon is of considerable importance for the interpretation of fluxes in epithelia and across cells bounded by asymmetrical membranes. For further discussion, see Curran and Schultz (1968).

(*f*) *Isotope interaction.* Since unidirectional fluxes are measured with radioactively labeled ions, interactions between the flow of the abundant species and the tracer species can account for deviations of the flux ratio equation like those discussed under (*c*) and (*d*). This was demonstrated by Kedem and Essig (1965), who derived a general flux ratio equation considering isotope interaction and the effect of active transport.

It is clear from this discussion of the possible causes for deviations from Eq. A3.3 that measurements of the flux ratio alone do not suffice to identify a given mechanism responsible for a deviation of the flux ratio from its theoretical value. To analyze the problem further, the conductance of an ionic channel is often measured by electrical methods. This amounts to a determination of the membrane conductance and the choice of the appropriate means to separate the membrane conductance into the respective ionic conductances.

To understand how these electrical measurements can be used to complement tracer measurements, it is possible to derive the equation for an ionic conductance from the flux equation and then compare it with the electrically measured quantity. A current imposed on a membrane from an external source will create an additional passive net flux across it in a yet unspecified way. Therefore, Eqs. A3.1 will give the conductances of the individual ionic channels when converted to current density and differentiated with respect to the membrane potential:

$$\left.\begin{aligned} G^+ &= \frac{dJ^+}{dV} = \frac{F^3}{(RT)^2} P^+ V \frac{C_o^+ C_i^+}{C_o^+ - C_i^+} \\ G^- &= \frac{dJ^-}{dV} = \frac{F^3}{(RT)^2} P^- V \frac{C_o^- C_i^-}{C_i^- - C_o^-} \end{aligned}\right\} \tag{A3.9}$$

where G^+ and G^- are the differential (slope) conductances for the cations and anions, respectively. Once the differential conductances for the individual ionic channels are known, the differential membrane conductance can be obtained by summation:

$$G_m = \sum G^+ + \sum G^- \tag{A3.10}$$

In a given experimental situation, the quantity measured with the square-pulse method is the chord conductance (Table 2.5), which is not identical with the slope conductance because of the nonlinearity of the membrane conductance (see Sec. 1.5.1.1). G_m or, after appropriate current separation, G^+ or G^- can be obtained from a current–voltage curve based on a dynamic resistance measurement or on voltage-clamp data. If, in a square-pulse measurement, the current is kept small enough so that the potential and concentration profiles within the membrane are not appreciably disturbed, one obtains for a given steady-state membrane potential $g_m \approx G_m$.

On the other hand, it is possible to evaluate the ionic conductances from flux measurements. Multiplying Eqs. A3.1 by the Faraday constant and differentiating with respect to V, one obtains, using Eqs. A3.2 and A3.3, for the conductance of a given monovalent ion:

$$\left.\begin{aligned} G^+ &= \frac{F^2}{RT}\left\{ M^+ \left[\frac{RT}{F}\left(\frac{d\ln P^+}{dV}\right) - \frac{1}{1-\exp(-VF/RT)} + \frac{RT}{VF}\right] - M^+_{\text{out}} \right\} \\ G^- &= \frac{F^2}{RT}\left\{ M^- \left[\frac{RT}{F}\left(\frac{d\ln P^-}{dV}\right) - \frac{1}{1-\exp(-VF/RT)} + \frac{RT}{VF}\right] - M^-_{\text{in}} \right\} \end{aligned}\right\} \qquad \text{(A3.11)}$$

Equation A3.11 was derived for cations by Sjodin (1959) and was later used and modified by Brinley (1965) for single-file diffusion. Comparison of Eqs. A3.9 and A3.11 shows that Eqs. A3.11 are more complex because they take into account the change in permeability with the membrane potential. A simpler form of this equation employing constant permeability constants was derived by Keynes (1951).

As mentioned earlier, the electrically determined conductance depends only on the passive net flux, which is much less sensitive to the factors causing disagreement with the flux ratio equation than the unidirectional fluxes. On the other hand, Eqs. A3.11 depend on the unidirectional flows. This means that a disagreement in the values obtained by electrical and tracer methods can be used to postulate a mechanism responsible for the disagreement with the flux ratio equation; see, for example, Haas (1964), Brinley (1965), Hodgkin and Keynes (1955), Keynes (1954), Hodgkin and Horowicz (1959a, b), Weidmann (1952), Carmeliet (1961b), Haas and Glitsch (1962), and Haas, Glitsch, and Trautwein (1963).

References

Abe, Y., and T. Tomita. 1968. Cable properties of smooth muscle. J. Physiol. **196:** 87–100.

Abramowitz, M., and I. A. Stegun. 1967. Handbook of mathematical functions, with formulas, graphs, and mathematical tables, 6th printing. Dover, New York.

Aceves, J., and D. Erlij. 1971. Sodium transport across the isolated epithelium of the frog skin. J. Physiol. **212:** 195–210.

Achelis, J. D. 1932. Über die Polarisationskapazität (Permeabilität) des Skeletmuskels bei indirekter Reizung. Pfluegers Arch. Gesamt. Physiol. Mensch. Tiere **230:** 412–422.

Adam, K. R., H. Schmidt, R. Stämpfli, and C. Weiss. 1966. The effect of scorpion venom on single myelinated nerve fibres of the frog. Brit. J. Pharmacol. Chemother. **26:** 666–677.

Adam, K. R., and C. Weiss. 1959. Actions of scorpion venom on skeletal muscle. Brit. J. Pharmacol. Chemother. **14:** 334–339.

Adelman, W. J., Jr., F. M. Dyro, and J. P. Senft. 1966. Internally perfused axons: effects of two different anions on ionic conductance. Science **151:** 1392–1394.

Adelman, W. J., Jr., and Y. B. Fok. 1964. Internally perfused squid axons studied under voltage clamp conditions. II. Results. The effects of internal potassium and sodium on membrane electrical characteristics. J. Cell. Comp. Physiol. **64:** 429–443.

Adelman, W. J., Jr., and D. L. Gilbert. 1964. Internally perfused squid axons studied under voltage clamp conditions. I. Method. J. Cell. Comp. Physiol. **64:** 423–428.

Adelman, W. J., Jr., and J. P. Senft. 1966. Voltage clamp studies on the effect of internal cesium ion on sodium and potassium currents in the squid giant axon. J. gen. Physiol. **50:** 279–293.

Adrian, R. H. 1956. The effect of internal and external potassium concentration on the membrane potential of frog muscle. J. Physiol. **133:** 631–658.

———. 1960. Potassium chloride movement and the membrane potential of frog muscle. J. Physiol. **151:** 154–185.

———. 1964. The rubidium and potassium permeability of frog muscle membrane. J. Physiol. **175:** 134–159.

Adrian, R. H., W. K. Chandler, and A. L. Hodgkin. 1966. Voltage clamp experiments in skeletal muscle fibres. J. Physiol. **186:** 51–52P.

———. 1968. Voltage clamp experiments in striated muscle fibers. J. gen. Physiol. **51:** 188–192s.

———. 1969. The kinetics of mechanical activation in frog muscle. J. Physiol. **204:** 207–230.

———. 1970a. Voltage clamp experiments in striated muscle fibres. J. Physiol. **208:** 607–644.

———. 1970b. Slow changes in potassium permeability in skeletal muscle. J. Physiol. **208:** 645–668.

Adrian, R. H., L. L. Costantin, and L. D. Peachey. 1969. Radial spread of contraction in frog muscle fibers. J. Physiol. **204:** 231–257.

Adrian, R. H., and W. H. Freygang. 1962a. The potassium and chloride conductance of frog muscle membrane. J. Physiol. **163:** 61–103.

———. 1962b. Potassium conductance of frog muscle membrane under controlled voltage. J. Physiol. **163:** 104–114.

Adrian, R. H., and L. D. Peachey. 1965. The membrane capacity of frog twitch and slow muscle fibres. J. Physiol. **181:** 324–336.

Adrian, R. H., and C. L. Slayman. 1966. Membrane potential and conductance during transport of sodium, potassium and rubidium in frog muscle. J. Physiol. **184:** 970–1014.

Agin, D. P. 1969. Electrochemical properties of glass microelectrodes, p. 62–75. *In* M. Lavallée, O. F. Schanne, and N. C. Hebert (Eds.), Glass microelectrodes. Wiley, New York.

Almers, W. 1972a. Potassium conductance changes in skeletal muscle and the potassium concentration in the transverse tubules. J. Physiol. **225:** 33–56.

———. 1972b. The decline of potassium permeability during extreme hyperpolarization in frog skeletal muscle. J. Physiol. **225:** 57–83.

Amatniek, E. 1958. Measurement of bioelectric potentials with microelectrodes and neutralized input capacity amplifiers. IRE Trans. Med. Electron. **10:** 3–14.

Amatniek, E., W. Freygang, H. Grundfest, G. Kiebel, and A. Shanes. 1957. The effect of temperature, potassium and sodium on the conductance change accompanying the action potential in the squid giant axon. J. gen. Physiol. **41:** 333–342.

Anderson, N. C., Jr. 1969. Voltage-clamp studies on uterine smooth muscle. J. gen. Physiol. **54:** 145–165.

Anderson, N. C., F. Ramon, and J. W. Moore. 1970. Effects of Na^+ and Ca^{++} on uterine smooth muscle excitation under current-clamp and voltage-clamp conditions. Fed. Proc. **29:** 261.

Anderson, N. C., F. Ramon, and A. Snyder. 1971. Studies on calcium and sodium in uterine smooth muscle excitation under current-clamp and voltage-clamp conditions. J. gen. Physiol. **58:** 322–339.

Antoni, H., J. Töppler, and N. Krause. 1970. Polarization effects of sinusoidal 50-cycle alternating current on membrane potential of mammalian cardiac fibres. Pfluegers Arch. Europe. J. Physiol. **314:** 274–291.

Araki, T., and T. Otani. 1955. Response of single motoneurons to direct stimulation in toad's spinal cord. J. Neurophysiol. **18:** 472–485.

Araki, T., T. Otani, and T. Furukawa. 1953. The electrical activities of single motoneurons in toad's spinal cord, recorded with intracellular electrodes. Jap. J. Physiol. **3:** 254–267.

Araki, T., and C. A. Terzuolo. 1962. Membrane currents in spinal motoneurons associated with the action potential and synaptic activity. J. Neurophysiol. **25:** 772–789.

Armstrong, C. M. 1971. Interaction of tetraethylammonium ion derivatives with the potassium channel of giant axons. J. gen. Physiol. **58:** 413–437.

Armstrong, C. M., and L. Binstock. 1964. The effects of several alcohols on the properties of the squid giant axon. J. gen. Physiol. **48:** 265–277.

Atwater, I., F. Bezanilla, and E. Rojas. 1969. Sodium influxes in internally perfused squid giant axon during voltage clamp. J. Physiol. **201:** 657–664.

Atwood, H. L. 1963. Differences in muscle fibre properties as a factor in "fast" and "slow" contraction in *Carcinus*. Comp. Biochem. Physiol. **10:** 17–32.

Baker, P. F., A. L. Hodgkin, and H. Meves. 1964. The effect of diluting the internal solution on the electrical properties of a perfused giant axon. J. Physiol. **170:** 541–560.

Baker, P. F., A. L. Hodgkin, and E. B. Ridgway. 1970. Two phases of calcium entry during the action potential in giant axons of *Loligo*. J. Physiol. **208:** 80–82P.

Baker, P. F., A. L. Hodgkin, and T. I. Shaw. 1961. Replacement of the protoplasm of a giant nerve fiber with artificial solutions. Nature **190:** 885–887.

———. 1962. The effects of changes in internal ionic concentrations on the electrical properties of perfused giant axons. J. Physiol. **164:** 335–374.

Baker, R. D., M. J. Wall, and J. L. Long. 1971. Intestinal transmural electrical activity: selective effects of mucosal and serosal anaerobiosis. Biochim. Biophys. Acta **225:** 392–396.

Balk, O., and H. Müller-Mohnssen. 1966. Die stationäre Strom-Spannungs-Charakteristik des Ranvierschen Schnürrings. Z. Vgl. Physiol. **52:** 56–78.

———. 1968. Experimental demonstration of relations between the action potential and the stationary current-voltage characteristic of the Ranvier node. Biophysik **5:** 126–131.

Bani, G. 1966. Osservazioni sulla ultrastruttura della cute di *Bufo bufo* (L.) e modificazioni a livello dell'epidermide, in relatione a differenti condizioni ambientali. Monit. Zool. Ital. **74:** 93–112.

Barbashova, Z. I., and Y. Y. Moskalenko. 1961. Changes in electrical parameters of skeletal muscle tissue of animals acclimatized to hypoxia. Biophysics (Engl. Transl.) **6:** 365–368.

Barr, L. 1961. Transmembrane resistance of smooth muscle cells. Am. J. Physiol. **200:** 1251–1255.

———. 1963. Propagation in vertebrate visceral smooth muscle. J. Theor. Biol. **4:** 73–85.

Barr, L., and W. Berger. 1964. The role of current flow in the propagation of cardiac muscle action potentials. Pfluegers Arch. Gesamt. Physiol. Mensch. Tiere **279:** 192–194.

Barr. L., W. Berger, and M. M. Dewey. 1968. Electrical transmission at the nexus between smooth muscle cells. J. gen. Physiol. **51:** 347–368.

Barr, L., M. M. Dewey, and W. Berger. 1965. Propagation of action potentials and the structure of the nexus in cardiac muscle. J. gen. Physiol. **48:** 797–823.

Barry, P. H., J. M. Diamond, and E. M. Wright. 1971. The mechanism of cation permeation in rabbit gallbladder. Dilution potentials and bionic potential. J. Memb. Biol. **4:** 358–394.

Beeler, G. W., and H. Reuter. 1970a. Voltage clamp experiments on ventricular myocardial fibres. J. Physiol. **207:** 165–190.

———. 1970b. Membrane calcium current in ventricular myocardial fibres. J. Physiol. **207:** 191–209.

———. 1970c. The relation between membrane potential, membrane currents and activation of contraction in ventricular myocardial fibres. J. Physiol. **207:** 211–229.

Bennett, M. R. 1967a. The effect of cations on the electrical properties of the smooth muscle cells of the guinea-pig vas deferens. J. Physiol. **190:** 465–479.

———. 1967b. The effect of intracellular current pulses in smooth muscle cells of the guinea-pig vas deferens at rest and during transmission. J. gen. Physiol. **50:** 2459–2475.

Bennett, M. R., and G. Burnstock. 1966. Application of the sucrose-gap method to determine the ionic basis of the membrane potential of smooth muscle. J. Physiol. **183:** 637–648.

Bennett, M. R., and N. C. R. Merrillees. 1966. An analysis of the transmission of excitation from autonomic nerves to smooth muscle. J. Physiol.**185:** 520–535.

Bennett. M. V. L. 1966. Physiology of electrotonic junctions. Ann. N.Y. Acad. Sci. **137:** 509–539.

Bennett, M. V. L., S. M. Crain, and H. Grundfest. 1959. Electrophysiology of supramedullary neurons in *Spheroides maculatus*. II. Properties of the electrically excitable membrane. J. gen. Physiol. **43:** 189–219.

Berger, W. 1963. Die Doppelsaccharosetrennwandtechnik. Eine Methode zur Untersuchung des Membranpotentials und der Membraneigenschaften glatter Muskelzellen. Pfluegers Arch. Gesamt. Physiol. Mensch. Tiere **277:** 570–576.

Bergman, C., and R. Stämpfli. 1966. Différence de perméabilité des fibres nerveuses myélinisées sensorielles et motrices à l'ion potassium. Helv. Physiol. Pharmacol. Acta **24:** 247–258.

Bernhardt, J. 1972. Properties of a three-terminal bridge for dielectric measurements of biological materials with a well-defined sample field from 20 Hz to 500 kHz. Biophysik **8:** 151–157.

Besseau, A., and Y. M. Gargouïl. 1969. Ionic currents in rat ventricular heart fibres: voltage-clamp experiments using double sucrose-gap technique. J. Physiol. **204:** 95–96P.

Besseau, A., C. Léoty, et Y. M. Gargouïl. 1969. Etude en courant ou voltage imposé, de préparations ventriculaires cardiaques de rat par la technique de double séparation de saccharose. Compt. Rend. Acad. Sci. Ser. D. **268:** 2714–2717.

Bezanilla, F., C. Caputo, H. Gonzalez-Serratos, and R. A. Venosa. 1972. Sodium dependence of the inward spread of activation in isolated twitch muscle fibres of the frog. J. Physiol. **223:** 507–523.

Bianchi, M., G. Torelli, F. Celentiano, and G. Cortili. 1969. Misure della correnti di "corto circuito" su pelle di rana in situ. Boll. Soc. Ital. Biol. Sper. **45:** 385–388.

Biber, T. U. L., R. A. Chez, and P. F. Curran. 1966. Na transport across frog skin at low external Na concentrations. J. gen. Physiol. **49:** 1161–1176.

Biber, T. U. L., L. Cruz, and P. F. Curran. 1972. Sodium influx at the outer surface of frog skin. Evaluation of different extracellular markers. J. Memb. Biol. **7:** 365–376.

Biber, T. U. L., and P. F. Curran. 1970. Direct measurement of uptake of sodium at the outer surface of the frog skin. J. gen. Physiol. **56:** 83–99.

Binstock, L., and L. Goldman. 1967. Giant axon of *Myxicola*: some membrane properties as observed under voltage clamp. Science **158:** 1467–1469.

———. 1969. Current and voltage clamped studies in *Myxicola* giant axons; effect of tetrodotoxin. J. gen. Physiol. **54:** 730–740.

Blaustein, M. P. 1968a. Action of certain tropine esters on voltage-clamped lobster axon. J. gen. Physiol. **51:** 309–319.

———. 1968b. Barbiturates block sodium and potassium conductance increases in voltage-clamped lobster axons. J. gen. Physiol. **51:** 293–307.

Blaustein, M. P., and D. E. Goldman. 1966. Competitive action of calcium and procaine on lobster axon. J. gen. Physiol. **49:** 1043–1063.

Blinks, L. R. 1930. The direct current resistance of *Valonia*. J. gen. Physiol. **13:** 361–378.

Bode, H. W. 1945. Network analysis and feedback amplifier design. Van Nostrand Reinhold, New York.

Boistel, J., and F. Fatt. 1958. Membrane permeability change during inhibitory transmitter action in crustacean muscle. J. Physiol. **144:** 176–191.

Bothwell, T. P., and H. P. Schwan. 1956. Electrical properties of the plasma membrane of erythrocytes at low frequencies. Nature **178:** 265–266.

Bothwell, T. P., H. P. Schwan, and F. J. Wiercinski. 1954. Electric properties of beef erythrocyte suspensions at low frequencies. Fed. Proc. **13:** 15–16.

Boulpaep, E. L., and J. F. Seely. 1971. Electrophysiology of proximal and distal tubules in the autoperfused dog kidney. Am. J. Physiol. **221:** 1084–1096.

Boyd, I. A., and A. R. Martin. 1959. Membrane constants of mammalian muscle fibres. J. Physiol. **147:** 450–457.

Bozler, E. 1942. The initiation of impulses in cardiac muscle. Am. J. Physiol. **138:** 273–282.

Bozler, E., and K. S. Cole. 1935. Electric impedance and phase angle of muscle in rigor. J. Cell. Comp. Physiol. **6:** 229–241.

Brading, A., E. Bülbring, and T. Tomita. 1969. The effect of temperature on the membrane conductance of the smooth muscle of the guinea pig *taenia coli*. J. Physiol. **200:** 621–635.

Bradl, M. 1967. Die Verteilung von Chlor- und Kaliumionen zwischen dem Axon und der Membran markhaltiger Nervenfasern. Acta Biol. Med. Ger. **19:** 73–90.

———. 1970. Zur Elektrophysiologie zytoplasmatischer Membranen. 2. Über den Einfluss von Kalium- und Chlorionen auf den Membranwiderstand markhaltiger Nervenfasern. Acta Biol. Med. Ger. **25:** 379–388.

Brady, A. J. 1964. Excitation and excitation-contraction coupling in cardiac muscle. Ann. Rev. Physiol. **26:** 341–356.

———. 1966. Ionic movements and the transmembrane potential, pp. 35–46. *In* L. S. Dreifus and W. Likoff (Eds.), Mechanisms and Therapy of Cardiac Arrhythmias. Grune and Stratton, New York.

Brady, A. J., and J. W. Woodbury. 1960. The sodium-potassium hypothesis as the basis of electrical activity in frog ventricle. J. Physiol. **154:** 385–407.

Brandt, P. W., J. P. Reuben, C. Girardier, and H. Grundfest. 1965. Correlated morphological and physiological studies on isolated single muscle fibers. I. Fine structure of the crayfish muscle fiber. J. Cell. Biol. **25:** 233–260.

Brandt, P. W., J. P. Reuben, and H. Grundfest. 1968. Correlated morphological and physiological studies on isolated single muscle fibers. II. The properties of the crayfish transverse tubular system: localization of the sites of reversible swelling. J. Cell. Biol. **38:** 115–129.

Bravený, P., and J. Šumbera. 1970. Electromechanical correlations in the mammalian heart muscle. Pfluegers Arch. Europe. J. Physiol. **319:** 36–48.

Brinley, F. J. 1965. Sodium, potassium and chloride concentrations and fluxes in the isolated giant axon of *Homarus*. J. Neurophysiol. **28:** 742–772.

Brock, L. G., J. S. Coombs, and J. C. Eccles. 1952. The recording of potentials from motoneurons with an intracellular electrode. J. Physiol. **117:** 431–460.

Brodwick, M. S., and D. Junge. 1972. Post-stimulus hyperpolarization and slow potassium conductance increase in *Aplysia* giant neurone. J. Physiol. **223:** 549–570.

Bromm, B., and H. Lullies. 1966. Über den Mechanismus der Reizwirkung mittelfrequenter Wechselströme auf die Nervenmembran. Pfluegers Arch. Gesamt. Physiol. Mensch. Tiere **289:** 215–226.

Bromm, B., and R. Simon. 1971. Inward rectification in frog skeletal muscle membrane during alternating current stimulation. Pfluegers Arch. Europe. J. Physiol. **328:** 155–169.

Brooks, S. C. 1923. Conductivity as a measure of vitality and death. J. gen. Physiol. **5:** 365–381.

Brown, A. C. 1962. Current and potential of frog skin in vivo and in vitro. J. Cell. Comp. Physiol. **60:** 263–269.

Brown, A. M., and P. R. Berman. 1970. Mechanism of excitation of *Aplysia* neurons by carbon dioxide. J. gen. Physiol. **56:** 543–558.

Brown, A. M., L. Walker, and R. B. Sutton. 1970. Increased chloride conductance as the proximate cause of hydrogen ion concentration effects in *Aplysia* neurons. J. gen. Physiol. **56:** 559–582.

Brown, H. F., E. Clark, and S. J. Noble. 1972. Pacemaker current in frog atrium. Nat. New Biol. **235:** 30–31.

Brown, H. F., and S. J. Noble. 1969a. Membrane currents underlying delayed rectification and pacemaker activity in frog atrial muscle. J. Physiol. **204:** 717–736.

———. 1969b. A quantitative analysis of the slow component of delayed rectification in frog atrium. J. Physiol. **204:** 737–747.

Bruggeman, D. A. G. 1935. Berechenung verschiedener physikalisher Konstanten von heterogenen Substanzen. I. Dielektrizitätskonstanten und Leitfähigkeiten der Mischkörper aus isotropen Substanzen. Ann. Phys. **24:** 636–664.

Buchthal, F. 1934. Widerstandsmessungen an einzelnen, quergestreiften Muskelfasern in Ruhe und während der Kontraktion. Skand. Arch. Physiol. **70:** 199–226.

Bueno, E. J., and L. Corchs. 1968. Induced pacemaker activity on toad skin. J. gen. Physiol. **51:** 785–801.

Bülbring, E., and T. Tomita. 1968. The effect of catecholamines on the membrane resistance and spike generation in the smooth muscle of guinea-pig taenia coli. J. Physiol. **194:** 74–76P.

———. 1969a. Effect of calcium, barium and manganese on the action of adrenaline in smooth muscle of the guinea-pig taenia coli. Proc. Roy. Soc. B. **172:** 121–136.

———. 1969b. Increase of membrane conductance by adrenaline in the smooth muscle of guinea-pig taenia coli. Proc. Roy. Soc. B. **172:** 89–102.

Bureš, J., M. Petráň, and J. Zachar. 1967. Electrophysiological methods in biological research. Academic, New York, 2nd edition.

Burg, M. B., L. Isaacson, J. Grantham, and J. Orloff. 1968. Electrical properties of isolated perfused rabbit renal tubules. Am. J. Physiol. **215:** 788–794.

Burke, W., and D. L. Ginsborg. 1956a. The electrical properties of the slow muscle fibre membrane. J. Physiol. **132:** 586–598.

———. 1956b. The action of the neuromuscular transmitter on the slow fibre membrane. J. Physiol. **132:** 599–610.

Burlakova, Y. V., B. N. Veprintsev, and I. T. Rass. 1959. Lag of the impedance effect behind the action potential in frog nerve trunk. Biophysics (Engl. Transl.) **4:** 127–131.

Burnstock, G., and C. L. Prosser. 1960. Conduction in smooth muscles: comparative electrical properties. Am. J. Physiol. **199:** 553–559.

Caillé, J. P. 1969. Perméabilités ioniques et potentiel de repos des membranes excitables et non-excitables. Thèse de Ph.D. (Biophysique) Université de Sherbrooke, Sherbrooke.

Caillé, J. P., and O. F. Schanne. 1972. Nombres de transport, t_K, t_{Na} et t_{Cl} pour la membrane de la cellule hépatique in vivo. Can. J. Physiol. Pharmacol. **50:** 416–422.

Candia, O. A. 1966. An automatic voltage-clamp recorder for inexcitable membranes. IEEE Trans. Biomed. Eng. **13:** 189–194.

———. 1970. The hyperpolarization region of the current-voltage curve in frog skin. Biophys. J. **10:** 323–344.

Candia, O. A., and J. A. Zadunaisky. 1964. Potassium uptake and sodium transport at the inside surface of isolated frog skin. Physiologist **7:** 98.

———. 1972. Potassium flux and sodium transport in the isolated skin. Biochim. Biophys. Acta **255:** 517–529.

Carlisky, N. J., and V. L. Lew. 1970. Bicarbonate secretion and non-Na component of the short-circuit current in the isolated colonic mucosa of *Bufo arenarum*. J. Physiol. **206:** 529–541.

Carmeliet, E. E. 1961a. Chloride ions and the membrane potential of Purkinje fibres. J. Physiol. **156:** 375–388.

———. 1961b. Chloride and potassium permeability in cardiac Purkinje fibres. Presses Académiques Européennes, Brussels.

Carmeliet, E., and P. van Bogaert. 1969. Strontium action potentials in cardiac Purkynjē fibers. Arch. Int. Physiol. Biochim. **77:** 134–135.

Carmeliet, E., and J. Vereecke. 1969. Adrenaline and the plateau phase of the cardiac action potential. Importance of Ca^{++}, Na^{+} and K^{+} conductance. Pfluegers Arch. Europe. J. Physiol. **313:** 300–315.

Carmeliet, E., and J. Willems. 1971. The frequency dependent character of the membrane capacity in cardiac Purkinje fibres. J. Physiol. **213:** 85–93.

Carslaw, H. S., and J. C. Jaeger. 1959. p. 078. Conduction of heat in solids. Clarendon Press, Oxford, 2nd ed.

Carstensen, E. L.. 1967. Passive electrical properties of micro-organisms. II. Resistance of the bacterial membrane. Biophys. J. **7:** 493–503.

Carstensen, E. L., H. A. Cox, W. B. Mercer, and L. A. Natale. 1965. Passive electrical properties of microorganisms. I. Conductivity of *Escherichia coli* and *Micrococcus lysodeikticus*. Biophys. J. **5:** 289–300.

Carstensen, E. L., and R. E. Marquis. 1968. Passive electrical properties of microorganisms. III. Conductivity of isolated bacterial cell walls. Biophys. J. **8:** 536–548.

Casteels, R. 1969. Calculation of the membrane potential in smooth muscle cells of the guinea-pig's taenia coli by the Goldman equation. J. Physiol. **205:** 193–208.

Casteels, R., G. Droogmans, and H. Hendrickx. 1971a. Membrane potential of smooth muscle cells in K-free solution. J. Physiol. **217:** 281–295.

———. 1971b. Electrogenic sodium pump in smooth muscle cells of the guinea-pig's taenia coli. J. Physiol. **217:** 297–313.

Cereijido, M., and P. Curran. 1965. Intracellular electrical potentials in frog skin. J. gen. Physiol. **48:** 543–557.

Cereijido, M., I. Reisin, and C. A. Rotunno. 1968. The effect of sodium concentration on the content and distribution of sodium in the frog skin. J. Physiol. **196:** 237–253.

Cereijido, M., and C. A. Rotunno. 1967. Transport and distribution of sodium across frog skin. J. Physiol. **190:** 481–497.

———. 1968. Fluxes and distribution of sodium in frog skin: a new model. J. gen. Physiol. **51:** 280–289.

———. 1971. The effect of antidiuretic hormone on Na movement across frog skin. J. Physiol. **213:** 119–133.

Ceretti, E. R. P., J. L. Webb, and O. F. Schanne: Interatrial conduction in the rat. Am. J. Physiol. **215**: 671–676.

Chailakhian, L. M., and S. A. Iur'ev. 1957. An investigation of the time relations of the action potential and impedance changes on excitation in the frog nerve. Biophysics (Engl. Transl.) **2**: 415–424.

Chalazonitis, N., G. Romey, and A. Arvanitaki. 1967. Résistance de la neuromembrane en fonction de la temperature (neurones d'*Aplysia* et d'*Helix*). Compt. Rend. Soc. Biol. **161**: 1625–1628.

Challice, C. E. 1971. Functional morphology of the specialized tissues of the heart. Methods Achiev. Exp. Pathol. **5**, 121–172.

Chamberlain, S. G., and G. A. Kerkut. 1967. Voltage clamp studies on snail (*Helix aspersa*) neurones. Nature **216**: 89.

———. 1969. Voltage clamp analysis of the sodium and calcium inward currents in snail neurons. Comp. Biochem. Physiol. **28**, 787–801.

Chandler, W. K., A. L. Hodgkin, and H. Meves. 1965. The effect of changing the internal solution on sodium inactivation and related phenomena in giant axons. J. Physiol. **180**: 821–836.

Chandler, W. K., and H. Meves. 1965. Voltage clamp experiments on internally perfused giant axons. J. Physiol. **180**: 788–280.

———. 1970a. Sodium and potassium currents in squid axons perfused with fluoride solutions. J. Physiol. **211**: 623–652.

———. 1970b. Evidence for two types of sodium conductance in axons perfused with sodium fluoride solution. J. Physiol. **211**: 653–678.

———. 1970c. Rate constants associated with changes in sodium conductance in axons perfused with sodium fluoride. J. Physiol. **211**: 679–705.

———. 1970d. Slow changes in membrane permeability and long-lasting action potentials in axons perfused with fluoride solutions. J. Physiol. **211**: 707–728

Chowdhury, T. K., and F. M. Snell. 1965. A microelectrode study of electrical potentials in frog skin and toad bladder. Biochim. Biophys. Acta **94**: 461–471.

Civan, M. M. 1970. Effects of active sodium transport on current–voltage relationship of toad bladder. Am. J. Physiol. **219**: 234–245.

Civan, M. M., and H. S. Frazier. 1968. The site of the stimulatory action of vasopressin on sodium transport in toad bladder. J. gen. Physiol. **51**: 589–605.

Civan, M. M., O. Kedem, and A. Leaf. 1966. Effect of vasopressin on toad bladder under conditions of zero net sodium transport. Am. J. Physiol. **211**: 569–575.

Cohen, L. B., B. Hille and R. D. Keynes. 1969. Light scattering and birefringence changes during activity in the electric organ of *electrophorus electricus*. J. Physiol. **203**: 489–509.

———. 1970. Changes in axon birefringence during the action potential. J. Physiol. **211**: 495–515.

Cohen, L. B., B. Hille, R. D. Keynes, D. Landowne, and E. Rojas. 1971. Analysis of the potential-dependent changes in optical retardation in the squid giant axon. J. Physiol. **218**: 205–237.

Cohen, L. B., and R. D. Keynes. 1969. Optical changes in the voltage-clamped squid axon. J. Physiol. **204**: 100–101P.

———. 1971. Changes in light scattering associated with the action potential in crab nerves. J. Physiol. **212**: 259–275.

Cohen, L. B., R. D. Keynes, and B. Hille. 1968. Light scattering and birefringence changes during nerve activity. Nature **218**: 438–441.

Cohen, L. B., R. D. Keynes, and D. Landowne. 1972a. Changes in light scattering that accompany the action potential in squid giant axons: potential-dependent components. J. Physiol. **224**: 701–725.

———. 1972b. Changes in axon light scattering that accompany the action potential: current-dependent components. J. Physiol. **224**: 727–752.

Cole, K. S. 1928a. Electric impedance of suspensions of *Arbacia* eggs. J. gen. Physiol. **12**: 37–54.

———. 1928b. Electric impedance of suspensions of spheres. J. gen. Physiol. **12**: 29–36.

———. 1932. Electric phase angle of cell membranes. J. gen. Physiol. **15**: 641–649.

———. 1933. Electric conductance of biological systems. Cold Spring Harbor Symp. Quant. Biol. **1**: 107–116.

———. 1935. Electric impedance of *Hipponoë* eggs. J. gen. Physiol. **18**: 877–887.

———. 1937. Electric impedance of marine egg membranes. Trans. Faraday Soc. **33**: 966–972.

———. 1938. Electric impedance of marine egg membranes. Nature **141**: 79.

———. 1947. Dielectric polarization and cell membrane, pp. 20–41. Four lectures on biophysics, Instituto de Biofisica, Rio de Janeiro, Brasil.

———. 1949. Dynamic electrical characteristics of the squid axon membrane. Arch. Sci. Physiol. **3**: 253–258.

———. 1961. Non-linear current–potential relations in an axon membrane. J. gen. Physiol. **44**: 1055–1057.

———. 1968. Membranes, ions and impulses. A chapter of Classical biophysics. University of California Press, Berkeley.

Cole, K. S., and R. F. Baker. 1941. Longitudinal impedance of the squid giant axon. J. gen. Physiol. **24**: 771–788.

Cole, K. S., and R. H. Cole. 1936. Electrical impedance of *Asterias* eggs. J. gen. Physiol. **19**: 609–623.

———. 1941. Dispersion and absorption in dielectrics. I. Alternating current characteristics. J. Chem. Phys. **9**: 341–351.

Cole, K. S., and H. J. Curtis. 1936. Electric impedance of nerve and muscle. Cold Spring Harbor Symp. Quant. Biol. **4**: 73–89.

———. 1937. Wheatstone bridge and electrolytic resistor for impedance measurements over a wide frequency range. Rev. Sci. Instrum. **8**: 333–339.

———. 1938a. Electric impedance of *Nitella* during activity. J. gen. Physiol. **22**: 37–64.

———. 1938b. Electrical impedance of nerve during activity. Nature **142**: 209–210.

———. 1938c. Electric impedance of single marine eggs. J. gen. Physiol. **21**: 591–599.

———. 1939. Electric impedance of the squid axon during activity. J. gen. Physiol. **22**: 649–670.

———. 1941. Membrane potential of the squid giant axon during current flow. J. gen. Physiol. **24**: 551–563.

———. 1950. Bioelectricity: Electric physiology, pp. 82–90. *In* O. Glasser (Ed.), Medical physics. Year Book Publishers, Chicago, vol. 2.

Cole, K. S., and R. M. Guttman. 1942. Electric impedance of the frog egg. J. gen. Physiol. **25**: 765–775.

Cole, K. S., and A. L. Hodgkin. 1939. Membrane and protoplasm resistance in the squid giant axon. J. gen. Physiol. **22**: 671–687.

Cole, K. S., and T. L. Jahn. 1937. The nature and permeability of grasshopper egg membranes. IV: The alternating current impedance over a wide frequency range. J. Cell. Comp. Physiol. **10**: 265–275.

Cole, K. S., and G. Marmont. 1942. The effect of ionic environment upon the longitudinal impedance of the squid giant axon. Fed. Proc. **1**: 15–16.

Cole, K. S., and J. Moore. 1960. Ionic current measurements in the squid giant axon membrane. J. gen. Physiol. **44**: 123–167.

Cole, K. S., and J. M. Spencer. 1938. Electric impedance of fertilized *Arbacia* egg suspensions. J. gen. Physiol. **21**: 583–590.

Cole, R. H. 1965. Relaxation processes in dielectrics. J. Cell. Comp. Physiol. **66**, suppl. 2: 13–20.

Cole, R. H., and P. N. Gross. 1949. A wide range capacitance-conductance bridge. Rev. Sci. Instrum. **20**: 252–260.

Connor, J. A., and C. F. Stevens. 1971a. Inward and delayed outward membrane currents in isolated neural somata under voltage clamp. J. Physiol. **213**: 1–19.

———. 1971b. Voltage clamp studies of a transient outward membrane current in gastropod neural somata. J. Physiol. **213**: 21–30.

———. 1971c. Prediction of repetitive firing behaviour from voltage clamp data on an isolated neurone soma. J. Physiol. **213**: 31–53.

Conti, F., and G. Eisenman. 1965. The steady state properties of ion exchange membranes with fixed sites. Biophys. J. **5**: 511–530.

Conti, F., and G. Palmieri. 1968. Nerve fiber behaviour in heavy water under voltage clamp. Biophysik **5**: 71–77.

Conti, F., and I. Tasaki. 1970. Changes in extrinsic fluorescence in squid axons during voltage-clamp. Science **169**: 1322–1324.

Coombs, J. S., D. R. Curtis, and J. C. Eccles. 1956. Time courses of motoneural responses. Nature **178**: 1049–1050.

———. 1959. The electrical constants of the motoneurone membrane. J. Physiol. **145**: 505–528.

Coombs, J. S., J. C. Eccles, and P. Fatt. 1955. The electrical properties of the motoneurone membrane. J. Physiol. **130**: 291–325.

Coraboeuf, E. 1969. Resistance measurements by means of microelectrodes in cardiac tissue, pp. 224–271. *In* M. Lavallée, O. F. Schanne, and N. C. Hebert (Eds.), Glass microelectrodes. Wiley, New York.

Coraboeuf, E., P. Distel, and J. Boistel. 1955. Electrophysiologie élémentaire des tissus cardiaques, pp. 123–145. *In* Colloque international de microphysiologie du Conseil National de Recherches Scientifique. Paris-Gif sur Yvette, CNRS.

Coraboeuf, E., G. Galand, G. Wallon, and Y. M. Gargouïl. 1959. Effets de milieux hyposodique sur l'activité électrique du coeur isolé de mammifère. Compt. Rend. Soc. Biol. **153**: 1837.

Coraboeuf, E., and P. Guilbault. 1960. Action comparée des milieux hyposodiques sur le coeur isolé de rat et de cobaye. Compt. Rend. Soc. Biol. **154**: 1053.

Coraboeuf, E., and G. Vassort. 1967. Effets de la tétrodotoxine, du tétraéthylammonium et du manganèse sur l'activité du myocarde de rat et de cobaye. Compt. Rend. Acad. Sci. Ser. D. **264**: 1072–1075.

Coraboeuf, E., and S. Weidmann. 1949. Potentiels d'action du muscle cardiaque obtenus à l'aide de microélectrodes intracellulaires. Présence d'une inversion de potential. Compt. Rend. Soc. Biol. **143**: 1360–1361.

Coraboeuf, E., F. Zacouto, Y. M. Gargouïl, and J. Laplaud. 1958. Mesure de la résistance membranaire du myocarde ventriculaire de mammifères au cours de l'activité. Compt. Rend. Acad. Sci. **246**: 2934–2937.

Costantin, L. L. 1968. The effect of calcium on contraction and conductance thresholds in frog skeletal muscle. J. Physiol. **195**: 119–132.

———. 1970. The role of sodium current in the radial spread of contraction in frog muscle fibers. J. gen. Physiol. **55**: 703–715.

Cranefield, P. F., J. A. E. Eyster, and W. E. Gilson. 1951. Electrical characteristics of injury potentials. Am. J. Physiol. **167**: 450–456.

Creed, K. E.. 1971a. Membrane properties of the smooth muscle membrane of the guinea-pig urinary bladder. Pfluegers Arch. Europe. J. Physiol. **326**: 115–126.

———. 1971b. Effects of ions and drugs on the smooth muscle cell membrane of the guinea-pig urinary bladder. Pfluegers Arch. Europe. J. Physiol. **326**: 127–141.

Creese, R., J. B. Dillon, J. Marshall, P. B. Sabawala, D. J. Schneider, D. B. Taylor, and D. E. Zinn. 1957. The effect of neuromuscular blocking agents on isolated human intercostal muscles. J. Pharmacol. Exp. Ther. **119**: 485–494.

Crill, W. E., R. E. Rumery, and J. W. Woodbury. 1959. Effects of membrane current on transmembrane potentials of cultured chick embryo heart cells. Am. J. Physiol. **197**: 733–735.

Curran, P. F., and S. C. Schultz. 1968. Transport across membranes, general principles, pp. 1217–1243. *In* C. F. Code (Ed.), Handbook of physiology. Section 6: Alimentary canal, vol. 3, Intestinal absorption. Amer. Physiol. Soc., Washington, D.C.

Curran, P. F., and G. F. Schwartz. 1960. Na, Cl and water transport by rat colon. J. gen. Physiol. **43**: 555–571.

Curtis, D. R. 1964. Microelectrophoresis, pp. 144–190. *In* W. L. Nastuk (Ed.), Physical techniques in biological research, vol. 5. Academic, New York.

Curtis, D. R., and J. C. Eccles. 1959. Time courses of excitatory and inhibitory synaptic actions. J. Physiol. **145**: 529–546.

Cuthbert, A. W., and E. Painter. 1968. Independent action of antidiuretic hormone, theophylline and cyclic 3′,5′-adenosine monophosphate on cell membrane permeability in frog skin. J. Physiol. **199**: 593–612.

———. 1969a. The action of antidiuretic hormone on cell membranes. Voltage transient studies. Brit. J. Pharmacol. Chemother. **35**: 29–50.

———. 1969b. Capacitance changes in frog skin caused by theophylline and antidiuretic hormone. Brit. J. Pharmacol. Chemother. **37**: 314–324.

Danielli, J. F. 1939. The resistance of nerve in relation to interpolar length. J. Physiol. **96**: 65–73.

Danisi, G., and F. L. Vieira. 1971. Energetica do transporte de sodio. Acta Cient. Venez. **22**, suppl. 2: 27R.

Dänzer, H. 1938. Theorie des Verhaltens biologischer Körper im Hochfrequenzfeld, pp. 191–231. *In* H. Dänzer, H. E. Hollman, B. Rajewsky, H. Schaefer, and E. Schliephake (Eds.), Ultrakurzwellen in ihren medizinisch-biologischen Anwendungen. Georg Thieme Verlag, Leipzig.

Davson, H. 1970. A textbook of general physiology, vol. 2, 4th ed. Williams and Wilkins, Baltimore.

Debye, P. 1929. Polar molecules. Chemical Catalog, New York.

Deck, K. A., R. Kern, and W. Trautwein. 1964. Voltage clamp technique in mammalian cardiac fibres. Pfluegers Arch. Gesamt. Physiol. Mensch. Tiere **280**: 50–62.

Deck, K. A., and W. Trautwein. 1964. Ionic currents in cardiac excitation. Pfluegers Arch. Gesamt. Physiol. Mensch. Tiere **280**: 63–80.

DeFelice, L. J., and C. E. Challice. 1969. Anatomical and ultrastructural study of the electrophysiological atrioventricular node of the rabbit. Circ. Res. **24**: 457–474.

Deguchi, T., and T. Narahashi. 1971. Effects of procaine on ionic conductances of end-plate membranes. J. Pharmacol. Exp. Ther. **176**: 423–433.

Del Castillo, J., and B. Katz. 1954. The membrane change produced by the neuromuscular transmitter. J. Physiol. **125**: 546–565.

Del Castillo, J., J. Y. Lettvin, W. S. McCulloch, and W. Pitts. 1957. Membrane currents in clamped vertebrate nerve. Nature **180**: 1290–1291.

Del Castillo, J., and X. Machne. 1953. Effect of temperature on the passive electrical properties of the muscle fibre membrane. J. Physiol. **120**: 431–434.

Del Castillo, J., and L. Stark. 1952. The effect of calcium ions on the motor end-plate potentials. J. Physiol. **116**: 507–515.

De Hemptinne, A. 1971a. Properties of the outward currents in frog atrial muscle. Pfluegers Arch. Europe. J. Physiol. **329**: 321–331.

———. 1971b. The frequency dependence of outward current in frog auricular fibres. An experimental and theoretical study. Pfluegers Arch. Europe. J. Physiol. **329**: 332–340.

Dhalla, N. S., D. B. McNamara, and P. V. Sulakhe. 1970. Excitation-contraction coupling in heart. V. Contribution of mitochondria and sarcoplasmic reticulum in the regulation of calcium concentration in the heart. Cardiology **55**: 178–191.

Diamond, J. M., and E. M. Wright. 1969. Molecular forces governing non-electrolyte permeation through cell membrane. Proc. Roy. Soc. B. **172**: 273–316.

Diecke, F. P. J., M. E. Westecker, and R. Vogt. 1971. The effect of streptomycin on sodium and potassium currents in myelinated nerve. Arch. Int. Pharmacodyn. Ther. **193**: 5–13.

Dierolf, B. M., and H. S. McDonald. 1969. Effects of temperature acclimation on electrical properties of earthworm giant axons. Z. Vgl. Physiol. **62**: 284–290.

Dodge, F. A., and B. Frankenhaeuser. 1958. Membrane currents in isolated frog nerve fibre under voltage clamp conditions. J. Physiol. **143**: 76–90.

———. 1959. Sodium currents in the myelinated nerve fibre of *Xenopus laevis* investigated with the voltage clamp technique. J. Physiol. **148**: 188–200.

Dowben, R. M. 1969. General physiology. a molecular approach. Harper & Row, New York.

Draper, M. H., and S. Weidmann. 1951. Cardiac resting and action potentials recorded with an intracellular electrode. J. Physiol. **115**: 74–94.

Dubois, J. M., and C. Bergman. 1971a. Variation de la conductance sodium de la membrane nodale en fonction de la concentration en ions Na^+. Compt. Rend. Acad. Sci. Ser. D. **272**: 2796–2799.

———. 1971b. Conductance sodium de la membrane nodale: inhibition compétitive calcium-sodium. Compt. Rend. Acad. Sci. Ser. D. **272**: 2924–2927.

Du Bois-Reymond, E. 1849. Untersuchungen über thierische Elektrizität, vol. 2. Reiner, Berlin.

Dudel, J., M. Morad, and R. Rüdel. 1968. Contractions of single crayfish muscle fibers induced by controlled changes of membrane potential. Pfluegers Arch. Gesamt. Physiol. Mensch. Tiere **299**: 38–51.

Dudel, J., K. Peper, R. Rüdel, and W. Trautwein. 1966. Excitatory membrane current in heart muscle (Purkinje fibers). Pfluegers Arch. Gesamt. Physiol. Mensch. Tiere **292**: 255–273.

———. 1967a. The dynamic chloride component of membrane current in Purkinje fibers. Pfluegers Arch. Gesamt. Physiol. Mensch. Tiere **295**: 197–212.

———. 1967b. The effect of tetrodotoxin on the membrane current in cardiac muscle (Purkinje fibers). Pfluegers Arch. Gesamt. Physiol. Mensch. Tiere **295**: 213–226.

———. 1967c. The potassium component of membrane current in Purkinje fibers. Pfluegers Arch. Gesamt. Physiol. Mensch. Tiere **296**: 308–327.

Dudel, J., K. Peper, and W. Trautwein. 1966. The contribution of Ca^{++} ions to the current voltage relation in cardiac muscle (Purkinje fibers). Pfluegers Arch. Gesamt. Physiol. Mensch. Tiere **288**: 262–281.

Dudel, J., and R. Rüdel. 1970. Voltage and time dependence of excitatory sodium current in cooled sheep Purkinje fibres. Pfluegers Arch. Europe. J. Physiol. **315**: 136–155.

Ebashi, S., and M. Endo. 1968. Calcium ion and muscle contraction. Prog. Biophys. Mol. Biol. **18**: 123–183.

Eccles, J. C. 1966. The physiology of nerve cells. Johns Hopkins Press, Baltimore.

Edmonds, C. J., and J. Marriott. 1968. Electrical potential and short circuit current of an in vitro preparation of rat colon mucosa. J. Physiol. **194**: 479–494.

Edmonds, C. J., and O. E. Nielsen. 1968. Transmembrane electrical potential differences and ionic composition of mucosal cells of rat colon. Acta Physiol. Scand. **72**: 338–349.

Einolf, C. W., and E. L. Carstensen. 1969. Passive electrical properties of microorganisms. IV. Studies of the protoplasts of *Microccus lysodeikticus*. Biophys. J. **9**: 634–643.

Eisenberg, B., and R. S. Eisenberg. 1968. Selective disruption of the sarcotubular system in frog sartorius muscle. A quantitative study with exogenous peroxidase as a marker. J. Cell. Biol. **39**: 451–467.

Eisenberg, R. S. 1967. The equivalent circuit of single crab muscle fibers as determined by impedance measurements with intracellular electrodes. J. gen. Physiol. **50**: 1785–1806.

Eisenberg, R. S., and L. L. Costantin. 1971. The radial variation of potential in the transverse tubular system of skeletal muscle. J. gen. Physiol. **58**: 700–701.

Eisenberg, R. S., and P. W. Gage. 1967. Frog skeletal muscle fibers: changes in electrical properties after disruption of transverse tubular system. Science **158**: 1700–1701.

———. 1969. Ionic conductance of the surface and transverse tubular membrane of frog sartorius fibers. J. gen. Physiol. **53**: 279–297.

Eisenberg, R. S., and E. A. Johnson. 1970. Three dimensional electrical field problems in physiology. Prog. Biophys. Mol. Biol. **20**: 1–65.

Eisenman, G. 1962. Cation selective glass electrodes and their mode of operation. Biophys. J. **2**: 259–323.

Elmquist, D., T. R. Johns, and S. Thesleff. 1960. A study of some electrophysiological properties of human intercostal muscle. J. Physiol. **154**: 602–607.

Engel, E., V. Barcillon, and R. S. Eisenberg. 1972. The interpretation of current–voltage relations recorded from a spherical cell with a single microelectrode. Biophys. J. **12**: 384–403.

Engelmann, T. W. 1877. Vergleichende Untersuchungen zu der Lehre von der Muskel und

Nervenelektrizität. Arch. Gesamt. Physiol. Mensch. Tiere **15**: 116–148.

Essig, A., and S. R. Caplan. 1968. Energetics of active transport processes. Biophys. J. **8**: 1434–1457.

Eyster, J. A. E., and W. E. Gilson. 1947. Electrical characteristics of injuries to heart muscle. Am. J. Physiol. **150**: 572–579.

Eyster, J. A. E., and W. J. Meek. 1921. The origin and conduction of the heart beat. Physiol. Rev. **1**: 1–43.

Falk, G. 1968. Predicted delays in the activation of the contractile system. Biophys. J. **8**: 608–625.

Falk, G., and P. Fatt. 1964. Linear electrical properties of striated muscle fibres observed with intracellular electrodes. Proc. Roy. Soc. B. **160**: 69–123.

———. 1968a. Passive electrical properties of rod outer segments. J. Physiol. **198**: 627–646.

———. 1968b. Conductance changes produced by light in rod outer segments. J. Physiol. **198**: 647–699.

Fatt, P. 1957. Sequence of events in synaptic activation of a motoneurone. J. Neurophysiol. **20**: 61–80.

———. 1961. Intracellular microelectrodes. *In* J. H. Quastel (Ed.), Methods in medical research, vol. 9, pp. 381–404. Year Book Medical Publishers, Chicago.

———. 1964. An analysis of the transverse electrical impedance of striated muscle. Proc. Roy. Soc. B. **159**: 606–651.

Fatt, P., and B. L. Ginsborg. 1958. The ionic requirements for the production of action potentials in crustacean muscle fibers. J. Physiol. **142**: 516–543.

Fatt, P., and B. Katz. 1951. An analysis of the end-plate potential recorded with an intracellular electrode. J. Physiol. **115**: 320–370.

———. 1953a. The electrical properties of crustacean muscle fibres. J. Physiol. **120**: 171–204.

———. 1953b. The effect of inhibitory nerve impulses on a crustacean muscle fibre. J. Physiol. **121**: 374–389.

Ferreira, H. G., F. A. Harrison, and R. D. Keynes. 1966. The potential and short-circuit current across isolated rumen epithelium of the sheep. J. Physiol. **187**: 631–644.

Ferris, C. D. 1974. Introduction to bioelectrodes. Plenum Press, New York.

Fessard, A., and L. Tauc. 1956. Capacité, résistance et variations actives d'impédance d'un soma neuronique. J. de Physiol. **48**: 541–544.

Field, M., D. Fromm, and I. McColl. 1971. Ion transport in rabbit ileal mucosa. I. Na and Cl fluxes and short-circuit current. Am. J. Physiol. **220**: 1388–1396.

Finkelstein, A. 1964. Electrical excitability of isolated frog skin and toad bladder. J. gen. Physiol. **47**: 545–565.

Finkelstein, A., and A. Mauro. 1963. Equivalent circuits as related to ionic systems. Biophys. J. **3**: 215–237.

Finn, A. L. 1971. The kinetics of sodium transport in toad bladder. II. Dual effects of vasopressin. J. gen. Physiol. **57**: 349–362.

Finn, A. L., and M. L. Rockoff. 1971. The kinetics of sodium transport in the toad bladder. I. Determination of the transport pool. J. gen. Physiol. **57**: 326–348.

Firth, D. R., and L. J. DeFelice. 1971. Electrical resistance and volume flow in glass microelectrodes. Can. J. Physiol. Pharmacol. **49**: 436–447.

Fishman, H. M. 1969. Direct recording of K and Na current-potential characteristics of squid axon membrane. Nature **224**: 1116–1118.

———. 1970. Direct and rapid description of the individual ionic current currents of squid axon membrane by ramp potential control. Biophys. J. **10**: 799–817.

Fishman, H. M., and K. S. Cole. 1969. On line measurement of squid axon current-potential characteristics. Fed. Proc. **28**: 333.

Fishman, H. M., and R. I. Macey. 1968. Calcium effects in the electrical excitability of "split" frog skin. Biochim. Biophys. Acta **150**: 482–487.

———. 1969a. The N-shaped current-potential characteristic in frog skin. I. Time development during step voltage clamp. Biophys. J. **9**: 127–139.

———. 1969b. The N-shaped current–potential characteristic in frog skin. II. Kinetic behavior during ramp voltage clamp. Biophys. J. **9**: 140–150.

———. 1969c. The N-shaped current–potential characteristic in frog skin. III. Ionic dependence. Biophys. J. **9**: 151–163.

Fitzhugh, R. 1960. Thresholds and plateaus in the Hodgkin-Huxley nerve equations. J. gen. Physiol. **43**: 867–896.

———. 1966. Theoretical effect of temperature on threshold in the Hodgkin-Huxley nerve model. J. gen. Physiol. **49**: 989–1005.

Flemström, G. 1971. Na^+ transport and impedance properties of the isolated frog gastric mucosa at different O_2 tensions. Biochim. Biophys. Acta **225**: 35–45.

Fozzard, H. A. 1966. Membrane capacity of the cardiac Purkinje fibre. J. Physiol. **182**: 255–267.

Fozzard, H. A., and G. Dominguez. 1969. Effect of formaldehyde and glutaraldehyde on electrical properties of cardiac Purkinje fibers. J. gen. Physiol. **53**: 530–540.

Fozzard, H. A., and D. C. Hellam. 1968. Relationship between membrane voltage and tension in voltage-clamped cardiac Purkinje fibres. Nature **218**: 588–589.

Fozzard, H. A., and W. Sleator. 1967. Membrane ionic conductances during rest and activity in guinea pig atrial muscle. Am. J. Physiol. **212**: 945–951.

Frank, K., and M. C. Becker. 1964. Microelectrodes for recording and stimulation, pp. 23–87. *In* W. L. Nastuk (Ed.), Physical techniques in biological research, vol. 5. Academic, New York.

Frank, K., and M. G. F. Fuortes. 1956. Stimulation of spinal motoneurones with intracellular electrodes. J. Physiol. **134**: 451–470.

Frank, K., M. G. F. Fuortes, and P. G. Nelson. 1959. Voltage clamp of motoneuron soma. Science **130**: 38–39.

Frank, K., and L. Tauc. 1964. Voltage clamp studies of molluscan neuron membrane properties, pp. 113–135. *In* J. F. Hoffman (Ed.), The cellular functions of membrane transport. Prentice Hall, Englewood Cliffs, N.J.

Frankenhaeuser, B. 1965. Computed action potential in nerve from *Xenopus laevis*. J. Physiol. **180**: 780–787.

Frankenhaeuser, B., and A. L. Hodgkin. 1956. The after-effects of impulses in the giant nerve fibres of *Loligo*. J. Physiol. **131**: 341–376.

———. 1957. The action of calcium on the electrical properties of squid axons. J. Physiol. **137**: 218–224.

Frankenhaeuser, B., and A. F. Huxley. 1964. The action potential in the myelinated nerve fibre of *Xenopus laevis* as computed on the basis of voltage clamp data. J. Physiol. **171**: 302–315.

Frankenhaeuser, B., B. D. Lindley, and R. S. Smith. 1966. Potentiometric measurement of membrane action potentials in frog muscle fibres. J. Physiol. **183**: 152–166.

Frankenhaeuser, B., and L. E. Moore. 1963. The effect of temperature on the sodium and potassium permeability changes in myelinated nerve fibres of *Xenopus laevis*. J. Physiol. **169**: 431–437.

Frankenhaeuser, B., and A. B. Vallbo. 1969. Measurement and control of membrane potential in myelinated nerve fibers, pp. 217–223. *In* M. Lavallée, O. F. Schanne, and N. C. Hebert (Eds.), Glass microelectrodes. Wiley, New York.

Frankenhaeuser, B., and B. Waltman. 1959. Membrane resistance and conduction velocity of large myelinated nerve fibres from *Xenopus laevis*. J. Physiol. **148**: 677–682.

Freygang, W. H., Jr., D. A. Goldstein, D. C. Hellam, and L. D. Peachey. 1964. The relation between the late after-potential and the size of the transverse tubular system of frog muscle. J. gen. Physiol. **48**: 235–263.

Freygang, W. H., Jr., S. I. Rapoport, and L. D. Peachey. 1967. Some relations between changes in the linear electrical properties of striated muscle fibers and changes in ultrastructure. J. gen. Physiol. **50**: 2437–2458.

Freygang, W. H., and W. Trautwein. 1969. Theorie und experimentelle Bestätigung eines neuen Ersatzschaltbildes des Purkinje-Fadens. Pfluegers Arch. Europe. J. Physiol. **307**: R35.

———. 1970. The structural implications of the linear electrical properties of cardiac Purkinje strands. J. gen. Physiol. **55**: 524–547.

Fricke, H. 1924. A mathematical treatment of the electrical conductivity and capacity of disperse systems. I. The electric conductivity of a suspension of homogenous spheroids. Phys. Rev. **24**: 575–587.

———. 1925a. The electrical capacity of suspensions of red corpuscles of a dog. Phys. Rev. **26**: 682–687.

———. 1925b. A mathematical treatment of the electrical conductivity and capacity of disperse systems. II. The capacity of a suspension of conducting spheroids surrounded by a non-conducting membrane for a current of low frequency. Phys. Rev. **26**: 678–681.

———. 1931. The electric conductivity and capacity of disperse systems. Physics **1**: 106–115.

———. 1932. The theory of electrolytic polarization. Phil. Mag. **14**: 310–318.

———. 1933. The electric impedance of suspensions of biological cells. Cold Spring Harbor Symp. Quant. Biol. **1**: 117–124.

———. 1953a. The Maxwell-Wagner dispersion in a suspension of ellipsoids. J. Phys. Chem. **57**: 934–937.

———. 1953b. The electric permittivity of a dilute suspension of a membrane-covered ellipsoids. J. Appl. Phys. **24**: 644–646.

———. 1953c. Relation of the permittivity of biological cell suspensions to fractional cell volume. Nature **172**: 731–732.

———. 1955. The complex conductivity of a suspension of stratified particles of spherical or cylindrical form. J. Phys. Chem. **59**: 168–170.

Fricke, H., and H. J. Curtis. 1934a. Specific resistance of the interior of the red blood corpuscle. Nature **133**: 651.

———. 1934b. Electric impedance of suspensions of yeast cells. Nature **134**: 102–103.

———. 1935. Electric impedance of suspensions of leucocytes. Nature **135**: 436.

———. 1936. The determination of surface conductance from measurements on suspensions of spherical particles. J. Phys. Chem. **40**: 715–722.

Fricke, H., and S. Morse. 1926. The electric resistance and capacity of blood for frequencies between 800 and $4\frac{1}{2}$ million cycles. J. gen. Physiol. **9**: 153–167.

Fricke, H., H. P. Schwan, K. Li, and V. Bryson. 1956. A dielectric study of the low-conductance surface membrane in *E. coli*. Nature **177**: 134–135.

Frölich, H. 1958. Theory of dielectrics. Oxford University Press, London, 2nd ed.

Frömter, E. 1972. The route of passive ion movement through the epithelium of *Necturus* gall bladder. J. Memb. Biol. **8**: 259–301.

Frömter, E., and J. Diamond. 1972. Route of passive ion permeation in epithelia. Nat. New Biol. **235**: 9–13.

Frumento, H. S. 1965. The electrical effects of an ionic pump. J. Theor, Biol. **9**: 253–262.

Funder, J., H. H. Ussing, and J. O. Wieth. 1967. The effects of CO_2 and hydrogen ions on active Na transport in the isolated frog skin. Acta Physiol. Scand. **71**: 65–76.

Fuoss, R. M., and F. Accascina. 1959. Electrolytic conductance. Interscience, New York.

Furchgott, R. F., and T. De Gubareff. 1956. Does contractile force of cardiac muscle depend on the rate of an "activation" process between beats? J. Pharmacol. Exp. Ther. **116**: 21–22.

Furshpan, E. J., and D. D. Potter. 1959. Transmission at the giant motor synapses of the crayfish. J. Physiol. **145**: 289–325.

Gage, P. W., and C. M. Armstrong. 1968. Miniature end-plate currents in voltage clamped muscle fibre. Nature **218**: 363–365.

Gage, P. W., and R. S. Eisenberg. 1969. Capacitance of the surface and transverse tubular membrane of frog sartorius fibers. J. gen. Physiol. **53**: 265–278.

Gagné, S. 1970. Etude sur la nature des paramètres électriques d'une microélectrode de verre à bout ouvert: un modèle pour une membrane à pore unique. Thèse de Ph.D. (Biophysique), Université de Sherbrooke, Sherbrooke.

Galeotti, G. 1907. Ricerche di elettrofisiologia secondo i criteri dell' elettrochimica. Z. Allg. Physiol. **6**: 99–118.

Galler, H. 1913. Über den elektrischen Leitungswiderstand des tierischen Körpers. Pfluegers Arch. Gesamt. Physiol. Mensch. Tiere **149**: 156–174.

Garcia Romeu, F., A. Salibián, and S. Pezzani-Hernández. 1969. The nature of the in vivo sodium and chloride uptake mechanisms through the epithelium of the chilean frog *Calyptocephalella gayi* (Dum. et Bibr., 1841). Exchanges of hydrogen against sodium and of bicarbonate against chloride. J. gen. Physiol. **53**: 816–835.

Gargouïl, Y. M., R. Tricoche, D. Fromenty, and E. Coraboeuf. 1958. Effets de l'adrénaline sur l'activité électrique du coeur de mammifères. Compt. Rend. Acad. Sci. **246**: 334–336.

Garnier, D., O. Rougier, Y. M. Gargouïl, and E. Coraboeuf. 1969. Analyse électrophysiologique du plateau des réponses myocardiques, mise en évidence d'un courant lent entrant en absence d'ions bivalents. Pfluegers Arch. Europe. J. Physiol. **313**: 321–342.

Geddes, L. A. 1972. Electrodes and the measurement of bioelectric events. Wiley, New York.

Geddes, L. A., and L. E. Baker. 1967. The specific resistance of biological material—a compendium of data for the biomedical engineer and physiologist. Med. Biol. Eng. **5**: 271–293.

———. 1968. Principles of applied biomedical instrumentation. Wiley, New York.

Geduldig, D. 1968. Analysis of membrane permeability coefficient ratios and internal ion concentrations from a constant field equation. J. Theor. Biol. **19**: 67–78.

Geduldig, D., and R. Gruener. 1970. Voltage clamp of the *Aplysia* giant neurone: early sodium and calcium currents. J. Physiol. **211**: 217–244.

Geduldig, D., and D. Junge. 1968. Sodium and calcium components of action potentials in *Aplysia* giant neurone. J. Physiol. **199**: 347–365.

Gentile, D. E., and W. D. Brodsky. 1969. Effect of ambient pH on sodium transport across isolated turtle bladders. Am. J. Physiol. **217**: 652–660.

George, E. P. 1961. Resistance values in a syncytium. Aust. J. Exp. Biol. Med. Sci. **39**: 267–274.

Gestland, R. C., B. Howland, Y. Lettvin, and W. H. Pitts. 1959. Comments on microelectrodes. Proc. IRE. **47**: 1856–1862.

Gibbons, W. R., and H. A. Fozzard. 1971. Voltage dependence and time dependence of contraction in sheep cardiac Purkinje fibers. Circ. Res. **28**: 446–460.

Giebisch, G., and S. Weidmann. 1967. Membrane currents in mammalian ventricular heart muscle fibres using a "voltage-clamp" technique. Helv. Physiol. Pharmacol. Acta **25**: 189–190 CR.

Gildemeister, M. 1928. Die passivelektrischen Erscheinungen im Tier-und Pflanzenreich, pp. 657–702. In A. Bethe (Ed.), Handbuch die normalen und pathologishen Physiologie, vol 8 (2).

Gingell, D., and J. F. Palmer. 1968. Changes in membrane impedance associated with a cortical contraction in the egg of *Xenopus laevis*. Nature **217**: 98–102.

Ginsborg, B. L. 1967. Ion movements in junctional transmission. Pharmacol. Rev. **19**: 289–316.

Ginzburg, B. Z., and J. Hogg. 1967. What does a short-circuit current measure in biological systems? J. Theor. Biol. **14**: 316–322.

Girardier, L., J. P. Reuben, P. W. Brandt, and H. Grundfest. 1963. Evidence for anion-perm-selective membrane in crayfish muscle fibers and its possible role in excitation-contraction coupling. J. gen Physiol. **47**: 189–214.

Gola, M. 1967. Effets de la pO_2 sur la résistance électrique de la neuromembrane d'*Helix pomatia*. Compt. Rend. Soc. Biol. **161**: 1634–1638.

Gola, M., and G. Romey. 1971. Réponses anormales à des courants sous-liminaires de certaines membranes somatiques (neurones géants d'*Helix pomatia*). Analyse par la méthode du voltage imposé. Pfluegers Arch. Europe. J. Physiol. **327**: 105–131.

Goldman, D. E. 1943. Potential, impedance and rectification in membranes. J. gen. Physiol. **27**: 37–60.

Goldman, L., and L. Binstock. 1969a. Current separations in *Myxicola* giant axons. J. gen. Physiol. **54**: 741–754.

———. 1969b. Leak current rectification in *Myxicola* giant axons. Constant field and constant conductance components. J. gen. Physiol. **54**: 755–764.

Gonzalez, C. F., Y. E. Shamoo, and W. A. Brodsky. 1967. Electrical nature of active chloride transport across short circuited turtle bladders. Am. J. Physiol. **212**: 641–650.

Gonzalez, C. F., Y. E. Shamoo, H. R. Wyssbrod, R. E. Solinger, and W. A. Brodsky. 1967. Electrical nature of sodium transport across the isolated turtle bladder. Am. J. Physiol. **213**: 333–340.

Gonzalez-Serratos, H. 1966. Inward spread of contraction during a twitch. J. Physiol. **185**: 20–21P.

———. 1967. Studies on the inwards spread of activation in isolated muscle fibers. Ph.D. thesis, University of London, England.

Goodford, P. J., and K. Hermansen. 1961. Sodium and potassium movements in the unstriated muscle of the guinea-pig taenia coli. J. Physiol. **158**: 426–448.

Gorman, A. L. F., and M. F. Marmor. 1970. Contributions of the sodium pump and ionic gradients to the membrane potential of a molluscan neurone. J. Physiol. **210**: 897–917.

Goto, M. 1971. Voltage-clamp experiments on the excitation-contraction coupling in the bullfrog ventricle. *In* F. F. Kao, K. Koizumi, and M. Vassalle (Eds.), Research in physiology. Aulo Gaggi, Bologna.

Goto, M., Y. Kimoto, and Y. Kato. 1971. A study on the excitation-contraction coupling of the bullfrog ventricle with voltage clamp technique. Jap. J. Physiol. **21**: 159–173.

Goto, M., Y. Kimoto, M. Saito, and Y. Wada. 1972. Tension fall after contraction of bullfrog atrial muscle examined with the voltage-clamp technique. Jap. J. Physiol. **22**: 637–650.

Goto, M., Y. Kimoto, and Y. Suetsugu. 1972. Membrane currents responsible for contraction and relaxation of the bullfrog ventricle. Jap. J. Physiol. **22**: 315–331.

Gougerot, L., and M. Foucher. 1972. La membrane de l'hématie est-elle un diélectrique parfait? Ann. Phys. Biol. Méd. **6**: 17–42.

Graham, J., and R. W. Gerard. 1946. Membrane potentials and excitation of impaled single muscle fibres. J. Cell. Comp. Physiol. **28**: 99–117.

Green, R. G., and W. P. Larson. 1922. Conductivity of bacterial cells. J. Infect. Dis. **30**: 550–558.

Grundfest, H., R. Guttman, C. D. Hendley, and I. B. Wilson. 1952. Membrane resistance changes in the course of axonal spikes modified by low Na^+ concentration. Science **115**: 522–523.

Grundfest, H., A. M. Shanes, and W. Freygang. 1953. The effect of sodium ions on the impedance change accompanying the spike in the squid giant axon. J. gen. Physiol. **37**: 25–37.

Guilbault, P., J. Delahayes, and M. Paillard. 1966. Résistance membranaire de la fibre ventriculaire de cobaye. Influence des ions calcium et magnesium. Compt. Rend. Soc. Biol. **160**: 2031–2035.

Guld, C. 1964. A glass-covered platinum microelectrode. Med. Electron. Biol. Eng. **2**: 317–324.

Guttman, R. 1939. The electrical impedance of muscle during the action of narcotics and other agents. J. gen. Physiol. **22**: 567–591.

Haas, H. G. 1964. Ein Vergleich zwischen Fluxmessungen und elektrischen Messungen am Myokard. Pfluegers Arch. Gesamt. Physiol. Mensch. Tiere **281**: 271–281.

Haas, H. G., and H. G. Glitsch. 1962. Kalium-Fluxe am Vorhof des Froschherzens. Pfluegers Arch. Gesamt. Physiol. Mensch. Tiere **275**: 358–375.

Haas, H. G., H. G. Glitsch, and W. Trautwein. 1963. Natrium-Fluxe am Vorhof des Froschherzens. Pfluegers Arch. Gesamt. Physiol. Mensch. Tiere **277**: 36–47

Haas, H. G., R. Kern, and H. M. Einwächter. 1970. Electrical activity and metabolism in cardiac tissue: an experimental and theoretical study. J. Memb. Biol. **3**: 180–209.

Haas, H. G., R. Kern, H. M. Einwächter, and M. Tarr. 1971. Kinetics of Na inactivation in frog atria. Pfluegers Arch. Europe. J. Physiol. **323**: 141–157.

Haas, H. G., R. Kern, and E. Lack. 1971. Die Voltage-Clamp-Technik als pharmakologische Testmethode: Der antifibrillatorische Effekt eines Neuroleptikums (Droperidol). Med. Res. **25**: 110–125.

Haas, H. G., and M. Tarr. 1969. Membrane currents in frog atria as affected by metabolic inhibition. Pfluegers Arch. Europe. J. Physiol. **307**: 31S.

Hagiwara, S., S. Chichibu, and K. Naka. 1964. The effects of various ions on resting and spike potentials of barnacle muscle fibers. J. gen. Physiol. **48**: 163–179.

Hagiwara, S., H. Hayashi, and K. Takahashi. 1969. Calcium and potassium currents of the membrane of a barnacle muscle fibre in relation to the calcium spike. J. Physiol. **205**: 115–129.

Hagiwara, S., and K. Naka. 1964. The initiation of spike potential in barnacle muscle fibers under low internal Ca^{++}. J. gen. Physiol. **48**: 141–162.

Hagiwara, S., and S. Nakajima. 1966. Differences in Na and Ca spikes as examined by application of tetrodotoxin, procaine and manganese ions. J. gen. Physiol. **49**: 793–806.

Hagiwara, S., and N. Saito. 1957. Mechanism of action potential production in the nerve cell of a puffer. Proc. Jap. Acad. **33**: 682–685.

———. 1959a. Membrane potential change and membrane current in supramedullary nerve cell of puffer. J. Neurophysiol. **22**: 204–221.

———. 1959b. Voltage current relations in nerve cell membrane of *Onchidium verruculatum*. J. Physiol. **148**: 161–179.

Hagiwara, S., and K. Takahashi. 1967. Surface density of calcium ions and calcium spikes in the barnacle muscle fiber membrane. J. gen. Physiol. **50**: 583–601.

Hagiwara, S., K. Takahashi, and D. Junge. 1968. Excitation-contraction coupling in a barnacle muscle fiber as examined with voltage clamp technique. J. gen. Physiol. **51**: 157–175.

Hagiwara, S., and I. Tasaki. 1958. A study on the mechanism of impulse transmission across the giant synapse of the squid. J. Physiol. **143**: 114–137.

Hagiwara, S., A. Watanabe, and N. Saito. 1959. Potential changes in syncytial neurons of lobster cardiac ganglion. J. Neurophysiol. **22**: 554–572.

Hall, A. E., O. F. Hutter, and D. Noble. 1963. Current-voltage relations of Purkinje fibers in sodium-deficient solutions. J. Physiol. **166**: 225–240.

Halle, W. 1967. Beitrag zur Wirkung verschiedener Ionen und Pharmaka auf die Automatic von Herzventrikel-und Amnionmuskelzellen in der Zellkultur. Naunyn-Schmiedebergs Arch. Pharmakol. Exp. Pathol. **256**: 322–332.

Harris, E. J. 1972. Transport and accumulation in biological systems. 3rd ed., University Park Press, Baltimore and Butterworth, London.

Harris, E. J., and S. Ochs. 1966. Effects of sodium extrusion and local anaesthetics on muscle membrane resistance and potential. J. Physiol. **187**: 5–21.

Harris, E. J., and R. A. Sjodin. 1961. Kinetics of exchange and net movement of frog muscle potassium. J. Physiol. **155**: 221–245.

Hartree, W., and A. V. Hill. 1921. The specific electrical resistance of frog's muscle. Biochem. J. **15**: 379–382.

Hashimoto, Y., M. E. Holman, and J. Tille. 1966. Electrical properties of the smooth muscle membrane of the guinea-pig vas deferens. J. Physiol. **186**: 27–41.

Hauswirth, O., D. Noble, and R. W. Tsien. 1972a. The dependence of plateau currents in cardiac Purkinje fibres on the interval between action potentials. J. Physiol. **222**: 27–49.

———. 1972b. Separation of the pacemaker and plateau components of delayed rectification in cardiac Purkinje fibers. J. Physiol. **225**: 211–235.

Hays, E. A., M. A. Lang, and H. Gainer. 1968. A re-examination of the Donnan distribution as a mechanism for membrane potentials and potassium and chloride ion distributions in crab muscle fibers. Comp. Biochem. Physiol. **26**: 761–792.

Hazlewood, C. F. 1973. The physicochemical state of ions and water in living tissues and model systems. Ann. N. Y. Acad. Sci. **204**: 1–631.

Heaviside, O. 1899. Electromagnetic theory. vol. 2. The Electrician Printing and Publishing Co., London.

Hebert, N. C. 1969. Properties of microelectrode glasses, pp. 25–31. *In* M. Lavallée, O. F. Schanne, and N. C. Hebert (Eds.), Glass microelectrodes. Wiley, New York.

Hecht, H. H., O. F. Hutter, and D. W. Lywood. 1964. Voltage-current relation of short Purkinje fibres in sodium-deficient solutions. J. Physiol. **170**: 5–7P.

Hegel, U., E. Frömter, and T. Wick. 1967. Der elektrische Wandwiderstand des proximalen Konvolutes der Rattenniere. Pfluegers Arch. Physiol. Mensch. Tiere **294**: 274–290.

Helfferich, F. G. 1962. Ion exchange. McGraw-Hill, New York.

Helman, S. I., J. J. Grantham, and M. B. Burg. 1971. Effect of vasopressin on electrical resistance of renal cortical collecting tubules. Am. J. Physiol. **220**: 1825–1832.

Henderson, P. 1907. Zur Thermodynamik der Flüssigkeitsketten. Z. Phys. Chem. **59**: 118–127.

———. 1908. Zur Thermodynamik der Flüssigkeitsketten. Z. Phys. Chem. **63**: 325–345.

Heppner, O. B., and R. Plonsey. 1970. Simulation of electrical interaction of cardiac cells. Biophys. J. **10**: 1057–1075.

Heppner, R. L., S. Weidmann, and E. H. Wood. 1966. Positive and negative inotropic effects by constant electric currents or current pulses applied during the cardiac action potential. Helv. Physiol. Pharmacol. Acta **24**: 94–96C.

Hermsmeyer, K., R. Rulon, and N. Sperelakis. 1972. Loss of the plateau of the cardiac action potential in hypertonic solutions. J. gen. Physiol. **59**: 779–793.

Hermsmeyer, K., and N. Sperelakis. 1970. Decrease in K^+ conductance and depolarization of frog cardiac muscle produced by Ba^{++}. Am. J. Physiol. **219**: 1108–1114.

Hidaka, T., and N. Toida. 1969. Biophysical and mechanical properties of red and white muscle fibres in fish. J. Physiol. **201**: 49–59.

Hille, B. 1968a. Pharmacological modifications of the sodium channels of frog nerve. J. gen. Physiol. **51**: 199–219.

———. 1968b. Charges and potentials at the nerve surface: divalent ions and pH. J. gen. Physiol. **51**: 221–236.

Hingson, D. J., and J. M. Diamond. 1972. Comparison of non electrolyte permeability patterns in several epithelia. J. Memb. Biol. **10**: 93–135.

Hinkle, M., P. Heller, and W. Van der Kloot. 1971. The influence of potassium and chloride ions on the membrane potential of single muscle fibers of the crayfish. Comp. Biochem. Physiol. **40**: 181–201.

Hirschhorn, N., and H. S. Frazier. 1971. Intracellular electrical potential of the epithelium of turtle bladder. Am. J. Physiol. **220**: 1158–1161.

Höber, R. 1910. Eine Methode die elektrische Leitfähigkeit im Innern von Zellen zu messen. Arch. Gesamt. Physiol. Mensch. Tiere **133**: 237–259.

———. 1912. Ein zweites Verfahren die Leitfähigkeit im Innern von Zellen zu messen. Arch. Gesamt. Physiol. Mensch. Tiere **148**: 189–221.

———. 1913. Messungen der inneren Leitfähigkeit von Zellen III. Arch. Gesamt. Physiol. Mensch. Tiere **150**: 15–45.

Hodgkin, A. L. 1947. The membrane resistance of a non-medullated nerve fibre. J. Physiol. **106**: 305–318.

———. 1951. The ionic basis of electrical activity in nerve and muscle. Biol. Rev. **26**: 339–409.

———. 1958. Ionic movements and electrical activity in giant nerve fibres. Proc. Roy. Soc. B. **148**: 1–37.

Hodgkin, A. L., and P. Horowicz. 1959a. Movements of Na and K in single muscle fibres. J. Physiol. **145**: 405–432.

———. 1959b. The influence of potassium and chloride ions on the membrane potential of single muscle fibres. J. Physiol. **148**: 127–160.

———. 1960. The effect of sudden changes in ionic concentrations on the membrane potential of single muscle fibres. J. Physiol. **153**: 370–385.

Hodgkin, A. L., and A. F. Huxley. 1939. Action potentials recorded from inside a nerve fibre. Nature **144**: 710–711.

———. 1945. Resting and action potentials in single nerve fibers. J. Physiol. **104**: 176–195.

———. 1952a. Currents carried by sodium and potassium ions through the membrane of the giant axon of *Loligo*. J. Physiol. **116**: 449–472.

———. 1952b. The components of membrane conductance in the giant axon of *Loligo*. J. Physiol. **116**: 473–496.

———. 1952c. The dual effect of membrane potential on sodium conductance in the giant axon of *Loligo*. J. Physiol. **116**: 497–506.

———. 1952d. A quantitative description of membrane current and its application to conduction and excitation in nerve. J. Physiol. **117**: 500–544.

Hodgkin, A. L., A. F. Huxley, and B. Katz. 1952. Measurement of current-voltage relations in the membrane of the giant axon of *Loligo*. J. Physiol. **116**: 424–448.

Hodgkin, A. L., and B. Katz. 1949. The effect of sodium ions on the electrical activity of the giant axon of the squid. J. Physiol. **108**: 37–77.

Hodgkin, A. L., and R. D. Keynes. 1955. The potassium permeability of a giant nerve fibre. J. Physiol. **128**: 61–88.

Hodgkin, A. L., and S. Nakajima. 1972a. The effect of diameter on the electrical constants of skeletal frog muscle fibres. J. Physiol. **221**: 105–120.

———. 1972b. Analysis of the membrane capacity in frog muscle. J. Physiol. **221**: 121–136.

Hodgkin, A. L., and W. A. H. Rushton. 1946. The electrical constants of a crustacean nerve fibre. Proc. Roy. Soc. B. **133**: 444–479

Hogben, C. A. M. 1955. Active transport of chloride by isolated frog gastric epithelium. Origin of the gastric mucosal potential. Am. J. Physiol. **180**: 641–649.

Hollander, P. B. and J. L. Webb.. 1955. Cellular membrane potentials and contractility of normal rat atrium and the effects of temperature, tension and stimulus frequency. Circ. Res. **3**: 604–612.

Hoshi, T., and F. Sakai. 1967. A comparison of the electrical resistances of the surface cell membrane and cellular wall in the proximal tubule of the newt kidney. Jap. J. Physiol. **17**: 627–637.

Hoshiko, T. 1961. Electrogenesis in frog skin, pp. 31–47. *In* A. M. Shanes (Ed.), Biophysics of physiological and pharmacological actions. American Association for the Advancement of Science, Washington, D. C.

Howell, J. N., and D. J. Jenden. 1967. T-tubules of skeletal muscle: morphological alterations which interrupt excitation contraction coupling. Fed. Proc. **26**: 553.

Hoyle, G., and C. A. G. Wiersma. 1958a. Excitation at neuromuscular junctions in crustacea. J. Physiol. **143**: 403–425.

———. 1958b. Inhibition at neuromuscular junctions in crustacea. J. Physiol. **143**: 426–440.

Hubbard, M. J., and Lord Rothschild. 1939. Spontaneous rhythmical impedance changes in the trout's egg. Proc. Roy. Soc. B. **127**: 510–526.

Hubbard, S. J. 1963. The electrical constants and the component conductances of frog skeletal muscle after denervation. J. Physiol. **165**: 443–456.

Huf, E. 1935. Versuche über den Zusammenhang zwischen Stoffwechsel, Potentialbildung und Funktion der Froschhaut. Pfluegers Arch. Gesamt. Physiol. Mensch. Tiere **235**: 655–673.

Hutter, O. F. 1969. Potassium conductance of skeletal muscle treated with formaldehyde. Nature **224**: 1215–1217.

Hutter, O. F., and D. Noble. 1959. The influence of anions on impulse generation and membrane conductance in Purkinje and myocardial fibres. J. Physiol. **147**: 16–17P.

———. 1960a. The chloride conductance of frog skeletal muscle. J. Physiol. **151**, 89–102.

———. 1960b. Rectifying properties of heart muscle. Nature **188**: 495.

———. 1961. Anion conductance of cardiac muscle. J. Physiol. **157**: 335–350.

Hutter, O. F., and S. M. Padsha. 1959. Effect of nitrate and other anions on the membrane resistance of frog skeletal muscle. J. Physiol. **146**: 117–132.

Hutter, O. F., and H. E. Warner. 1967. The pH sensitivity of the chloride conductance of frog skeletal muscle. J. Physiol. **189**: 403–425.

Huxley, A. F., and R. Stämpfli. 1949. Evidence for saltatory conduction in peripheral myelinated nerve fibers. J. Physiol. **108**: 315–339.

———. 1951. Direct determination of membrane resting potential and action potential in single myelinated nerve fibres. J. Physiol. **112**: 476–495.

Huxley, H. E. 1964. Evidence for continuity between the central elements of the triads and extracellular space in frog sartorius muscle. Nature **202**: 1067–1071.

Iida, T. T. 1943. Effects of diluted sea water on membrane capacitance of sea urchin eggs. J. Fac. Sci. Imp. Univ. Tokyo Sect. 4. **6**: 165–173.

Ildefonse, M., and O. Rougier. 1968. Activation et inactivation du courant potassium de la fibre musculaire squelettique. Compt. Rend. Acad. Sci. Ser. D. **267**: 2344–2347.

———. 1969. Sodium and potassium components of the membrane current in twitch skeletal muscle fibres investigated with voltage-clamp technique. J. Physiol. **205**: 97P.

———. 1972. Voltage-clamp analysis of the early current in frog skeletal muscle fibre using the double sucrose-gap method. J. Physiol. **222**: 373–395.

Inoue, F. 1971. Membrane ionic current of the propagated action potential, a new method. Jap. J. Physiol. **21**: 601–606.

Isenberg, G. 1971. Voltage-Clamp Experimente an isolierten Skelettmuskelfasern unter Kombination von Microelektroden- und Doppelsaccharosetrennwandtechnik. Acta Biol. Med. Ger. **26**: 311–321.

Isenberg, G., and G. Küchler. 1970a. Zur Ableitung bioelektrischer Potentiale von einzelnen Skelettmuskelfasern mit der Saccharosetrennwandmethode. III. Membranwiderstands-messungen. Acta Biol. Med. Ger. **24**: 623–638.

———. 1970b. Zur Ableitung bioelektrischer Potentiale von einzelnen Skelettmuskelfasern mit der Saccharosetrennwandmethode. IV. Strom-Spannungsbeziehungen unter "current clamp" in Kalium-reichen Lösungen. Acta Biol. Med. Ger. **24**: 639–656.

Ishiko, N., and M. Sato. 1960. The effect of stretch on the electrical constants of muscle fibre membrane. Jap. J. Physiol. **10**: 194–203.

Ito, M., and T. Oshima. 1965. Electrical behaviour of the motoneurone membrane during intracellularly applied current steps. J. Physiol. **180**: 607–635.

Ito, Y., and H. Kuriyama. 1971. Membrane properties of the smooth muscle fibres of the guinea-pig portal vein. J. Physiol. **215**: 427–441.

Ives, D. J. G., and G. J. Janz (Eds.). 1961a. Reference electrodes. Theory and practice. Academic, New York.

Ives, D. J. G., and G. J. Janz. 1961b. General and theoretical introduction, pp. 1–70. *In* D. J. G. Ives and G. J. Janz (Eds.), Reference electrodes. Theory and practice. Academic, New York.

Jack, J. J. B., and S. J. Redman. 1971. An electrical description of the motoneurone and its application to the analysis of synaptic potentials. J. Physiol. **215**: 321–352.

Jahn, T. L. 1936. Studies on the nature and permeability of the grasshopper egg. III. Changes in electrical properties of the membranes during development. J. Cell. Comp. Physiol. **8**: 289–300.

Jamakosmanović, A., and W. R. Loewenstein. 1968. Cellular uncoupling in cancerous thyroid epithelium. Nature, **218**: 775.

James, T. N. 1967. Anatomy of the cardiac conduction system in the rabbit. Circ. Res. **20**: 638–648.

James, T. N., L. Sherf, G. Fine, and A. R. Morales. 1966. Comparative ultrastructure of the sinus node in man and dog. Circulation **34**: 139–163.

Janáček, K., and R. Rybová. 1970. Nonpolarized frog bladder preparation. The effects of oxytocin. Pfluegers Arch. Europe. J. Physiol. **318**: 294–304.

Janáček, K., R. Rybová, and M. Slaviková. 1971. Independent stimulation of sodium entry and sodium extrusion in frog urinary bladder by aldosterone. Pfluegers Arch. Europe. J. Physiol. **326**: 316–323.

Jenerick, H. P. 1953. Muscle membrane potential, resistance, and external potassium chloride. J. Cell. Comp. Physiol. **42**: 427–448.

———. 1959. The control of membrane ionic currents by the membrane potentials of muscle. J. gen. Physiol. **42**: 923–930.

———. 1961. Ionic currents in membrane of active muscle fibre. Nature **191**: 1074–1076.

———. 1963. Phase plane trajectories of the muscle spike potential. Biophys. J. **3**: 363–377.

———. 1964. An analysis of the striated muscle fiber action current. Biophys. J. **4**: 77–91.

Johnson, E. A., and M. Lieberman. 1971. Heart: excitation and contraction. Annu. Rev. Physiol. **33**: 479–532.

Johnson, E. A., P. A. Robertson, and J. J. Tille. 1958. Purkinje and ventricular membrane resistance during the rising phase of the action potential. Nature **182**: 1161–1162.

Johnson, E. A., and J. Tille. 1960. Changes in polarization resistance during the repolarisation phase of the rabbit ventricular action potential. Aust. J. Exp. Biol. Med. Sci. **38**: 509–514.

———. 1961. Investigations of the electrical properties of cardiac muscle fibers with the aid of intracellular double-barreled electrodes. J. gen. Physiol. **44**: 443–467.

———. 1964. The repolarization phase of the cardiac ventricular action potential: a time-dependent system of membrane conductances. Biophys. J. **4**: 387–399.

Johnson, E. A., and L. Wilson. 1962. Membrane ionic permeabilities during the cardiac action potential. Aust. J. Exp. Biol. Med. Sci. **40**: 93–104.

Johnson, F. H., H. Eyring, and M. J. Pollisar. 1963. The kinetic basis of molecular biology, 2nd printing. Wiley, New York.

Johnson, S. L., and J. W. Woodbury. 1964. Membrane resistance of human red cells. J. gen. Physiol. **47**: 827–837.

Jones, A. W., and T. Tomita. 1967. The longitudinal tissue resistance of the guinea-pig taenia coli. J. Physiol. **191**: 109–110P.

Jongsma, H. J., and H. E. van Rijn. 1972. Electrotonic spread of current in monolayer cultures of neonatal rat heart cells. J. Memb. Biol. **9**: 341–360.

Julian, F. J., J. W. Moore, and D. E. Goldman. 1962a. Membrane potentials of the lobster giant axon obtained by use of the sucrose gap technique. J. gen. Physiol. **45**: 1195–1216.

———. 1962b. Current voltage relations in the lobster giant axon membrane under voltage clamp conditions. J. gen. Physiol. **45**: 1217–1238.

Junge, D., and G. P. Moore. 1966. Interspike-interval fluctuations in *Aplysia* pacemaker neurons. Biophys. J. **6**: 411–434.

Kamiyama, A., and K. Matsuda. 1966. Electrophysiological properties of the canine ventricular fiber. Jap. J. Physiol. **16**: 407–420.

Kandel, E. R., and L. Tauc. 1965. Mechanism of heterosynaptic facilitation in the giant cell of the abdominal ganglion of *Aplysia depilans*. J. Physiol. **181**: 28–47.

———. 1966. Anomalous rectification in the metacerebral giant cells and its consequence for synaptic transmission. J. Physiol. **183**: 287–304.

Kanno, Y., and W. R. Loewenstein. 1966. Cell-to-cell passage of large molecules. Nature **212**: 629–630.

Kanno, Y., and Y. Matsui. 1968. Cellular uncoupling in cancerous stomach epithelium. Nature **218**: 775–776.

Kao, C. Y., and P. R. Stanfield. 1968. Actions of some anions on electrical properties and mechanical threshold of frog twitch muscle. J. Physiol. **198**: 291–309.

———. 1970. Actions of some cations on the electrical properties and mechanical threshold of frog sartorius muscle fibers. J. gen. Physiol. **55**: 620–639.

Katchalsky, A., and P. F. Curran. 1965. Nonequilibrium thermodynamics in biophysics. Harvard University Press, Cambridge, Mass.

Katz, B. 1942. Impedance changes in frog's muscle associated with electrotonic and "end-plate" potentials. J. Neurophysiol. **5**: 169–184.

———. 1948. The electric properties of the muscle fibre membrane. Proc. Roy. Soc. B. **135**: 506–534.

———. 1949. Les constantes électriques de la membrane du muscle. Arch. Sci. Physiol. **3**: 285–300.

Katz, G. M., and T. L. Schwartz. 1967. Some problems and criteria for voltage clamping of excitable cells, p. 360. In Digest of the 7th International Conference on Medical and Biological Engineering, Stockholm.

Kavaler, F. 1959. Membrane depolarization as a cause of tension development in mammalian ventricular muscle. Am. J. Physiol. **197**: 968–970.

Kawata, H. 1965. Der Innenwiderstand des Froschventrikels. Pfluegers Arch. Gesamt. Physiol. Mensch. Tiere **285**: 222–228.

Kedem, O. 1961. Criteria of active transport, pp. 87–93. *In* A. Kleinzeller and A. Kotyk (Eds.), Membrane transport and metabolism. Academic, New York.

Kedem, O., and A. Essig. 1965. Isotope flows and flux ratios in biological membranes. J. gen. Physiol. **48**: 1047–1070.

Kern, R. 1966. Untersuchungen über die elektrischen Eigenschaften von Glasmikroelektroden. Ph.D. thesis. University of Heidelberg, Germany.

Kern, R., and H. M. Einwächter. 1969. Influence of droperidol on membrane currents and fluxes in frog atria. Pfluegers Arch. Europe. J. Physiol. **307**: 32S.

Kern, R., H. M. Einwächter, H. G. Haas, and E. G. Lack. 1971. Cardiac membrane currents as affected by an neuroleptic agent: droperidol. Pfluegers Arch. Europe. J. Physiol. **325**: 262–278.

Kernell, D. 1966. Input resistance, electrical excitability and size of ventral horn cells in cat spinal cord. Science **152**: 1637–1640.

Keynes, R. D. 1951. The ionic movements during nervous activity. J. Physiol. **114**: 119–150.

———. 1954. The ionic fluxes in frog muscle. Proc. Roy. Soc. B. **142**: 359–382.

Kidder, G. W., III, M. Cereijido, and P. F. Curran. 1963. Transient changes in electrical potential differences across frog skin. Am. J. Physiol. **207**: 933–940.

King, R. W. P. 1965. Transmission line theory. Dover, New York.

Kirkwood, J. G. 1954. Transport of ions through biological membranes from the standpoint of irreversible thermodynamics, pp. 119–127. *In* H. T. Clarke and D. Nachmanson (Eds.), Ion transport across membranes. Academic, New York.

Kishimoto, U. 1965. Voltage clamp and internal perfusion studies on *Nitella* internodes. J. Cell. Comp. Physiol. **66**: 43–54.

Kiyohara, T., and M. Sato. 1967. Membrane constants of red and white muscle fibers in the rat. Jap. J. Physiol. **17**: 720–725.

Klahr, S., and N. S. Bricker. 1964. Na transport by isolated turtle bladder during anaerobiosis and exposure to KCN. Am. J. Physiol. **206**: 1333–1339

Knutsson, E. 1961. Effects of ethanol on membrane potential and membrane resistance of frog muscle fibres. Acta Physiol. Scand. **52**: 242–253.

Knutsson, E., and S. Katz. 1967. The effect of ethanol on the membrane permeability to sodium and potassium ions in frog muscle fibres. Acta Pharmacol. Toxicol. **25**: 54–64.

Kobayashi, M. 1971. Relationship between membrane potential and spike configuration recorded by sucrose gap method in the ureter smooth muscle. Comp. Biochem. Physiol. **38**: 301–308.

Kobayashi, M., C. L. Prosser, and T. Nagai. 1967. Electrical properties of intestinal muscle as measured intracellularly and extracellularly. Am. J. Physiol. **213**: 275–286.

Koefoed-Johnsen, V., and H. H. Ussing. 1953. The contributions of diffusion and flow to the passage of D_2O through living membranes. Acta Physiol. Scand. **28**: 60–76.

———. 1958. The nature of the frog skin potential. Acta Physiol. Scand. **42**: 298–308.

Kohlhardt, M., B. Bauer, H. Krause, and A. Fleckenstein. 1972a. New selective inhibitors of the transmembrane Ca conductivity in mammalian myocardial fibres. Studies with the voltage clamp technique. Experientia **28**: 288–289.

———. 1972b. Differentiation of the transmembrane Na and Ca channels in mammalian cardiac fibres by the use of specific inhibitors. Pfluegers Arch. Europe. J. Physiol. **335**: 309–322.

Kohlrausch, F. 1953. Praktische physik. Vol. 2. 3rd printing. B. G. Teubner Verlagsgesellschaft, Leipzig.

Koide, F. T. 1967. Determination of intercellular bridge resistance in smooth muscle by measurement of injury and sucrose-gap potentials. J. Theor. Biol. **16**: 268–279.

Koppenhöfer, E., and H. Schmidt. 1968a. Die Wirkung von Skorpiongift auf die Ionenströme des Ranvierschen Schnürrings. I. Die Permeabilitäten P_{Na} und P_K. Pfluegers Arch. Europe. J. Physiol. **303**: 133–149.

———. 1968b. Die Wirkung von Skorpiongift auf die Ionenströme des Ranvierschen Schnürrings. II. Unvollständige Natriuminaktivierung. Pfluegers Arch. Europe. J. Physiol. **303**: 150–161.

Kordaš, M. 1969. The effect of membrane polarization on the time course of the end-plate current in frog sartorius muscle. J. Physiol. **204**: 493–502.

———. 1972a. An attempt at an analysis of the factors determining the time course of the end-plate current I. The effects of prostigmine and of the ratio of Mg^{2+} to Ca^{2+}. J. Physiol. **224**: 317–332.

———. 1972b. An attempt at an analysis of the factors determining the time course of the end-plate current II. Temperature. J. Physiol. **224**: 333–348.

Kornacker, K. 1969 Physical principles of active transport and electrical excitability, pp. 39–57. *In* R. M. Dowben (Ed.), Biological membranes. Little, Brown, Boston.

Kostyuk, P. G., Z. A. Sorokina, and Y. D. Kholodova. 1969. Measurement of activity of hydrogen, potassium and sodium ions in striated muscle fibers and nerve cells, pp. 322–348. *In* M. Lavallée, O. F. Schanne, and N. C. Hebert (Eds.), Glass microelectrodes. Wiley, New York.

Kotyk, A., and K. Janáček. 1970. Cell membrane transport, principles and techniques. Plenum Press, New York.

Kriebel, M. E. 1968. Electrical characteristics of tunicate heart cell membranes and nexuses. J. gen. Physiol. **52**: 46–59.

Krischer, C. C. 1969. Theoretical treatment of ohmic and rectifying properties of electrolyte filled micropipettes. Z. Naturforsch. **24**: 151–155.

Krishtal, O. A., and I. S. Magura. 1970. Calcium ions as inward current carriers in mollusc neurones. Comp. Biochem. Physiol. **35**: 857–866.

Krupa, J., B. Kwiatkowski, and J. Terlecki. 1972. Methode zur Leitfähigkeitsbestimmung des Inneren der menschlichen Erythrozyten auf der Grundlage der Messung elektrischer Grössen der Suspension. Biophysik **8**: 227–236.

Küchler, G. 1964. Zur Frage der Übertragungseigenschaften von Glasmikroelectroden bei der intracellulären Membranpotentialmessung. Pfluegers Arch. Gesamt. Physiol. Mensch. Tiere **280**: 210–223.

Kuffler, S. W., and E. M. Vaughan-Williams. 1953. Properties of the "slow" skeletal muscle fibres of the frog. J. Physiol. **121**: 318–340.

Kumamoto, M., and L. Horn. 1970. Voltage clamping of smooth muscle from taenia coli. Microvasc. Res. **2**: 188–201.

Kurihara, S., and K. E. Creed. 1972. Changes in the membrane potential of the smooth muscle cells of the guinea pig urinary bladder in various environments. Jap. J. Physiol. **22**: 667–683.

Kuriyama, H., K. Ohshima, and Y. Sakamoto. 1971. The membrane properties of the smooth muscle of the guinea-pig portal vein in isotonic and hypertonic solutions. J. Physiol. **217**: 179–199.

Kuriyama, H., T. Osa, and H. Tasaki. 1970. Electrophysiological studies of the antrum muscle fibers of the guinea pig stomach. J. gen. Physiol. **55**: 48–62.

Kuriyama, H., T. Osa, and N. Toida. 1967a. Electrophysiological study of the intestinal smooth muscle of the guinea-pig. J. Physiol. **191**: 239–255.

———. 1967b. Membrane properties of the smooth muscle of guinea-pig ureter. J. Physiol. **191**: 225–238.

Kuriyama, H., and T. Tomita. 1965. The responses of single smooth muscle cells of guinea-pig taenia coli to intracellularly applied currents and their effect on the spontaneous electrical activity. J. Physiol. **178**: 270–289.

———. 1970. The action potential in the smooth muscle of the guinea-pig taenia coli and ureter studied by the double sucrose-gap method. J. gen. Physiol. **55**: 147–162.

Lachance, M. 1972. L'état des ions dans le cytoplasme des globules rouges. Thèse de maîtrise, Université de Sherbrooke, Sherbrooke.

Lachance, M., and O. F. Schanne. 1972. The state of ions in the cytoplasm: a confirmation of the theory of H. Pauly and H. P. Schwan. Biophys. Soc. Annu. Meet. Abstr. **12**: 193a.

Lammel, E., and K. Golenhofen. 1971a. Erregbarkeit und Ca-Austausch intestinaler glatter Muskulatur (Taenia Coli des Meerschweinchens) unterhalb der kritischen Temperatur für die Spontanaktivität. Pfluegers Arch. Europe. J. Physiol. **329**: 258–268.

———. 1971b. Messungen der ^{45}Ca-Aufnahme an intestinaler glatter Muskulatur zur Hypothese eines von Ca-Ionen getragenen Aktionsstromes. Pfluegers Arch. Europe. J. Physiol. **329**: 269–282.

Landolt-Börnstein. 1960. Zahlenwerte und Funktionen. Teil 7, Elektrische Eigenschaften II. Springer, Berlin.

Langer, C. A. 1968. Ion fluxes in cardiac excitation and contraction and their relation to myocardial contractility. Physiol. Rev. **48**: 708–757.

Lanthier, R., and O. F. Schanne. 1966. Change of microelectrode resistance in solutions of different resistivities. Naturwiss. **53**: 430.

Larsen, H. E. 1971. The relative contributions of sodium and chloride ions to the conductance of toad skin in relation to shedding of the stratum corneum. Acta Physiol. Scand. **81**: 254–263.

Lassen, U. V., and O. Sten-Knudsen. 1968. Direct measurement of membrane potential and membrane resistance of human red cells. J. Physiol. **195**: 681–696.

Lavallée, M. 1964. Intracellular pH of rat atrial muscle fibers measured by glass micropipette electrodes. Circ. Res. **15**: 185–193.

Lavallée, M., O F. Schanne, and N. C. Hebert. 1969. Glass microelectrodes. Wiley, New York.

Lavallée, M., and G. Szabo. 1969. The effect of glass surface conductivity phenomena on the tip potential of micropipette electrodes, pp. 95–110. *In* M. Lavallée, O. F. Schanne, and N. C. Hebert (Eds.), Glass microelectrodes. Wiley, New York.

Law, P. K., and H. L. Atwood. 1971. Membrane resistance change induced by nitrate and other anions in long and short sarcomere muscle fibres of crayfish. Comp. Biochem. Physiol. **40**: 265–271.

———. 1972. Nonequivalence of surgical and natural denervation in dystrophic mouse muscles. Exp. Neurol. **34**: 200–209.

Leaf, A., J. Anderson, and L. B. Page. 1958. Active sodium transport by the isolated toad bladder. J. gen. Physiol. **41**: 657–668.

Leaf, A., L. B. Page, and J. Anderson. 1959. Respiration and active sodium transport by isolated toad bladder. J. Biol. Chem. **234**: 1625–1629.

Leaf, A., and A. Renshaw. 1957. Ion transport and respiration of isolated frog skin. Biochem. J. **65**: 82–90.

Lecar, H., G. Ehrenstein, L. Binstock, and R. E. Taylor. 1967. Removal of potassium negative resistance in perfused squid giant axons. J. gen. Physiol. **50**: 1499–1515.

Leicht, R., H. Meves, and H. Wellhöner. 1970. Voltage-Clamp Versuche an Riesenzellen der Weinbergschnecke *Helix pomatia*. Pfluegers Arch. Europe. J. Physiol. **316**: 66–67R.

Léoty, C., and G. Raymond. 1972. Mechanical activity and ionic currents in frog atrial trabeculae. Pfluegers Arch. Europe. J. Physiol. **334**: 114–128.

Léoty, C., G. Raymond, and Y. M. Gargouïl. 1970a. Evolution du courant calcique transmembranaire et de la contraction sur la fibre myocardique de grenouille. J. de Physiol. **62**, suppl. 1: 181.

———. 1970b. Tension phasique et tonique de la fibre myocardique de grenouille et les courants ioniques transmembranaires; étude en voltage imposé et par microphotométrie. Compt. Rend. Acad. Sci. Ser. D. **271**: 1545–1548.

———. 1971. La réponse contractile du myocarde sino-auriculaire de grenouille: influence de l'amplitude et de la durée de la dépolarisation transmembranaire. J. de Physiol. **63**: 68–69.

Lev, A. A. 1969. Electrochemical properties of "incompletely sealed" cation-sensitive microelectrodes, pp. 76–94. *In* M. Lavallée, O. F. Schanne, and N. C. Hebert (Eds.), Glass microelectrodes. Wiley, New York.

Levi, H., and H. H. Ussing. 1948. The exchange of sodium and chloride ions across the fibre membrane of the isolated frog sartorius. Acta Physiol. Scand. **16**: 232–249.

Levine, L. 1966. An electrophysiological study of chelonian skeletal muscle. J. Physiol. **183**: 683–713.

Lew, V. L. 1970. Short-circuit current and ionic fluxes in the isolated colonic mucosa of *Bufo arenarum*. J. Physiol. **206**: 509–528.

Lewis, R. 1971. Potassium inactivation in frog nerve membrane. Life Sci. **10**: 251–258.

Lewis, R., and F. P. J. Diecke. 1968. Interactions of rubidium fluxes in frog nerve membrane. Life Sci. **7**: 429–436.

Lieberman, M. 1967. Effects of cell density and low K on action potentials of cultured chick heart cells. Circ. Res. **21**: 879–888.

Linderholm. H. 1952. Active transport of ions through frog skin with special reference to the action of certain diuretics. Acta Physiol. Scand. **27**, suppl. 97: 1–144.

Lindermann, B., and U. Thorns. 1967. Fast potential spike of frog skin generated at the outer surface of the epithelium. Science **158**: 1473–1477.

Lindley, B. D., Bartsch, G. E., and Eberle, B. J. 1967. Determination of relative permeabilities for the constant field equation from experimental data. Math. Biosci. **1**:515–543.

Ling, G., and R. W. Gerard. 1949. The normal membrane potential of frog sartorius fibers. J. Cell. Comp. Physiol. **34**: 383–396.

Loewenstein, W. R. 1966. Permeability of membrane junctions. Ann. N. Y. Acad. Sci. **137**: 441–472.

———. 1967a. On the genesis of cellular communication. Dev. Biol. **15**: 503–520.

———. 1969b. Cell surface membranes in close contact. Role of calcium and magnesium ions. J. Colloid Sci. **25**: 34–46.

Loewenstein, W. R., and Y. Kanno. 1964. Studies on an epithelial (gland) cell junction. I. Modifications of surface membrane permeability. J. Cell Biol. **22**: 565–586.

Loewenstein, W. R., M. Nakas, and S. J. Socolar. 1967. Junctional membrane uncoupling. Permeability transformations at a cell membrane junction. J. gen. Physiol. **50**: 1865–1891.

Loewenstein, W. R., and R. D. Penn. 1967. Intercellular communication and tissue growth. II. Tissue regeneration. J. Cell Biol. **33**: 235–242.

Loewenstein, W. R., S. J. Socolar, S. Higashino, Y. Kanno, and N. Davidson. 1965. Intercellular communication: renal, urinary bladder, sensory and salivary gland cells. Science **149**: 275–298.

Lorente de Nó, R. 1947. Nature of the electrotonic potential, pp. 390–441. *In* A study of nerve physiology. Vol. 131, Rockefeller Institute for Medical Research, New York.

Lorkovič, H., and C. Edwards. 1968. Threshold for contracture and delayed rectification in muscle. Life Sci. **7**: 367–370.

Lullies, H. 1930. Uber die Polarisation in Geweben. II. Mitteilung. Die Polarisation im Nerven. I. Pfluegers Arch. Gesamt. Physiol. Mensch. Tiere **225**: 69–86.

Lüttgau, H.-Ch. 1960. Das Kalium-Transportsystem am Ranvier-Knoten isolierter markhaltiger Nervenfasern. Pfluegers Arch. Gesamt. Physiol. Mensch. Tiere **271**: 613–633.

Lux, H. D. 1967. Eigenschaften eines Neuron-Modells mit Dendriten begrenzter Länge. Pfluegers Arch. Gesamt. Physiol. Mensch. Tiere **297**: 238–255.

Machne, X., and R. Orozco. 1970. Electrical properties of axons of *Callinectes sapidus*. Am. J. Physiol. **219**: 1147–1153.

MacInnes, D. A. 1961. The principles of electrochemistry. Dover, New York.

Maeno, T. 1966. Analysis of sodium and potassium conductances in the procaine end-plate potential. J. Physiol. **183**: 592–606.

Magaribuchi, T., and H. Kuriyama. 1972. Effects of noradrenaline and isoprenaline on the electrical and mechanical activities of guinea-pig depolarized taenia coli. Jap. J. Physiol. **22**: 253–270.

Magleby, K. L., and C. F. Stevens. 1972a. The effect of voltage on the time course of end-plate currents. J. Physiol. **223**: 151–171.

———. 1972b. A quantitative description of end-plate currents. J. Physiol. **223**: 173–197.

Maiskii, V. A. 1963. Changes in the electrical characteristics of muscle fibres with increase in the concentration of potassium in the surrounding medium. Biophysics (Engl. Transl.) **8**: 649–659.

———. 1964. Electrical characteristics of surface membrane of the giant nerve cells of *Helix pomatia*. Fed. Proc. **23**, part II, Transl. Suppl.: T1173–1176.

Mandel, L. J., and P. F. Curran. 1972. Response of the frog skin to steady-state voltage clamping. I. The shunt pathway. J. gen. Physiol. **59**: 503–518.

Mandrino, M. 1971. Relation entre l'inactivation du courant sodium de la fibre musculaire striée de grenouille et une accumulation de potassium dans le système tubulaire transverse. Compt. Rend. Soc. Biol. **165**: 460–465.

Manthey, A. A. 1966. The effect of calcium on the desensitization of membrane receptors at the neuromuscular junction. J. gen. Physiol. **49**: 963–976.

Marey, E. J. 1876. Excitations électriques du coeur, physiologie expérimentale, p. 63. *In* Travaux de laboratoire de M. Marey, vol. 2. Masson et Cie, Paris.

Marmont, G. 1949. Studies on axon membrane. I. A new method. J. Cell. Comp. Physiol. **34**: 351–382.

Marmor, M. F. 1970. Anomalous rectification and electrogenic sodium transport in a molluscan neurone. Nature **226**: 1252–1253.

———. 1971a. The effects of temperature and ions on the current-voltage relation and electrical characteristics of a molluscan neurone. J. Physiol. **218**: 573–598.

———. 1971b. The independence of electrogenic sodium transport and membrane potential in a molluscan neurone. J. Physiol. **218**: 599–608.

Marmor, M. F., and A. L. F. Gorman. 1970. Membrane potential as the sum of ionic and metabolic components. Science **167**: 65–67.

Marmor, M. F., and G. C. Salmoiraghi. 1970. Electrogenic sodium transport and current-voltage relations in a molluscan neuron. Fed. Proc. **29**: abs. 834.

Maruyama, T., and T. Hoshi. 1972. The effect of D-glucose on the electrical potential profile across the proximal tubule of newt kidney. Biochem. Biophys. Acta **282**: 214–225.

Mascher, D., and K. Peper. 1969. Two components of inward current in myocardial muscle fibers. Pfluegers Arch. Europe. J. Physiol. **307**: 190–203.

Mashima, H., and H. Washio. 1968. The changes in membrane potential produced by alternating current or repetitive square pulses in the frog skeletal muscle fibres. Jap. J. Physiol. **18**: 403–416.

Masszi, G., and A. Tigyi-Sebes. 1962. The state of potassium in muscle investigated by high frequency. Acta Physiol. Acad. Sci. Hung. **22**: 272–280.

Matsubara, I., and K. Matsuda. 1969. Contribution of calcium current to the ventricular action potential of dog. Jap. J. Physiol. **19**: 814–823.

Matsuda, K. 1960. Some electrophysiological properties of terminal Purkinje fibers of heart. *In* Y. Katsuki (Ed.), Electrical activity of single cells. Igaku Shoin, Tokyo.

Matsumoto, N., I. Inoue, and U. Kishimoto. 1970. The electric impedance of the squid axon membrane measured between internal and external electrodes. Jap. J. Physiol. **20**: 516–526.

Matsumura, M. 1972a. Electromechanical coupling in crayfish muscle fibers examined by the voltage clamp method. Jap. J. Physiol. **22**: 53–69.

———. 1972b. The effects of metal ions and caffeine on electromechanical coupling in crayfish muscle fibers. Jap. J. Physiol. **22**: 71–85.

Mauro, A. 1961. Anomalous impedance, a phenomenological property of time-variant resistance. Biophys. J. **1**: 353–372.

Maxwell, J. C. 1873. Treatise on electricity and magnetism. Clarendon Press, Oxford.

McAllister, R. E. 1968. Computed action potentials for Purkinje fiber membrane with resistance and capacitance in series. Biophys. J. **8**: 951–954.

———. 1969. Computed membrane currents in cardiac Purkinje fibers during voltage clamps. Biophys. J. **9**: 571–585.

McAllister, R. E., and D. Noble. 1966a. Delayed rectification in cardiac Purkinje fibers. J. Physiol. **182**: 36–37P.

———. 1966b. The time constants of the slow outward current in cardiac Purkinje fibres. J. Physiol. **183**: 23–25P.

———. 1966c. The time and voltage dependence of the slow outward current in cardiac Purkinje fibres. J. Physiol. **186**: 632–662.

———. 1967. The effect of subthreshold potentials on the membrane current in cardiac Purkinje fibers. J. Physiol. **190**: 381–387.

McCann, F. V. 1963. Electrophysiology of an insect heart. J. gen. Physiol. **46**: 803–821.

———. 1964. The effect of anion substitution on bioelectric potentials in the moth heart. Comp. Biochem. Physiol. **13**: 179–188.

———. 1966. The effect of intracellular pulses on membrane potentials in the moth heart. Comp. Biochem. Physiol. **17**: 599–608.

McClendon, J. F. 1926. Colloidal properties of the surface of the living cell. II. Electric conductivity and capacity of blood to alternating currents of long duration and varying in the frequency from 260 to 2,000,000 cycles per second. J. Biol. Chem. **69**: 733–754.

McComas, A. J., K. Mrozek, D. Gardner-Medwin, and W. H. Stanton. 1968. Electrical properties of muscle fibre membranes in man. J. Neurol. Neurosurg. Psychiatry **31**: 434–440.

McDougal, B., and L. P. Sullivan. 1971. The effect of varying O_2 tensions on Na transport in the toad bladder. Proc. Soc. Exp. Biol. Med. **136**: 871–873.

McGuigan, J. A. S. 1968. Tension in ventricular fibres during a voltage clamp. Helv. Physiol. Pharmacol. Acta **26**: CR 362–363.

Mekata, F. 1971. Electrophysiological studies of the smooth muscle cell membrane of the rabbit common carotid artery. J. gen. Physiol. **57**: 738–751.

Merideth, J., C. Mendez, J. Mueller, and G. K. Moe. 1968. Electrical excitability of atrioventricular nodal cells. Circ Res. **23**: 69–85.

Meves, H. 1968. The ionic requirements for the production of action potentials in *Helix pomatia* neurones. Pfluegers Arch. Europe. J. Physiol. **304**: 215–241.

Meves, H., and K. G. Völkner. 1958. Die Wirkung von CO_2 auf das Ruhemembranpotential und die elektrischen Konstanten der quergestreiften Muskelfaser. Pfluegers Arch. Gesamt. Physiol. Mensch. Tiere **265**: 457–476.

Meves, H., and H. Wellhöner. 1969. Die Wirkung von Veratridin auf die Riesennervenzellen der Weinbergschnecke. Pfluegers Arch. Europe. J. Physiol. **312**: R 100–101.

Michalke, W., and W. R. Loewenstein. 1971. Communication between cells of different type. Nature **232**: 121–122.

Minor, A. V., and V. V. Maksimov. 1969. Passivnye electrischeskie svoĭsta modeli ploskoĭ kletki. Biofizika **14**: 328–335.

Minorsky, N. 1947. Introduction to non-linear mechanics. J. W. Edwards, Ann Arbor, Michigan.

Mironneau, J., and J. Lenfant. 1971. Analyse des réponses électriques de la fibre musculaire lisse d'utérus de ratte: mise en évidence d'un courant lent calcico-sodique. Compt. Rend. Acad. Sci. Ser. D **272**: 436–439.

———. 1972a. Activité électrique répétitive et courants transmembranaires du faisceau musculaire lisse de l'utérus; action de l'ocytocine. Compt. Rend. Acad. Sci. Ser. D. **274**: 3269–3272.

———. 1972b. Activité électrique du faisceau musculaire lisse de l'utérus. Mise en évidence et analyse d'un courant ionique sortant. J. de Physiol. **64**: 97–105.

Mironneau, J., J. Lenfant, and Y. Gargouïl. 1971. L'activité électrique de la fibre musculaire lisse de l'utérus: la repolarisation et les processus d'activation et d'inactivation du courant ionique sortant. Compt. Rend. Acad. Sci. Ser. D. **273**: 2290–2293.

Miyamoto, V. K., and T. E. Thompson. 1967. Some electrical properties of lipid bilayer membranes. J. Colloid Sci. **25**: 16–25.

Mobley, B. A., and E. Page. 1972. The surface area of sheep cardiac Purkinje fibres. J. Physiol. **220**: 547–563.

Moore, J. W., and K. S. Cole. 1963. Voltage clamp techniques, pp. 263–321. *In* W. L. Nastuk (Ed.), Physical techniques in biological research, vol. 6. Academic, New York.

Moore, J. W., T. Narahashi, and W. Ulbricht. 1964. Sodium conductance shift in an axon internally perfused with sucrose and low-potassium solution. J. Physiol. **172**: 163–173.

Moore, J. W., W. Ulbricht, and M. Takata. 1964. Effect of ethanol on the sodium and potassium conductances of the squid axon membrane. J. gen. Physiol. **48**: 279–295.

Moore, L. E. 1967. Membrane currents at large positive internal potentials in single myelinated nerve fibres of *Rana pipiens*. J. Physiol. **193**: 433–442.

———. 1971. Effect of temperature and calcium ions on rate constants of myelinated nerve. Am. J. Physiol. **221**: 131–137.

———. 1972. Voltage-clamp experiments on single muscle fibers of *Rana pipiens*. J. gen. Physiol. **60**: 1–19.

Morad, M., and W. Trautwein. 1968. The effect of the duration of the action potential on contraction in the mammalian heart muscle. Pfluegers Arch. Gesamt. Physiol. Mensch. Tiere **299**: 66–82.

Moreno, J. H., I. L. Reisin, E. R. Boulan, C. A. Rotunno, and M. Cereijido. 1973. Barriers to sodium movement across frog skin. J. Memb. Biol. **11**: 99–115.

Morlock, N. L., D. A. Benady, and H. Grundfest. 1968. Analysis of spike electrogenesis of eel electroplaques with phase plane and impedance measurements. J. gen. Physiol. **52**: 22–45.

Mounier, Y., and P. Guilbault. 1970. Influence du chlore sur les phénomènes électriques de repos de la fibre musculaire striée de crabe. Compt. Rend. Acad. Sci. Ser. D. **271**: 415–418.

Mounier, Y., J. P. Vilain, and P. Guilbault. 1970. Influence du pH externe sur la perméabilité ionique membranaire de la fibre musculaire de crabe (*Carcinus maenas*). Compt. Rend. Soc. Biol. **164**: 2566–2570.

Mozhayev, G. A., G. N. Mozhayeva, and I. I. Marakhova. 1963. Change in the conductivity of the membrane during the action potential of the isolated nerve fibre of the grass crab. Biophysics (Engl. Transl.) **8**: 288–297.

Mueller, P., and D. O. Rudin. 1969. Bimolecular lipid membranes: techniques of formation, study of electrical properties and induction of ionic gating phenomena, pp. 141–156. *In* H. Passow and R. Stämpfli (Eds.), Laboratory techniques in membrane biophysics. Springer, Berlin.

Muir, A. R. 1957a. An electron microscope study of the embryology of the intercalated disc in the heart of the rabbit. J. Biophys. Biochem. Cytol. **3**: 193–202.

———. 1957b. Observations on the fine structure of the Purkinje fibres in the ventricles of the sheep's heart. J. Anat. **91**: 251–258.

Müller-Mohnnsen, H., and O. Balk. 1965. Relations between stationary and dynamic properties of Ranvier nodes. Nature **207**: 1255–1257.

———. 1966. Zur Entstehung des K-Aktions potentials. Naturwiss. **53**: 130–131.

Mullins, L. J., and K. Noda. 1963. The influence of sodium-free solutions on the membrane potential of frog muscle fibres. J. gen. Physiol. **47**: 117–132.

Munk, L., and E. P. George. 1968. A mathematical model for the cardiac action potential based on slow inactivation of sodium conductance. Aust. J. Biol. Sci. **21**: 37–43.

Nachmansohn, D., and I. B. Wilson. 1955. Molecular basis for generation of bioelectric potentials, pp. 167–186. *In* T. Shedlovsky (Ed.), Electrochemistry in biology and medicine. Wiley, New York.

Nagai, T., and C. L. Prosser. 1963a. Electrical parameters of smooth muscle cells. Am. J. Physiol. **204**: 915–924.

———. 1963b. Patterns of conduction in smooth muscle. Am. J. Physiol. **204**: 910–914.

Nagasawa, J., and T. Suzuki. 1970. Ca flux and action potential in smooth muscle of guinea-pig taenia coli. Tohoku J. Exp. Med. **102**: 1–12.

Nagel, W., and A. Dörge. 1971. A study of the different sodium compartments and the transepithelial sodium fluxes of the frog skin with the use of ouabain. Pfluegers Arch. Europe. J. Physiol. **324**: 267–278.

Nakagawa, K., S. Klahr, and N. S. Bricker. 1967. Sodium transport by isolated epithelium of the urinary bladder of the fresh-water turtle. Am. J. Physiol. **13**: 1565–1569.

Nakajima, S., and A. L. Hodgkin. 1970. Effect of diameter on the electrical constants of frog skeletal muscle fibres. Nature **227**: 1053–1055.

Nakajima, S., S. Iwasaki, and K. Obata. 1962. Delayed rectification and anomalous rectification in frog's skeletal muscle membrane. J. gen. Physiol. **46**: 97–115.

Nakamura, Y., S. Nakajima, and H. Grundfest. 1965a. Analysis of spike electrogenesis and depolarizing K inactivation in electroplaques of *Electrophorus electricus*, L. J. gen. Physiol. **49**: 321–349.

———. 1965b. The action of tetrodotoxin on electrogenic components of squid giant axons. J. gen. Physiol. **48**: 985–996.

Narahashi, T. 1963a. The properties of insect axons. Adv. Insect Physiol. **1**: 175–256.

———. 1963b. Dependence of resting and action potentials on internal potassium in perfused squid giant axons. J. Physiol. **169**: 91–115.

———. 1964. Restoration of action potential by anodal polarization in lobster giant axons. J. Cell. Comp. Physiol. **64**: 73–96.

Narahashi, T., T. Deguchi, and E. X. Albuquerque. 1971. Effects of batrachotoxin on nerve membrane potential and conductances. Nature (New Biol.) **229**: 221–222.

Narahashi, T., and H. G. Haas. 1968. Interaction of DDT with the components of lobster nerve membrane conductance. J. gen. Physiol. **51**: 177–197.

Narahashi, T., and J. W. Moore. 1968. Neuroactive agents and nerve membrane conductances. J. gen. Physiol. **51**: 93–101S.

Narahashi, T., J. W. Moore, and D. T. Frazier. 1969. Dependence of tetrodotoxin blockage of nerve membrane conductance on external pH. J. Pharmacol. Exp. Ther. **169**: 224–228.

Narahashi, T., J. W. Moore, and W. R. Scott. 1964. Tetrodotoxin blockage of sodium conductance increase in lobster giant axon. J. gen. Physiol. **47**: 965–974.

Nastuk, W. L., and A. L. Hodgkin. 1950. The electrical activity of single muscle fibers. J. Cell. Comp. Physiol. **35**: 39–73.

Nastuk, W. L., and R. L. Parsons. 1970. Factors in the inactivation of postjunctional membrane-receptors of frog skeletal muscle. J. gen. Physiol. **56**: 218–249.

Neher, E. 1971. Two fast transient current components during voltage clamp on snail neurons. J. gen. Physiol. **58**: 36–53.

Neher, E., and H. D. Lux. 1969. Voltage clamp on *Helix pomatia* neuronal membrane; current measurement over a limited area of the soma surface. Pfluegers Arch. Europe. J. Physiol. **311**: 272–277.

———. 1971. Properties of somatic membrane patches of snail neurons under voltage clamp. Pfluegers Arch. Europe. J. Physiol. **322**: 35–38.

Nelson, P. G., and K. Frank. 1964. La production du potentiel d'action étudiée par la tęchnique du voltage imposé sur le motoneurone du chat. Actual. Neurophysiol. **15**: 15–35.

———. 1967. Anomalous rectification in cat spinal motoneurons and effect of polarizing currents on excitatory post-synaptic potential. J. Neurophysiol. **30**: 1097–1113.

Nelson, P. G., and H. D. Lux. 1970. Some electrical measurements of motoneuron parameters. Biophys, J. **10**: 55–73.

Nicholls, J. G. 1956. The electrical properties of denervated skeletal muscle. J. Physiol. **131**: 1–12.

Noble, D. 1962a. The voltage dependence of the cardiac membrane conductance. Biophys. J. **2**: 381–393.

———. 1962b. A modification of the Hodgkin-Huxley equations applicable to Purkinje fibre action and pacemaker potentials. J. Physiol. **160**; 317–352.

———. 1966. Applications of Hodgkin-Huxley equations to excitable tissues. Physiol. Rev. **46**; 1–50.

Noble, D., and R. W. Tsien. 1968. The kinetics and rectifier properties of the slow potassium current in cardiac Purkinje fibers. J Physiol. **195**; 185–214.

———. 1969a. Outward membrane currents activated in the plateau range of potentials in cardiac Purkinje fibres. J. Physiol. **200**: 205–231.

———. 1969b. Reconstruction of the repolarization process in cardiac Purkinje fibres based on voltage clamp measurements of membrane current. J. Physiol. **200**: 233–254.

Nonner, W. 1969. A new voltage clamp method for Ranvier nodes. Pfluegers Arch. Europe. J. Physiol. **309**: 176–192.

Nonomura, Y., Y. Hotta, and H. Ohashi. 1966. Tetrodotoxin and manganese ions: effects on electrical activity and tension in taenia coli of guinea pig. Science **152**: 97–99.

Ochi, R. 1970. The slow inward current and the action of manganese ions in guinea-pig's myocardium. Pfluegers Arch. Europe. J. Physiol. **316**: 81–94.

Ochi, R., and W. Trautwein. 1971. The dependence of cardiac contraction on depolarization and slow inward current. Pfluegers Arch. Europe. J. Physiol. **323**, 187–203.

Ochs, A. L. 1967. Changes in membrane properties with hyperpolarization in snail neurons. Am. J. Physiol. **213**: 16–20.

Oikawa, T., C. S. Spyropoulos, I. Tasaki, and T. Teorell. 1961 Methods for perfusing the giant axons of *Loligo pealii*. Acta Physiol. Scand. **52**: 195–196.

Oker-Blom, M. 1900. Thieriesche Säfte und Gewebe in physikalisch-chemischer Beziehung. Arch. Gesamt. Physiol. Mensch. Tiere **79**: 510–533.

O'Lague, P., H. Delen, H. Rubin, and C. Tobias. 1970. Electrical coupling: low resistance junctions between mitotic and interphase fibroblasts in tissue culture. Science **170**: 464–466.

Oliveira-Castro, G. M., and W. R. Loewenstein. 1971. Junctional membrane permeability. Effects of divalent cations. J. Memb. Biol. **5**: 51–77.

Oncley, J. L. 1943. E. J. Cohn and J. T. Edsall (Eds.) Proteins, aminoacids and peptides as ions and dipolar ions. Reinhold, New York, pp. 543–568. *In*

Oomura, Y., and T. Tomita. 1960. Study on properties of neuromuscular junction, pp. 181–205. *In* Y. Katsuki (Ed.), Electrical activity of single cells. Igaku Shoin, Tokyo.

Orentlicher, M., and J. P. Reuben. 1971. Localization of ionic conductances in crayfish muscle fibers. J. Memb. Biol. **4**: 209–226.

Otsuka, M. 1958. Die Wirkung von Adrenalin auf Purkinje-Fasern von Säugetierherzen. Pfluegers Arch. Gesamt. Physiol. Mensch. Tiere **266**: 512–517.

Paes de Carvalho, A., B. F. Hoffman, and M. De Paula Carvalho. 1969. Two components of the cardiac action potential. I. Voltage-time course and the effect of acetylcholine on atrial and nodal cells of the rabbit heart. J. gen. Physiol. **54**: 607–635.

Page, E., B. Power, H. A. Fozzard, and D. A. Medooff. 1969. Sarcoleminal evaginations with knob-like or stalked projections in Purkinje fibers of the sheep's heart. J. Ultrastruct. Res. **28**: 288–300.

Palti, Y., and W. J. Adelman, Jr. 1969. Measurement of axonal membrane conductances and capacity by means of a varying potential control voltage clamp. J. Memb. Biol. **1**: 431–458.

Pappano, A. J., and N. Sperelakis. 1969a. Spike electrogenesis in cultured heart cells. Am. J. Physiol. **217**: 615–624.

———. 1969b. Low K^+ conductance and low resting potentials of isolated single cultured heart cells. Am. J. Physiol. **217**: 1076–1082.

Patlak, C. S. 1960. Derivation of an equation for the diffusion potential. Nature **188**: 944–945.

Pauly, H. 1959. Electrical conductance and dielectric constant of the interior of erythrocytes. Nature **183**: 333–334.

———. 1962. Electrical properties of the cytoplasmic membrane and the cytoplasm of bacteria and of protoplasts. IRE Trans. Med. Electron. **9**: 93–95.

———. Über die elektrische Kapazität der Zellmembran und die Leitfähigkeit des Zytoplasmas von Ehrlich-Aszitestumorzellen. Biophysik **1**: 143–153.

Pauly, H., L. Packer, and H. P. Schwan. 1960. Electrical properties of mitochondrial membranes. J. Biophys. Biochem. Cytol. **7**: 589–601.

Pauly, H., and H. P. Schwan. 1959. Über die Impedanz einer Suspension von kugelförmigen Teilchen mit einer Schale. Z. Naturforsch. **14b**: 125–131.

———. 1966. Dielectric properties and ion mobility in erythrocytes. Biophys. J. **6**: 621–639.

Peachey, L. D. 1965a. Transverse tubules in excitation-contraction coupling. Fed. Proc. **24**: 1124–1134.

———. 1965b. The sarcoplasmic reticulum and transverse tubules of the frog's sartorius. J. Cell. Biol. **25**: 209–231.

———. 1967. Membrane systems of crab fibers. Am. Zool. **7**: 505–513.

Peachey, L. D., and A. F. Huxley. 1964. Transverse tubules in crab muscle. J. Cell. Biol. **23**: 70–71A.

Peachey, L. D., and R. F. Schild. 1968. The distribution of the *T*-system along sarcomeres of frog and toad sartorius. J. Physiol. **194**: 249–258.

Peiffert, J. 1943. Impedanzmessungen des Nerven mit Hochfrequenz. Verleich mit dem Widerstand für Gleichstrom. J. Radiol. Electrol. Med. Nucl. **25**: 147–148.

Penn, R. D. 1966. Ionic communication between liver cells. J. Cell. Biol. **29**: 171–174.

Peper, K., and W. Trautwein. 1967. The effect of aconitine on the membrane current in cardiac muscle. Pfluegers Arch. Gesamt. Physiol. Mensch. Tiere **296**: 328–336.

———. 1968. A membrane current related to the plateau of the action potential of Purkinje fibers. Pfluegers Arch. Europe. J. Physiol. **303**: 108–123.

———. 1969. A note on the pacemaker current in Purkinje fibers. Pfluegers Arch. Europe. J. Physiol. **309**: 356–361.

Pflüger, E. 1859. Physiologie des Elektrotonus. Hirschwald, Berlin.

Philippson, M. 1921. Les lois de la résistance électrique des tissus vivants. Bull. Acad. Roy. Belg. Cl. Sci. **7**: 387–405.

Pichon, Y. 1968. Nature des courants membranaires dans une fibre nerveuse d'insecte: l'axone géant de *Periplaneta americana*. Compt. Rend. Soc. Biol. **162**: 2233–2240.

———. 1969a. Aspects électriques ioniques du fonctionnement nerveux chez les insectes, cas particulier de la chaîne nerveuse abdominale d'une blatte (*Periplaneta americana* L.). Thèse de doctorat d'état, Université de Rennes, France.

———. 1969b. Effets de la tetrodotoxine (T. T. X.) sur les caractéristiques de perméabiliteé membranaire de la fibre nerveuse isolée d'insecte. Compt. Rend. Acad. Sci. Ser. D. **268**: 1095–1097.

———. 1969c. Effets des ions tetraethylammonium (TEA) sur la membrane de l'axone géant d'insecte. Compt. Rend. Soc. Biol. **163**: 952–958.

———. 1969d. Effets du D.D.T. sur la fibre nerveuse isolée d'insecte. Etude en courant et en voltage imposés. J. de Physiol. **61**, suppl. 1: 162–163.

———. 1970a. Ionic content of haemolymph in the cockroach *Periplaneta americana*. A critical analysis. J. Exp. Biol. **53**: 195–209.

———. 1970b. Voltage clamp study of the cockroach giant axon: a simple method. J. Physiol. **210**: 86–88P.

Pichon, Y., and J. Boistel. 1967. Current-voltage relations in the isolated giant axon of the cockroach under voltage-clamp conditions. J. Exp. Biol. **47**: 343–355.

Plonsey, R., and D. G. Fleming. 1969. Bioelectric phenomena. McGraw-Hill, New York.

Politoff, A. L., S. J. Socolar, and W. R. Loewenstein. 1969. Permeability of a cell membrane junction. Dependence on energy metabolism. J. gen. Physiol. **53**: 498–515.

Portela, A., J. Vaccari, R. J. Perez, A. Ardizzone, J. C. Perez, and P. Stewart. 1970. Electrical membrane constants of sartorius muscle fibers from the South American frog *Leptodactylus ocellatus*. Experientia **26**: 957–958.

Porter, K. R. 1961. The sarcoplasmic reticulum: its recent history and present status. J. Biophys. Biochem. Cytol. 10: 219–226.

Poussart, D. J. M. 1969. Nerve membrane current noise: direct measurements under voltage clamp. Proc. Natl. Acad. Sci. **64**: 95–99.

———. 1971. Membrane current noise in lobster axon under voltage clamp. Biophys. J. **11**: 211–234.

Proske, U., and P. Vaughan. 1968. Histological and electrophysiological investigation of lizard skeletal muscle. J. Physiol. **199**: 495–509.

Prosser, C. L., G. Burnstock, and J. Kahn. 1960. Conduction in smooth muscle: comparative structural properties. Am. J. Physiol. **199**: 545–552.

Prudnikova, I. F. 1959. The effects of eserine on the action potential and on the impedance spike in frog nerve. Biophysics (Engl. Transl.) **4**: 23–32.

Pugsley, I. D. 1966. Contribution of the transverse tubular system to the membrane capacitance of striated muscle of the toad (*Bufo marinus*). Aust. J. Exp. Biol. Med. Sci. **44**: 9–22.

Quensel, W. 1932. Über die Polarisationskapazität (Permeabilität) des Froschmuskels in Abhängigkeit vom Stoffwechsel. Pfluegers Arch. Gesamt. Physiol. Mensch. Tiere **230**: 423–433.

Rajewsky, B. 1938. Biophysikalische Grundlagen der Ultrakurzwellen-Wirkung im lebenden Gewebe. *In* H. Dänzer, H. E. Hollman, B. Rajewsky, H. Schaefer, and E. Schliephake (Eds.), Ultrakurzwellen in ihren Medizinisch-Biologischen Anwendungen. Georg Thieme Verlag, Leipzig.

Rall, W. 1957. Membrane time constant of motoneurons. Science **126**: 454.

———. 1959. Branching dendritic trees and motoneuron membrane resistivity. Exp. Neurol. **1**: 491–527.

———. 1960. Membrane potential transients and membrane time constant of motoneurons. Exp. Neurol. **2**: 503–532.

———. 1964. Theoretical significance of dendritic trees for neuronal input-output relations, pp. 73–97. *In* R. F. Reiss (Ed.), Neural theory and modeling. Stanford University Press, Palo Alto.

———. 1967. Distinguishing theoretical synaptic potentials computed for different soma dendritic distributions of synaptic input. J. Neurophysiol. **30**: 1138–1168.

———. 1969a. Time constants and electrotonic length of membrane cylinders and neurons. Biophys. J. **9**: 1483–1508.

———. 1969b. Distributions of potential in cylindrical coordinates and time constants for a membrane cylinder. Biophys. J. **9**: 1509–1541.

Rapport, D., and G. B. Ray. 1927. Changes of electrical conductivity in the beating tortoise ventricle. Am. J. Physiol. **80**: 126–139.

Rasmussen, H. 1970. Cell communication, calcium ion and cyclic adenosine monophosphate. Science **170**: 404–412.

Rawlins, F., L. Mateu, F. Fragachan, and G. Whittembury. 1970. Isolated toad skin epithelium: transport characteristics. Pfluegers Arch. Europe. J. Physiol. **316**: 64–80.

Rayleigh, Lord. 1892. On the influence of obstacles arranged in rectangular order upon the properties of a medium. Philos. Mag. **34**: 481–502.

Rehm, W. S., and M. E. Lefevre. 1965. Effect of dinitrophenol on potential, resistance, and H^+ rate of frog stomach. Am. J. Physiol. **208**: 922–930.

Reiter, M. 1966. Der Einfluss der Natriumionen auf die Beziehung zwischen Frequenz und Kraft der Kontraktion des isolierten Meerschweinchenmyokards. Arch. Pharmakol. Exp. Pathol. **254**: 261–286.

Reuben, J. P., and H. Gainer. 1962. Membrane conductance during depolarizing postsynaptic potentials of crayfish muscle fibres. Nature **193**: 142–143.

Reuter, H. 1967. The dependence of slow inward current in Purkinje fibres on the extracellular calcium concentration. J. Physiol. **192**: 479–492.

———. 1968. Slow inactivation of currents in cardiac Purkinje fibres. J. Physiol. **197**: 233–253.

Reuter, H. and G. W. Beeler, Jr. 1969a. Sodium current in ventricular myocardial fibers. Science **163**: 397–399.

———. 1969b. Calcium current and activation of contraction in ventricular myocardial fibers. Science **163**: 399–401.

Reuter, H. and N. Seitz. 1968. The dependence of calcium efflux from cardiac muscle on temperature and external ion composition. J. Physiol. **195**: 451–470.

Richardson, I. W. 1972. Multiple membrane systems as biological models. J. Memb. Biol. **8**: 219–236.

Rojas, E., F. Bezanilla, and R. E. Taylor. 1970. Demonstration of sodium and potassium conductance changes during a nerve action potential. Nature **225**: 747–748.

Rojas, E., and R. E. Taylor. 1970. Calcium influxes in perfused squid giant axons during voltage clamp. J. Physiol. **210**: 135P.

Rose, B. 1970. Junctional membrane permeability: restoration by repolarizing current. Science **169**: 607–609.

Rose, R. C., and S. G. Schultz. 1971. Studies on the electrical potential profile across rabbit ileum. Effects of sugars and amino acids on transmural and transmucosal electrical potential differences. J. gen. Physiol. **57**: 639–663.

Rosenberg, T. 1948. On accumulation and active transport in biological systems. I. Thermodynamic considerations. Acta Chem. Scand. **2**: 14–33.

Rothschild, Lord. 1946. The theory of alternating current measurements in biology and its application to the investigation of the biophysical properties of the trout egg. J. Exp. Biol. **23**: 77–99.

———. 1957. The membrane capacitance of the sea urchin egg. J. Biophys. Biochem. Cytol. **3**: 103–110.

Rothschuh, K. E., and H. Meier. 1954. Experimentelle Untersuchungen über das Membranruhepotential und das Kernhüllenwiderstandsverhältnis am M. sartorius. Z. Biol. **107**: 264–274.

Rotunno, C. A., M. I. Pouchan, and M. Cereijido. 1966. Location of the mechanism of active transport of sodium across the frog skin. Nature **210**: 597–599.

Rotunno, C. A., F. A. Vilallonga, M. Fernandez, and M. Cereijido. 1970. The penetration of sodium into the epithelium of the frog skin. J. gen. Physiol. **55**: 716–735.

Rougier, O., M. Ildefonse, and Y. M. Gargouïl. 1966. Application de la technique du double "sucrose-gap" à l'étude électrophysiologique du muscle cardiaque. Compt. Rend. Acad. Sci. Ser. D. **263**: 1482–1485.

Rougier, O., G. Vassort, D. Garnier, Y. M. Gargouïl, and E. Coraboeuf. 1968. Données nouvelles concernant le rôle des ions Na^+ et Ca^{++} sur les propriétés électrophysiologiques des membranes cardiaques; existence d'un canal lent. Compt. Rend. Acad. Sci. Ser. D. **266**: 802–805.

———. 1969. Existence and role of a slow inward current during the frog atrial action potential. Pfluegers Arch. Europe. J. Physiol. **308**: 91–110.

Rougier, O., G. Vassort, and M. Ildefonse. 1968. Analyse qualitative en voltage imposé du courant de membrane de la fibre musculaire squelettique. Compt. Rend. Acad. Sci. Ser. D. **266**: 1754–1757.

Rougier, O., G. Vassort, and R. Stämpfli. 1967. Expérience de "voltage clamp" de fibres musculaires cardiaques à l'aide de la technique du "sucrose-gap." J. de Physiol. **59**: 490.

———. 1968. Voltage clamp experiments on frog atrial heart muscle fibers with the sucrose gap technique. Pfluegers Arch. Europe. J. Physiol. **301**: 91–108.

Rubio, G., and G. Zubieta. 1961. The variation of the electric resistance of microelectrodes during the flow of current. Acta Physiol. Lat. Am. **11**: 91–94.

Ruhenstroth-Bauer, G., and K. Zeininger. 1956. Die Membranen von Leberzellen und deren geformte Bestandteile. Naturwiss. **43**: 426.

Ruiz-Ceretti, E., and A. Ponce Zumino. 1976. Action potential changes under varied $[Na^+]_o$ and $[Ca^{2+}]_o$ indicating the existence of two inward currents in cells of the rabbit atrioventricular node. Circ. Res.**39**: 326–336.

Ruiz de Ceretti, E., A. Ponce Zumino, and I. M. Parisii. 1971. Resolution of two components in the upstroke of the action potential in atrioventricular fibers of the rabbit heart. Can. J. Physiol. Pharmacol. **49**: 642–648.

Ruiz-Ceretti, E., O. F. Schanne, and J. L. Bonnardeaux. 1976. Effects of amphotericin *B* on isolated rabbit hearts. J. Mol. Cell. Cardiol. **8**: 77–88.

Rush, E., E. Lepeschkin, and H. O. Brooks. 1968. Electrical and thermal properties of double-barreled ultra microelectrodes. IEEE Trans. Biomed. Eng. **15**: 80–93.

Rushton, W. A. H. 1927. The effect upon the threshold for nervous excitation of the length of nerve exposed and the angle between current and nerve. J. Physiol. **63**: 357–377.

———. 1934. A physical analysis of the relation between threshold and interpolar length in the electric excitation of medullated nerve. J. Physiol. **82**: 332–352.

Sakamoto, Y. 1969. Membrane characteristics of the canine papillary muscle fiber. J. gen. Physiol. **54**: 765–781.

———. 1971. Electrical activity of guinea-pig taenia coli in calcium Locke solution. Jap. J. Physiol. **21**: 295–306.

Sanders, S. S., J. F. Kurfees, and W. S. Rehm. 1968. Estimation of the electromotive force for the H^+ mechanism of the frog stomach by voltage-clamping. Arch. Biochem. Biophys. **124**: 493–496.

Schaefer, H. 1940. Elektrophysiologie. Vol 1. F. Deuticke, Vienna.

———. 1942. Elektrophysiologie. Vol 2. F. Deuticke, Vienna.

Schanne, O. F. 1969. Measurement of cytoplasmic resistivity by means of the glass microelectrode, pp. 299–321. *In* M. Lavallée, O. F. Schanne, and N. C. Hebert (Eds.), Glass microelectrodes. Wiley, New York.

Schanne, O. F., and E. R. P. de Ceretti. 1965. Simultaneous changes of membrane resistance and of the resistivity of the myoplasm in frog sartorius muscle. Proc. 23rd Int. Congr. Physiol. Sci. Tokyo, Abstr. **156**: 89.

———. 1971. Measurement of input impedance and cytoplasmic resistivity with a single microelectrode. Can. J. Physiol. Pharmacol. **49**: 713–716.

Schanne, O. F., E. R. P. de Ceretti, and C. Rivard. 1973. Glass microelectrode resistance in solutions of different resistivities. Biophys. Soc. Abstr. 300a.

Schanne, O. F., H. Kawata, B. Schäfer, and M. Lavallée. 1966. A study on the electrical resistance of the frog sartorius muscle. J. gen. Physiol. **49**: 897–912.

Schanne, O. F., M. Lavallée, R. Laprade, and S. Gagné. 1968. Electrical properties of glass microelectrodes. Proc. IEEE **56**: 1072-1082.

Schanne, O. F., Ph. Soulier, E. R. P. Ceretti, and J. M. Demers. 1972. Action of streptomycin on the ionic currents of frog atrium. Proc. Can. Fed. Biol. Soc. **15**: 63.

Schanne, O. F., Ph. Soulier, E. R. P. Ceretti, J. M. Demers, and J. Bonnardeaux. 1973. L'influence de l'amphotéricine B sur les courants ioniques des trabécules auriculaires de grenouilles. J. de Physiol. **67**: 308A.

Schanne, O. F., L. J. Thomas, and E. Ceretti. 1966. The input resistance of the rat atria. Fed. Proc. **25**: Abs. 2515.

Scherf, D. 1947. Studies on auricular tachycardia caused by aconitine administration. Proc. Soc. Exp. Biol. Med. **64**: 233–239.

Schmidt, H., and R. Stämpfli. 1966. Die Wirkung von Tetraäthylammonium chlorid auf den einzelnen Ranvierschen Schnürring. Pfluegers Arch. Gesamt. Physiol. Mensch. Tiere **287**: 311–325.

Schmidt, R. F. 1960. Versuche mit Aconitin zum Problem der spontanen Erregungsbildung im Herzen. Pfluegers Arch. Gesamt. Physiol. Mensch. Tiere **271**: 526–556.

Schmitt, O. H. 1955. Dynamic negative admittance components in statically stable membranes, pp. 91–120. *In* T. Shedlovsky (Ed.), Electrochemistry in biology and medicine. Wiley, New York.

Schmitt, O., and H. Schmidt. 1972. Influence of calcium ions on the ionic currents of nodes of Ranvier treated with scorpion venom. Pflueger Arch. Europe. J. Physiol. **333**: 51–61.

Schneider, M. F. 1970. Linear electrical properties of the transverse tubules and surface membrane of skeletal muscle fibers. J. gen. Physiol. **56**: 640–671.

Schultz, S. G. 1969. Mechanisms of absorption, pp. 59–108. *In* R. M. Dowben (Ed.), Biological membranes. Little, Brown, Boston.

———. 1972. Electrical potential differences and electromotive forces in epithelial tissues. J. gen. Physiol. **59**: 794–798.

Schultz, S. G., P. F. Curran, and E. M. Wright. 1967. Interpretation of hexose-dependent electrical potential differences in small intestine. Nature **214**: 509–510.

Schultz, S. G., and R. Zalusky. 1964. Ion transport in isolated rabbit ileum. I. Short-circuit current and Na fluxes. J. gen. Physiol. **47**: 567–584.

Schwan, H. P. 1951. Elektrodenpolarisation und ihr Einfluss auf die Bestimmung dielektrischer Eigenschaften von Flüssigkeiten und biologischem Material. Z. Naturforsch. **6b**: 121–129.

———. 1954. Die elektrischen Eigenschaften von Muskelgewebe bei Niederfrequenz. Z. Naturforsch. **9b**: 245–251.

———. 1957. Electrical properties of tissue and cell suspensions, pp. 147–209. *In* J. H. Lawrence and C. A. Tobias (Eds.), Advances in biological and medical physics, vol. **5**. Academic, New York.

———. 1963. Determination of biological impedances, pp. 323–407. *In* W. L. Nastuk (Ed.), Physical techniques in biological research, vol. **6**. Academic, New York.

———. 1965. Electrical properties of bound water. Ann. N. Y. Acad. Sci. **125**: 344–354.

Schwan, H. P., and E. L. Carstensen. 1957. Dielectric properties of the membrane of lysed erythrocytes. Science **125**: 985–986.

Schwan, H. P., and C. F. Kay. 1957. The conductivity of living tissues. Ann. N. Y. Acad. Sci. **65**, 1007–1013.

Schwan, H. P., and K. Li. 1955. Measurements of materials with high dielectric constants and conductivity at ultrahigh frequencies. AIEE Trans. **73**, part I: 603–607.

Schwan, H. P., and H. J. Morowitz. 1962. Electrical properties of the membranes of the pleuropneumonia-like organism A 5969. Biophys. J. **2**: 395–407.

Schwan, H. P., G. Schwarz, J. Maczuk, and H. Pauly. 1962. On the low-frequency dielectric dispersion of colloidal particles in electrolyte solution. J. Physiol. Chem. **66**: 2626–2635.

Schwan, H. P., and K. Sittel. 1953. Wheatstone bridge for admittance determinations of highly conducting materials at low frequencies. AIEE Trans. May, 114–121.

Schwartz, T. L. 1971. Direct effects on the membrane potential due to pumps that transfer no net charge. Biophys. J. **11**: 944–960.

Schwarz, G. 1962. A theory of the low-frequency dielectric dispersion of colloidal particles in electrolyte solution. J. Phys. Chem. **66**: 2636–2642.

Schwarz, J. R., and W. Vogel. 1971a. Potassium inactivation in myelinated nerve fibres. Experientia **27**: 397–398.

———. 1971b. Potassium inactivation in single myelinated nerve fibres of *Xenopus laevis*. Pfluegers Arch. Europe. J. Physiol. **330**: 61–73.

Seely, J. F., and E. L. Boulpaep. 1971. Renal function studies on the isobaric autoperfused dog kidney. Am. J. Physiol. **221**: 1075–1083.

Shanes, A. M. 1958. Electrochemical aspects of physiological and pharmacological action in excitable cells. Part II. The action potential and excitation. Pharmacol. Rev. **10**: 165–273.

Shanes, A. M., W. H. Freygang, H. Grundfest, and E. Amatniek. 1959. Anesthetic and calcium action in the voltage clamped squid axon. J. gen. Physiol. **42**: 793–802.

Shanes, A. M., H. Grundfest, and W. Freygang. 1953. Low level impedance changes following the spike in the squid giant axon before and after treatment with "veratrine" alkaloids. J. gen. Physiol. **37**: 39–51.

Shaw, T. I. 1964. Core conductor properties of tissue reticulum, pp. 159–163. *In* A collection of papers presented to Sir Charles Lovatt Evans.

Shearer, C. 1919. Studies on the action of electrolytes on bacteria. I. The action of monovalent and divalent salts on the conductivity of bacterial emulsions. J. Hyg. **18**: 337–360.

Sheridan, J. D. 1966. Electrophysiological study of special connections between cells in the early chick embryo. J. Cell Biol. **31**: 1–5.

Shiba, H. 1971. Heaviside's "Bessel cable" as an electric model for flat simple epithelial cells with low resistive junctional membranes. J. Theor. Biol. **30**: 59–68.

Shiba, H., and Y. Kanno. 1971. Further study of the two-dimensional cable theory: An electric model for a flat thin association of cells with a directional intercellular communication. Biophysik **7**: 295–301.

Shigenobu, K., and N Sperelakis. 1971. Development of sensitivity to tetrodotoxin of chick embryonic hearts with age. J. Mol. Cell. Cardiol. **3**: 271–286.

Siegenbeek van Heukelom, J. 1971. Cell communication in epithelial systems. Ph.D. thesis. University of Utrecht.

Siegenbeek van Heukelom, J., J. J. Denier Van der Gon, and F. J. A. Prop. 1970. Epithelial monolayers: A study for cell communications. Biochim. Biophys. Acta **211**: 98–100.

Simpson, F. O., and S. J. Oertelis. 1962. The fine structure of sheep myocardial cells; sarcolemmal invaginations and the transverse tubular system. J. Cell Biol. **12**: 91–100.

Sjodin, R. A. 1959. Rubidium and cesium fluxes in muscle as related to the membrane potential. J. gen. Physiol. **42**: 983–1003.

———. 1961. Some cation interactions in muscle. J. gen. Physiol. **44**: 929–962.

Sjöstrand, F. S., and E. Andersson. 1954. Electron microscopy of the intercalated discs of cardiac muscle tissue. Experientia **10**: 369–370.

Smith, D. S. 1966. The organization and function of the sarcoplasmic reticulum and T-system of muscle cells. Prog. Biophys. Mol. Biol. **16**: 107–142.

Smith, T. G., R. B. Wuerker, and K. Frank. 1967. Membrane impedance changes during synaptic transmission in cat spinal motoneurons. J. Neurophysiol. **30**: 1072–1096.

Snell, F. M. 1969. Some electrical properties of fine-tipped pipette microelectrodes, pp. 111–123. *In* M. Lavallée, O. F. Schanne, and N. C. Hebert (Eds.), Glass microelectrodes. Wiley, New York.

Socolar, S. J., and A. L. Politoff. 1971. Uncoupling cell junctions in a glandular epithelium by depolarizing current. Science **172**: 492–494.

Solomon, A. K. 1960. Red cell membrane structure and ion transport. J. gen. Physiol. **43**, suppl. 1: 1–15.

Solomon, S. 1969. Failure of tetrodotoxin to affect sodium transport by frog skin and proximal tubule of rat kidney. Life Sci. **8**, part 1: 397–400.

Sperelakis, N. 1969a. Lack of electrical coupling between contiguous myocardial cells in vertebrate hearts, pp. 135–165. *In* F. V. McCann (Ed.), Comparative physiology of the heart: Current trends. Birkhäuser Verlag, Basel and Stuttgart.

———. 1969b. Changes in conductances of frog sartorius fibers produced by CO_2, ReO_4^- and temperature. Am. J. Physiol. **217**: 1069–1075.

Sperelakis, N., and T. Hoshiko. 1961. Electrical impedance of cardiac muscle. Circ. Res. **9**: 1280–1283.

Sperelakis, N., T. Hoshiko, and R. M. Berne. 1960. Non-syncytial nature of cardiac muscle: Membrane resistance of single cells. Am. J. Physiol. **198**: 531–536.

Sperelakis, N., and D. Lehmkuhl. 1964. Effect of current on transmembrane potentials in cultured chick heart cells. J. gen. Physiol. **47**: 895–927.

———. 1966. Ionic interconversion of pacemaker and non-pacemaker cultured chick heart cells. J. gen. Physiol. **49**: 867–895.

Sperelakis, N., and A. J. Pappano. 1969. Increase in P_{Na} and P_K of cultured heart cells produced by veratridine. J. gen. Physiol. **53**: 97–114.

Sperelakis, N., and R. Rubio. 1971. An orderly lattice of axial tubules which interconnect adjacent transverse tubules in guinea-pig ventricular myocardium. J. Mol. Cell. Cardiol. **2**: 211–220.

Sperelakis, N., and M. F. Schneider. 1968. Membrane ion conductances of frog sartorius fibers as a function on tonicity. Am. J. Physiol. **215**: 723–729.

Sperelakis, N., M. F. Schneider, and E. J. Harris. 1967. Decreased K^+ conductance produced by Ba^{++} in frog sartorius fibers. J. gen. Physiol. **50**: 1565–1583.

Sperelakis, N., and K. Shigenobu. 1972. Changes in membrane properties of chick embryonic hearts during development. J. gen. Physiol. **60**: 430–453.

Sperelakis, N., and M. Tarr. 1965. Weak electrotonic interaction between neighboring visceral smooth muscle cells. Am. J. Physiol. **208**: 737–747.

Spira, A. W. 1971. The nexus in the intercalated disc of the canine heart: Quantitative data for an estimation of its resistance. J. Ultrastruct. Res. **34**: 409–425.

Spring, K. R., and C. V. Paganelli. 1972. Sodium flux in *Necturus* proximal tubule under voltage clamp. J. gen. Physiol. **60**: 181–201.

Stämpfli, R. 1954. A new method for measuring membrane potentials with external electrodes. Experientia **10**: 508–509.

———. 1958. Die Strom-Spannungs-Charakteristik der erregbaren Membran eines einzelnen Schnürrings und ihre Abhängigkeit von der Ionenkonzentration. Helv. Physiol. Acta **16**: 127–146.

———. 1959. Is the resting potential of Ranvier nodes a potassium potential? Ann. N. Y. Acad. Sci. **81**: 265–284.

———. 1963. Die doppelte Saccharosetrennwandmethode zur Messung von elektrischen Membraneigenschaften mit extracellulären Elektroden. Helv. Physiol. Acta **21**: 189–204.

Stanfield, P. R. 1970. The differential effects of tetraethylammonium and zinc ions on the resting conductance of frog skeletal muscle. J. Physiol. **209**: 231–256.

Stefani, E., and A. B. Steinbach. 1969. Resting potential and electrical properties of frog slow muscle fibres. Effect of different external solutions. J. Physiol. **203**: 383–401.

Steinbach, H. B., S. Spiegelman, and N. Kawata. 1944. The effects of potassium and calcium on the electrical properties of squid axons. J. Cell. Comp. Physiol. **24**: 147–154.

Stevens, Ch. F. 1969. Voltage clamp analysis of a repetitively firing neuron, pp. 76–82. *In* H. H. Jaspers, A. A. Ward, and A. Pope (Eds.), Basic mechanisms of the epilepsies. Little, Brown, Boston.

Stewart, G. N. 1899. The relative volume or weight of corpuscles and plasma in blood. J. Physiol. **24**: 356–373.

Straub, R. W. 1963. Bestimmung elektrischer Konstanten von C-Fasern. Pfluegers Arch. Gesamt. Physiol. Mensch. Tiere **278**: 108.

Strickholm, A. 1961. Impedance of a small electrically isolated area of the muscle cell surface. J. gen. Physiol. **44**: 1073–1088.

———. 1962. Excitation currents and impedance of a small electrically isolated area of the muscle cell surface. J. Cell. Comp. Physiol. **60**: 149–160.

———. 1963. Membrane current of crab muscle. Nature **198**: 393–394.

Suarez-Kurtz, G., J. P. Reuben, P. W. Brandt, and H. Grundfest. 1972. Membrane calcium activation in excitation-contraction coupling. J. gen. Physiol. **59**: 676–688.

Takata, M., J. W. Moore, C. Y. Kao, and F. P. Furhman. 1966. Blockage of sodium conductance increase in lobster giant axon by tetrodotoxin. J. gen. Physiol. **49**: 977–988.

Takata, M., W. F. Pickard, J. Y. Lettvin, and J. W. Moore. 1966. Ionic conductance changes in lobster axon membrane when lanthanum is substituted for calcium. J. gen. Physiol. **50**: 461–471.

Takeda, K. 1967. Permeability changes associated with the action potential in procaine-treated crayfish abdominal muscle fibers. J. gen. Physiol. **50**: 1049–1074.

Takeda, K., and Y. Oomura. 1968. Conductance increase causing anomalous rectification in frog muscle in fluoride-rich solution. Proc. Jap. Acad. **44**: 285–289.

———. 1969. Two component anomalous rectification in frog muscle fibers. Proc. Jap. Acad. **45**: 814–819.

———. 1970. Regenerative response in sarcotubular system of frog muscle fibers in F-rich solution. Proc. Jap. Acad. **46**: 1046–1050.

———. 1971. Enhancement by EDTA of sarcotubular regenerative response produced in F-rich solution. Proc. Jap. Acad. **47**: 732–735.

———. 1972. Sarcotubular regenerative response induced by EDTA in propionate solution. Proc. Jap. Acad. **48**: 753–757.

Takeuchi, A., and N. Takeuchi. 1959. Active phase of frog's end-plate potential. J. Neurophysiol. **22**: 395 411.

———. 1960a. An analysis of end-plate potential, pp. 207–216. *In* Y. Katsuki (Ed.), Electrical activity of single cells. Igaku Shoin, Tokyo.

———. 1960b. On the permeability of end-plate membrane during the action of transmitter. J. Physiol. **154**: 52–67.

Takeuchi, N. 1963. Effects of calcium on the conductance change of the end-plate membrane during the action of transmitter. J. Physiol. **167**: 141–155.

Tamasige, M. 1950. Membrane and sarcoplasm resistance in an isolated frog muscle fibre. Annot. Zool. Jap. **23**: 125–134.

Tanaka, I. 1959. Apparent membrane resistance changes during repolarization of the toad atrium. Fed. Proc. **18**: 156.

Tanaka, I., and Y. Sasaki. 1966. On the electrotonic spread in cardiac muscle of the mouse. J. gen. Physiol. **49**: 1089–1110.

Tarr, M. 1967. Electrotonus in cardiac and smooth muscle. Ph. D. thesis. Western Reserve University, Cleveland.

———. 1971. Two inward currents in frog atrial muscle. J. gen. Physiol. **48**: 523–543.

Tarr, M., and N. Sperelakis. 1964. Weak electrotonic interaction between contiguous cardiac cells. Am. J. Physiol. **207**: 691–700.

Tarr, M., and J. Trank. 1971. Equivalent circuit of frog atrial tissue as determined by voltage clamp-unclamp experiments. J. gen. Physiol. **58**: 551–522.

Tasaki. I. 1949. Collision of two nerve impulses in the nerve fibre. Biochim. Biophys. Acta **3**: 494–497.

———. 1955. New measurements of the capacity and the resistance of the myelin sheath and the nodal membrane of the isolated frog nerve fiber. Am. J. Physiol. **181**: 639–650.

———. 1959a. Conduction of the nerve impulse, pp. 75–121. *In* J. Field, H. W. Magoun, and V. E. Hall (Eds.), Handbook of physiology, vol. 1. American Physiological Society, Washington, D. C.

———. 1959b. Demonstration of two stable states of the nerve membrane in potassium-rich media. J. Physiol. **148**: 306–331.

———. 1968. Nerve excitation. A macromolecular approach. Charles C. Thomas, Springfield, Illinois.

Tasaki, I., and A. F. Bak. 1957. Oscillatory membrane currents of squid axon under voltage-clamp. Science **126**: 696–697.

Tasaki, I., and A. F. Bak. 1958a. Current-voltage relations of single nodes of Ranvier as examined by voltage-clamp technique. J. Neurophysiol. **21**: 124–131.

———. 1958b. Discrete threshold and repetitive responses in the squid axon under voltage clamp. Am. J. Physiol. **193**: 301–308.

Tasaki, I., and W. H. Freygang, Jr. 1955. The parallelism between the action potential, action current, and membrane resistance at a node of Ranvier. J. gen. Physiol. **39**: 211–223.

Tasaki, I., and S. Hagiwara. 1957a. Capacity of muscle fiber membrane. Am. J. Physiol. **188**: 423–429.

———. 1957b. Demonstration of two stable potential states in the squid giant axon under tetraethylammonium chloride. J. gen. Physiol. **40**: 859–885.

Tasaki, I., and K. Mizuguchi. 1949. The changes in the electric impedance during activity and the effect of alkaloids and polarization upon the bioelectric processes in the myelinated nerve fibre. Biochim. Biophys. Acta **3**: 484–493.

Tasaki, I., and M. Shimamura. 1962. Further observations on resting and action potential of intracellularly perfused squid axon. Proc. Natl. Acad. Sci. **48**: 1571–1577.

Tasaki, I., I. Singer, and T. Takenaka. 1965. Effects of internal and external ionic environment on excitability of squid giant axon: A macromolecular approach. J. gen. Physiol. **48**: 1095–1123.

Tasaki, I., I. Singer, and A. Watanabe. 1966. Excitation of squid giant axons in sodium-free external media. Am. J. Physiol. **211**: 746–754.

———. 1967. Cation interdiffusion in squid giant axons. J. gen. Physiol. **50**: 989–1007.

Tasaki, I., and C. S. Spyropoulos. 1958a. Nonuniform response in the squid axon membrane under voltage-clamp. Am. J. Physiol. **193**: 309–317.

———. 1958b. Membrane conductance and current-voltage relation in the squid axon under "voltage clamp." Am. J. Physiol. **193**: 318–327.

Tasaki, I., T. Takenaka, and S. Yamagishi. 1968 Abrupt depolarization and bi-ionic action potentials in internally perfused squid giant axons. Am. J. Physiol. **215**: 152–159.

Tasaki, I., A. Watanabe, and I. Singer. 1966. Excitability of squid giant axons in the absence of univalent cations in the external medium. Proc. Natl. Acad. Sci. **56**: 1116–1122.

Tasaki, K., Y. Tsukahara, and S. Ito. 1968. A simple, direct and rapid method for filling microelectrodes. Physiol. Behav. **3**: 1009–1010.

Tauc, L., and H. Gerschenfeld. 1960. Acetylcholine as a possible transmitter of synaptic inhibiton in *Aplysia*. Compt. Rend. Acad. Sci. **251**: 3076–3078.

Tauc, L., and E. R. Kandel. 1964. An anomalous form of rectification in a molluscan central neurone. Nature **202**: 1339–1346.

Taylor, R. E. 1959. Effect of procaine on electrical properties of squid axon membrane. Am. J. Physiol. **196**: 1071–1078.

———. 1965. Impedance of the squid axon membrane. J. Cell. Comp. Physiol. **66**, suppl. 2: 21–26.

Taylor, R. E., J. W. Moore, and K. S. Cole. 1960. Analysis of certain errors in squid axon voltage clamp measurements. Biophys. J. **1**: 161–202.

Teorell, T. 1946. Application of "square wave analysis" to bioelectric studies. Acta Physiol. Scand. **12**: 235–254.

———. 1948. Membrane electrophoresis in relation to bioelectrical polarization effects. Nature **162**: 961.

———. 1949. Membrane electrophoresis in relation to bioelectrical polarization effects. Arch. Sci. Physiol. **3**: 205–219.

———. 1951. Zur quantitativen Behandlung der Membranpermeabilität. Z. Elektrochem. **55**: 460–469.

———. 1953. Transport processes and electrical phenomena in ionic membranes. Prog. Biophys. Biophys. Chem. **3**: 305–369.

———. 1959a. Electrokinetic membrane processes in relation to properties of excitable tissues. I. Experiments on oscillatory transport phenomena in artificial membranes. J. gen. Physiol. **42**: 831–845.

———. 1959b. Electrokinetic membrane processes in relation to properties of excitable tissues. II. Some theoretical considerations. J. gen. Physiol. **42**: 847–863.

Terzuolo, C. A., and T. Araki. 1959. Voltage clamp of the post-synaptic and spike potentials in spinal motoneurons. Fed. Proc. **18**: Abs. 158.

———. An analysis of intra- versus extracellular potential changes associated with activity of single spinal motoneurons. Ann. N. Y. Acad. Sci. **94**: 547–558.

Thomas, R. C. 1969. Membrane current and intracellular sodium changes in a snail neurone during extrusion of injected sodium. J. Physiol. **201**: 495–514.

Tille, J. 1966. Electrotonic interaction between muscle fibers in the rabbit ventricle. J. gen. Physiol. **50**: 189–202.

Tomita, T. 1956. The nature of action potentials in the lateral eye of the horseshoe crab as revealed by simultaneous intra- and extracellular recording. Jap. J. Physiol. **6**: 327–340.

———. 1966a. Electrical response of smooth muscle to external stimulation in hypertonic solution. J. Physiol. **183**: 450–468.

———. 1966b. Membrane capacity and resistance of mammalian smooth muscle. J. Theor. Biol. **12**: 216–227.

———. 1967. Current spread in the smooth muscle of the guinea-pig vas deferens. J. Physiol. **189**: 163–176.

———. 1969a. Single and coaxial microelectrodes in the study of the retina, pp. 124–153. *In* M. Lavallée, O. F. Schanne, and N. C. Hebert (Eds.), Glass microelectrodes. Wiley, New York.

———. 1969b. The longitudinal tissue impedance of the smooth muscle of guinea-pig taenia coli. J. Physiol. **201**: 145–159.

———. 1970. Electrical properties of mammalian smooth muscle, pp. 197–243. *In* E. Bülbring, A. F. Brading, A. W. Janes, and T. Tomita (Eds.), Smooth muscle. Edward Arnold Publishers, London.

Trautwein, W., K. A. Deck, and R. Kern. 1963. Voltage-clamp-Experimente an Herzmuskelfasern. Pfluegers Arch. Gesamt. Physiol. Mensch. Tiere **278**: 13.

Trautwein, W., and J. Dudel. 1958. Zum Mechanismus der Membranwirkung des Acetylcholin an der Herzmuskelfaser. Pfluegers Arch. Gesamt. Physiol. Mensch. Tiere **266**: 324–334.

Trautwein, W., J. Dudel, and K. Peper. 1965. Stationary S-shaped current voltage relation and hysteresis in heart muscle fibers: Excitatory phenomena in Na^+ free bathing solution. J. Cell. Comp. Physiol. **66**: 79–90.

Trautwein, W., and D. G. Kassebaum. 1961. On the mechanism of spontaneous impulse generation in the pacemaker of the heart. J. gen. Physiol. **45**: 317–330.

Trautwein, W., S. W. Kuffler, and C. C. Edwards. 1956. Changes in membrane characteristics of heart muscle during inhibition. J. gen. Physiol. **40**: 135–145.

Troshin, A. S. 1961. Sorption properties of protoplasm and their role in cell permeability, pp. 45–53. *In* A. Kleinzeller and A. Kotyk (Eds.), Membrane transport and metabolism. Academic, New York.

———. 1966. Problems of cell permeability, pp. 224–225. Pergamon, New York.

Ulbricht, W. 1965. Voltage clamp studies on veratrinized frog nodes. J. Cell. Comp. Physiol. **66**, suppl. 2: 91–98.

———. 1969. The effect of veratridine on excitable membranes of nerve and muscle. Ergeb. Physiol. Biol. Chem. Exp. Pharmakol. **61**: 18–71.

Ussing, H. H. 1948. The use of tracers in the study of active ion transport across animal membranes. Cold Spring Harbor Symp. Quant. Biol. **13**: 193–200.

———. 1949a. The distinction by means of tracers between active transport and diffusion. Acta Physiol. Scand. **19**: 43–56.

———. 1949b. The active ion transport through the isolated frog skin in the light of tracer studies. Acta Physiol. Scand. **17**: 1–37.

———. 1960. The alkali metal ions in isolated systems and tissues, pp. 1–195. *In* O. Eichler and A. Farah (Eds.), Handbuch der experimentellen Pharmakologie. Springer, Berlin.

———. 1961. Experimental evidence and biological significance of active transport, pp. 1–11. *In* Biochemie des aktiven Transports, 12. Colloquium der Gesellschaft für Physiologische Chemie. Springer, Berlin.

Ussing, H. H., and E. E. Windhager. 1964. Nature of shunt path and active sodium transport path through frog skin epithelium. Acta Physiol. Scand. **61**: 484–504.

Ussing, H. H., and K. Zerahn. 1951. Active transport of sodium as a source of electric current in the short-circuited isolated frog skin. Acta Physiol. Scand. **23**: 110–127.

Van der Kloot, W. G., and B. Dane. 1964. Conduction of the action potential in the frog ventricle. Science **146**: 74–75.

Vassalle, M. 1966. Analysis of cardiac pacemaker potential using a "voltage clamp" technique. Am. J. Physiol. **210**: 1335–1341.

Vassort, G., and O. Rougier. 1972. Membrane potential and slow inward current dependence of frog cardiac mechanical activity. Pfluegers Arch. Europe. J. Physiol. **331**: 191–203.

Vassort, G., O. Rougier, D. Garnier, M. P. Sauviat, E. Coraboeuf, and Y. M. Gargouïl. 1968. Effets de l'adrénaline sur les courants entrants transmembranaires au cours de l'activité cardiaque. Compt. Rend. Acad. Sci. Ser. D. **267**: 1762–1765.

———. 1969. Effects of adrenaline on membrane inward currents during the cardiac action potential. Pfluegers Arch. Europe. J. Physiol. **309**: 70–81.

Velick, S., and M. Gorin. 1940. The electrical conductance of suspensions of ellipsoids and its relation to the study of avian erythrocytes. J. gen. Physiol. **23**: 753–771.

Vereecke, J., and E. Carmeliet. 1971a. Sr action potentials in cardiac Purkyně fibres. I. Evidence for a regenerative increase in Sr conductance. Pfluegers Arch. Europe. J. Physiol. **322**: 60–72.

———. 1971b. Sr action potentials in cardiac Purkyně fibres. II. Dependence of the Sr conductance on the external Sr concentration and Sr-Ca antagonism. Pfluegers Arch. Europe. J. Physiol. **322**: 73–82.

Verveen, A. A., and H. E. Derksen. 1969. Amplitude distribution of axon membrane noise voltage. Acta Physiol. Pharmacol. Neerl. **15**: 353–379.

Verveen, A. A., H. E. Derksen, and K. L. Schick. 1967. Voltage fluctuations of neural membrane. Nature **216**: 588–589.

Vieira, F. L., S. R. Caplan, and A. Essig. 1972a. Energetics of sodium transport in frog skin. I. Oxygen consumption in the short-circuited state. J. gen. Physiol. **59**: 60–76.

———. 1972b. Energetics of sodium transport in frog skin. II. The effects of electrical potential on oxygen consumption. J. gen. Physiol. **57**: 77–91.

Vitek, M., and W. Trautwein. 1971. Slow inward current and action potential in cardiac Purkinje fibres. The effect of Mn^{++}-ions. Pfluegers Arch. Europe. J. Physiol. **323**: 204–218.

Wagner, K. W. 1913. Zur Theorie der unvollkommenen Dielektrika. Ann. Phys. **40**: 817–855.

Warburg, E. 1899. Über das Verhalten sogenannter unpolarisierbarer Elektroden gegen Wechselstrom. Ann. Phys. Chem. **67**: 493–499.

Wardell, W. M., and T. Tomita. 1967. Coupling resistance of double barrelled microelectrodes. Nature **216**: 1007–1008.

Watanabe, A., S. Obarà, and T. Akiyama. 1967. Pacemaker potentials for the periodic burst discharge in the heart ganglion of a stomatopod, *Squilla oratoria*. J. gen. Physiol. **50**: 839–862.

Weidmann, S. 1951a. Effect of current flow on the membrane potential of cardiac muscle. J. Physiol. **115**: 227–236.

———. 1951b. Electrical characteristics of sepia axons. J. Physiol. **114**: 372–381.

———. 1952. The electrical constants of Purkinje fibers. J. Physiol. **118**: 348–360.

———. 1955a. Effects of calcium ions and local anaesthetics on electrical properties of Purkinje fibres. J. Physiol. **129**: 568–582.

———. 1955b. The effect of the cardiac membrane potential on the rapid availability of the sodium-carrying system. J. Physiol. **127**: 213–224.

———. 1955c. Rectifier properties of Purkinje fibers. Am. J. Physiol. **183**: 671.

———. 1956. Elektrophysiologie der Herzmuskelfaser. Huber, Bern.

———. 1965. The functional significance of the intercalated discs, pp. 149–152. *In* B. Taccardi and G. Marchetti (Eds.), Electrophysiology of the heart. Pergamon, London.

———. 1966. The diffusion of radiopotassium across intercalated discs of mammalian cardiac muscle. J. Physiol. **187**: 323–342.

———. 1970. Electrical constants of trabecular muscle from mammalian heart. J. Physiol. **210**: 1041–1054.

West, T. C. 1955, Ultramicroelectrode recording from the cardiac pacemaker. J. Pharmacol. Exp. Ther. **115**: 283–290.

Whittembury, G. 1964. Electrical potential profile of the toad skin epithelium. J. gen. Physiol. **47**: 795–808.

Willems, J., and E. E. Carmeliet. 1969. Membrane capacity of cardiac Purkyně fibre in sodium- and strontium-Tyrode. Arch. Int. Physiol. Biochim. **77**: 943–944.

Windhager, E. E., E. L. Boulpaep, and G. Giebisch. 1966. Electrophysiological studies on single nephrons. Proc. Int. Congr. Nephrol. **1**: 35–47.

Winegrad, S. 1971. Studies of cardiac muscle with a high permeability to calcium produced by treatment with ethylenediaminetetraacetic acid. J. gen. Physiol. **58**: 71–93.

Wood, E. H., R. L. Heppner, and S. Weidmann. 1969. Inotropic effects of electric currents. I. Positive and negative effects of constant electric current or current pulses applied during cardiac action potentials. II. Hypotheses: calcium movement, excitation-contraction coupling and inotropic effects. Circ. Res. **24**: 409–445.

Woodbury, J. W. 1961. Voltage and time-dependent membrane conductance changes in cardiac muscle, pp. 501–528. *In* A. M. Shanes (Ed.), Biophysics of physiological and pharmacological actions. American Association for the Advancement of Science, Publ. 69. Washington, D. C.

———. 1962. Cellular electrophysiology of the heart, pp. 237–286. *In* W. F. Hamilton and P. Dow (Eds.), Handbook of physiology, Sect. 2: Circulation, vol. I. American Physiological Society, Washington, D. C.

Woodbury, J. W., and W. E. Crill. 1961. On the problem of impulse conduction in the atrium, pp. 124–135. *In* E. Florey (Ed.), Nervous inhibition. Pergamon, New York.

———. 1970. The potential in the gap between two abutting cardiac muscle cells. A closed solution. Biophys. J. **10**: 1076–1083.

Woodbury, J. W., and A. M. Gordon. 1965. The electrical equivalent circuit of heart muscle. J. Cell. Comp. Physiol. **66**: 35–42.

Woodbury, L. A., J. W. Woodbury, and H. H. Hecht. 1950. Membrane resting and action potentials of single cardiac muscle fibers. Circulation **1**: 264–266.

Wright, D. W., and R. S. Snart. 1971. Simultaneous measurement of the effect of vasopressin on sodium and water transport across toad bladder. Life Sci. **10**: 301–308.

Wright, E. M., and J. M. Diamond. 1968. Effects of pH and polyvalent cations on the selective permeability of gall-bladder epithelium to monovalent ions. Biochim. Biophys. Acta **163**: 57–74.

———. 1969a. An electrical method of measuring non-electrolyte permeability. Proc. Roy. Soc. B. **172**: 203–225.

———. 1969b. Patterns of non-electrolyte permeability. Proc. Roy. Soc. B. **172**: 227–271.

Yamagishi, S., and H. Grundfest. 1971. Contributions of various ions to the resting and action potentials of crayfish medial giant axons. J. Memb. Biol. **5**: 345–365.

Yelnosky, J., R. Katz, and E. V. Dietrich. 1964. A study of some of the pharmacologic actions of Droperidol. Toxicol. Appl. Pharmacol. **6**: 37–47.

Yonath, J., and M. M. Civan. 1971. Determination of the driving force of the Na^+ pump in toad bladder by means of vasopressin. J. Memb. Biol. **5**: 366–385.

Yonemura, K. 1967. Resting and action potentials in red and white muscles of the rat. Jap. J. Physiol. **17**: 708–719.

Young, J. Z. 1936. The structure of nerve fibres in cephalopods and crustacea. Proc. Roy. Soc. B. **121**: 319–337.

Zablow, L. 1958. Microelectrode resistance. Appendix to E. Amatniek, Measurement of bioelectric potentials with microelectrodes and neutralized input capacity amplifiers. IRE Trans. Med. Electron. **10**: 3–14.

Zachar, J. 1971. Electrogenesis and contractility in skeletal muscle cells. University Park Press, Baltimore.

Zadunaisky, J. A., O. A. Candia, and D. J. Chiarandini. 1963. The origin of the short-circuit current in the isolated skin of the South American frog *Leptodactylus ocellatus*. J. gen. Physiol. **47**: 393–402.

Zerahn, K. 1956. Oxygen consumption and active sodium transport in the isolated and short-circuited frog skin. Acta Physiol. Scand. **36**: 300–318.

———. 1961. Active sodium transport across the isolated frog skin in relation to metabolism, pp. 237–255. *In* A. Kleinzeller and A. Kotyk (Eds.), Membrane transport and metabolism. Academic, New York.

———. 1969. Nature and localization of the sodium pool during active transport in the isolated frog skin. Acta Physiol. Scand. **77**: 272–281.

Zivy, D., F. M. Lévy, and B. Hénon. 1968. La résistivité électrique des suspensions de mycobacteries. Etude comparative sur diverses espèces. Applications taxonomiques. Pathol. Microbiol. **31**: 337–344.

Zobel, O. J. 1923. Theory and design of uniform and composite electric wave-fitters. Bell Syst. Tech. J. **11**: 1–46.

Zolovick, A. J., R. L. Norman, and M. R. Fedde. 1970. Membrane constants of muscle fibers of rat diaphragm. Am. J. Physiol. **219**: 654–657.

Zoond, A. 1927. The interpretation of changes in electrical resistance accompanying the death of bacterial cells. J. Bacteriol. **14**: 279–286.

Index